Geophysicists, astronomers (especially those with an interest in history), historians and orientalists should all find Richard Stephenson's work fascinating. The culmination of many years of research, it discusses, in depth, ancient and medieval eclipse observations and their importance in studying Earth's past rotation. This is the first major book on this subject to have appeared in the last 20 years. The author has specialised in the interpretation of early astronomical records and their application to problems in modern astronomy for many years. The book contains an in-depth discussion of numerous eclipse records from Babylon, China, Europe and the Arab lands. Translations of almost every record studied are given. It is shown that although tides play a dominant long-term role in producing variations in Earth's rate of rotation – causing a gradual increase in the length of the day – there are significant, and variable, non-tidal changes in opposition to the main trend.

HISTORICAL ECLIPSES AND EARTH'S ROTATION

HISTORICAL ECLIPSES AND EARTH'S ROTATION

F. Richard Stephenson
University of Durham

CAMBRIDGE UNIVERSITY PRESS
Cambridge, New York, Melbourne, Madrid, Cape Town, Singapore, São Paulo

Cambridge University Press
The Edinburgh Building, Cambridge CB2 8RU, UK

Published in the United States of America by Cambridge University Press, New York

www.cambridge.org
Information on this title: www.cambridge.org/9780521461948

First published 1997
This digitally printed version 2008

A catalogue record for this publication is available from the British Library

Library of Congress Cataloguing in Publication data

Stephenson, F. Richard (Francis Richard), 1941–
Historical eclipses and Earth's rotation / F. Richard Stephenson.
p. cm.
Includes bibliographical references and index.
ISBN 0 521 46194 4
1. Eclipses. 2. Earth–Rotation. I. Title.
QB541.S663 1997
523.3′8′09–dc20 96-20968 CIP

ISBN 978-0-521-46194-8 hardback
ISBN 978-0-521-05633-5 paperback

To my wife Ellen for her love, encouragement and patience

Contents

Principal Symbols

C	gravitational lunar acceleration on GMT/UT
$C*$	lunar sidereal acceleration on GMT/UT
c	lunar sidereal acceleration on GMT/UT with gravitational term removed
c'	solar sidereal acceleration on GMT/UT
cy	centuries (Julian)
GMT	Greenwich mean time
E	time in Julian centuries from the epoch 2000.0 (= J2000)
ET	ephemeris time
h	hours
J2000	epoch 2000.0
L	lunar mean longitude on GMT/UT
L'	solar mean longitude on GMT/UT
LMT	local mean time
LOD	length of day
LT	local apparent time
ms	milliseconds
n	lunar mean orbital angular velocity
n'	solar mean orbital angular velocity (apparent)
$\dot{n}$	lunar orbital acceleration on TT
p.e.	probable error (= $0.67 \times$ s.e.)
q	acceleration of Earth's spin (= $\Delta T/t^2$)
r	lunar mean distance from Earth
s.e.	standard error
SI	Systeme Internationale (series of units)
T	time in Julian centuries from the epoch 1900.0
t	time in Julian centuries from the epoch 1800.0 (or 1820.0)
TAI	international atomic time

TT	terrestrial time
UT	universal time
y	years
z	equation of time
ΔT	Earth's rotational clock error (= TT – UT)
θ	time in Julian centuries from an arbitrary epoch
λ	lunar mean longitude on TT
λ'	solar mean longitude on TT
Λ	geographic longitude (negative to the east of Greenwich)
ϕ	geographic latitude
ω	terrestrial angular velocity of rotation

1
Variations in the length of the day: a historical perspective

1.1 Introduction

The main purpose of this book is to investigate in detail long-term variations in the length of the day, or equivalently, changes in the Earth's rate of rotation, using pre-telescopic observations of eclipses. Such variations are mainly produced by lunar and solar tides, but non-tidal mechanisms are also significant. Despite its historical bias, this subject has become an important topic in modern geophysics.

In studying changes in the length of the day which have occurred in recent centuries (since the invention of the telescope), more accurate data than eclipses are available. These include occultations of stars by the Moon up to 1955 and systematic monitoring of the Earth's rotation relative to the atomic time-scale since that date. However, observations of eclipses provide by far the most consistent and reliable source of information on variations in the terrestrial rate of rotation during the pre-telescopic period.

Although numerous early accounts of eclipses are to be found in astronomical works, many others are scattered in a variety of writings – for example the Greek and Latin classics, imperial annals of China and monastic chronicles of Europe. Observations recorded in these works are often extremely crude by modern standards, but the lengthy time-scale covered is highly suited to the detection of long-term trends which are not apparent from more recent data. The investigation of this diverse material adds a whole new dimension to what might otherwise have been a somewhat narrow scientific discipline.

This book has as its nucleus an extensive paper by Stephenson and Morrison (1995) entitled 'Long-term fluctuations in the Earth's rotation: 700 BC to AD 1990'. The present work is a considerable expansion of this paper and contains detailed discussion of almost every observation used – as well as an in-depth examination of calendar conversion and early methods of measuring time. Although in general much the same ancient and medieval eclipse observations are analysed, a significant number of medieval Chinese data have now been added (see chapter 9).

In the present chapter, the gradual recognition of the existence of changes in the length of the day will be outlined, commencing with ancient writers and extending through recent centuries to the present. The lunar acceleration will also be considered; from a historical viewpoint this is closely linked with changes in the length of the day. Other related topics include the introduction of modern time-systems which are independent of changes in the Earth's rate of rotation, and the Julian and Gregorian calendars.

It should be noted that in quoting the names of authors born before about 1900, both surname and given name(s) will be cited throughout this chapter. However, the publication of so many multi-authored papers in recent decades would make continuation of this practice tedious.

NB throughout this book I have deliberately avoided the use of accents above and below individual letters. This choice should considerably simplify the text for the reader who is not a linguist, without – I hope – causing too much inconvenience to the reader who has a linguistic background.

1.2 The solar day

The period of rotation of the Earth has been regarded as one of the fundamental units of time since remote antiquity. Although the full realisation that our planet does not rotate at an absolutely uniform rate is a relatively modern discovery, several preliminary stages in this recognition can be discerned. These relate exclusively to *apparent* changes in the length of the day which are caused by both the tilt of the Earth's axis and the ellipticity of the terrestrial orbit around the Sun.

Seasonal changes in the length of daylight and darkness were probably recognised at a very archaic stage in history. Even in relatively low latitudes (*c.* 30–35 deg N) where the great ancient civilisations of the Fertile Crescent, Indus Valley and China developed, the length of daylight ranges over a ratio of as much as 1.4:1 from summer to winter. In Egypt, the night was divided into 12 hours at least as early as 2150 BC, but a similar division of the period of daylight cannot be traced until *c.* 1450 BC (Parker, 1974). By the twelfth century BC, evidence for the use of 24 equal hours in a full solar day can also be found. A papyrus of that date lists the number of hours of daylight and darkness for each month of the year, the total always being 24. This system spread to Greece and ultimately to the modern world.

At least as early as the eighth century BC, the astronomers of Babylon divided the solar day into 12 equal double hours, each of which was subdivided into 30 equal parts; the basic unit thus corresponded to 4 minutes. Babylonian arithmetic was sexagesimal and the origin of our division of the hour into 60 minutes and the minute into 60 seconds can ultimately be traced to this system of reckoning.

The ancient Greeks employed both seasonal and equal hours (each 24 in number) side by side. On the former scheme, day and night were each divided into 12 hours, each ranging from about 0.8 to 1.2 of our present hours, depending on the season. Although seasonal hours were more relevant to everyday life, equal hours were of greater value to astronomers in studying the motion of the Moon, Sun and planets. This dual practice was followed in the medieval Arab world and subsequently in Europe.

The great Alexandrian astronomer Claudius Ptolemy (*c.* AD 150) was well aware that the length of the solar day – which may be defined as the interval between two successive transits of the Sun across an observer's meridian – varied slightly during the course of a year. In a section of his *Almagest* entitled 'On the inequality in the days' (III, 9) he gave an essentially correct explanation of the difference between the *anomalistic* or irregular solar day and the mean solar day (although, of course, he assumed that the Sun revolved around the Earth and that its motion was epicyclic).

The annual variation which Ptolemy describes arises from two separate causes: the ellipticity of the Earth's orbit and the tilt of the terrestrial axis (the obliquity of the ecliptic). The former effect, which is also responsible for the inequality of the seasons, produces a cyclical change in the apparent daily solar motion in longitude (currently from 1.020 deg in January to 0.954 deg in July). Since the Sun (apparently) moves in the ecliptic rather than along the celestial equator, the obliquity is responsible for an additional variation of the daily motion of the Sun in right ascension.

As a result of the superposition of these two components, during the course of the year the solar day can be up to about 30 sec longer or shorter than the average value of 86 400 sec (the mean solar day) – see figure 1.1. Ptolemy was unable to detect such a small variation, but he could readily discern the cumulative effect. He remarked:

> Neither of these effects causes a perceptible difference between the mean and the anomalistic return (of the Sun) over a single solar day, but the accumulated difference over a number of solar days is quite noticeable.
>
> [Trans. Toomer (1984, p. 170).]

On account of the build-up of individual small changes in the length of the solar day, a sundial (or other device keeping solar time – usually known as apparent local time) may be appreciably fast or slow relative to a clock measuring mean time (as defined by the average daily motion of the Sun). The difference between local apparent time (subsequently

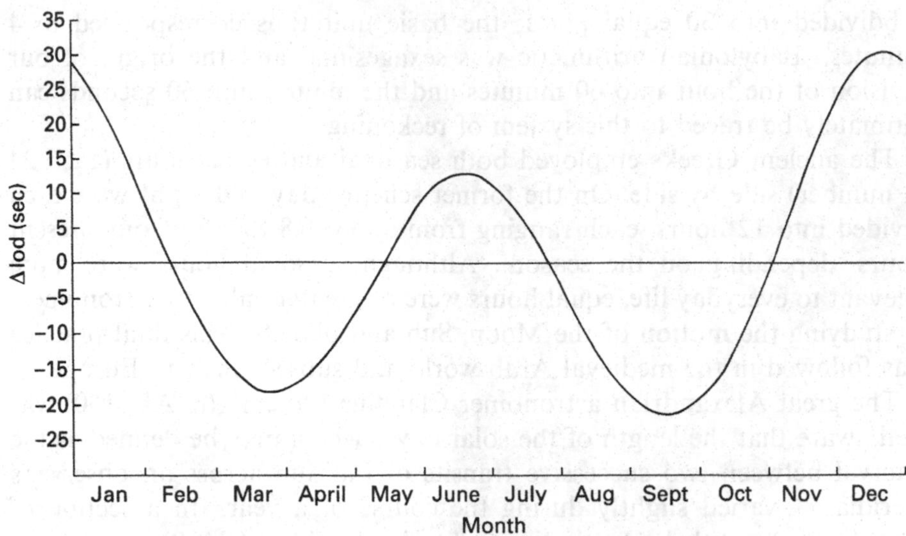

Fig. 1.1 Annual variation in the length of the apparent solar day (Δlod) arising from the tilt of the Earth's axis and the eccentricity of the orbit.

abbreviated to LT) and local mean time (LMT) at any moment is known as the equation of time (z). Extreme values of z are currently attained in mid-February (−0.24 h) and early November (+0.27 h). In February, the Sun crosses the observer's meridian significantly after mean noon, while in November the reverse is true. The form of this annual variation has slowly altered down the centuries, largely on account of the gradual advance of the Earth's perihelion caused by planetary action. Figures 1.2a and 1.2b compare the present curve with that calculated for 700 BC (the approximate date of the earliest reliable astronomical observations from any part of the world). For a series of equation of time curves at 1000-year intervals from 4000 BC to AD 4000, see Hughes *et al.* (1989).

Despite the early recognition of the apparent variation in the length of the solar day, mean solar time remained an essentially theoretical concept for more than 1500 years. As a result, all timings of eclipses (and other celestial events) from any part of the ancient and medieval world are expressed either directly or indirectly in terms of LT. Until the introduction of pendulum clocks – the first by Christiaan Huyghens in 1656 – the inaccuracy of primitive chronometers meant that the equation of time was of little practical significance. However, this parameter gradually became of more importance as attempts were made to use pendulum clocks to mark LMT. This was first achieved by regulating the clocks with the aid of a sundial and a conversion table from LT to LMT. In about AD 1700, such tables for setting clocks and watches began to be commercially

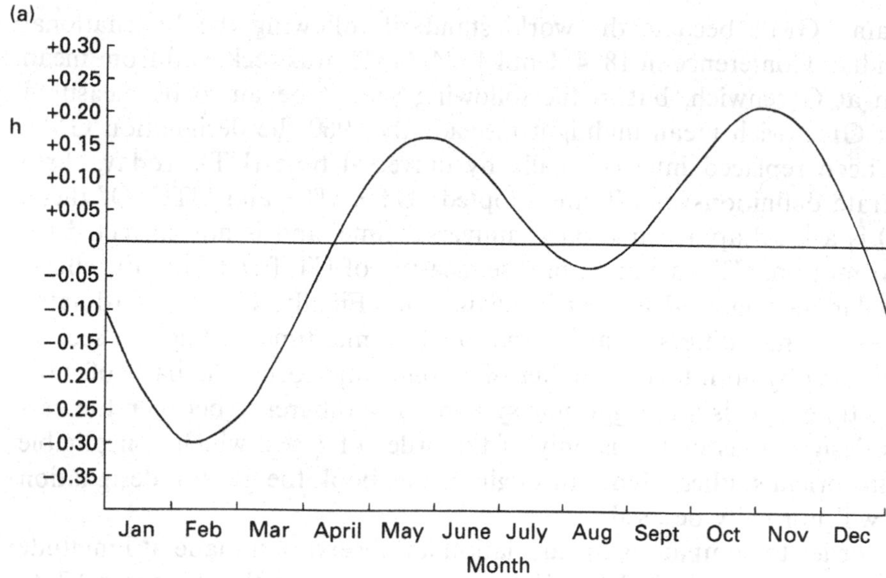

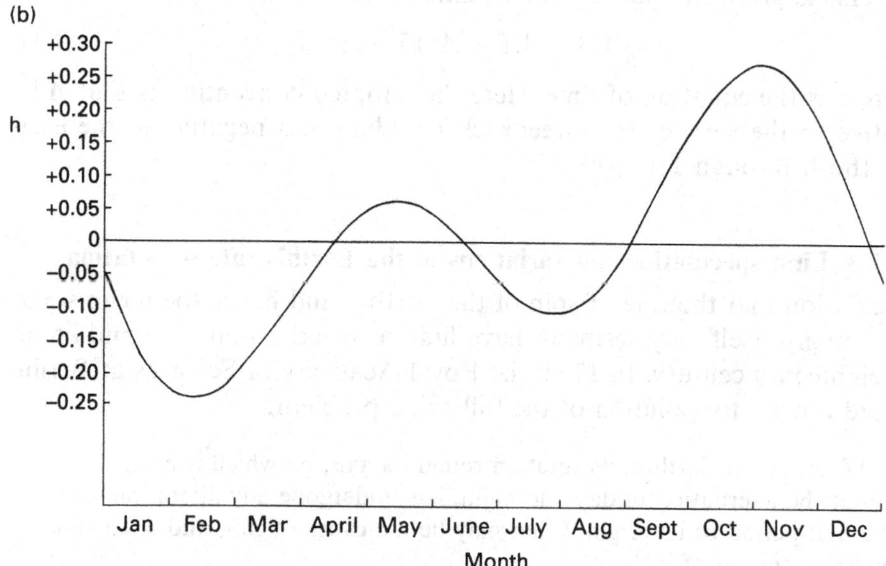

Fig. 1.2 Changes in the form of the equation of time curve between 700 BC (a) and the present (b).

available (Macey, 1994, p. 443), while almost from its inception in 1767 the *Nautical Almanac* tabulated this quantity to the nearest 0.1 sec.

The data in the various national ephemerides did not begin to be specified on LMT until 1834. Commencing in 1840, the railways were largely instrumental in the adoption of Greenwich mean time (GMT) throughout

Britain. GMT became the world standard following the International Meridian Conference in 1884. Until 1924, GMT was reckoned from mean noon at Greenwich, but in the following year it began to be measured from Greenwich mean midnight instead. By 1960 the designation GMT had been replaced internationally by universal time (UT). Today, three separate definitions of UT are adopted: UT0, UT1 and UTC. Of these, UT0 is a local approximation to universal time, and is not corrected for polar motion. UT1 is a more precise measure of UT, freed from the effects of polar motion, and is used in navigation. Finally, UTC (co-ordinated universal time) differs from international atomic time (defined in section 1.8 below) by an integral number of seconds and forms the basis of most radio time-signals and legal time systems. The difference between the various designations of UT is only of the order of 1 sec, which is negligible for historical studies. Hence throughout this book the general designation UT will normally be used.

In order to compare with calculation an observation made at longitude Λ deg and expressed in LT, it is necessary to convert the measured LT to UT. This is given in hours by the equation

$$UT = LT + \Lambda/15 - z, \tag{1.1}$$

where z is the equation of time. Here the adopted convention is that Λ is positive to the west of the Greenwich meridian and negative to the east of it (both through 180 deg).

1.3 First speculations on variations in the Earth's rate of rotation

Speculation that the rate of spin of the Earth – and hence the mean solar day – might itself vary seems to have first occurred around the middle of the eighteenth century. In 1754, the Royal Academy of Sciences at Berlin offered a prize for solution of the following problem:

> Whether the Earth in its rotation round its axis, by which it brings about the alternation of day and night, has undergone any alteration since the first period of its origin. What may be the cause of this, and what can make us certain of it?
>
> [Ley (1968, p. 157).]

According to Ley (1968, p. xi), Leonhard Euler may have first mentioned the problem to the secretary of the Royal Academy in Berlin. Ley notes that Euler, in a letter to the Bishop of Bergen (Henrik Pontoppidan), written not long before the prize was offered, had stated that he suspected that the day was growing shorter.

Immanuel Kant suggested, on the contrary, a cause that would lengthen the day. In a paper whose title may be translated 'Whether the Earth has undergone any alteration of its axial rotation', Kant (1754) reasoned that the combined attraction of the Moon and Sun, by moving the waters of the Earth in a westerly direction, would oppose the diurnal rotation of our planet. Estimating the velocity of ocean movement at '1 foot per second', Kant considered the loss of momentum on the east coast of North and South America – because of the large extension in latitude. He calculated that in two million years the 'whole motion of the Earth' would be exhausted, corresponding to a mean increase in the length of the day by about 4 sec per century. Although this result is about 2000 times the true figure, the basic principle was sound.

Kant further concluded that on account of the tidal retardation of its spin, in the remote future the Earth would ultimately turn the same face towards the Moon, the terrestrial surface then being at rest relative to the source of disturbance. He also correctly explained the fact that the Moon always turns the same face towards the Earth as being due to the retardation of the lunar rotation by tides raised by the Earth. Tidal friction is now recognised as the major long-term cause of variation in the length of the mean solar day (usually abbreviated to LOD). Remarkably, Kant later came to the conclusion that the Earth might be contracting owing to gradual cooling, which would cause it to spin faster in opposition to the effect of the tides.

Some 25 years after Kant formulated his hypothesis of the tidal retardation of the Earth's spin, Sir William Herschel considered the question of changes in the length of the day from an observational basis. In a letter published in 1781, he remarked that no astronomer had then investigated whether the Earth's diurnal rotation was constant or not, adding that this was 'probably [on account of] the difficulty of finding a proper standard to measure it by; since it is itself used as the standard by which we measure all the other motions'.

Herschel made the further perceptive remark:

> It is perhaps not altogether impossible but that inequalities may exist in the (diurnal) motion which, in an age where observations are carried to such a degree of refinement, may be of some consequence.

Because even the best of contemporary clocks were too imprecise to demonstrate the existence of such 'inequalities', Herschel envisaged a long-term plan. He recommended determination of the rotation periods of the other planets – especially Mars – as accurately as possible relative to the LOD, in the hope that future astronomers would be able to decide whether or not these periods had maintained their relative values. This was to be a vain hope. However, nearly a century beforehand, the analysis of ancient observations of eclipses by Edmond Halley had provided the first suggestion of a seemingly independent phenomenon: an acceleration of the Moon. Ultimately, this proved to be partly real and partly apparent, the cause of the apparent component being a gradual increase in the adopted unit of time – i.e. the Earth's period of rotation.

1.4 Discovery of the lunar secular acceleration

As noted in section 1.1, although ancient observations are of low precision compared with telescopic measurements, the greatly enhanced time-scale renders them much more suitable for determining an accelerative term. By Halley's epoch, the investigation of ancient eclipses to obtain improved knowledge of the lunar motion was already a long-established tradition. Both Hipparchus (*c.* 150 BC) and Ptolemy (*c.* AD 150) had utilised Babylonian eclipse observations from several centuries previously for this purpose (see chapter 4 below). Long afterwards, the great Chinese astronomer of the thirteenth century Kuo Shou-ching had used a series of eclipse observations extending over more than a thousand years to test the accuracy of contemporary eclipse tables (see chapter 9). Nevertheless, until Halley's researches, the *mean* motion of the Moon (i.e. with the various short-term fluctuations averaged out) had been assumed to be uniform.

In 1695, towards the end of a paper entitled 'Some account of the ancient state of the city of Palmyra...', Halley recommended that the longitudes of Baghdad, Aleppo and Alexandria be carefully determined (using forthcoming lunar eclipses), explaining his motive in the following words:

> For in and near these Places were made all the Observations whereby the Middle Motions of the Sun and Moon are limited: And I could then pronounce in what Proportion the Moon's motion does Accelerate; which that it does, I think I can demonstrate, and shall (God Willing) one day make it appear to the Publick.

Two years earlier, Halley (1693) had investigated a series of celestial observations made by the medieval Muslim astronomer al-Battani (AD 850–929). Al-Battani, whose latinised name was Albategnius, observed both at Raqqa (in present-day Syria) and Antakya (Turkey) – see also chapter 13. When Halley undertook his investigation, only an inaccurate Latin translation of al-Battani's writings, made in the twelfth century, was available. Halley's objective can best be summarised in the title of his paper (also in Latin), which may be rendered as follows: 'Corrections and comments on the ancient astronomical observations of Albategnius in order to restore the solar and lunar tables of that author'. Although this paper was purely an historical exercise, Halley's research may well have inspired him to test contemporary lunar tables against the observations made by al-Battani and other early astronomers. However, apart from the above brief reference in his 1695 paper his publications remain silent on the question of the lunar acceleration.

In 1749, the Reverend Richard Dunthorne not only provided firm evidence for the existence of a lunar acceleration but also obtained a result which was not significantly improved upon until the present century. Using lunar and solar tables which he had himself constructed, Dunthorne analysed a selection of six ancient and medieval eclipse timings from the following years: 721 and 313 BC (both from Babylon and recorded by Ptolemy in his *Almagest*); 201 BC and AD 364 (both from Alexandria); and AD 977 and 978 (both from Cairo). Details of these and other roughly contemporary observations are given in chapters 4, 10 and 13 below. In each case Dunthorne found that the computed time was earlier than the observed time, the discrepancy being greatest for the most ancient observations. He inferred an acceleration of the Moon, and concluded:

> If we take this Acceleration (of the Moon) to be uniform, as the Observations whereupon it is grounded are not sufficient to prove the contrary, the Aggregate of it will be as the Square of the Time: And if we suppose it to be 10″ in 100 years...it will best agree with the above mentioned Observations.

Dunthorne here implied a coefficient of θ^2 of 10″ in the Moon's longitude (where θ is measured in Julian centuries, each of 36 525 mean solar days from an arbitrary epoch). Of course, this result, in common with all similar deductions made until comparatively recent times, was based on the assumption of a fixed length of day. It became customary to refer to the coefficient of θ^2 in the lunar mean longitude – expressed purely in arcseconds – as the Moon's *secular acceleration* – rather than quoting twice this figure. Interestingly, the observation in 721 BC still remains one of the very earliest reports of an eclipse from any part of the world, even though numerous ancient records unknown to Dunthorne have since come to light. Because the 721 BC eclipse occurred nearly 25 centuries before Dunthorne's own time, a secular acceleration of 10″ implied a discrepancy between observation and theory of as much as 1.7 deg. The Moon takes fully 3 hours to move through this angle, a time-interval which could be measured with tolerable precision by the astronomers of Babylon (see chapter 4).

In his paper, Dunthorne also stated that he had investigated several eclipse observations made by Tycho Brahe in the late sixteenth century and by Bernard Walther and Johann Muller (the latter better known as Regiomontanus) in the late fifteenth century. Unfortunately he only published very brief details. On comparing Tycho's measurements with his own eclipse tables, Dunthorne had found that observation and theory agreed 'full as well as could be expected considering the imperfection of his (Tycho's) clocks...'. For the observations by Walther and Regiomontanus, Dunthorne's computed lunar longitudes proved to be mainly some 5 arcmin ahead of their positions as derived from measurement. However, in this latter case he considered the mutual discord between the individual results too great for an acceleration to be inferred with certainty.

Not long after Dunthorne's work, Tobias Mayer (1753) made a similar investigation of historical eclipses. He obtained a significantly lower result for the lunar secular acceleration of $6''.7$, but subsequently revised it to $9''$. At about this same time, Joseph-Jerome Lalande (1757) obtained a result of $10''$. Hence by early in the second half of the eighteenth century, the existence of the secular acceleration of the Moon's mean motion was firmly established on observational grounds.

1.5 Early attempts to explain the Moon's secular acceleration

The significant acceleration of the Moon derived from ancient data invited theoretical explanation, and commencing in 1770 several prizes were offered for its solution by the Academy of Sciences at Paris. Following unsuccessful attempts by Leonhard Euler and Joseph-Louis Lagrange, what in retrospect proved to be a *partially* successful explanation was obtained by Pierre-Simon Laplace (1787a and b). Laplace attributed the lunar acceleration to the effect of the gradual diminution in the eccentricity of the Earth's orbit caused by planetary action. This produces a long-term decrease in the mean action of the Sun on the Moon, resulting in a slow increase in the lunar mean motion. The variation in the Earth's orbital eccentricity – and hence the change in the average motion of the Moon – is actually cyclic with a periodicity of the order of 100 000 years (Berger, 1977). However, over the relatively short time-scale covered by historical observations the lunar acceleration arising from this cause has consistently been positive – see chapter 2.

Laplace approximated the change in eccentricity (e') of the terrestrial orbit by a linear equation of the form

$$e' = e'_0 - ay. \tag{1.2}$$

Here e'_0 is the eccentricity at epoch 1850 and a is a constant; y is reckoned in Julian years from 1850. Laplace deduced that the lunar secular acceleration (C) is given by

$$C = \tfrac{3}{2}m^2 e'_0 n' a, \tag{1.3}$$

where m is the ratio of the mean motions of the Sun and Moon (0.0747) and n' the mean solar motion. Laplace derived a secular acceleration of $11''.135$, which appeared to be in excellent accord with observation.

Consequently he rejected the possibility that variation in the LOD was responsible. He remarked that since the time of Hipparchus, almost 2000 years previously, the LOD 'had not varied by $\frac{1}{100}$ of a second'. Some years later, in 1792, Lagrange closely confirmed Laplace's value for the lunar acceleration and it thus seemed that the problem was solved.

Laplace's apparently satisfactory result was not disputed until 1853 when John Couch Adams pointed out an error in his computations. Laplace and Lagrange had integrated the necessary differential equations as though the eccentricity of the Earth's orbit were constant, substituting its variable value in the results. Adams showed that although such a procedure was permissible when only a first approximation was required, for higher precision it was necessary to introduce the variability of e' into the differential equations. He at first replaced $\frac{3}{2}m^2$ in equation (1.3) by $\frac{3}{2}m^2 - (3771/64)m^4$, thus arriving at $C = 8''.3$. Adams subsequently reduced this result to $5''.78$ by evaluating the first few terms in a slowly converging power series (to m^7). The sum of the individual terms for C was

$$C = 10''.66 - 2''.34 - 1''.58 - 0''.71 - 0''.25 = 5''.78.$$

Adams' revised value, first announced in a letter to Charles-Eugene Delaunay, was published in 1859. It was disputed by such distinguished mathematicians as Peter Andreas Hansen, Giovanni Plana and Philippe de Pontecoulant, who each closely supported Laplace's figure. However, following an extensive analysis, in which he evaluated as many as 42 terms, Delaunay (1859) obtained a secular acceleration of $6''.11$ – only a slight revision of Adams' value. To quote Simon Newcomb (1898, p. 98):

> Although their antagonists never formally surrendered, they tacitly abandoned the field, leaving Adams and Delaunay in its undisturbed possession.

Twentieth century investigations have closely confirmed Delaunay's deduction. In his exhaustive investigation of the lunar theory, Ernest W. Brown (1915a) derived a value for C of $6''.03$. More recently, Chapront-Touze and Chapront (1983) made a new determination of this parameter. Their result comprises the following components: Earth figure ($0''.1925$); planetary perturbations ($0''.0025$); Earth's orbital eccentricity ($5''.8864$); lunar figure ($-0''.0151$). The sum, $6''.0463$, is almost identical to Brown's value.

William Ferrel (1865) and Delaunay (1865) independently revived the hypothesis of tidal retardation of the Earth's spin to account for the observed excess acceleration of the Moon. Ferrel derived a result for the apparent acceleration of the Moon on the simplifying assumptions that the surface of the Earth was covered by oceans and that the tidal wave took the form of a prolate spheroid. He also made the further rather arbitrary suppositions that the vertices of this spheroid were displaced by friction through 30 deg (relative to the line of centres of the Earth and Moon) and that the height of the tidal wave above low water was 2 feet (0.6 metres) at the equator. Ferrel calculated that the effect of both lunar and solar tides would be to retard the rotatory motion of the Earth at the equator by 'about forty-four miles in a century'. Since the Moon moves through 84″ in its orbit while the terrestrial equator turns through this distance, Ferrel concluded that the apparent secular acceleration of the Moon would be 84″. This was roughly an order of magnitude greater than the observed acceleration and Ferrel concluded that the effect of tides must be largely 'counteracted by a corresponding contraction of the Earth's volume from a gradual loss of heat' – essentially the explanation offered by Kant.

The equator rotates through 44 miles (70 km) in about 150 sec. As a (Julian) century contains 36 525 days, it can be seen that Ferrel's result would correspond to a rate of increase in the LOD by about 8 milliseconds per century (ms/cy). A similar figure was reported by Sir William Thompson – later Lord Kelvin – at a lecture delivered in 1866 (Kelvin, 1894, p. 68). Delaunay (1865) did not attempt to deduce the rate of lengthening of the day due to tides. However, on the basis of the difference between the observed secular acceleration of the Moon and its gravitational equivalent (amounting to some 6″), he calculated that the equivalent secular retardation of the terrestrial rotation would be $27\frac{1}{3}$ times as great or 164″. Since the Earth turns through this angle in about 11 sec, the corresponding rate of increase in the LOD would be about 0.6 ms/cy. This is, in fact, about one-quarter of the actual amount (see chapter 2).

Neither Ferrel nor Delaunay – nor Kelvin at this early date – seem to have considered the reciprocal effect of lunar tidal friction on the Moon's orbit. However, Emmanuel Liais (1866), who was of the opinion that magnetism might play a key role in the interaction between the Earth and Moon, stressed that the gain in angular momentum by the Moon at the expense of the Earth would produce an orbital acceleration. Liais concluded:

> By the laws of mechanics, a reduction in the speed of a celestial body causes a recession such that the 'angular movement' [*mouvement angulaire*] is increased. Thus there exist (both) a real acceleration of the motion of the Moon and an apparent acceleration arising from the increase in the length of the day.

A satellite which gains angular momentum both recedes from its primary and loses angular velocity, as can be seen from the following equations, which are based on the approximation of circular motion. If A represents the orbital angular momentum of the Moon, r the distance from the Earth, and n the angular velocity,

$$dA/dr = Kr^{-1/2}, \tag{1.4}$$

$$dA/dn = -kn^{-5/3}, \tag{1.5}$$

where both K and k are positive constants.

Liais effectively implied that the observed acceleration of the Moon would be the sum of the apparent and real components, which is a clear statement of the true situation.

Sir George Darwin (1880) and also Lord Kelvin and Peter Guthrie Tait (1883) discussed the effect of tidal friction in detail. However, they were well aware that the approximations which they made were necessarily crude. Kelvin later (1894, p. 69n) stressed that direct evaluation was a complex task. He remarked:

> It seems hopeless, without waiting for some centuries, to arrive at any approach to an exact determination of the the amount of the actual retardation of the Earth's rotation by tidal friction, except by extensive and accurate observation of the amounts and times of the tides on the shores of continents and islands in all seas, and much assistance from *true* dynamical theory to estimate these elements all over the sea. But supposing them known for every part of the sea, the retardation of the Earth's rotation could be calculated by quadratures.

Not until 1920 was the first reasonably successful calculation of this kind made (by Sir Harold Jeffreys), but this empirical method has been superseded by studies of the orbits of artificial satellites and lunar laser ranging – see chapter 2.

Darwin emphasised that in the geological past the Moon must have been much closer to the Earth than at present. He showed that since the lunar orbital angular momentum is about 5 times that of the Earth about its axis of rotation, if the Earth and Moon were once a single fluid body, the rotation period would be about 4 hours. This led to his 'resonance theory' of the origin of the Moon, in which solar tides raised on the primeval Earth were amplified by resonance to such an extent that the planet broke into two. Although this theory is no longer regarded as tenable, it was briefly revised in modified form some 25 years ago (Wise, 1969; O'Keefe, 1970).

The effects of tidal friction on the rotation of the Earth and motion of the Moon may be briefly outlined as follows; for a more detailed discussion, see Jeffreys (1976, pp. 316ff.). The gravitational attraction of the Moon on the Earth will produce diametrically opposite tidal bulges – as illustrated schematically in figure 1.3. At bulge X, which is slightly nearer the Moon, the lunar gravitational force will be greater than at the

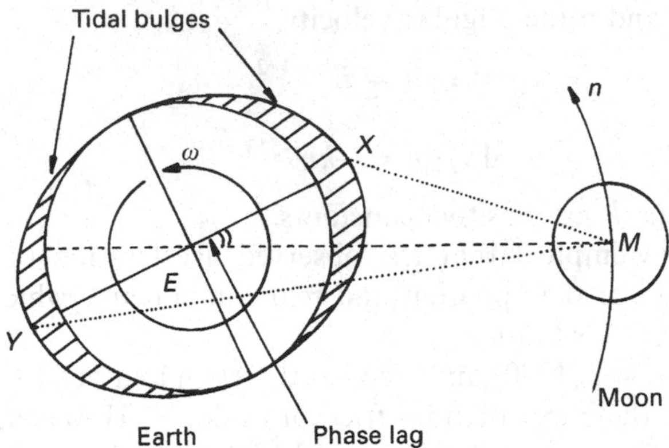

Fig. 1.3 Tidal interaction between the Moon and the Earth.

centre of the Earth, while at Y it will be weaker. On account of friction in the oceans (and in the interior of the Earth owing to the imperfect elasticity of the material) there will be a phase lag between the passage of the Moon across any particular meridian of longitude and the associated high tide. The two bulges will thus be out of alignment with the line of centres of the Earth and Moon. Hence the gravitational attraction of the Moon will have a component in the opposite sense to the direction of rotation of the Earth – which will thus be retarded.

In the reciprocal case of the attraction of the bulges X and Y on the Moon, the force produced by X is greater than that due to Y since the distance XM is less than YM and the angle XME is greater than YME. There will thus be a net force acting on the Moon with a component in the direction of revolution. Following the principle of conservation of angular momentum, the angular momentum lost by the Earth will be gained by the Moon.

Despite the enormous mass of the Sun, solar tides are significantly less effective in retarding the Earth's rotation – by a factor of about 4 – because of the vastly greater mean distance.

Because the angular momentum in the Earth's orbit is so huge, the Sun can have negligible real acceleration due to inverse tidal action. Nevertheless, a gradual decrease in the length of the day caused by tides would produce an *apparent* solar acceleration. Until early in the present century there was no observational support for any solar acceleration and this had somewhat weakened the arguments put forward in favour of the tidal retardation of the Earth's spin. Eventually, evidence was to come from a study of ancient eclipses – just as it had in the case of the lunar acceleration.

1.6 Discovery of the solar secular acceleration

Interest in historical eclipses waned towards the end of the eighteenth century, but it was revived by Sir George Airy (1853, 1857). *Untimed* reports of total obscurations of the Sun now began to attract special attention, the basic principle being that since the lunar umbra sweeps out a very narrow path across the Earth's surface (see chapter 3), small changes in the adopted value for the lunar acceleration would materially alter the maximum degree of obscuration of the Sun at any particular site. Other nineteenth century authors who analysed this type of material included Giovanni Celoria (1877a and b), and Friedrich Ginzel (1884a and b). In general, results for the lunar secular acceleration fairly close to 10″ were derived – essentially the value obtained by Dunthorne (1749) and significantly greater than the theoretical figure.

By the mid-nineteenth century, a small secular acceleration of the equinox (the reference point for measuring celestial co-ordinates) had been discovered. Hansen (1857) calculated that the equinox had a secular acceleration (coefficient of θ^2) due to planetary action of 1″.1, while Newcomb (1895b) obtained 1″.11. (Recent computations by Laskar (1986) yield an almost identical value of 1″.112.) Observational results for the lunar acceleration inevitably included the effect of precession, and when this was subtracted it became customary to refer to the remainder as the *sidereal* acceleration of the Moon. Results for the sidereal acceleration were a little closer to the theoretical result for C (approximately 6″.05), which was calculated relative to an inertial frame. However, there was still a significant discrepancy of several arcsec/cy^2.

Throughout his career, Simon Newcomb considered historical allusions to *total* solar eclipses as of very dubious worth. In his own studies, starting in 1878, Newcomb relied instead on timed measurements. These originated from Babylon and Greece between 721 BC and AD 136 (as quoted in Ptolemy's *Almagest*) and from Baghdad and Cairo between AD 829 and 1004. (For details of these various observations, see chapters 4, 10 and 13.) Newcomb obtained a result for the lunar sidereal acceleration of 8″.8. As will be apparent from chapter 10 below, Newcomb was fully justified in his rejection of the ancient European observations of total solar eclipses which his contemporaries analysed. Often both the date and place of observation were in doubt, while in most cases it was not even clear whether the Sun was indeed completely obscured. However, medieval observations of this kind have proved to be of considerable utility – see chapter 11.

In 1905, Philip H. Cowell found evidence from ancient eclipses for a significant solar acceleration. Only a minute secular acceleration of the Sun (amounting to −0″.02) was expected on gravitational theory. This was purely apparent and arose from a slow increase in the mean distance of the Earth from the Sun caused by planetary perturbations. Cowell had previously made several analyses of ancient eclipse records – both timed and untimed – in order to derive a value for the lunar coefficient of θ^2. He now concluded that to best satisfy the observations it was necessary to assume a secular acceleration of the Sun as well as of the Moon. His results for the sidereal lunar and solar accelerations (which will be denoted by C^* and c' respectively) were $C^* = 10''.9$ and $c' = 4''.1$. At first Cowell attributed the observed solar acceleration to the resistance of the Earth's orbital motion by the 'ether'. Soon afterwards (1906a) he reverted to the hypothesis of a steady increase in the unit of time – the mean solar day – produced by tidal action.

Cowell noted that the ratio which he derived for C^*/c' (i.e. 2.7:1) was much less than that of the mean motions of the Moon and Sun (13.37:1). He correctly explained this difference as arising from the gradual diminution in the angular velocity of the Moon as its orbit expanded under the reciprocal action of the tides. In essence, Cowell showed that if ω represents the angular velocity of the Earth's rotation and n and n' the mean angular velocity of revolution of the Moon and Sun, then, with θ in centuries as above,

$$c' = 0.5\left(-\frac{n'}{\omega}\frac{\mathrm{d}\omega}{\mathrm{d}\theta}\right), \tag{1.6}$$

$$C^* = 0.5\left(\frac{\mathrm{d}n}{\mathrm{d}\theta} - \frac{n}{\omega}\frac{\mathrm{d}\omega}{\mathrm{d}\theta}\right). \tag{1.7}$$

In equation (1.7), $\mathrm{d}n/\mathrm{d}\theta$ represents the real tidal acceleration of the Moon (negative) and $(n/\omega)(\mathrm{d}\omega/\mathrm{d}\theta)$ its apparent acceleration due to the increase in the adopted unit of time. As noted earlier (section 1.5), tides have negligible effect on the motion of the Sun so that any acceleration must be purely apparent. The factor 0.5 is introduced in both equations because the secular accelerations C^* and c' were customarily regarded as coefficients of θ^2.

In general, if one takes the mean solar motion as $0''.0411$ per second of time, it is readily shown that the rate of lengthening of the day corresponding to a selected value for c' is given by 1.33 c' ms/cy. From his value for c' of $4''.1$, Cowell derived a result for the rate of increase in the LOD of 5 ms/cy, which is rather more than twice the current estimate.

Cowell's detection of the apparent solar acceleration, although based on rather flimsy historical evidence, represented a pioneering step. His conclusion that the apparent acceleration of the Sun arose from a diminution of the Earth's rate of rotation was rejected by Newcomb and other authors (see Cowell, 1907). However, Cowell's work inspired other astronomers – notably John K. Fotheringham – to solve directly for both

C^* and c' when analysing ancient observations. In a protracted series of papers, published between 1908 and 1935, Fotheringham analysed numerous early observations of eclipses, occultations and equinoxes. Most of his timed data were taken from the *Almagest*, but he also investigated several allusions to eclipses in ancient historical and poetical works. He derived (1920b) values for C^* and c' of respectively 10″.6 and 1″.5, remarking that 'all classes of ancient observations' were adequately satisfied by these figures. Fotheringham's result for the solar acceleration, corresponding to a rate of increase in the LOD of 2.0 ms/cy, is remarkably close to the figure deduced from present-day researches. Although both he and Cowell used a miscellany of observations of varying degrees of reliability, Fotheringham placed greater emphasis on the more viable timed data.

Willhelm de Sitter (1927) revised Fotheringham's analysis of ancient observations. In quoting the lunar acceleration he subtracted the gravitational contribution derived by Brown (6″.03), obtaining a result which we shall term c. De Sitter obtained values for c and c' (the solar acceleration) of respectively 5″.22 ± 0″.30 (p.e.) and 1″.80 ± 0.16 (p.e.). Fotheringham's results, reinterpreted in this way, were to play an important part in the recognition of short-term fluctuations in the Earth's rate of rotation as revealed by telescopic observations. It is to this issue that we now turn.

1.7 Fluctuations in the mean motions of the Moon, Sun and planets

During his extensive researches on the lunar theory, Hansen discovered two inequalities of long period in the Moon's mean longitude which he attributed to the action of Venus. These terms, whose derived amplitudes were respectively 15″.3 and 21″.5 and periods 273 years and 239 years, were included in his *Tables de la Lune*, published in 1857. At the time, Hansen's tables were believed to represent closely all observations of the Moon since 1750. In particular, they were adopted for computing the lunar ephemerides printed in the *Nautical Almanac* from 1862.

Of the two inequalities, the amplitude and period of the former were closely confirmed by Delaunay (1860a and b). The term became known as the 'Great Venus Term'. According to Brown (1919, p. 27), its amplitude is 14″.27 while its period is 271 years. However, the latter inequality was shown by Delaunay (1860a and b) to have no theoretical justification. This was the first reliable indication of major irregularities in the lunar motion which could not be satisfactorily explained by gravitational theory. By 1870, Newcomb had noted that comparisons between observations of the position of the Moon and predictions based on Hansen's theory revealed significant deviations. As these discrepancies could not be explained by observational errors, it was evident that they either arose from deficiencies in Hansen's theory or fluctuations in the motion of the Moon of non-gravitational origin.

In 1878, Newcomb analysed numerous telescopic observations of the Moon, some – uncovered as the result of his own literature searches – dating as far back as 1645. He demonstrated that the fluctuations in the lunar motion could be represented approximately by an empirical oscillatory term of period some 260 years and amplitude 12″ on which were superposed additional fluctuations on a time-scale of decades and of typical amplitude 3 or 4 sec. Eventually Newcomb (1909) mapped these fluctuations in great detail – see figure 1.4. He named the principal fluctuation the 'Great Empirical Term', assigning it a period of 275 years and amplitude 12″.95.

In 1878 Newcomb made the following discerning comment:

> Now it is most remarkable that observations of transits of Mercury agree with those of the Moon, and with those of the first satellite of Jupiter, in indicating that this apparent inequality (in the Moon's motion) was in part at least due to the Earth's rotation.

Lord Kelvin, at a lecture delivered in 1876, had also ascribed the fluctuations to variations in the Earth's rate of rotation (Kelvin, 1894, pp. 271–2). However, it was many years before this conclusion was generally accepted. Not long before his death, Newcomb (1909) was more cautious, asserting:

> I regard these fluctuations as the most enigmatic phenomenon presented by the celestial motions, being so difficult to account for by the action of any known causes, that we cannot but suspect them to arise from some action in nature hitherto unknown.

Although Newcomb again considered the possibility that 'the inequalities are only apparent, being perhaps due to fluctuations in the Earth's speed of rotation, and therefore in our measure of time', this remained no more than a hypothesis to him. In the concluding remarks to his *Researches on the Motion of the Moon and Related Astronomical Elements*, published posthumously in 1912, he stated:

> The most unsatisfactory feature of the conclusion of the entire work as carried through by the author is that, until the matter is cleared up, it will be impossible to predict the Moon's longitude with the precision required for astronomical purposes. We shall be obliged to correct the Moon's mean longitude from time to time, perhaps at intervals of 10 or 20 years, from observations.

This was an unhappy conclusion to many years of research.

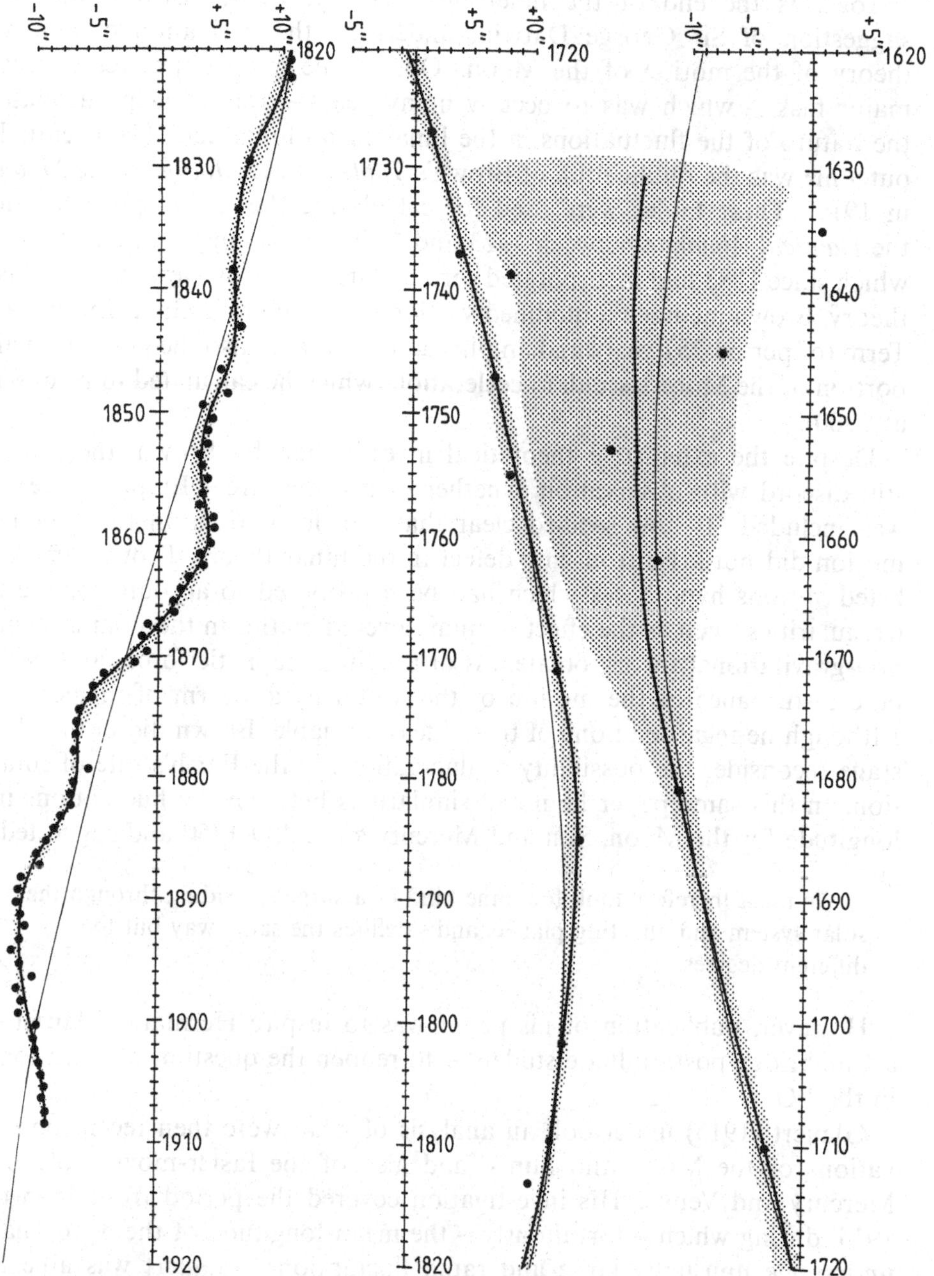

Fig. 1.4 Fluctuations in the lunar mean longitude from AD 1630 to 1909 as derived by Newcomb (1909).

Towards the end of the nineteenth century, Ernest Brown, on the suggestion of Sir George Darwin, undertook the derivation of a new theory of the motion of the Moon. One of the main objectives of this major task – which was to occupy many years – was to help elucidate the nature of the fluctuations in the lunar mean longitude. The eventual outcome was the publication of Brown's *Tables of the Motion of the Moon* in 1919. These tables were used for calculating the lunar ephemeris in the *Nautical Almanac* between 1923 and 1959 – replacing Hansen's tables which since 1883 had incorporated certain corrections by Newcomb. In his theory, Brown included a modified version of Newcomb's Great Empirical Term (of period 257 years and amplitude 10″.71) and also the gravitational portion of the Moon's secular acceleration (which he calculated to be 6″.03 at 1900).

Despite the exhaustive theoretical investigation by Brown, there was still discord with observation, whether or not the Great Empirical Term was included. It now seemed clear that the fluctuations in the Moon's motion did not arise from any defect in the lunar theory. Brown (1915b) listed various hypotheses which had been proposed to account for these irregularities, such as the effect of undiscovered matter in the solar system, non-gravitational forces, oblateness of the Sun, magnetic forces and periodic disturbance of the motion of the Moon by a swarm of meteorites. Although he regarded none of these ideas as viable, Brown did not at this stage reconsider the possibility of fluctuations in the Earth's rate of rotation. In this same paper he noted similarities between the fluctuations in longitude for the Moon, Sun and Mercury since AD 1750 and suggested:

> We must therefore look for some kind of a surge spreading through the solar system and affecting planets and satellites the same way but to different degrees.

However, publication of his paper was to inspire Hermann Glauert – a Cambridge postgraduate student – to reopen the question of variations in the LOD.

Glauert (1915) undertook an analysis of what were then recent observations of the Moon and Sun – and also of the faster-moving planets Mercury and Venus. His investigation covered the period from 1865 to 1914, during which – fortuitously – the mean longitude of the Moon had undergone unusually large and rapid fluctuations. Glauert was able to show that the discrepancies in mean longitude for the Sun, Mercury and Venus fairly closely resembled those for the Moon. He concluded:

> It appears that the errors in the longitude of the Moon, and the three bodies considered in this paper may be accounted for by a rather irregular variation in the rotation of the Earth...

Although Glauert's result was inconclusive, since he was not able to prove that the deviations were in the ratio of the mean motions of the celestial bodies, he had paved the way for a return to the idea that the fluctuations were only apparent and had their origin in the variations of the adopted unit of time. Other partially successful investigations of telescopic observations were carried out by Robert T. A. Innes (1925), Brown (1926), Sir Harold Spencer Jones (1926, 1932) and de Sitter (1927). However, not until a later paper by Jones (1939) was it clearly demonstrated that changes in the Earth's rate of rotation were fully responsible for the observed anomalies. De Sitter, and subsequently Jones, expressed the problem as follows. Let B be defined as the fluctuation in the Moon's mean longitude. Then the difference (δL) between observational and gravitational theory (Brown's theory with the Great Empirical Term removed) for the lunar mean longitude can be represented by an equation of the form

$$\delta L = a + bT + cT^2 + B, \tag{1.8}$$

where T is expressed in Julian centuries from the epoch 1900.0. For the Sun, the corresponding expression – the difference between observation and Newcomb's (1895a) theory – is:

$$\delta L' = a' + b'T + c'T^2 + QBn'/n. \tag{1.9}$$

Here n'/n is the ratio of the mean solar and lunar motions, while Q is a factor to be determined (as are the coefficients a, b, a', b' and c'). Similar expressions apply to Mercury and Venus. In the case of the Sun, $n'/n = 0.0747$, while for Mercury and Venus the corresponding figures are 0.310 and 0.112. It was necessary to use a value for c, the lunar coefficient of T^2, derived from ancient observations in order to solve for Q. As noted above, de Sitter revised Fotheringham's result for c, obtaining $5''.22$. If the observed fluctuations in the mean longitude were entirely Earth-based, Q would have the value unity. Although de Sitter was able to show that, within the limits of error, Q had the same value for the Sun, Mercury and Venus, his result was considerably greater then unity: 1.25 ± 0.03 (p.e.). Shortly afterwards, Jones (1932) reinvestigated the problem and found $Q = 1.19 \pm 0.05$ (p.e.). Hence neither solution excluded the possibility that part of the observed lunar fluctuations was real.

In 1939, Jones made an exhaustive analysis of the following series of observations: occultations of stars by the Moon and other lunar observations since AD 1680; solar declinations since 1760; solar right ascensions (RA) since 1835; Mercury transits since 1677; and measurements of the RA of Venus since 1835. His principal conclusions can be divided into two parts:

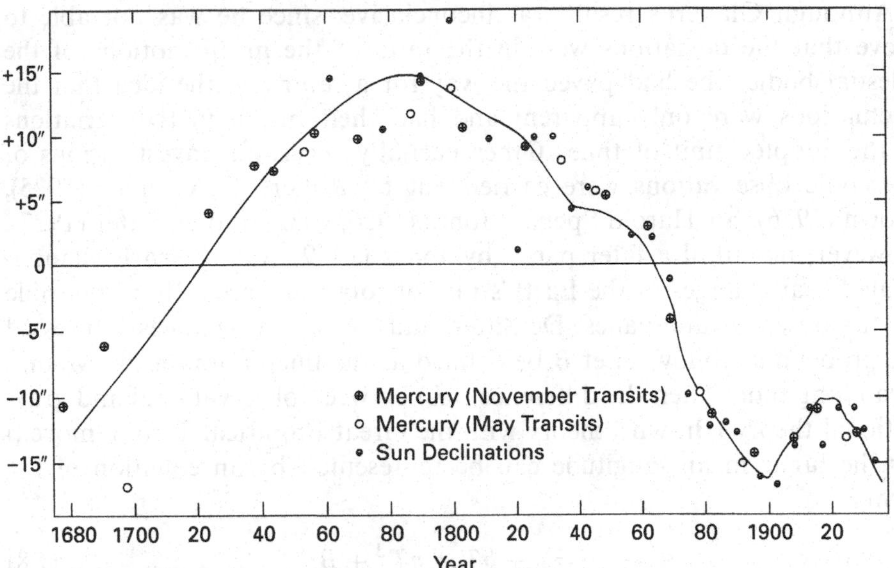

Fig. 1.5 Fluctuations in the lunar mean longitude from AD 1680 to 1939 as derived by Jones (1939).

(i) For the Sun, Moon, Mercury and Venus, $Q = 1.025 \pm 0.050$ (p.e.). Hence:

> The fluctuations in the mean longitudes of the Sun, Moon, Mercury and Venus are proportional to the respective mean motions of these bodies and can be attributed to variations in the rate of rotation of the Earth caused by changes in the moment of inertia [*sic*].

(ii) On the assumption of a lunar secular acceleration of 5″.22, the secular acceleration of the Sun was found to be 1″.23 ± 0″.04 (p.e.), while for Mercury and Venus the accelerations were respectively 5″.10 and 2″.00. Thus:

> The secular accelerations of the Sun, Mercury and Venus are proportional to their mean motions and can be accounted for by retardation of the Earth's rotation by tidal friction.

Figure 1.5, which is taken from figure 1 of Jones' (1939) paper, shows the fluctuations in the mean longitude of the Moon which he determined from observations of the Moon (solid line), Sun and Mercury. The results for the Sun are scaled up by a factor of 13.37 and for Mercury by 3.23 – the reciprocal ratios of the mean motions relative to the Moon. Although the scatter among the data for the Sun and Mercury is fairly large, there is good general accord with the lunar curve.

Jones made the comment that observations of the Sun and Mercury agree in indicating that the average effects of tidal friction during the past 250 years are smaller than the average effects over the past 2000 years. He particularly had in mind the difference between his result for the solar acceleration derived from telescopic data (1″.23) and that obtained by de Sitter from ancient observations (1″.80). Tidal friction has probably remained virtually constant over the last two or three millennia (see chapter 2). Although Jones assumed that the variations in the rate of rotation of the Earth as observed from the telescopic observations were caused by changes in the terrestrial moment of inertia, this is only one of several possible causes. Today, fluctuations in the Earth's rate of rotation at the millisecond level have been detected on a wide range of time-scales from days to millennia. These are superposed on the main tidal slowing down and are ascribed to a variety of mechanisms: e.g. varying trade wind patterns, core–mantle coupling, sea-level changes and post-glacial uplift – see also section 1.9.

1.8 The introduction of a theoretically invariant time-system

As early as 1929, Andre-Louis Danjon recognised that mean solar time (i.e. GMT) was unsatisfactory for astronomical purposes. A new time-system which was independent of variations in the Earth's rate of rotation was required instead. Danjon proposed the adoption of a theoretically uniform time-system, based on Newtonian dynamics. He wrote:

> We should not delay longer in adopting a new practical definition of time...in order to fix the time by the position of the planets in their orbits it is necessary to make the calculations in celestial mechanics accounting to the laws of nature.
>
> [Trans. Clemence (1971).]

Little interest was shown in Danjon's proposal for nearly 20 years. However, the corrections to the mean longitudes of the Moon and Sun derived by Jones (1939) were to form the basis for both the definition of a Newtonian time-system and its measurement (using lunar observations).

When Jones produced his paper in 1939, the standard expressions for the mean longitudes of the Moon (L) and Sun (L') relative to the mean equinox of date were as follows:

$$L = 973\,571''.71 + 1732\,564\,406''.06T + 7''.14T^2 + 0''.0068T^3 + \text{GET}, \quad (1.10)$$

$$L' = 1006\,908''.04 + 129\,602\,768''.13T + 1''.089T^2. \quad (1.11)$$

Here (as above) T is measured in Julian centuries of 36 525 mean solar days from the epoch 1900.0 (Julian day number 2415020.0) and GET denotes the Great Empirical Term. These equations had been derived respectively by Brown (1915a) and Newcomb (1895b). Both of the above expressions include linear and quadratic terms deduced by Newcomb (1895b) to allow for precession – the latter of coefficient 1″.11.

As noted above, Jones (1939) adopted de Sitter's result for the lunar acceleration c of 5″.22. In order to secure close agreement with modern observations he also added a constant (+4″.65) and a linear term (+12″.96 T). The resulting correction to Brown's expression is therefore

$$\delta L = +4''.65 + 12''.96T + 5''.22T^2 - \text{GET}. \tag{1.12}$$

In the case of the Sun, Jones derived the following amendment to Newcomb's expression for L':

$$\delta L' = +1''.00 + 2''.97T + 1''.23T^2. \tag{1.13}$$

In 1948, Gerald M. Clemence – who at the time was unaware of Danjon's suggestion – used Jones' correction to Newcomb's mean solar longitude (equation (1.13)) to define 'Newtonian time'. This definition was based on the mean motion of the Sun in longitude: 1 arcsec in 24.349 sec of time. Denoting ΔT as the difference between GMT and the proposed new time-system (in the sense Newtonian time – GMT), Clemence derived the equation

$$\Delta T = 24.349 + 72.3165T + 29.949T^2 + 1.821B \text{ sec.} \tag{1.14}$$

Subsequently at the International Astronomical Union (IAU) General Assembly held at Rome in 1952 it was recommended that:

> In all cases where the mean solar second is unsatisfactory as a unit of time by reason of its variability, the unit adopted should be the sidereal year at 1900.0; that the time reckoned in these units be designated *Ephemeris Time*; that the change of mean solar time to ephemeris time be accomplished by the following correction:
>
> $$\Delta T = 24.349 + 72.3165T + 29.949T^2 + 1.821B \text{ sec,}$$
>
> where T is reckoned in Julian centuries from 1900.0 January 0 Greenwich Mean Noon and B has the meaning given by SPENCER JONES in *Monthly Notices*, R.A.S **99**, 541 (1939) and that the above formula also define the second.
>
> [IAU (1954).]

The name 'ephemeris time' (usually abbreviated to ET) had been suggested by Dirk Brouwer.

The unit of ET was thus chosen to be the same as that of Newcomb's solar ephemeris. In this ephemeris, the mean motion of the Sun was expressed in terms of the average length of the second – regarded as 1/86 400 of the mean solar day – during the latter half of the eighteenth century and most of the nineteenth. This resulted from the fitting of his theory to observations ranging in date from 1750 to 1892 (Newcomb, 1895b). Thus the average LOD on the GMT/UT scale during this interval (mean epoch *c.* 1820) became the standard LOD on the ET scale. At the IAU General Assembly at Dublin in 1954, the duration of the second was fixed on the basis of the above definition as 1/31 556 925.975 of the length of the tropical year for 1900.

On account of its slow motion and extreme brilliance, the Sun was by no means a suitable object for measuring ET. The much more rapidly moving Moon was well suited for this purpose. A revised formula was thus needed for the lunar mean longitude. Clemence (1948) showed that since the mean motion of the Moon is 13.37 times as rapid as that of the Sun, equation (1.13) is equivalent to the following correction to the mean lunar longitude:

$$\Delta L = 13''.37 + 39''.71T + 16''.41T^2. \qquad (1.15)$$

In order to convert Brown's expression for L to Newtonian Time (i.e. ET), the correction $(\delta L - \Delta L)$, as obtained from equations (1.12) and (1.15), was required. Representing the lunar mean longitude on the new system by λ,

$$\lambda = L + \delta L - \Delta L, \qquad (1.16)$$

or, inserting numerical values,

$$\lambda = L - 8''.72 - 26''.75T - 11''.22T^2 - \text{GET}, \qquad (1.17)$$

one could thus derive ΔT directly by comparing longitudes of the Moon computed from the lunar theory with observation.

The above amendment to the lunar mean longitude (equation (1.17)) was also recommended at the 1952 IAU Rome meeting 'in order to bring the lunar ephemeris into accordance with the solar ephemeris'. With a minor alteration to the coefficient of T (to 26''.74), it was incorporated in the *Improved Lunar Ephemeris* (ILE, 1954) and subsequently the lunar ephemeris $j = 2$ (IAU, 1968).

From section 1.6, twice the coefficient of T^2 in the above equation (i.e. -22.44 arcsec/cy^2) represents the estimated non-gravitational acceleration of the Moon (dn/dt or $\dot{n}$) caused by tides. This gradual retardation is accompanied by a slow recession of the Moon from the Earth. As will be discussed in chapter 2, the most reliable modern investigations, using a variety of techniques, indicate a value for $\dot{n}$ close to -26 arcsec/cy^2.

In 1960, the International Conference on Weights and Measures approved the definition of the second as:

the fraction 1/31 556 925.9747 of the tropical year 1900 January 0 at 12 hours ephemeris time.

In the same year, ET became the standard time-system adopted in the *Astronomical Ephemeris*. By this date the numerical coefficients for ΔT had been very slightly altered and the finally adopted formula was

$$\Delta T = 24.349 + 72.318T + 29.950T^2 + 1.82144B \text{ sec.} \quad (1.18)$$

Ephemeris time was to have a relatively short active life. As early as 1945, Isidor Isaac Rabi had suggested the idea of using atomic resonances in a chronometer. In 1955, the first atomic clock – the caesium beam resonator – was developed by L. Essen and J. V. L. Parry (Essen and Parry, 1957). This device had a stability of the order of 1 part in 10^{13}, losing or gaining less than 10^{-8} sec in a day. It was thus far superior to any solar system clock and – in particular – was ideally suited to the accurate determination of fluctuations in the LOD. By July 1955, atomic time-scales became available in several countries. At the International Conference on Weights and Measures in 1967, the second was redefined as:

The duration of 9192 631 770 periods of the radiation corresponding to the transition between the two hyperfine levels of the ground state of the caesium atom 133.

This unit, which is the fundamental unit of time in the international system (SI), was chosen to be as close to the length of the second of the ET scale as measurement would allow (Markowitz *et al.*, 1958).

The new standard became known as international atomic time (TAI) in 1971. As there had been no discontinuity in the adopted time-scale since 1955, the designation TAI can be effectively used as far back as that date. The origin of TAI was arbitrarily chosen so that the TAI and UT1 readings were identical at 0 h on Jan 1 in 1958. Since the value of ΔT was 32.184 sec at this moment, the difference ET–TAI was also 32.184 sec. Currently (1997) ΔT has a value of close to 60 sec.

In 1984, ET was replaced by terrestrial dynamical time (TDT) for astronomical purposes and the fundamental geocentric ephemerides were defined in terms of TDT. In order to provide continuity with ET, the difference between TDT and TAI was set as 32.184 sec. No significant variations between TDT – now simply known as terrestrial time (TT) – and TAI have been detected. The designation TT will be used in place of both TDT and ET throughout most of the remainder of this book.

1.9 Short-term fluctuations in the Earth's rotation

Although the fundamental aim of this book is the investigation of variations in the Earth's rate of rotation on a millennial time-scale, some brief remarks on shorter-term fluctuations are also appropriate. These fluctuations – which cannot be resolved using pre-telescopic observations – may be divided into two main groups, depending on their periodicity. The categories are: (i) sub-annual and inter-annual variations; and (ii) decade fluctuations.

Variations in category (i), which are of approximate amplitude 1 millisecond (1 ms), are of too short a periodicity to be in any way relevant to the theme of the present book. It is well established that they are produced by exchange of angular momentum between the atmosphere and the surface of the Earth, alterations in the force of the winds against mountain ranges being largely responsible. (For reference, the principal moment of inertia of the atmosphere is about 1.8×10^{-6} of that of the Earth itself.) Significant peaks at about 6 months, 1 year, 2.2 years and 4 years have been detected (e.g. Hide *et al.*, 1980; Jordi *et al.*, 1994). The higher frequency oscillations can only be mapped in detail (using techniques such as very long base-line interferometry) since 1955, when the TAI time-scale was introduced. However, the inter-annual variations can be detected as far back as 1890 by analysing timings of lunar occultations of stars.

The fluctuations which occur on a time-scale of decades – category (ii) – are of typical amplitude 3 ms. These variations can be traced over the last 350 years or so, mainly using recorded timings of occultations of stars by the Moon (Stephenson and Morrison, 1984). The principal causal mechanism would appear to be angular momentum transfer between the outer fluid core of the Earth and the surrounding solid mantle by the process of electromagnetic coupling. However, small alterations in global sea-level owing to freezing or melting of polar ice may also be partly responsible.

Over a period of several decades, the cumulative effect of systematic changes in the LOD at the millisecond level is very significant. To give a fairly typical example, over the past 40 years (some 15 000 days) the average LOD has been approximately 2 ms longer than the standard value of 86 400 SI sec. Hence the accumulated clock error TT − UT (i.e. ΔT) during that interval has amounted to some 30 sec. Even fairly crude telescopic timings of occultations of stars (which are effectively instantaneous events) can be used to trace similar decade fluctuations in the past in some detail. Resolution of these fluctuations is very good as far back as about AD 1860, but progressively deteriorates in earlier centuries (Stephenson and Morrison, 1984).

Figure 1.6, which is based on figure 1 of Stephenson and Morrison (1995), depicts the form of the ΔT curve since AD 1620 by a dotted line. If only tides and other long-term effects producing a fairly steady change in the LOD were significant, the basic form of this curve would have been parabolic. However, on a time-scale of centuries the decade fluctuations have tended to obscure the long-term trend (which is represented by a

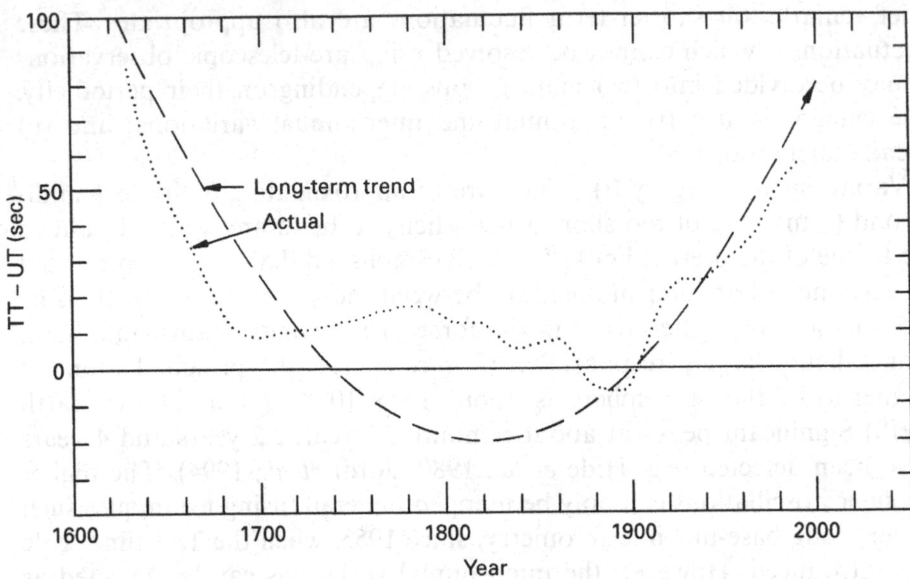

Fig. 1.6 The form of the ΔT curve since AD 1620 (shown by a dotted line), as depicted by Stephenson and Morrison (1995). The long-term parabolic trend (represented by a dashed line) is also depicted. The cusp of this mean parabola is placed close to AD 1820.

dashed line in the diagram). The cusp of this mean parabola is placed close to AD 1820 in figure 1.6. As discussed in section 1.8, this is the approximate mean epoch of the observations analysed by Newcomb (1895b) which were subsequently used to define the standard LOD on the ET scale. From Stephenson and Morrison (1995), the approximate equation of this parabola is $-20 + 31t^2$, where t is expressed in centuries relative to AD 1820 (see also chapter 14).

Figure 1.7 shows the first time-derivative of the modern ΔT curve, as derived by Morrison and Stephenson (1981). This diagram delineates actual changes in the LOD relative to the standard length of 86 400 SI sec. Power spectral analysis reveals a broad peak around a periodicity of 30 years. Calculations indicate that torques of the order of 10^{18} Nm would be required to produce the observed variations. Such torques are much greater than those produced by lunar and solar tides – which amount to about 5×10^{16} Nm (see chapter 2).

In analysing the various eclipse observations later in this book, the primary objective will be to determine a series of ΔT results by comparing individual values of UT derived from observation with their computed equivalents on TT. For each measurement,

$$\Delta T = \mathrm{TT} - \mathrm{UT}. \qquad (1.19)$$

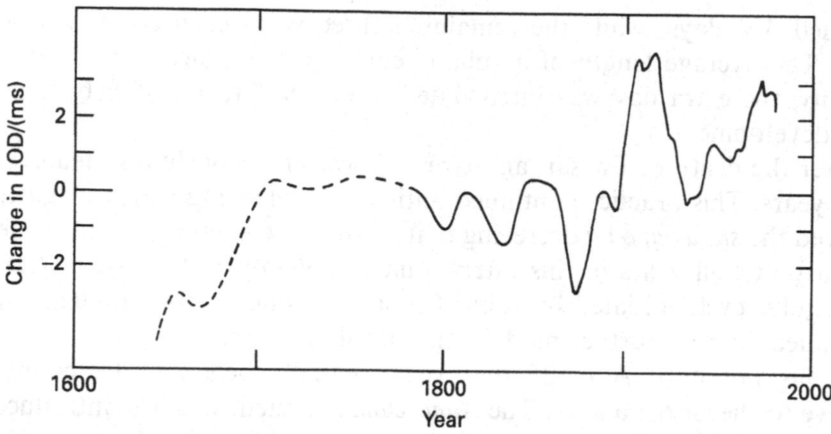

Fig. 1.7 The first time-derivative of the modern ΔT curve, as derived by Morrison and Stephenson (1981); changes in the LOD relative to the standard mean solar day of length of 86 400 SI sec are shown.

1.10 The Julian and Gregorian calendars

It seems appropriate to conclude this introductory chapter with some remarks on the Julian and Gregorian calendars, to which all eclipse dates are reduced throughout this book.

The fundamental calendrical periods are the tropical year of 365.2422 mean solar days (which fixes the seasons) and the synodic month of 29.5306 days (which determines the availability of moonlight). Early calendars followed several basic forms. Some were purely solar (e.g. that of ancient Egypt or the Julian system), others were exclusively lunar (e.g. the Muslim calendar), while a number of calendars were luni-solar (e.g. those of Babylon and China). Over the centuries, a variety of epochs have been adopted for counting years – for example the epoch of the Olympiads (776 BC), the traditional date of the foundation of Rome (753 BC), the Seleucid era (311 BC), the era of Diocletian (AD 284), and the era of the Hijra (AD 622) – in addition to the Christian era. Conversion between the various early calendars and their Julian or Gregorian equivalents will be discussed in the relevant chapters of this book.

The Julian calendar was instituted by Julius Caesar in 45 BC with the advice of the Alexandrian astronomer Sosigenes. This calendar replaced the luni-solar system of the Roman republic which, owing to faulty intercalation, became increasingly out of phase with the seasons. For example, by 50 BC the vernal equinox was falling in mid-May rather than in late March. Caesar directed that the year which is now designated 46 BC should contain an extra 90 days; starting with the following year a regular four-year cycle was to be adopted. In each cycle, the first year was assigned 366 days, while the remaining three years each contained 365 days. The average length of a Julian year was thus exactly 365.25 days. Initially, the extra day was intercalated after Feb 24; use of Feb 29 is a later development.

After the death of Caesar, an extra day was erroneously inserted every *three* years. This practice continued until 9 BC, following which Augustus restored the *status quo* by decreeing that further intercalation be suspended for 16 years, all years in this interval having 365 days. Finally, in AD 8 the regular cycle initiated by Julius Caesar was once more adopted. This continued without further modification until AD 1582.

It was not until AD 525 that years actually began to be counted relative to the Christian era. The *Anno Domini* system was first introduced by the monk Dionysius Exiguus, who employed it in his Easter tables. In England, these tables, and with them the Christian era, were adopted at the Synod of Whitby in AD 664. Not long afterwards, the Venerable Bede (AD 672–735) systematically employed *Anno Domini* in his important chronological writings, which gradually led to its popularisation. Use of the Christian Era became widespread in Europe by the eleventh century.

Dionysius had derived an incorrect date of 753 AUC (from the foundation of the city of Rome) for the birth of Christ; this event probably took place about 4 or 5 years previously. Although Dionysius' result was questioned by Bede, it was not altered by him. In the late ninth century, the German monk Regino rejected the equation 753 AUC = AD 1, but by then Dionysius' choice was too well established for any further alteration to be made. The use of BC dates (in addition to AD) did not find general favour until AD 1650, following the extensive chronological researches of Dionysius Petavius (AD 1583–1652).

Today, historians adopt a standardised Julian calendar which follows the same basic rules for all BC and AD dates up to AD 1582. For any date after the Christian era (up to AD 1582), all year numbers which are divisible by 4 – including century years – are bissextile or leap years. Since there is no year zero on this system (1 BC is immediately followed by AD 1), 4 BC is not a leap year, but 5 BC, 9 BC and so on are bissextile. In order to avoid a discontinuity of a year, historians of astronomy frequently adopt a variant of the BC/AD scheme in which years are numbered as positive or negative. On this system (which includes a year zero), years represented by positive numbers (e.g. +1133) correspond exactly with their AD equivalents. However, years denoted by negative numbers differ by unity from their BC alternatives; thus –430 corresponds to 431 BC. In particular, the year 0 is equivalent to to 1 BC. This choice simplifies the enumeration of intervals between selected years. Throughout the present book, all dates derived from historical researches will be expressed on the BC/AD system. However, when ΔT results are tabulated – as in the

Appendix – positive and negative year numbers will normally be used instead since a continuous system is necessary when analysing the results.

As an average Julian year is slightly longer than a tropical year (365.25 days compared with 365.2422 days), the start of each year on the Julian calendar became progressively delayed relative to the seasons. The difference amounted to one day every 128 years. Over the centuries, this defect produced a marked change in the vernal equinox (and thus in the date of Easter). Eventually it led to the reformation of the calendar by Pope Gregory XIII in AD 1582. For this task, Pope Gregory was advised by the Jesuit astronomer Christopher Clavius (AD 1537–1612).

In essence, there were two main steps in the adoption of the Gregorian calendar:

(i) Ten days in AD 1582 were omitted from calendar reckoning. The date 1582 Oct 4 (a Thursday), the last day of the Julian system, was immediately followed by 1582 Oct 15 (a Friday), the first day of the Gregorian scheme. In consequence, the date of the vernal equinox was restored to Mar 21, as it had been at the time of the Council of Nicea in AD 325. The sequence of weekdays was not interrupted, however.

(ii) In century years which were not divisible by 400 (e.g. AD 1700, 1800 and 1900) the intercalary day was now omitted. All other years which were divisible by 4 (including 1600 and 2000) remained as leap years.

Although the Gregorian calendar was at once officially adopted by most Roman Catholic countries, other countries delayed in its acceptance. For example in England the change was not made until as late as 1752; by then, since 1700 was not a leap year on the Gregorian reckoning, it was necessary to omit 11 days. Thus the day after 1752 Sep 3 (a Wednesday) was Sep 14 (a Thursday). The mean length of the Gregorian year is 365.2425 days, only 0.0003 days longer than the tropical year. The discrepancy amounts to no more than one day in about 3300 years.

On a systematic solar calendar – such as the Julian or Gregorian system – it is a relatively straightforward matter to calculate the exact number of days between any two selected epochs. Most computations of this sort make use of the Julian day number (JDN), devised by Joseph Justus Scaliger (AD 1540–1609). This scheme is based on a cycle of 7980 Julian years, commencing in 4713 BC. The precise epoch from which Julian days are numbered is Greenwich noon on Jan 1 (Julian calendar) in that year. A major reference epoch is 1900.0 (i.e. Greenwich noon on 1899 Dec 31), for which the JDN is 2415020.0. Since most modern astronomical computations make use of terrestrial time, it has become customary to use the Julian ephemeris date (JED) where necessary; this begins at 12h TT.

Various algorithms have been developed to convert from either the Julian or Gregorian calendar to the JDN, among the simplest being those deduced by Muller (1975), which make full use of the rules of FORTRAN. These are given in equations (1.20) and (1.21). Here J is the JDN, Y the year, M the month and D the day of the month; all variables are treated as integer. For any date AD on the Gregorian calendar:

$$J=367*Y - 7*(Y+(M+Y)/12)/4 - 3*((Y+(M-9)/7)/100+1)/4 +275*M/9+D+1721029. \quad (1.20)$$

For any date on the Julian calendar the following simpler formula applies:

$$J=367*Y-7(Y+5001+(M-9)/7)/4+275*M/9+D+1729777. \quad (1.21)$$

I have employed these formulae in many of the computer programs which I have developed.

2

Tidal friction and the ephemerides of the Sun and Moon

2.1 Introduction

In order to determine the value of ΔT from an eclipse observation, it is necessary to be able to calculate accurately the positions of the Sun and Moon at any selected epoch. By definition, the Sun has negligible acceleration on TT. However, the longitude of the Moon contains an appreciable quadratic term – part of which (owing to the reciprocal action of the tides) can only be determined empirically. Since many of the eclipses recorded in history are remote from the present-day, the effect of the lunar accelerative term on their calculated visibility is substantial. Furthermore, knowledge of the tidal component of this acceleration (usually denoted by $\dot{n}$), leads directly to a determination of the effect of the tides on the Earth's spin – see section 2.4. Consequently, it is important to investigate both the numerical value of $\dot{n}$ and its constancy during the historical period. Each of these questions will be considered in the immediately following sections (2.2 and 2.3).

2.2 Evaluation of the lunar tidal acceleration on TT

As discussed in chapter 1, the gravitational component of the lunar acceleration (coefficient of T^2 equal to $6''.05$) is well established. Up to about 1970, all estimates of the non-gravitational lunar acceleration were based on analyses in a UT framework. Results for the secular acceleration of the Moon (c) and Sun (c') on UT can be easily converted to $\dot{n}$ using the formula:

$$\dot{n} = 2(c - (n/n')c') \text{ arcsec/cy}^2, \tag{2.1}$$

where n/n' is the ratio of the mean lunar and solar motions. The factor of 2 is introduced because whereas c and c' are coefficients of T^2, $\dot{n}$ is a true acceleration. Substituting the numerical value for n/n' of 13.37, we obtain

$$\dot{n} = 2c - 26.74c' \text{ arcsec/cy}^2. \tag{2.2}$$

Thus for example using de Sitter's (1927) results for c of $5''.22 \pm 0''.30$ (p.e.) and c' of $1''.80 \pm 0''.16$ (p.e.) leads to $\dot{n} = -37.7 \pm 6.4$ (s.e.) arcsec/cy^2. Replacing the above figure for c' by Jones' (1939) value of $1''.23 \pm 0''.04$ (p.e.) gives instead $\dot{n} = -22.44 \pm 1.8$ (s.e.) arcsec/cy^2. As noted in chapter 1, this latter value for $\dot{n}$ was adopted in both the *Improved Lunar Ephemeris* (ILE, 1954) and the lunar ephemeris $j = 2$ (IAU, 1968).

Commencing in 1970, several investigators attempted to utilise ancient eclipse observations to solve simultaneously for both $\dot{n}$ and the mean acceleration of the Earth's spin ($\dot{\omega}/\omega$). These authors included Newton (1970 and 1972b), Muller and Stephenson (1975) and Muller (1976), results for $\dot{n}$ ranging from -42 to -30 arcsec/cy^2. Such efforts proved to be of limited usefulness because in the analysis of eclipses the two parameters $\dot{n}$ and ($\dot{\omega}/\omega$) are highly correlated. In practice, only a linear combination of these quantities can be satisfactorily determined from eclipse data.

In recent years, investigators of early eclipse observations have assumed an independent value for $\dot{n}$ in order to solve exclusively for ΔT. Fortunately for Earth rotation studies, several precise determinations of $\dot{n}$ using a variety of other means have been made in recent years. Techniques include the following: transits of the planet Mercury across the solar disk, artificial satellite data, and lunar laser ranging. Each of these methods will be discussed briefly below.

2.2.1 *Transits of Mercury*

In 1975, Morrison and Ward determined $\dot{n}$ from a comparison over most of the telescopic period between the ΔT curves obtained from (i) transits of the planet Mercury across the solar disk and (ii) occultations of stars by the Moon. Both Mercury and the Sun have negligible acceleration on TT. The investigation by Morrison and Ward, which covered the period from 1677 to 1973, yielded a result for $\dot{n}$ of -26 ± 2 arcsec/cy^2. This value has been closely confirmed in recent years using both artificial satellites and lunar laser ranging.

2.2.2 *Artificial satellite data*

The trajectory of a close artificial satellite is significantly perturbed by the tidal deformation of the solid Earth and oceans. Analysis of orbital measurements yield results for the tidal phase lag, leading to a value for $\dot{n}$.

Early results for $\dot{n}$ obtained using this technique include: -27.3 ± 5.2 arcsec/cy^2 (Lambeck, 1977) and -27.4 ± 3 (Goad and Douglas, 1978). Most recently, Christodoulidis *et al.* (1988) deduced -25.27 ± 0.61. In particular, these last authors determined the individual contributions to $\dot{n}$ from long-period, diurnal and semi-diurnal tides to be as follows (in arcsec/cy^2): −0.69 ± 0.36 (long-period), −3.18 ± 0.25 (diurnal), and −21.40 ± 0.43 (semi-diurnal). By far the largest contribution (−20.00 ± 0.40) comes from a single tide – the M2 (principal lunar semi-diurnal).

2.2.3 *Lunar laser ranging*

The first laser retro-reflector array was installed on the lunar surface by the *Apollo 11* astronaut Neil Armstrong in 1969. There are now five instruments of this kind on the Moon – the last deployed by *Lunakhod 2* in 1973; the signals have been mainly transmitted and received at McDonald Observatory in Texas.

The first published measurements in 1978 by Calame and Mulholland and separately by Williams *et al.* gave respectively $\dot{n} = -24.6 \pm 5$ and -23.8 ± 4 arcsec/cy^2. (Calame and Mulholland estimated the formal standard deviation to be ± 1.6 arcsec/cy^2, but suggested that ± 5 was a more realistic error estimate.) Soon afterwards, Ferrari *et al.* (1980) obtained −23.8 ± 2.6, while Dickey *et al.* (1982) deduced −23.8 ± 1.5.

Since 1992, results for $\dot{n}$ close to −26 arcsec/cy^2 have been derived from lunar laser ranging, the most recent figure being −25.88±0.5 (Dickey *et al.*, 1994). James G. Williams, one of the laser ranging team at Jet Propulsion Laboratory, Pasadena, has informed me (personal communication: April 1994) that lately values for $\dot{n}$ in the range −25.8 to −26.0 arcsec/cy^2 'have consistently been obtained'.

The rather low results obtained in the early 1980s were to have adverse consequences for contemporary ephemerides; a value for $\dot{n}$ of close to $-23''.9$/cy^2 was incorporated in both the Jet Propulsion Laboratory numerically integrated ephemeris DE200/LE200 (Standish, 1982) and the semi-analytical ephemeris ELP 2000 (Chapront-Touze and Chapront, 1983). Fortunately, a much more viable value for $\dot{n}$ of −26.21 arcsec/cy^2 is implicit in the long-term Jet Propulsion Laboratory numerical integration known as DE102/LE51 (Newhall *et al.*, 1983). This ephemeris, which is very useful for historical researches, covers the entire period from 1411 BC to AD 3002.

2.2.4 *Selected value for $\dot{n}$*

The weighted mean of the latest published results for $\dot{n}$ obtained from artificial satellite data (−25.27±0.61) and lunar laser ranging (−25.88±0.5) is 25.63 ± 0.39 arcsec/cy^2. However, bearing in mind the comment by Williams quoted above, I have adopted a figure for $\dot{n}$ of −26.0 arcsec/cy^2. This was the choice of Stephenson and Morrison (1995).

It is readily shown that as the lunar orbit expands,

$$\frac{\dot{r}}{r} = -\frac{2}{3}\frac{\dot{n}}{n}, \tag{2.3}$$

where r is the mean lunar distance. Inserting the appropriate values for r (= 3.84×10^8 m) and n (= $1.732 \times 10^{9''}$/cy) gives $\dot{r} = 0.148\dot{n}$ m/cy. A lunar acceleration of $-26/\text{cy}^2$ is equivalent to a rate of retreat of the Moon ($\dot{r}$) of 3.86 m/cy. Of more direct relevance in the present context, the length of the synodic month can be calculated to be increasing at an average rate of 0.038 sec/cy. Hence since 700 BC, the mean length of the month has increased by only 1.0 sec.

2.3 The constancy of the lunar tidal acceleration

The degree of constancy of $\dot{n}$, the lunar tidal acceleration, during the interval of approximately 2700 years covered by reliable historical observations is clearly a matter of some concern. This acceleration is directly proportional to the lunar tidal torque and hence to the rate of dissipation of the Earth's rotational energy (see section 2.5).

Lunar tidal dissipation takes place mainly in the hydrosphere; bodily friction represents only a small proportion of the total – perhaps 10 per cent (e.g. Lambeck, 1980, p. 329). Hydrospheric dissipation occurs both in the deep oceans and the continental shelves, but the precise ratio of the contributions from these two sources is uncertain. On the time-scale covered by the present investigation, dissipation in the deep oceans ought to be sensibly constant since global parameter changes (e.g. ocean–continent distribution) should be negligible over a very few millennia. However, there is evidence that even on the shelf areas changes should be very small. There has not been any appreciable change in sea-floor topography and although significant localised sea-level variations have probably occurred, global alterations in sea-level have been minor. Studies of tide-gauge records by Nakiboglu and Lambeck (1991) indicate a global sea-level rise in recent decades by 1.15 ± 0.3 mm/y. Measurements of global sea-level variations on a time-scale of millennia, although less reliable, suggest roughly the present rate over the past 2700 years (Pirazzoli, 1991). A significantly more marked change would appear to be unlikely since the entire period is long after the last ice-age (which ended some 10 000 years ago). On this basis, it may be estimated that since 700 BC mean sea-level has probably not changed by more than about 3 m.

In general, the average depth of water on the continental shelves is about 100 m; a change in the global sea-level by 3 per cent (i.e. by 3 m) would probably alter the area of the shelves by a similar ratio. No model is available which can predict just how tidal dissipation would respond to variations in both area and water depth on the continental margins. However, the very small magnitude of these variations suggests that the effect on global dissipation will be negligible. In summary, the assumption of a sensibly constant value for $\dot{n}$ over the historical period seems justified. (I am grateful to Dr J. M. Huthnance of the Proudman Oceanographic Laboratory, Bidston, Merseyside for helpful discussion on this question.)

2.4 The tidal acceleration of the Earth's spin

Analysis of ancient and medieval eclipse observations leads to a result for the total rate of change in the LOD owing to both tides and other causes. Hence any residual non-tidal lengthening of the day (for example produced by post-glacial uplift, sea-level variations, etc.) may be obtained by subtracting the tidal component which is produced by the action of both the Moon and Sun.

Christodoulidis *et al.* (1988) analysed the perturbations of the orbits of near-Earth artificial satellites by tidal effects. On the basis of conservation of angular momentum in the Earth–Moon system, these authors obtained the following empirical relation between the observed tidal acceleration of the Moon and the acceleration of the Earth's spin caused by lunar and solar tides ($\dot{\omega}_T$):

$$\dot{\omega}_T = (+49 \pm 3)\, \dot{n}\, \mathrm{arcsec/cy}^2. \qquad (2.4)$$

This result was only marginally smaller than that obtained by Lambeck (1980, p. 337) by combining oceanic estimates, tidal solutions and astronomical estimates for $\dot{\omega}_T$ – i.e. $(+51 \pm 4)\ \dot{n}$ arcsec/cy^2.

Converting from arcsec/cy^2 to rad/sec^2 (the customary units for $\dot{\omega}_T$), by multiplying by the factor 4.868×10^{-25} gives

$$\dot{\omega}_T = (2.39 \pm 0.15) \times 10^{-23}\, \dot{n}\, \mathrm{rad/sec}^2. \qquad (2.5)$$

Inserting $\dot{n} = -26.0$ arcsec/cy^2 yields

$$\dot{\omega}_T = (-6.20 \pm 0.38) \times 10^{-22} \mathrm{rad/sec}^2.$$

This last result may be reduced to the rate of change in the LOD (Z_T) in ms/cy by multiplying by a factor 3.745×10^{21}:

$$Z_T = 2.3 \pm 0.1 \mathrm{ms/cy}.$$

This result will be assumed in subsequent investigations; see chapter 14.

If the LOD were to increase at a constant rate of 2.3 ms/cy, there would be a parabolic divergence between UT (as measured by the Earth's rotation) and TT. This would take the form

$$\Delta T_{\mathrm{tidal}} = \mathrm{TT} - \mathrm{UT} = q_T t^2, \qquad (2.6)$$

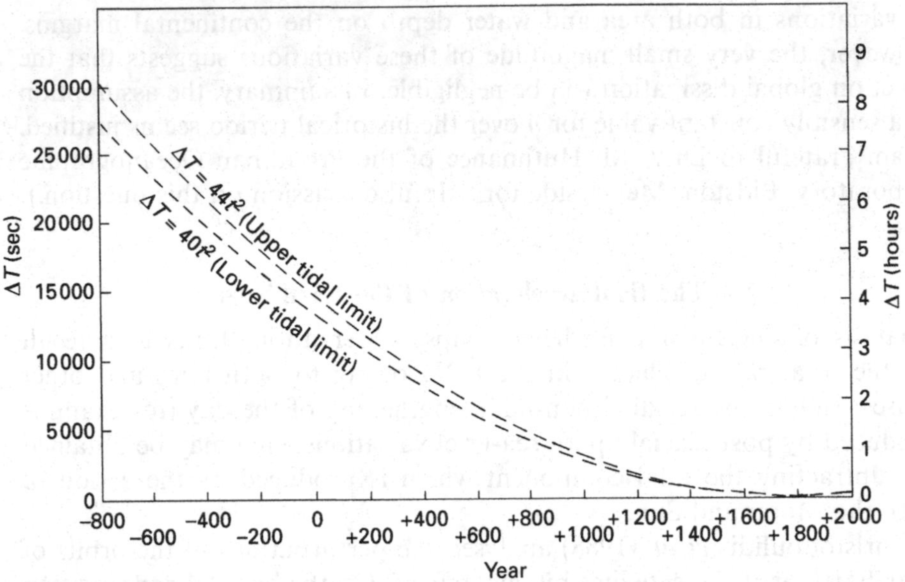

Fig. 2.1 The estimated range in $\Delta T_{\rm tidal}$ in sec as far back as 800 BC according to the results of Christodoulidis *et al.* (1988).

where $q_{\rm T}$ is a constant and t is reckoned from the epoch at which the unit of UT is equal to the unit of TT. From chapter 1, this epoch lies close to AD 1820. If ΔT is measured in seconds and t in Julian centuries of 36 525 ephemeris days, then $q_{\rm T}$ can be expressed in terms of $Z_{\rm T}$ (ms/cy) by

$$q_{\rm T} = 36.525 Z_{\rm T}/2. \tag{2.7}$$

With $Z_{\rm T} = 2.3 \pm 0.1$ ms/cy, then from equations (2.6) and (2.7)

$$\Delta T_{\rm tidal} = (42 \pm 2)t^2. \tag{2.8}$$

On this basis, the estimated range in $\Delta T_{\rm tidal}$ in sec as far back as 800 BC is depicted in figure 2.1. Since many ancient and medieval observations were timed to a small fraction of an hour, it can be seen that especially in earlier centuries any significant non-tidal variations from the mean ΔT parabola should be readily apparent.

2.5 Lunar and solar tidal dissipation

The torques which the Moon and Sun produce on the rotating Earth may be readily estimated. From Kepler's third law and the principle of conservation of angular momentum, the lunar tidal torque (F) is approximately given in terms of $\dot{n}$ by

$$F = -EMr^2\,\dot{n}/3(E+M), \tag{2.9}$$

where E and M are the masses of the Earth and Moon and r is the mean separation. Expressing $\dot{n}$ in terms of arcsec/cy^2 in the above equation and inserting numerical values for E ($= 5.58 \times 10^{24}$ kg), E/M ($= 81.3$) and r ($= 3.84 \times 10^8$ m) gives the mean lunar tidal torque F as

$$F \sim -1.74 \times 10^{15}\dot{n} \text{ Nm}. \tag{2.10}$$

With $\dot{n} = -26.0$ arcsec/cy^2, $F \sim +4.5 \times 10^{16}$ Nm.

Jeffreys (e.g. 1976, p. 319) showed that if friction in the oceans is assumed to be linear, the ratio of the solar tidal torque (F') to that of the Moon is equal to 1/4.9. At the other extreme, for quadratic friction, this ratio increases to 1/3.8 (Newton, 1968). Hence if one adopts a mean ratio for F'/F of 0.23, the combined lunar and solar torque amounts to some $+5.6 \times 10^{16}$ Nm. In opposition to the main deceleration, the semi-diurnal atmospheric tide is responsible for a small accelerative term. The torque produced by this mechanism is estimated by Volland (1990) as $(-3.2 \pm 0.3) \times 10^{15}$ Nm. Adding this to the above result of $+5.6 \times 10^{16}$ Nm gives $+5.3 \times 10^{16}$ Nm as a fair estimate of the tidal torque (F_T) causing a retardation of the Earth's spin.

The rate at which the tidal torque does work (P_T) is given by

$$P_T = F_T(\omega - n). \tag{2.11}$$

Inserting the appropriate values for ω ($= 7.29 \times 10^{-5}$ rad/sec) and n ($= 0.27 \times 10^{-5}$ rad/sec) yields

$$P_T \sim 3.7 \times 10^{12} \text{ W}.$$

The first estimate of global tidal dissipation based on oceanic studies was made by Jeffreys in 1920. Jeffreys considered that the contribution from the open oceans would be negligible, but he was of the opinion that shallow seas would be a major source of energy loss. Following a paper by Taylor (1919) which contained estimates of the rate of dissipation of energy by tides in the Irish Sea, Jeffreys (1920) obtained data on tidal ranges and currents for shallow seas in various parts of the Earth (mainly from the *Admiralty Pilot*). On this basis, he calculated a value for P_T of 1.1×10^{12} W. Soon afterwards, Heiskanen (1921) made similar computations, deriving $P_T = 1.9 \times 10^{12}$ W. Although both of these results were considerably smaller than the value derived above (i.e. 3.7×10^{12} W), they were at least of the correct order of magnitude.

In relatively recent years, several numerical models of global ocean tides have been published. These incorporate terms for ocean loading and gravitational self-attraction into the Laplace tidal equations. Perhaps the most comprehensive solution for P_T obtained in this way is that by Lambeck (1980, p. 330). By summing the contributions of the individual oceanic tides (both lunar and solar), Lambeck derived a value for P_T of

4.42×10^{12} W. This result is in adequately close accord with that obtained above from astronomical data.

2.6 Ephemerides of the Sun and Moon

All eclipse computations in this book have been made using specially designed programs – which are adaptations of standard ephemerides for the Sun and Moon. In computing solar positions I have used Newcomb's analytical ephemeris (Newcomb, 1895a). All Fourier terms of amplitude greater than 0.025 arcsec in longitude and 0.005 arcsec in latitude have been included. For lunar positions a modification of the analytical ephemeris $j = 2$ (IAU, 1968), in which $\dot{n} = -26$ arcsec/cy^2, has been adopted. In this case, all Fourier terms in longitude greater than 0.05 arcsec, and in latitude greater than 0.01 arcsec have been included. In testing the accuracy of these ephemerides, I am grateful to Dr E. M. Standish of Jet Propulsion Laboratory, Pasadena, for supplying comparison solar and lunar positions based on DE102/LE51. Further details are given below.

2.6.1 Solar ephemeris

Newcomb's ephemeris for the Sun – which effectively defines the TT scale – incorporates the following expression for the solar mean longitude relative to the mean equinox of date:

$$\lambda'_{1900} = 1006\,908''.04 + 129\,602\,768''.13T + 1''.089T^2, \tag{2.12}$$

where T is in Julian centuries of 36 525 days measured from the epoch 1900.0 (Julian day number 2415020.0). The corresponding expression for the epoch J2000 (i.e. 2000.0: Julian day number 2451545) is

$$\lambda'_{2000} = 1009\,677''.26 + 129\,602\,770''.31E + 1''.089E^2, \tag{2.13}$$

where E is also in Julian centuries.

For comparison, the linear term in the solar mean longitude derived by Bretagnon (1982) relative to J2000 has the coefficient 129 602 771″.36, while the parabolic term has coefficient +1″.093. (A minute cubic term is ignored.) The difference between the two parabolic terms ($0''.004E^2$) is negligible, amounting to no more than 2″ at 700 BC – the approximate date of the oldest reliable eclipse observations. Since the disparity between the linear terms (1″.05E) reaches some 30 arcsec at 700 BC, this would alter the computed times of eclipses or occultations by about 1 minute. Such a discrepancy is insignificant at this early period.

A comparison which I have made between solar longitudes obtained using Newcomb's ephemeris and those from the numerically integrated ephemeris DE102 for dates of historical eclipses yields deviations less than about 30″ throughout the entire period since 700 BC; these show a roughly linear trend.

2.6.2 Lunar ephemeris

The mean lunar longitude derived by Brown (1915a) relative to the mean equinox of date is

$$L = 973571''.71 + 1732\,564\,406''.06T + 7''.14T^2 + 0''.0068T^3. \quad (2.14)$$

As above, T is measured in Julian centuries from the epoch 1900.0. From chapter 1, the correction to the above expression necessary to convert from a UT to a TT framework – as in $j = 2$ – is

$$\lambda = L - 8''.72 - 26''.74T - 11''.22T^2, \quad (2.15)$$

where λ is the mean longitude of the Moon on TT.

The equivalent value of $\dot{n}$ is -22.44 arcsec/cy^2. A further amendment to $j = 2$ is necessary to incorporate a value for $\dot{n}$ of -26.0 arcsec/cy^2, as derived by Morrison (1979). From an analysis of numerous occultations reported between 1943 and 1974, Morrison showed that the following addition should be made in order to obtain the optimum accord with modern observations:

$$\delta\lambda = -1''.544 + 2''.330T - 1''.78T^2. \quad (2.16)$$

The sum of equations (2.15) and (2.16), i.e. $-10''.26 - 24''.41T - 13''.00T^2$, is the total correction which has been applied to Brown's expression for the lunar mean longitude. This yields a revised result at the epoch 1900.0 of

$$\lambda_{1900} = 973\,561''.45 + 1732\,564\,381''.65T - 5''.86T^2 + 0''.0068T^3. \quad (2.17)$$

When expressed in terms of the epoch J2000 this becomes:

$$\lambda_{2000} = 785\,937''.25 + 1732\,564\,369''.95E - 5''.88E^2 + 0''.0068E^3. \quad (2.18)$$

For comparison, the semi-analytical lunar ephemeris devised by Chapront-Touze and Chapront (1983), when adjusted for precession of $5029''.10E + 1''.11E^2$ (Laskar, 1986) incorporates a coefficient of E of $1732\,564\,372''.84$ and a quadratic term of $-4''.79E^2$. The latter corresponds to a value for $\dot{n}$ of -23.9 arcsec/cy^2. As noted earlier (section 2.2.3), this has in retrospect proved to be an unfortunate choice.

Over the period covered by historical observations, the main discrepancy between the adopted expression for the lunar mean longitude and that due to Chapront-Touze and Chapront (1983) lies in the difference between the quadratic terms – i.e. $1''.03E^2$. This amounts to as much as some 750″

around 700 BC. By comparison the linear terms differ by only about $3''E$, or less than 80″ at the above epoch.

A long-term comparison between the modified version of $j = 2$ used in this study – based on equation (2.18) – and DE102/LE51 (which incorporates $\dot{n} = -26.21$ arcsec/cy^2) yields much smaller deviations. This conclusion is expected since the difference between the quadratic terms in λ is only about $0''.11T^2$. Discrepancies in longitude between the two ephemerides reach about 100″ at 700 BC, while the maximum discrepancy in latitude is about 10″.

Most of the deviations in the lunar position between the adopted ephemeris and DE102/LE51 can be accounted for by the slight discord between the selected value for $\dot{n}$ (−26.0 arcsec/cy^2) and that implicit in DE102/LE51. When this parabolic difference is allowed for, the discrepancies in longitude at 700 BC reduce to no more than 25″ while latitude errors do not exceed 2″. Both of these deviations – which show a gradual decrease as E diminishes – are negligible for the present purpose.

2.7 Conclusion

Having considered the accuracy of the adopted solar and lunar ephemerides, we are now in a position to discuss the various techniques available for the analysis of solar and lunar eclipse observations. This will form the subject of chapter 3, after a preliminary discussion of the viability of alternative data.

3

Pre-telescopic eclipse observations and their analysis

3.1 Introduction

Despite their relatively low precision, pre-telescopic observations cover a sufficient time-span for long-term trends in the length of day (LOD) to become apparent. These trends cannot be discerned from modern measurements. Here we have the main reason why archaic observations are so important in the study of the Earth's past rotation.

The analysis of ancient and medieval eclipse records is just one of several techniques which have been utilised in recent years to investigate variations in the LOD over the historical past. Before discussing in detail the application of eclipse observations to this problem, it is necessary to briefly consider other available methods and to explain why eclipses are to be preferred to other types of data.

3.2 Observational requirements for determining ΔT in the pre-telescopic period

Any early astronomical observation which is of value in studying changes in the LOD in the past must satisfy a wide variety of criteria. These may be listed as follows:

(i) The observation must involve at least one of the brighter and more rapidly moving objects in the solar system (i.e. the Moon, Sun or one of the inner planets Mercury, Venus and Mars).

(ii) The exact Julian or Gregorian date of the observation must either be specified directly or be able to be determined unambiguously.

(iii) A reasonably accurate value of the UT must be reducible from the reported circumstances, or – as in the case of total solar eclipses – the rotational phase of the Earth must be able to be derived.

(iv) The corresponding TT must be able to be precisely calculated from the dynamical equations of motion of the celestial body or bodies concerned.

(v) In most cases the geographical position of the observer must be known with some precision – preferably to within a small fraction of a degree in both latitude and longitude.

Since careful measurements of the co-ordinates of the Sun, Moon and planets are fairly rare throughout the pre-telescopic period, studies of the Earth's past rotation have mainly concentrated on the observation of specific *events*, such as conjunctions between two celestial bodies. As will be discussed in detail below, numerous early records of both solar and lunar eclipses satisfy all of the above conditions and form an impressive body of data. Other types of observation are of much more limited utility.

3.3 Alternatives to eclipses in the pre-telescopic period

Potentially viable observations other than eclipses include: (i) equinox and related measurements; (ii) close planetary conjunctions; and (iii) occultations of stars and planets by the Moon. These will be considered in turn below.

3.3.1 Equinox and related measurements

Early astronomers not only recorded the dates of the vernal and autumnal equinoxes but also occasionally made careful estimates of the local times when the Sun crossed the celestial equator – for example by measuring the meridian altitude of the Sun on a pair of dates adjacent to either equinox. At the equinoxes, the declination of the Sun is changing by about 1 arcmin in an hour. This is close to the limit of resolution of the unaided eye so that in principle individual measurements are capable of fixing ΔT to the order of an hour. However, on account of instrumental defects and other factors, most early measurements do not attain this ideal. Preserved equinox and related observations (e.g. determinations of the meridian altitude of the Sun at other times of year) from the pre-telescopic period fall into two main groups: (i) ancient Greek measurements and (ii) medieval Arab records.

A series of 20 Greek equinox observations from the time of Hipparchus is reported by Ptolemy in his *Almagest* (III, 2). These range in date from 162 to 128 BC and most observations were probably made by Hipparchus himself. In each case the time when the Sun crossed the celestial equator is estimated to the nearest 6 hours. The most recent investigations of these measurements in the study of the Earth's past rotation are by Newton

(1970, pp. 9–16) and Muller (1976). Newton obtained a best estimate for ΔT of +4.2 ± 1.0 h (i.e. +15 100 ± 3600 sec) at epoch −143 (i.e. 144 BC), while Muller deduced +4.2 ± 0.8 h (+15 100 ± 2900 sec) at essentially the same epoch. As will be evident from subsequent chapters of this book, individual ancient eclipse observations frequently lead to uncertainties in ΔT smaller than those obtained from the whole series of equinox data.

In his *Kitab Tahdid Nikayat al-Amakin Li-Tashih Masafat al-Makasin* (The Determination of the Co-ordinates of Positions for the Correction of Distances between Cities), the great Muslim astronomer al-Biruni (AD 973–1048) recorded 12 equinox times determined between AD 829 and 1019. This treatise has been translated into English by Ali (1967). Most of these equinox times are quoted to the nearest hour. Al-Biruni also listed more than 40 measurements of the meridian altitude of the Sun between AD 832 and 1018, as well as a few miscellaneous observations. The equinox data will be referred to below as set (a), while the other observations will be denoted set (b).

The most comprehensive study of the medieval measurements is by Newton (1972a), who expressed his results in terms of ω, the rate of spin of the Earth. From set (a), Newton derived a value for 10^9 $(\dot{\omega}/\omega)$ of -18.3 ± 7.0 cy^{-1}. The corresponding value for ΔT at the mean epoch of AD 890 may be deduced as approximately +2500 ± 1000 sec. Newton did not quote a separate result for $(\dot{\omega}/\omega)$ from set (b) but he stated that his combined solution from (a) and (b) was 10^9 $(\dot{\omega}/\omega) = -26.5 \pm 5.8$ cy^{-1}. This implies a result for set (b) of -44.6 ± 10.4 cy^{-1}. At the mean epoch of AD 940, the equivalent value of ΔT would be about +5400 ± 1300 sec. It is thus clear that there is considerable discord between the individual solutions, the second being more than twice as large as the first. In particular, the standard errors would appear to have been considerably underestimated. It may be concluded that by the medieval period, equinox observations and meridian altitude measurements give only a general indication of the value of ΔT.

3.3.2 *Close planetary conjunctions*

Many close conjunctions between two planets or between one planet and a bright star are recorded in Babylonian, East Asian and Arab history. In some cases, it is specifically stated that a planet appeared to conceal another planet or star. Most recorded occurrences of this kind are untimed, but a number of such events observed by Arab astronomers between AD 858 and 1003 were timed to the nearest hour or so. These observations are reported by Ibn Yunus (d. AD 1009) in his *al-Zij al-Kabir al-Hakimi* (Astronomical Handbook dedicated to Caliph al-Hakim), sections of which have been translated into French by Caussin (1804). Most

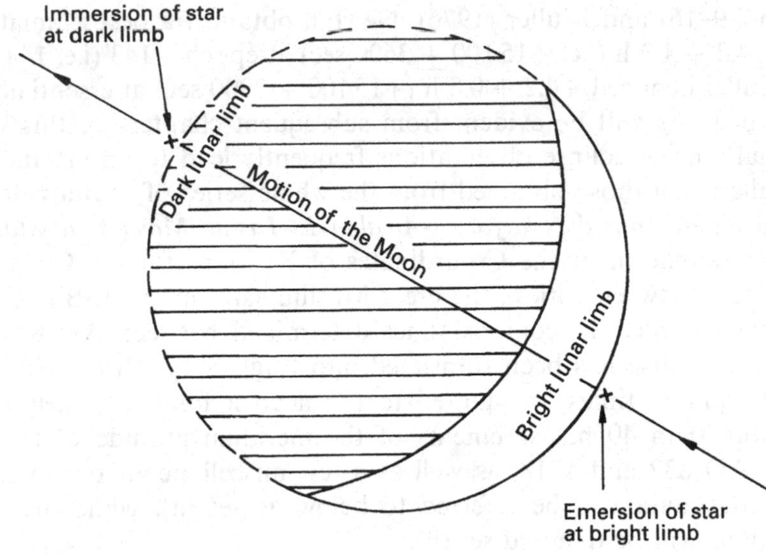

Fig. 3.1 Schematic representation of a lunar occultation.

conjunctions reported by Ibn Yunus involve the planet Venus. Although Venus moves fairly rapidly (mean motion 1.6 times as fast as that of the Sun), it is such a brilliant object that its light tends to overpower other planets or stars which are located in its immediate vicinity.

From an analysis of the medieval Arab data, Newton (1970, pp. 164–174 and 200–210) derived 10^9 $(\dot{\omega}/\omega) = -20 \pm 24$ cy^{-1}. At the mean epoch of AD 970 – roughly the same average date as for the Arab equinox observations – the corresponding value of ΔT is $+2300 \pm 2700$ sec. Newton expressed some disappointment at the large uncertainty in his result. Like the medieval equinox and related data, it would appear that close planetary conjunctions have little to offer in the determination of ΔT.

3.3.3 Occultations of stars and planets by the Moon

Figure 3.1 shows a schematic representation of a lunar occultation. First contact (immersion) occurs at the eastern limb of the Moon and last contact (typically about an hour later) at the western limb.

Although telescopic observations of occultations provide the best dataset for measurement of ΔT from soon after the introduction of the telescope to the advent of the TAI scale in 1955 (Stephenson and Morrison, 1984), most pre-telescopic reports of these events are of little use. Owing to the brilliance of the Moon – especially at large elongations from the Sun – it is difficult for the unaided eye to decide whether or not an occultation is actually taking place unless the occulted object is extremely bright (e.g. Venus or perhaps Jupiter). Further, the terminology in early texts is often vague.

Among ancient and medieval civilisations, only the Babylonians and the inhabitants of East Asia (Chinese, Koreans and Japanese) recorded occultations on a regular basis. Several tens of observations of this kind are described in Babylonian history, but East Asian history is replete with such reports. Nevertheless, very few times are noted; in most cases it is simply stated that an occultation of a certain star or planet occurred on a specified day. The occulted object is often described as 'entering' the Moon. Ancient Greek astronomers occasionally timed occultations of stars by the Moon but only two events of this kind involving a bright star are preserved (in the *Almagest*, VII, 3).

Hilton *et al.* (1992) analysed a series of 58 untimed Chinese observations of occultations of planets by the Moon between 68 BC and AD 575. Expressing ΔT in the form qt^2, where t is in Julian centuries measured from an epoch around AD 1800, they derived values for q by minimising the sum of the squares of the least separations between Moon and planet as a function of q. By applying equal weights to all observations, Hilton *et al.* obtained a result for q of $+12.6 \pm 10.2$ arcsec/cy^2. At the mean epoch – which was close to AD 450 – the approximate value for ΔT would thus be $+2400 \pm 1900$ sec. By adopting different weighting schemes, they derived results for q of around +35 arcsec/cy^2 ($\Delta T \sim +6600$ sec) but the uncertainties were very high – of the order of 20 or 30 arcsec/cy^2. Hilton *et al.* concluded that:

> Overall the check on the change in the rotation rate (of the Earth) is very weak, but it represents the limit of what can be done with known, untimed occultation records.

3.3.4 *Further remarks*

Ancient and medieval observations of equinoxes, close planetary conjunctions and occultations yield disappointing results for ΔT. It is indeed fortunate that a promising array of both timed and untimed observations of eclipses is available. Reliable records of these events commence around 700 BC and continue throughout most of the pre-telescopic period.

3.4 Historical eclipses

Both solar and lunar eclipses can often be very impressive events. As a result they have attracted considerable attention throughout much of recorded history. Indeed, several thousand individual records of eclipses are preserved from the pre-telescopic period in a variety of sources. However, only a small proportion of these early data proves to be of value in the study of the Earth's past rotation. Most useful observations, notably timings of individual phases, are the work of astronomers. Nevertheless, many useful accounts of eclipses – especially total obscurations of the Sun – were written by chroniclers and others who had no special interest in astronomy.

One of the principal reasons for early astronomers observing eclipses was to test the accuracy of existing almanacs and improve future predictions. In addition, medieval Arab stargazers measured the times of lunar eclipses in order to determine longitude on the Earth's surface. Especially in Babylon and China, the ultimate goal in watching for eclipses and other celestial phenomena was astrological. There can be little doubt that without the influence of astrology many of the astronomical observations which exist from the ancient and medieval world would have been neither made nor preserved. Here we have perhaps the only positive contribution which this pseudo-science has made to world culture!

In reporting eclipses, the main concern of chroniclers seems to have been to describe the phenomenon as a spectacle. Many of these events evidently produced a profound impression on eyewitnesses, whether or not they had any special interest in astronomy. Eclipses of both Sun and Moon are frequently noted in historical writings, but in such works only solar obscurations are of much value in the study of changes in the LOD.

Before discussing the different types of eclipse observation which are available from various parts of the ancient and medieval world, some remarks on the cause of eclipses are appropriate.

3.5 Cause of eclipses

From the point of view of Earth rotation studies, solar and lunar eclipses have a number of mutually distinct features. In the case of a solar obscuration, a shadow is cast by the Moon on the rotating Earth. As a result, the visibility of the various phases at any particular site depends very much on the value of the accumulated clock error ΔT at that date. A lunar eclipse occurs when the Moon enters the terrestrial shadow so that only the *local* time of each individual stage is a function of ΔT; parameters such as magnitude and duration are not usually affected by Earth's rotation. Nevertheless, early solar and lunar observations are sufficiently diverse as to complement one another in the investigation of long-term changes in the length of the day.

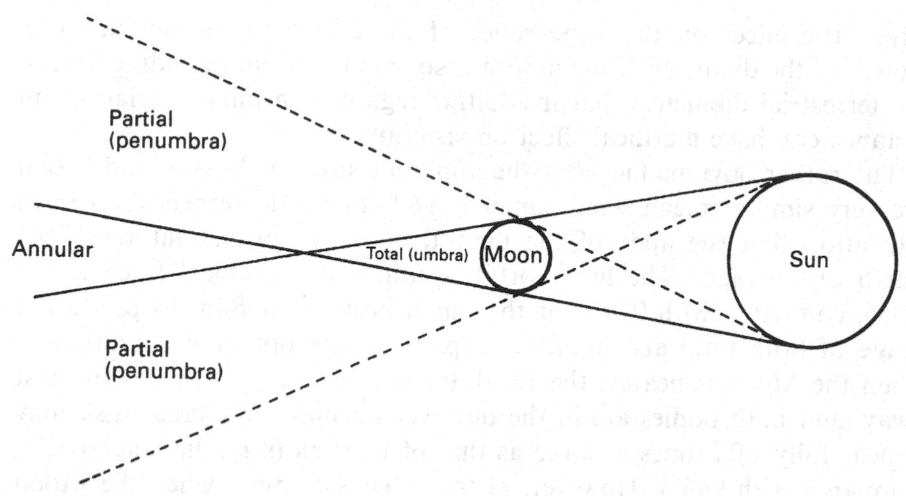

Fig. 3.2 Basic mechanism of solar eclipse formation.

3.5.1 Solar eclipses

Solar eclipses are produced by the direct interposition of the Moon between the Sun and Earth. As seen from our planet, the Moon is in conjunction with the Sun (astronomical new Moon) every synodic month – i.e. at mean intervals of 29.5306 days. Hence if the lunar and terrestrial orbits were in the same plane, there would be 12 or 13 solar eclipses each year. In fact, the two orbits are inclined at an angle of about 5.1 deg. Thus from the Earth's viewpoint, at most conjunctions the Moon passes either above or below the Sun and no eclipse is produced. A solar eclipse can only occur when at the time of conjunction the Moon is close to one of its nodes – the two points 180 deg apart where its orbital plane intersects that of the Earth. During an average century, there are approximately 238 solar eclipses visible on (at least some part of) the Earth's surface. This figure represents only about 19 per cent of the new Moons occurring during this same time. Nevertheless, on the Earth as a whole, eclipses of the Sun are still a fairly frequent phenomenon. In most years two or three of these events take place, but rarely this number can be as large as four or even *five* (as last took place in 1935).

The basic mechanism of solar eclipse formation is shown in figure 3.2. Within the penumbral region, the Sun is partially obscured by the Moon, but in the full umbral shadow the entire solar disk is hidden. Although the Moon is much smaller than the Earth (ratio of diameters 0.272), the penumbral shadow has a diameter rather more than half (approximately 0.54) of the Earth at mean distance. Hence a partial solar eclipse is usually visible over a wide area. Within the penumbra, variations in the distance of the Moon and Sun from the Earth owing to the ellipticity of the orbits have little effect on the appearance of an eclipse to the unaided eye. However, the diameter of the umbra is so small (seldom exceeding 0.02 of the terrestrial diameter) that inside this region even minor variations in distance can have a critical effect on visibility.

The rather obvious fact that the apparent sizes of the Sun and Moon are very similar (mean semi-diameters 960″ and 931″ respectively) is an indication that the apex of the umbral cone can just about reach the Earth on average. The lunar orbit is sufficiently elliptical (eccentricity 0.055, compared with 0.017 for the Earth around the Sun) to produce a range of both total and annular eclipses. Under optimum conditions – when the Moon is nearest the Earth (i.e. at perigee), the Sun is furthest away, and both bodies are in the observer's zenith – the lunar disk may appear fully 1.08 times as large as that of the Sun (semi-diameter 1023″, compared with 946″). However, at the other extreme – when the Moon is furthest from the Earth (at apogee), the Sun is closest to us, and both bodies are near the observer's horizon – the Moon may cover no more than about 0.90 of the Sun (semi-diameter 881″, compared with 978″). These two extremes are illustrated in figures 3.3a and 3.3b. Of the approximately 238 eclipses visible on the Earth during an average century, some 84 are partial, 77 are annular, 66 are total and 11 are combinations of annular and total (von Oppolzer, 1887). These frequencies are only slightly variable from one century to the next.

When the umbra extends fully to the Earth, the entire solar photosphere is screened from a small section of the surface of our planet for a short time. On account of the orbital motion of the Moon and rotation of the Earth, the umbral shadow sweeps progressively across the terrestrial surface in a narrow band; this is a section of a small circle – see figure 3.4, which shows the path of totality at the eclipse of AD 968 Dec 22 as calculated by Schroeter (1923). For observers in this zone, the sky suddenly grows very dark and often several stars become visible. The corona (the outer atmosphere of the Sun) also appears, but for some reason this is very rarely mentioned in early records. Under a vertical Sun the umbra can never exceed about 270 km in diameter. However, at low solar altitudes the elongated shadow of the Moon may be much wider than this, occasionally exceeding 500 km.

The change in light intensity during the various stages of a total solar eclipse is shown schematically in figure 3.5 (lower curve). During totality, most of the available light comes from scattered sunlight near the horizon where the eclipse is still only partial. When the atmosphere is relatively dust-free, the illumination may decrease by more than six orders of magnitude between the start of the eclipse and the onset of totality; most of the diminution in light level occurs in the last few seconds before the Sun is completely obscured.

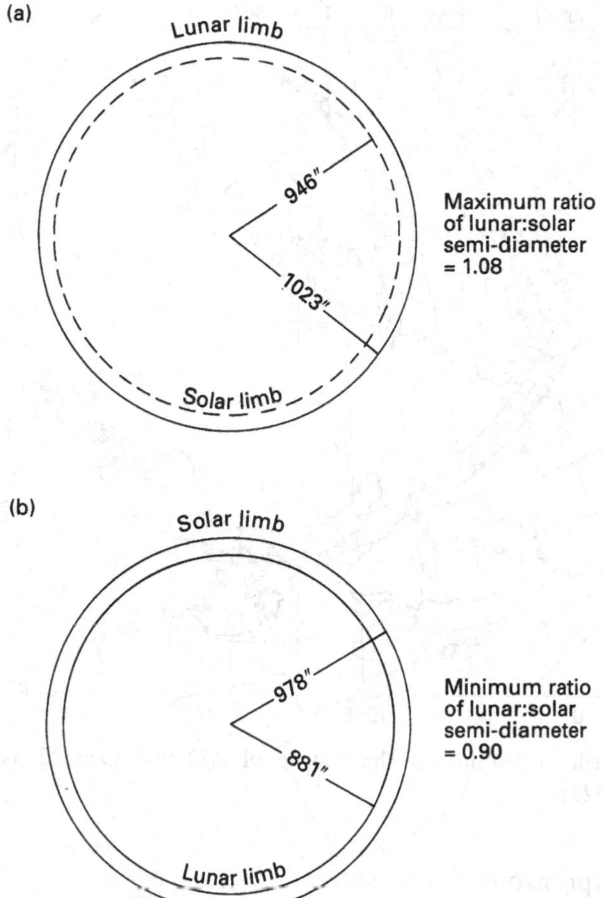

Fig. 3.3 Extreme total (a) and annular (b) obscurations of the Sun.

If the apex of the umbral cone stops short of the Earth, an annular or ring eclipse is formed instead. In most events of this kind, the loss of daylight is fairly small. This is evident from figure 3.5 (upper curve), where the light variation is calculated for a typical annular eclipse in which 0.95 of the solar diameter is obscured. In this diagram, approximate allowance is made for limb darkening around maximal phase.

When the apex of the umbral cone just touches the Earth, an eclipse may be marginally annular near the sunrise and sunset locations and marginally total near the noon position; at the former places an observer is significantly further from the Moon (by up to one Earth radius).

The *magnitude* (μ) of a partial solar eclipse is defined as the fraction of the solar diameter obscured by the Moon at the moment of greatest phase. In the case of a central eclipse, the magnitude (μ_c) is given by the

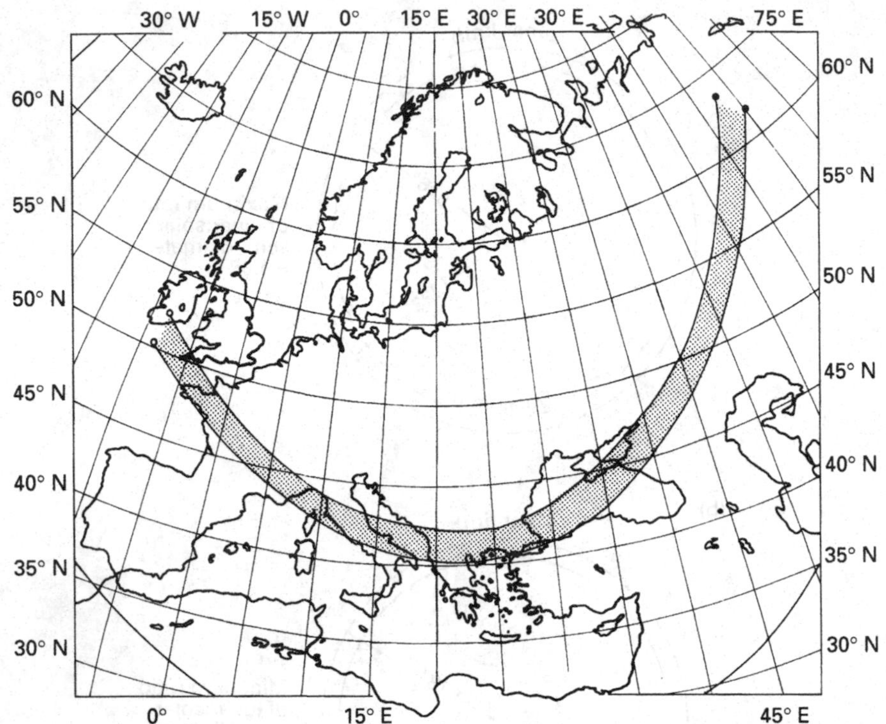

Fig. 3.4 Path of totality at the eclipse of AD 968 Dec 22 as calculated by Schroeter (1923).

following expression:

$$\mu_c = (P - U)/(P + U), \tag{3.1}$$

where P is the penumbral radius and U the umbral radius on a plane passing through the place of observation and perpendicular to the shadow axis (effectively the continuation of the line joining the centres of the Sun and Moon). Both P and U are customarily expressed in terms of the Earth's equatorial radius. The adopted convention is that $U < 0$ for total eclipses and $U > 0$ for annular obscurations; P is always positive. Since $(P - U)$ is constant, having a numerical value of 0.5464, equation (3.1) reduces to

$$\mu_c = 0.5464/(P + U). \tag{3.2}$$

Although the magnitude of a central annular eclipse is always less than unity, in the case of a total eclipse it exceeds unity. It should be noted that (ignoring minor local variations) the magnitude of a central eclipse is the same whether the place of observation lies near the middle of the shadow or close to its edge. As μ_c is also equal to the ratio of the apparent lunar

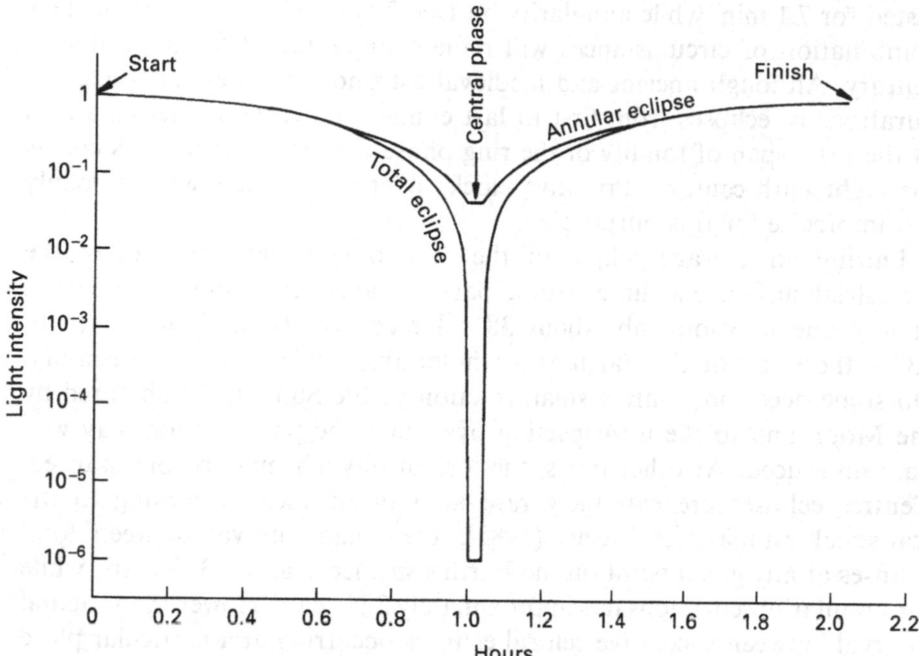

Fig. 3.5 Changes in light intensity during the various stages of typical annular and total solar eclipses.

and solar diameters, the extreme range in magnitude for a central eclipse extends from 0.90 to 1.08 (see above).

The Moon orbits the Earth at about 3400 km per hour. However, on account of the terrestrial rotation, the speed of the lunar shadow relative to a point on the ground is quite variable and is frequently much less than the above figure. As a result, the times of onset and the durations of both partial and central (i.e. total and annular) eclipses are much affected by geographical circumstances. When an eclipse occurs with the Moon overhead at the equator, the effect of the Earth's speed of rotation (some 1700 km per hour) is to roughly halve the ground velocity of the shadow. Near rising or setting, the corresponding contribution from the Earth's rotation is minimal so that the shadow travels much faster along the ground. Under ideal conditions, the interval from first to last contact (i.e. from the very beginning of the visible eclipse to its end) can exceed 4 hours. However, a duration as long as this is fairly rare.

For totality, the maximum duration (from second to third contact) is rather less than 8 min while for annularity the duration may exceed 12 min. The relatively long durations of annular eclipses arise largely from the low orbital velocity of the Moon when furthest from the Earth. Two solar eclipses of remarkable duration occurred in 1973. Totality on Jun 30 lasted for 7.1 min, while annularity on Dec 24 extended for 12.0 min. This combination of circumstances will remain unparalleled for more than a century. Although ancient and medieval astronomers often measured the durations of eclipses from first to last contact, no careful determinations of the brief span of totality or the ring phase are preserved until as late as the eighteenth century. Primitive clocks (clepsydras, etc.) were probably too imprecise for this purpose.

During an average eclipse of the Sun, only about one-sixth of the terrestrial surface encounters some part of the lunar shadow. As a result, at any one location only about 38 solar eclipses (roughly one-sixth of 238 – the total for the Earth as a whole) are visible in a typical century. On some occasions, only a small fraction of the Sun will be obscured by the Moon and to the unsuspecting bystander the phenomenon may well pass unnoticed. At other times, the loss of daylight may be pronounced. Central eclipses are extremely rare at a given site. According to the statistical estimate of Meeus (1982), the mean interval between total eclipses at any given point on the Earth's surface is about 375 years, while for annular obscurations it is approximately 224 years. However, the actual interval between successive central eclipses occurring at a particular place is very variable. Often, many centuries can elapse, while very occasionally two such events can take place in not much more than a year. For instance, both the eclipses of 1983 Jun 11 and 1984 Nov 22 were total at Port Moresby (Papua New Guinea) – a circumstance which aroused considerable local interest. One of the most remarkable examples in ancient and medieval history occurred at Split (Croatia) in the early thirteenth century. Both the eclipses of AD 1239 Jun 3 and 1241 Oct 6 were described by a chronicler of that city as causing intense darkness, and producing great terror among the populace (see chapter 11 for details).

So many careful observations of partial solar eclipses are preserved from antiquity that it seems likely that early astronomers must have often used some means to dim the brilliant disk of the Sun. It is known that medieval Arab astronomers were in the habit of viewing the eclipsed Sun by reflection in water (see chapter 13); it could then be scrutinised more easily. However, under these circumstances even a slight breeze can distort the solar image. The first century AD Roman author Seneca tells us that in his time more viscous liquids were utilised for this purpose:

> Whenever we want to watch an eclipse of the Sun we set out basins filled with oil or pitch, because the heavy liquid is not easily disturbed and so preserves the images it receives...
>
> [*Naturales Quaestiones*, I, 11; trans. Corcoran (1971, pp. 69–71).]

Possibly similar techniques were adopted by early astronomers in other parts of the world, but in general little is known about the methods used.

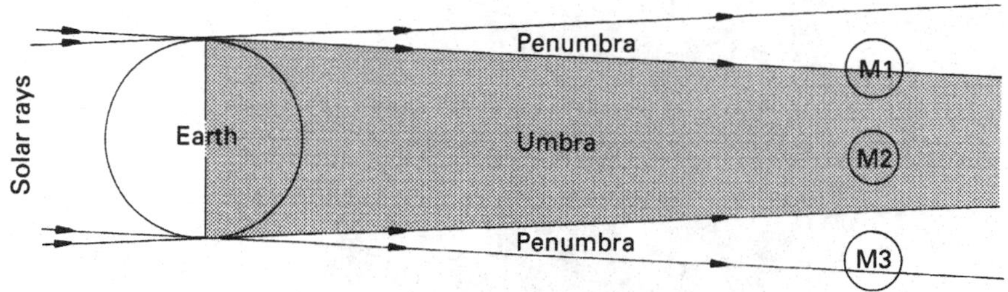

M1 = Moon partially in umbral shadow: partial eclipse
M2 = Moon fully in umbral shadow: total eclipse
M3 = Moon partially in penumbral shadow: penumbral eclipse

Fig. 3.6 Schematic representation of a lunar eclipse.

3.5.2 Lunar eclipses

A lunar eclipse takes place whenever the Moon encounters the terrestrial shadow, as depicted schematically in figure 3.6. At the Moon's distance, the diameter of the penumbral cone is about 4.8 times the lunar diameter, while the corresponding ratio for the umbral cone is approximately 2.7:1.

Figure 3.7, which is from the book *Astronomicum Caesareum*, published in 1540 by Petrus Apianus, shows the Moon, umbral shadow and penumbral shadow roughly to scale.

During a penumbral eclipse, there is a slight reduction in the amount of sunlight reaching the Moon, but this is scarcely noticeable to the unaided eye. As a result, records of these events from the pre-telescopic period are virtually unknown. An almost unique exception is the marginally penumbral eclipse of BC 188 Aug 1/2, which was observed in Babylon (see chapter 6). By contrast, umbral eclipses are readily visible and in general only these will be considered below.

For the Moon to enter the umbra either partially or totally, it must be close to one of the the two nodes of its orbit when at opposition (i.e. full Moon). In a typical century there are 154 lunar eclipses of which 70 are total and 84 are partial (Liu and Fiala, 1992). The combined figure represents only about 12 per cent of the number of full Moons during the same interval. Hence on the Earth as a whole, eclipses of the Moon occur significantly less frequently than their solar counterparts (only about two-thirds as often). Occasionally, there can be as many as three lunar obscurations in a single year. Very rarely, as last happened in 1982 and will not recur for more than 500 years!, three *total* eclipses can occur in one year.

When the Moon passes fairly centrally through the terrestrial umbra, the duration from start to finish (first contact to last contact), can reach about 3.7 hours, while totality (second contact to third contact) can last up to about 1.7 hours. As already noted, the rotation of the Earth has

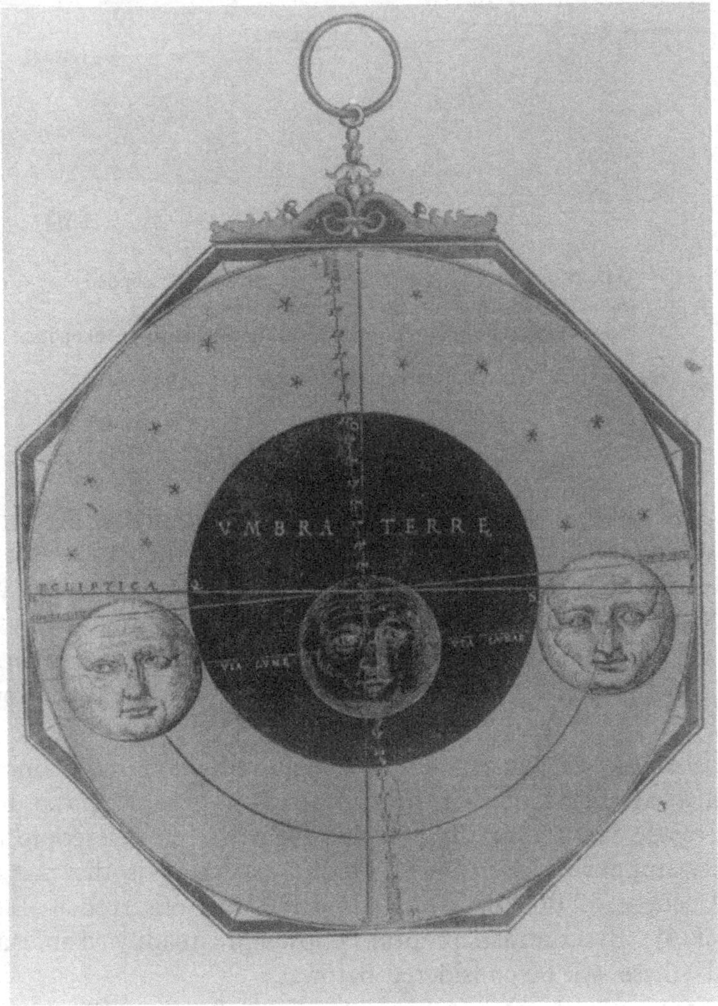

Fig. 3.7 The Moon, umbral shadow and penumbral shadow as depicted in the book *Astronomicum Caesareum* by Petrus Apianus (Ingolstadt, 1540).

no effect on these durations under normal circumstances. However, if the Moon happens to be near the observer's horizon, part of an eclipse may occur before local moonrise or after moonset, thus apparently reducing the durations.

For a partial eclipse, the magnitude is simply the fraction of the lunar diameter in shadow at maximal phase. In general, the magnitude may be derived from the relation:

$$\mu = (s + S - d)/2s. \tag{3.3}$$

Here s is the lunar semi-diameter, S the semi-diameter of the Earth's shadow at the Moon's distance and d the minimal angular distance of the Moon from the centre of the shadow. Mean values for s and S are respectively 933″ and 2472″. In the case of a total eclipse, μ can range from unity up to a maximum of 1.89, the latter value applying only when the Moon passes centrally through the umbra. However, only rarely does the magnitude exceed 1.80.

As the Moon is so much dimmer than the Sun, even small partial eclipses are relatively easy to observe with the unaided eye. For instance, a survey of lunar eclipses recorded by medieval Arab chroniclers has shown that although most observations relate to total eclipses, several obscurations of magnitude less than about 0.5 (including two as small as 0.28 and 0.12) were noted (Said and Stephenson, 1991).

Since the lunar parallax is fairly small (approximately 1.0 deg), no matter where an observer is situated on the Earth's surface the proportion of the lunar disk covered at any given instant is virtually the same provided that the Moon is above the horizon at the time. Similar remarks apply to the UT of the various phases; the local time of each contact depends directly on the geographical longitude of the place of observation. As a result, lunar eclipses have a lengthy history of application to longitude determination – extending at least as far back as the ninth century AD (see chapter 13). Solar eclipses are of little utility for this pursuit owing to the irregular movement of the Moon's shadow across the Earth's surface.

Although lunar obscurations are less frequent than solar eclipses on the Earth as a whole, the reverse is true at any specific place. This is because at any moment (weather permitting), the phenomenon is visible over the entire hemisphere of the Earth at which the Moon is above the horizon. In an average century as many as 95 lunar eclipses are visible at a given site, including those in which the Moon rises or sets whilst the phenomenon is is progress. This number, which represents more than 60 per cent of the quota for the entire Earth, is roughly $2\frac{1}{2}$ times the number of solar obscurations in the same time.

The Earth's atmosphere materially influences the appearance of a lunar eclipse. It effectively increases the diameter of the terrestrial umbra by about two per cent and also imparts a diffuse edge to the shadow. On account of refraction, the atmosphere allows some sunlight to reach the lunar disk even during totality. When there is minimal volcanic dust, etc. in the upper air, the totally eclipsed Moon takes on a deep red hue due to the preferential transmission of long-wavelength light. However, following a major volcanic eruption, there may be so much dust that the Moon may be virtually invisible during totality. Both red and very dark eclipses are recorded from time to time in history.

3.6 Eclipse observations which are of value in the investigation of ΔT

Observations of solar eclipses which are potentially viable for the study of Earth's past rotation include the following: (i) untimed total and annular eclipses; (ii) instances (also untimed) where totality or annularity is specifically denied; (iii) measurements of the local time of individual phases; (iv) estimates of eclipse magnitude (at greatest phase); and (v) cases where it is noted that the Sun rose or set whilst eclipsed. As will be discussed in section 3.8.6, reports that merely note the occurrence of an eclipse without further details are of minimal value.

To the above list may be added the following types of lunar eclipse observation: (vi) measurements of the local time of the various phases; (vii) examples where it is merely noted that the Moon rose or set whilst eclipsed; and (viii) specific estimates of the degree of obscuration of the Moon at moonrise or moonset. Since the peak magnitude of a lunar eclipse and the durations of individual phases are independent of ΔT, estimates of these quantities – or statements that the Moon was either totally or partially obscured – do not make any direct contribution to the study of the Earth's past rotation. However, as will be shown in sections 3.8.4, 3.9.1 and 3.9.3, analysis of recorded magnitudes and durations enables the material in some of the above categories to be better assessed.

Before considering the derivation of ΔT from observations in each of the above categories, the various historical sources will be briefly outlined. Fuller details, together with a discussion of the relevant calendar conversion, will be given in chapters 4 to 13.

3.7 Sources of data

Practically all of the ancient and medieval eclipse observations which are of value in the investigation of long-term changes in the length of the day originate from only four civilisations. These are: Babylon (*c.* 700 BC to 50 BC); China (*c.* 700 BC to AD 1500); Europe (*c.* 500 BC to AD 1600); and the Arab dominions (*c.* AD 800 to 1300). Somewhat surprisingly, there do not appear to be any useful records from ancient Egypt (apart from the Greek city of Alexandria), India or Central America, for example. According to the first century BC Greek historian Diodorus of Sicily (*Library of History*, book I, chapter 81), the Egyptians 'keenly observed the movements, orbits and stationary points of the planets'. However, this somewhat ambitious assertion is not supported by modern research into ancient Egyptian astronomy (Parker, 1974). Indian astronomers were much concerned with the prediction of eclipses and other aspects of mathematical astronomy (cf. Yabuuchi, 1979), but no early observations appear to be extant. Although the Mayans developed a highly sophisticated astronomy and were very interested in eclipse prediction (Thompson, 1974), no actual observations are known to survive. Many Mayan codices (possibly containing valuable astronomical records) were destroyed by the conquistadors in AD 1540.

Babylonian astronomers systematically observed eclipses along with many other celestial phenomena. Their records, inscribed on clay tablets, first came to light rather more than a century ago and are now largely in the British Museum. Translations of numerous astronomical texts have been published by Sachs and Hunger (1988, 1989, 1996). Only about ten per cent of the original material is extant, and nearly all surviving tablets are badly damaged. Nevertheless, nearly 150 important observations of eclipses – mainly of the Moon – are available; these are discussed in chapters 4 to 7 below. In arriving at this total, I have treated each individual measurement etc. as a separate observation, even if the same eclipse is counted several times. When dates are fully preserved, they are invariably accurately expressed in terms of the Babylonian luni-solar calendar. Precise conversion to the Julian calendar is straightforward (for details, see chapter 4). The place of observation is also well established as Babylon. There is nothing in the available records to indicate just when the cause of eclipses was known to the Babylonians. They probably obtained this knowledge from the Greeks during the Hellenistic period (late fourth century BC onwards). The first true explanation of eclipses appears to have been given by Anaxagoras of Athens *c.* 450 BC – see also chapter 10.

It was the practice of the astronomers of Babylon to estimate the magnitude of an eclipse (usually to the nearest twelfth of the lunar or solar diameter) and also to carefully time the various phases. Times were usually expressed to the nearest US (time-degree) equal to four minutes; these were probably determined with the aid of a *clepsydra* (water clock), but the records are silent on this question. Apart from the texts devoted to astronomy, Babylonian historical records make virtually no references to eclipses.

Chinese eclipse records (see chapters 8 and 9) are also dominated by the observations of astronomers, who maintained a regular watch of the sky for all kinds of celestial events. More than a thousand individual sightings of eclipses are reported in Chinese history, but most accounts simply note the occurrence of such an event without any other details. Only a relatively small proportion of these observations is of value in the investigation of ΔT, but this still represents a significant number of data (more than 100 in total). Practically all dates are precisely recorded (relative to the Chinese luni-solar calendar – see chapter 8) and the place of observation can usually be taken as the dynastic capital of the time. The correct explanation of eclipses was not given in China until as late as the first century AD – long after its realisation by the Greeks.

Chinese eclipse reports from before AD 400 almost exclusively relate to the Sun and there are several allusions to total or very large obscurations from this period. Between about AD 400 and 1300, many timings of both solar and lunar eclipses – determined with the aid of a water clock – are preserved. Most solar and lunar measurements were quoted to the nearest *k'o* ('mark'); one hundred of these units equalled a full day and night, so that each was equivalent to 0.24 h. However, some lunar times were expressed to the nearest fifth of a *keng* ('night watch') – a unit whose length varied with the seasons but was roughly equal to half an hour. After AD 1300, few careful observations are extant until the Jesuit era (seventeenth century). Chinese records of eclipses are largely found in the official dynastic histories (*cheng-shih*), all of which have been printed (and reprinted in recent years) and are thus readily accessible. An extensive and valuable compilation of eclipse and other astronomical records from China has been co-ordinated by Beijing Observatory (1988).

From about the middle of the first millennium AD, Korean and Japanese astronomers followed much the same observing pattern as their Chinese counterparts. However, few eclipse records from these countries appear to be sufficiently detailed and reliable to contribute effectively to the study of the Earth's past rotation.

Nearly all early European accounts of eclipses are the work of historians, chroniclers, etc. – rather than astronomers. With the exception of some ten observations made by Greek astronomers between about 200 BC and AD 130 (all contained in Ptolemy's *Almagest*), most *ancient* records of eclipses are of dubious reliability. These are largely found in the Greek and Latin Classics and are the subject of chapter 10 below. Ginzel (1899) made an extensive compilation of this material; see also Newton (1970). After about AD 800, some 50 qualitative but careful descriptions of total or very large solar obscurations are preserved in monastic and town chronicles (see chapter 11). Several independent accounts may often relate to the same event; for instance some ten separate reports of totality in AD 1239 are preserved from a variety of sites in the Mediterranean region. Despite the lack of measurements, and often only minimal understanding of the cause of eclipses, medieval European records are frequently of great value in the determination of ΔT. Dates are normally accurately specified (mostly in terms of the Julian calendar), while usually the observation can be assumed to have been made at the place where the chronicle was composed. Large numbers of medieval annals have been published in their original language (generally Latin) by editors such as Muratori (1723–) and Pertz (1826–), and are thus readily accessible. Celoria (1877a and 1877b) and Ginzel (1884a, 1884b and 1918) assembled numerous eclipse records from these sources, while many translations have been published by Newton (1972b). A few careful timings made by fourteenth and fifteenth century European astronomers are also extant – see chapter 11.

From about AD 800 to 1500, Arab chroniclers frequently documented the occurrence of both solar and lunar eclipses – as well as other striking celestial phenomena such as comets and meteors. Although times are only crudely estimated, there are several vivid accounts of total eclipses of the Sun between AD 900 and 1300. These are discussed in chapter 12. Over a relatively short period – between about AD 830 and 1020 – Arab astronomers recorded careful timings and magnitude estimates for both solar and lunar eclipses. More than 40 separate observations of this kind are preserved (see chapter 13). Individual eclipse phases were timed indirectly by measuring the altitude of the Sun, Moon or selected bright stars. In each account the place of observation is clearly stated and nearly all dates are carefully noted in terms of the Islamic lunar calendar. The major chronicles and astronomical treatises have been published in Arabic. Translations of several astronomical works are also available, notably the treatise of Ibn Yunus (d. AD 1009), which contains many careful observations of both solar and lunar eclipses (Caussin, 1804). Said *et al.* (1989) have published translations of a wide variety of solar eclipse records compiled from Arab chronicles.

Combining the above four sets of data, we have a remarkable series of more than 300 useful observations of both solar and lunar eclipses, extending from about 700 BC to the telescopic era. (As noted above, reports of different phases of the same eclipse are counted as separate data.) There are a few earlier eclipse records, notably from China *c.* 1000 BC (see chapter 8). However, the interpretation of these archaic texts is open to speculation, while the dates are uncertain. Most of the accessible sources from approximately 700 BC onwards have been thoroughly searched for references to eclipses. Although consultation of material as diverse as (i) unpublished Babylonian tablets, (ii) medieval European printed chronicles, and (iii) Arab manuscripts is currently in progress, the acquisition of further useful eclipse data seems likely to be fairly slow unless some fresh archive comes to light.

3.8 Solar eclipse observations and their reduction to ΔT

In section 3.6 above, five different categories of solar eclipse observation which are in principle of value in the calculation of ΔT are listed. The determination of ΔT from each type of observation will now be discussed, with examples. In each case it will be assumed that the only unknown quantity is ΔT itself. As noted in chapter 2, the various orbital parameters, including the lunar acceleration $\dot{n}$, are known with more than adequate precision.

3.8.1 Total and annular solar eclipses

Throughout this investigation it will be assumed that an eclipse was only fully total or annular if a record clearly describes either the complete disappearance of the Sun or the reduction of the solar disk to a ring of light. As discussed in chapter 11, other effects – such as darkness or the visibility of stars – are only general indications of a very large eclipse. If the sky is clear, the onset of totality should be extremely well defined, as is evident from figure 3.5. This is confirmed by the impressions of modern observers (e.g. Muller, 1975; see also the many eighteenth and nineteenth century reports compiled by Ranyard, 1879). An interesting illustration is provided by the eclipse of AD 1715 May 3, which was total in England. Edmond Halley circulated advance notification of this eclipse throughout much of England, specifically asking those who happened to be near the edges of the belt of totality to keep a special watch. As a result, Halley (1715) was able to obtain several careful unaided-eye observations from near the northern and southern limits of totality. These reports proved to be so self-consistent that it has proved possible to use them to determine the mean solar semi-diameter within 0.1 arcsec (Parkinson *et al.*, 1988).

The following account, relating to the eclipse of AD 1241 Oct 6, is fairly typical of medieval descriptions of the total phase. It originates from the monastery of Stade in north Germany (see chapter 11):

> 1241 ...There was an eclipse of the Sun ...on the day before the Nones of October (Oct 6), on Sunday some time after midday. Stars appeared and the Sun was *completely* hidden from our sight ...

Use of the first person plural affirms that totality was witnessed in Stade.

In the case of an annular eclipse, the ring phase is less marked since a portion of the Sun stays unobscured at all times. Hence the loss of daylight is usually relatively small. Although more than 40 careful accounts of totality are preserved from various parts of the world in the pre-telescopic era, no more than four direct allusions to annularity can be traced in this period (AD 873, 1147, 1292 and 1601). Evidently, most events of this kind passed unnoticed. The ring phase is clearly described in the following brief report from Ta-tu (the former name for Beijing) on a date corresponding to AD 1292 Jan 21 (see also chapter 8):

> Chih-yuan reign period, 29th year, first month, day *chia-wu* [31]. The Sun was eclipsed. A darkness invaded the Sun, which was not totally covered. It was like a golden ring ...

Even if the Sun is only marginally reduced to a complete circle of light, the effect of irradiation is to greatly enhance visibility of this ring, as has been noted by modern observers. Immediately outside the zone

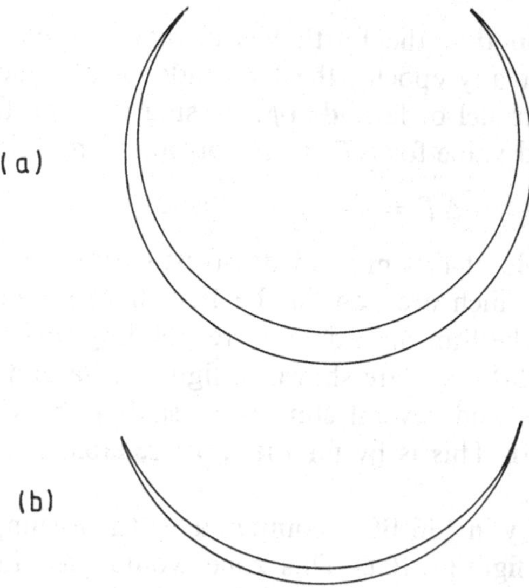

Fig. 3.8 Form of the crescent as seen just outside the central zone for a generally annular eclipse of typical central magnitude 0.95 (a) and a generally total obscuration of typical central magnitude 1.05 (b).

of annularity, the angle between the cusps of the crescent is small and remains less than 90 deg (i.e. making at least three-quarters of a full circle) even at some considerable distance from the central zone. This is illustrated in figure 3.8a, which is based on a typical central magnitude of 0.95. (By contrast, when a generally total obscuration is viewed from outside the umbral region this angle always exceeds 180 deg – see figure 3.8b.) Because of this feature, early descriptions of eclipses which were generally annular on the Earth's surface are often vague and difficult to interpret (e.g. 'a human head was seen in the Sun' – as reported from both Cologne and Hirschau in Germany in AD 1207 – see chapter 11). Hence unless a text clearly asserts that a full ring of sunlight was seen (or, on the contrary, likens the Sun to a crescent), the record is best rejected.

Provided it can be confidently established that a certain solar eclipse was fully total or annular on a definite date and at a specified place, measurement of the time of day when it occurred is usually superfluous. This results from the narrowness of the central zone on the Earth's surface. The effect of increasing ΔT is to displace the whole of the calculated zone of totality or annularity towards more easterly longitudes, the converse being true if ΔT is diminished. For any particular observation, made at a site with latitude ϕ deg and longitude Λ deg, an *approximate* solution for ΔT may be obtained in the following way. (NB the adopted sign conventions are: positive for latitudes north of the equator and positive for longitudes west of the Greenwich meridian up to 180 deg.)

On the assumption that the Earth has always rotated at its present rate (i.e. $\Delta T = 0$ at any epoch), the longitude Λ' at which the central line intersects the parallel of latitude (ϕ) passing through the site is first derived. The required value for ΔT is then given by the formula

$$\Delta T = 240(\Lambda' - \Lambda) \text{ sec.} \tag{3.4}$$

To give an example of this method of investigation, two separate late Babylonian texts – which are now in the British Museum – describe a complete eclipse of the Sun on a date corresponding to BC 136 Apr 15. (Photographs of both tablets are shown in figures 3.9a and 3.9b.) During totality, four planets and several stars were said to be visible (for full details see chapter 5). This is by far the most reliable ancient record of such a phenomenon.

The zone of totality in 136 BC – computed on the assumption that $\Delta T = 0$ – is shown in figure 3.10a. This zone would pass far to the west of Babylon ($\phi = +32.55$ deg, $\Lambda = -44.42$ deg), the central line crossing the parallel of latitude for the city at $\Lambda' = +4.3$ deg. Hence in order to obtain satisfactory agreement with the record, a correction in longitude of approximately +48.8 deg would be required. Such a correction implies a result for ΔT of around +11 700 sec (3.25 h). The computed central line with $\Delta T = +11\,700$ sec is depicted in figure 3.10b.

In practice, it is possible to set *firm* upper and lower bounds to the value of ΔT at the appropriate epoch by taking into account the width of the zone of totality or annularity. Thus, as shown in figure 3.11, for the southern edge of the umbral shadow to just reach Babylon, a value for ΔT of +11 210 sec (3.11 h) would be needed. For the northern edge to reach the site, the corresponding figure would be +12 140 sec (3.37 h). Hence to fully satisfy the observation, only values of ΔT between these two limits – i.e. $+11\,210 < \Delta T < +12\,140$ sec – are acceptable. The true value of ΔT in 136 BC has an equal probability of lying anywhere in this interval, but not outside it; this represents a tolerance of no more than about four per cent relative to the mean figure of +11 680 sec. It should be mentioned here that I have usually estimated ΔT limits obtained from observations of total and annular obscurations – and also for careful reports which specifically deny a central eclipse – to the nearest 10 sec. All other ΔT values or limits (e.g. obtained from contact timings or rising and setting phenomena) will be quoted to the nearest 50 sec – a precision which is still more than adequate for investigating long-term trends in ΔT.

Since for any given value of ΔT the computed central line in 136 BC crosses the parallel of latitude for Babylon at only one point along the entire track, no other solution range than the above will suffice. However, if a central line crosses the observer's latitude twice, as was true for places between latitudes 40 and 50 deg in AD 968 (see figure 3.4), an observation

Fig. 3.9 Two Late Babylonian tablets describing a complete eclipse of the Sun on a date corresponding to BC 136 Apr 15. (Courtesy: British Museum.)

(a)

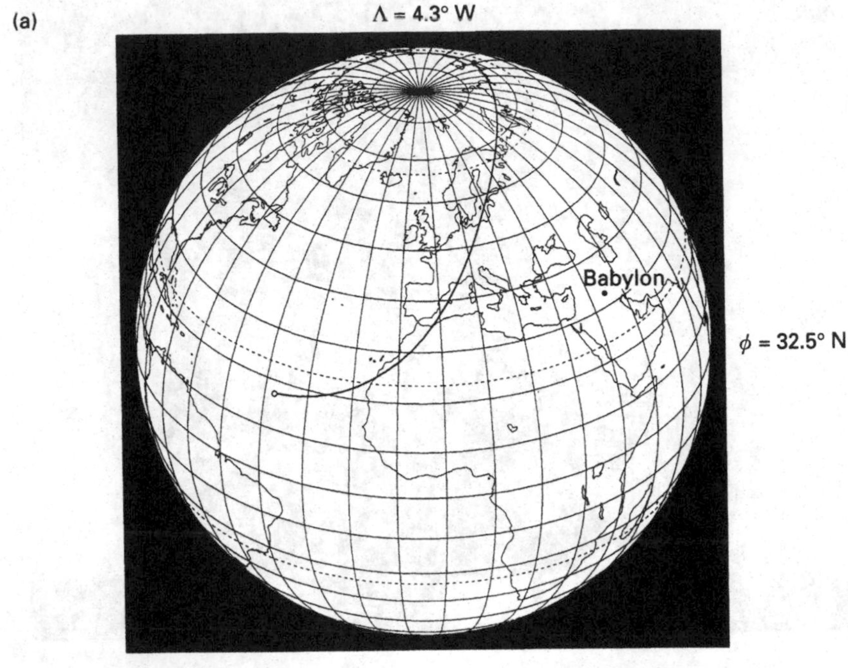

(b)

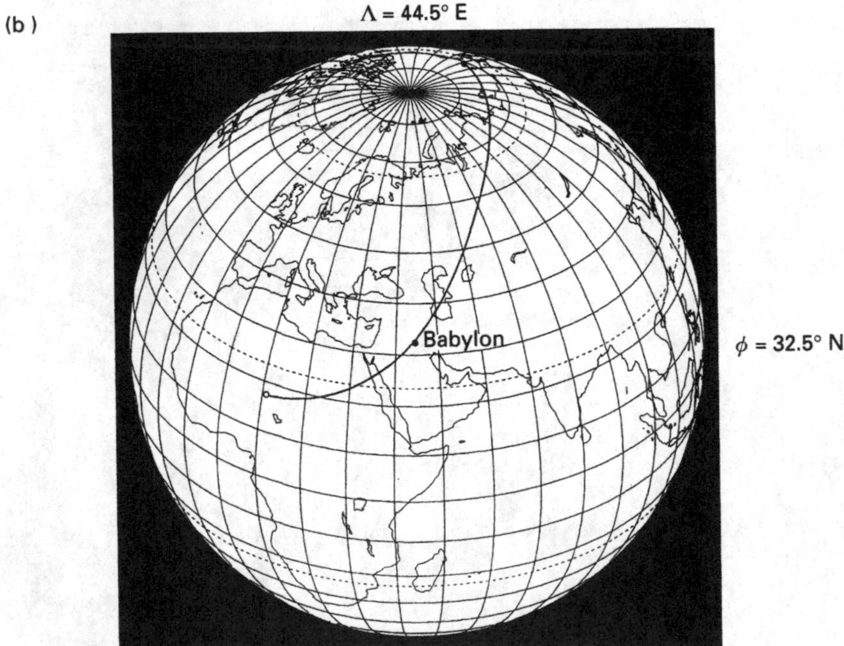

Fig. 3.10 Zone of totality in 136 BC, computed on the assumption that (a) $\Delta T = 0$ and (b) $\Delta T = +11\,700$ sec. The eclipse was observed to be complete at Babylon. (Courtesy: Dr S. Bell.)

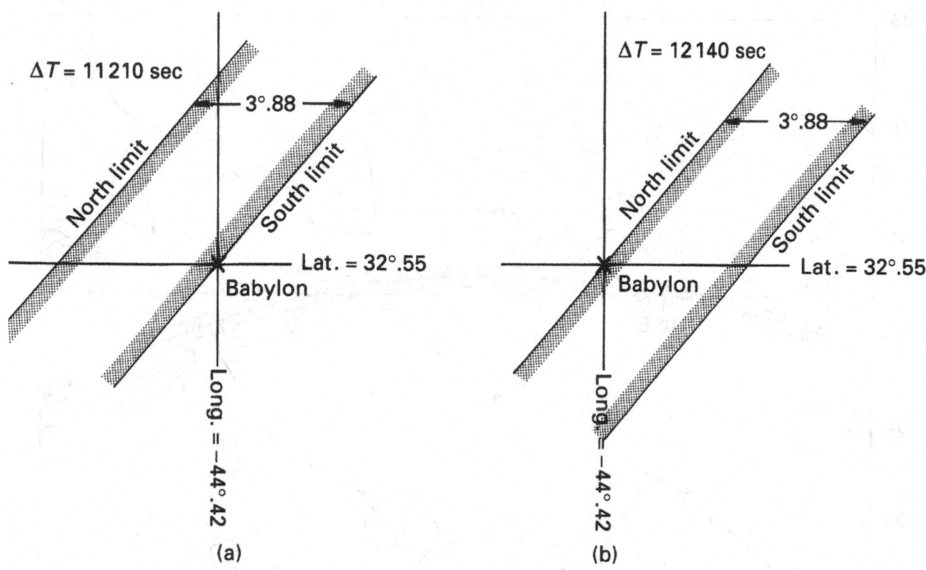

Fig. 3.11 Lower (a) and upper (b) bounds to the value of ΔT as derived from observations of the total solar eclipse of 136 BC at Babylon.

of a central eclipse will lead to two distinct ranges for ΔT. In principle, these can only be separated if a measurement (or at least an estimate) of the local time is recorded. However, in practice, the two derived ΔT ranges are often so far apart that only one solution can possibly be viable – as indicated by comparison with ΔT results obtained from roughly contemporaneous observations. An example in which the two sets of ΔT limits are unusually close is provided by the total solar eclipse of AD 454 Aug 10. As viewed from the capital of Chien-k'ang (Nanjing: $\phi = +32.03$ deg, $\Lambda = -118.78$ deg), this was said to be complete, and 'all the constellations were brightly lit' (see chapter 8). Unfortunately, the time of day is not specified.

As can be seen from figure 3.12, which is calculated using $\Delta T = 0$, the central line in AD 454 would cross the latitude of Nanjing twice (at $\Lambda' = -115$ deg and $\Lambda' = -90$ deg), reaching its maximum distance north of the equator ($\phi = +33.0$ deg) between these positions. Since these two longitudes are respectively some 4 and 29 deg to the west of Nanjing, the observation can thus be satisfied by two discrete ranges of ΔT: either around +1000 sec (for an early morning observation) or +7000 sec (for a mid-morning observation). Allowing for the width of the track of totality, the precise limits may be deduced; these are (i) $+70 < \Delta T < +1800$ sec and (ii) $+6130 < \Delta T < +7900$ sec. As will be discussed below (chapter 8), only the latter choice is supported by observations of similar date, so that set (i) can effectively be discounted. In most cases where two

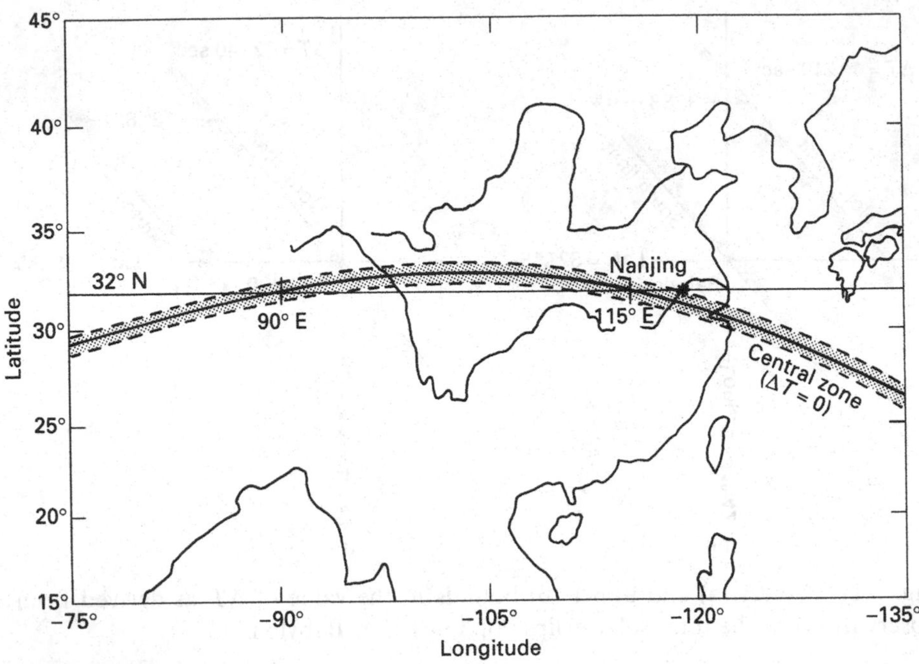

Fig. 3.12 Part of the central zone for the total solar eclipse of AD 454 Aug 10 computed using $\Delta T = 0$. Totality was observed at the Chinese capital of Nanjing.

ΔT ranges are indicated they are much further apart than this and an unambiguous solution can thus be readily obtained. A dual result will only be considered if both alternatives seem viable.

In 136 BC the zone of central eclipse was inclined at a rather large angle to the equator in the vicinity of Babylon. Hence although the track of totality was fairly wide (some 270 km) the resultant range of ΔT is reasonably narrow. However, if the path of the umbra ran almost parallel to the equator near the place of observation – as in AD 454 – the corresponding range of ΔT would be considerably increased. In AD 454 the breadth of the umbral zone was only 110 km but the derived ΔT range (i.e. $+6130 < \Delta T < +7900$ sec) is roughly twice as wide as that obtained in 136 BC In extreme cases the ΔT limits may be so wide apart as to render the observation redundant, even when a careful description of totality or the ring phase is preserved. This is well illustrated by the eclipse of AD 1185 May 1 (see figure 3.13, which is taken from Schroeter, 1923). On this date, the central zone (260 km in width) ran almost parallel to the equator across northern Russia. Detailed accounts of totality, mentioning visibility of the chromosphere, are preserved from Novgorod ($\phi = +58.50$ deg, $\Lambda = -31.33$ deg) – chapter 11. Although only a single range of ΔT can satisfy the observations made at this site, the limits are extremely

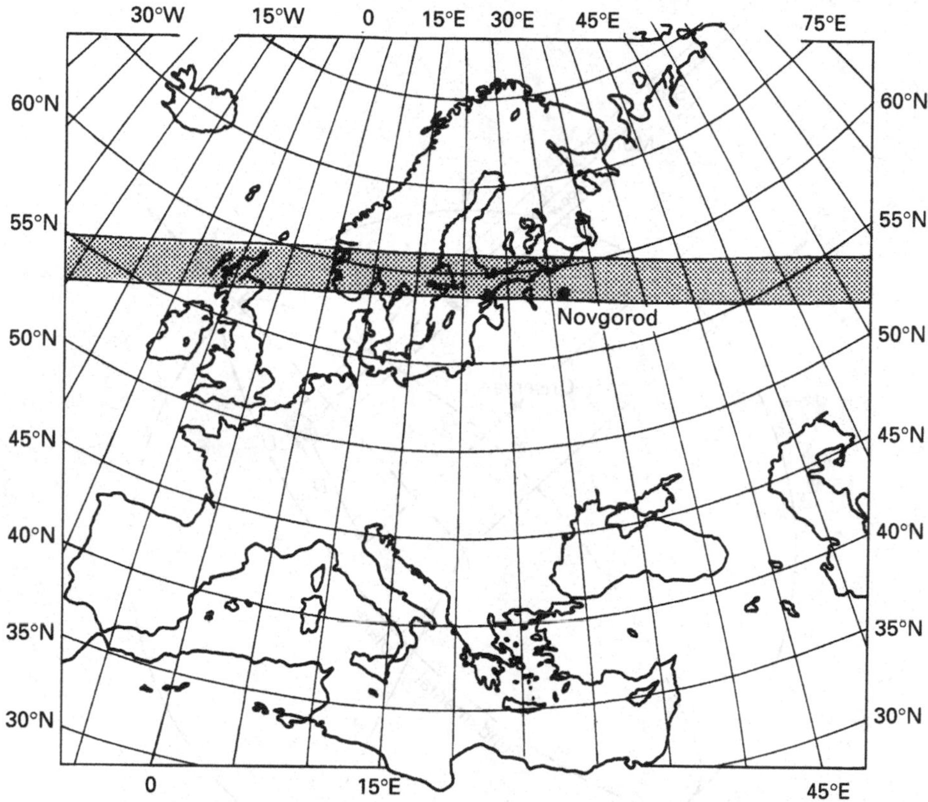

Fig. 3.13 Track of totality for the eclipse of AD 1185 May 1 according to Schroeter (1923). The eclipse was recorded as complete at Novgorod, the Russian capital.

wide: between −2200 and +10 500 sec. Roughly contemporaneous data indicate a true result for ΔT of close to +1200 sec at this fairly late date. Hence calculation can no more than confirm the reports of totality from Novgorod. Several similar examples could be given. Clearly the local geographical circumstances are of considerable importance in determining the utility of a particular observation.

Rather than map the belt of totality or annularity on each occasion, it is preferable to use a technique involving projection of the shadow on the 'fundamental plane' – the geocentric plane perpendicular to the axis of shadow and passing through the centre of the Earth (see figure 3.14). In this procedure, which has been adopted throughout this book, ΔT is varied until the modulus of the minimal distance ($|D|$) of the place of observation from the shadow axis equals that of the umbral radius ($|U|$) on the fundamental plane. The usual convention is to express both D and U in units of the Earth's equatorial radius. Limits to ΔT are obtained

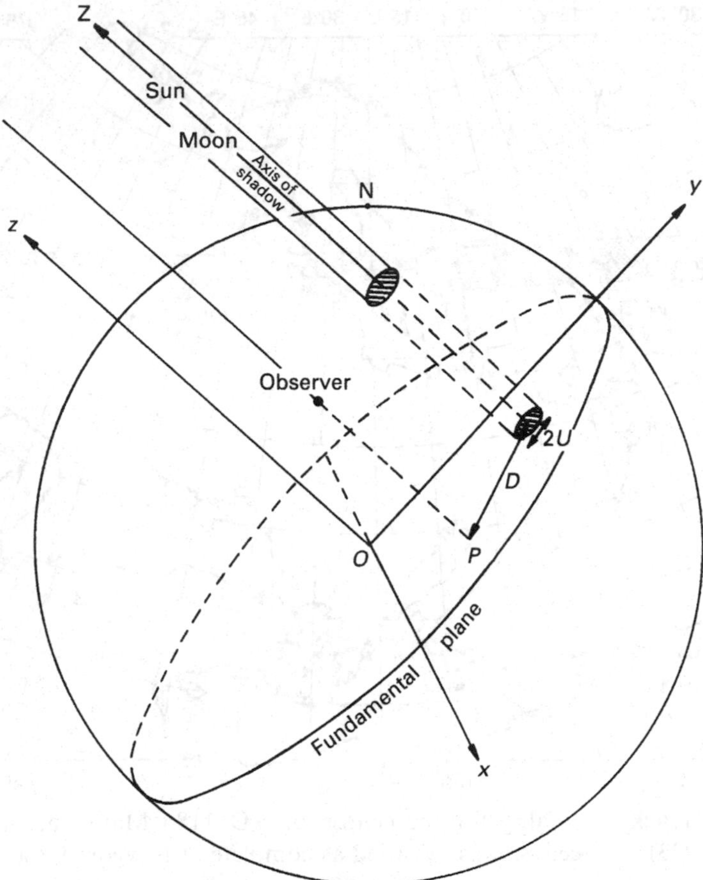

Fig. 3.14 Projection of the lunar shadow on the 'fundamental plane'.

corresponding to each edge of the central zone separately touching the place of observation. An iterative solution is necessary since D does not vary linearly with changes in ΔT.

3.8.2 Solar eclipses which, although large, were not quite central

Many solar eclipse reports deny that the central phase was witnessed, affirming instead that the Sun was reduced to a crescent (e.g. 'like a hook'). If a record states that the cusps of the crescent were pointing upwards, then it is clear that the eclipse was viewed from the south of the central zone and vice versa. Unfortunately, it is very rare for a text to be so specific and usually it is necessary to make two separate solutions – one on the assumption that the umbral region passed to the north of the observing site and the other to the south of it.

Computation proceeds very much as if the central phase was witnessed, but instead a range of ΔT – corresponding to the unobserved umbra – is *excluded.* Almost any reasonable value of ΔT on either side of the excluded range is usually acceptable since estimates of the magnitude of a large partial eclipse – if recorded at all – are often very crude. If the prohibited range of ΔT is very narrow (i.e. for an eclipse in which the apparent diameters of the Sun and Moon were nearly equal), almost any feasible solution space is permitted; the observation may thus prove to be redundant. An example of a more useful record is provided by the eclipse of AD 1004 Jan 24, in which the belt of annularity was unusually narrow. On this occasion, a very large partial obscuration of the Sun was reported from Cairo ($\phi = 30.05$ deg, $\Lambda = -31.25$ deg) by the astronomer Ibn Yunus in the following words (see chapter 13):

> The Sun was eclipsed until what remained of it resembled the crescent Moon on the first day of the month. I estimated the magnitude of the eclipse as 11 digits ...

In the vicinity of Cairo, the width of the central zone was only about 30 km. Calculations show that for any value of ΔT in the narrow range between +1770 and +1940 sec, the phase would have been annular at Cairo. Hence to satisfy the observation of a partial eclipse, either $\Delta T < +1750$ sec or $> +1920$ sec. Values of ΔT far beyond either of these limits would still produce a large partial eclipse at Cairo. As both estimates of magnitude can be considered merely approximate (and not even in good mutual accord – see section 3.8.4) the main conclusion from the above careful description is that the value of ΔT did *not* lie between +1750 and +1920 sec.

3.8.3 Timed solar eclipse contacts

Many measurements of solar eclipse timings are preserved in ancient and medieval history, notably from Babylon, China and the medieval Arab dominions. At each eclipse the astronomers typically determined three instants – the start (i.e. first contact), greatest phase and end (last contact) – unless the Sun was below the horizon or obscured by cloud for part of the eclipse duration. The various measurements, whether made directly (for example, with the aid of a water clock) or indirectly (in terms of solar altitude) are readily reducible to local apparent time and hence to UT. On only two occasions during the whole of the pre-telescopic period (at Babylon in 136 BC and at the Chinese capital of Ch'ang-an in AD 761) was the onset of totality carefully timed, but in neither case is there an estimate of the duration of the total phase.

Babylonian astronomers measured the time of first contact relative to the moment of sunrise or sunset, depending on which was nearer. Times of the later phases were then determined relative to the start of the eclipse. In computing the LT of rising or setting of the Sun or Moon, I have adopted the standard figure for mean horizontal refraction (0.57 deg). Except in extreme circumstances, corrections for temperature and pressure are unlikely to exceed 0.05 deg (Newton, 1972a) – a negligible amount. When the time of an eclipse contact was measured relative to sunrise or sunset, I have assumed that the upper limb of the Sun or Moon (semi-diameter roughly 0.26 deg) was on the visible horizon at these moments (a depression of the Sun's centre relative to the theoretical horizon of approximately 0.8 deg). There is some evidence that the Babylonians adopted this definition, which corresponds to either the very first (morning) or last (evening) gleam of sunlight (Stephenson, 1974).

In the vicinity of Babylon the terrain is so flat that the horizon is remarkably level. Hence in computing the LT of sunrise or sunset, no allowance need be made for horizon profile. However, since the walls of Babylon were some 15 m in height (Ravn, 1942, pp. 20 and 28), the observatory would have to be at least above this level. The depression of the horizon from this height would be 0.1 deg. According to Diodorus of Sicily (II, 9), the Babylonian astronomers used to survey the sky from the top of the ziggurat, a tower fully 90 m in height. From here the depression of the horizon would be as much 0.27 deg. Since Diodorus' story may be no more than hearsay, the figure of 0.1 deg (corresponding to some 15 m above ground level) will be adopted instead. When refraction and semi-diameter are included, this is equivalent to a solar depression of 0.9 deg at the moment of sunrise or sunset.

To give an example of a Babylonian timed observation, the solar eclipse of BC 254 Jan 31 was stated to begin 56 deg (= 3.73 h) before sunset, reaching maximal phase after a further 12 deg (0.80 h) and clearing after an additional 11 deg (0.73 h) – see chapter 5. In order to convert these measurements to LT, it is first necessary to compute the LT of sunset, then subtract the appropriate intervals. On this occasion, the Sun would attain a depression of 0.9 deg (i.e. assumed sunset) at 17.26 h; hence the LTs of the three observations would respectively be 13.53, 14.33 and 15.06 h. These times are all in hours and decimals – a system which is standard throughout this book.

It was the practice of Chinese astronomers to measure solar eclipse times in double hours and marks, the first of the 12 double hours commencing at an LT of 23 h. For instance, the eclipse of AD 1068 was reported to have begun at 8 marks in the hour of *szu* (9–11 h), reached its maximum after 5 marks in the hour of *wu* (11–13 h) and ended at 3 marks in the hour of *wei* (13–15 h). These measurements correspond to LTs of 10.96, 12.44 and 13.84 h (see chapter 9).

In order to fix the times of the various contacts for eclipses of the

Sun, medieval Arab astronomers systematically measured solar altitudes, afterwards reducing their results to local time with the aid of an astrolabe or tables. For example, at the eclipse of AD 923 Nov 11, the altitude of the Sun at mid-eclipse was estimated as 8 deg in the east, and at the end as 20 deg in the east (see chapter 13). The equivalent LTs are respectively 7.56 and 8.72 h. There is no mention of first contact in the text, but it may be inferred from the measured altitudes at mid-eclipse and last contact that the Sun probably rose eclipsed.

Once reduced to LT, the various measurements may be converted to UT by allowing for the equation of time (z) and the geographic longitude (Λ), as discussed in chapter 1 (equation (1.1)). In order to derive a figure for ΔT from any particular observation, it is necessary to use an iterative solution since the computed value of the TT for any phase of a solar eclipse is a function of ΔT itself. The Chinese observations of the eclipse of AD 1068 Feb 6 have already been discussed. This was recorded at the capital of Pien ($\phi = +34.78$ deg, $\Lambda = -114.33$ deg), measured times of the various phases leading to LTs of 10.96, 12.44 and 13.84 h. Adjusting for the equation of time ($z = -0.24$ h) and the longitude of Pien (equivalent to -7.62 h) the corresponding UTs are respectively 3.60, 5.07 and 6.48 h. Using iteration, the computed TTs are 4.05, 5.40 and 6.71 h, leading to ΔT values (to the nearest 50 sec) of +1600, +1200 and +850 sec. These three results are in fair accord with one another, bearing in mind that the measured times were only expressed to the nearest mark (i.e. 0.24 h or some 900 sec).

Despite the limited resolution of the unaided eye, I have normally assumed geometrical contacts for both the beginning and end of a solar eclipse. Since ancient and medieval methods of timing were fairly crude, determination of the LT of an actual contact between the lunar and solar limbs should only be marginally more precise than for the less well defined moment of maximal phase. In principle, apart from the rare measurements of the onset of totality, timings of last contact should be the most accurate since the observer merely needs to watch the indentation at the western limb of the Sun gradually disappear. However, among ancient and medieval astronomers it was common practice to roughly calculate in advance the time of onset for an eclipse so that observers would have some idea of when to look for first contact. In any given series of observations, I have thus not weighted observations according to phase.

On very rare occasions, a medieval Arab observer stated that he was late in detecting the start of an eclipse and had thus applied a small empirical correction to the measured time in order to estimate the moment of true first contact (see chapter 13). Under these circumstances, I have normally accepted the amended result.

Early astronomers occasionally reported the moment of mid-eclipse.

Unless it is clear from the text that this result was obtained by averaging the times of first and last contact, I have assumed that maximal phase was intended. For a solar obscuration, mid-eclipse cannot be identified directly; this moment can often be several minutes earlier or later than that of greatest phase owing to the uneven speed of the shadow across the ground. (In the case of a lunar eclipse, the moments of greatest phase and mid-eclipse are, of course, effectively identical.)

3.8.4 Solar eclipse magnitudes

It was the usual practice of Babylonian, Greek and Arab astronomers to estimate the degree of obscuration of the Sun or Moon at maximal phase to the nearest 'finger' or 'digit'. These units were each equal to one-twelfth of the apparent solar or lunar diameter. Chinese observers adopted a different system in which the basic unit of magnitude was originally one-fifteenth of the diameter of the Sun or Moon but after about AD 1000 was increased to one-tenth of the appropriate diameter. In most records of eclipse magnitude it is clear from the text that the luminary was above the horizon at greatest phase; if there is any doubt on this question, the observation will be rejected.

Each magnitude observation can lead to a discrete value for ΔT without the need for a measurement of time. However, there are several drawbacks in using this method. Preserved data from any historical source are relatively few in number. Further, not only do they represent crude unaided eye *estimates*, the computed magnitude of a solar eclipse is only a weak function of ΔT.

Partly to assess the potential of this method, Stephenson and Fatoohi (1994a) investigated the accuracy of a series of recorded estimates of *lunar* eclipse magnitudes from different parts of the world. For each individual observation, they compared the estimated magnitude with its computed equivalent (which is independent of ΔT). Stephenson and Fatoohi concluded that the mean uncertainties in such estimates are close to ten per cent. In the case of a *solar* eclipse, an error in the magnitude by this amount would lead to an uncertainty in the value of ΔT by some 2000 to 2500 sec. Hence compared with other eclipse data, observations of this type should give only poor resolution. Since it is seldom stated whether the upper or lower portion of the Sun was covered, each estimate of magnitude can – in principle – lead to up to four discrete values for ΔT. However, in practice more than one viable result is seldom obtained since, to obtain full accord with observation, calculation must place the Sun above the horizon at maximal phase.

An example is provided by an observation made at the Chinese capital of Lo-yang (ϕ = +34.75 deg, Λ= −112.47 deg) on a date equivalent to

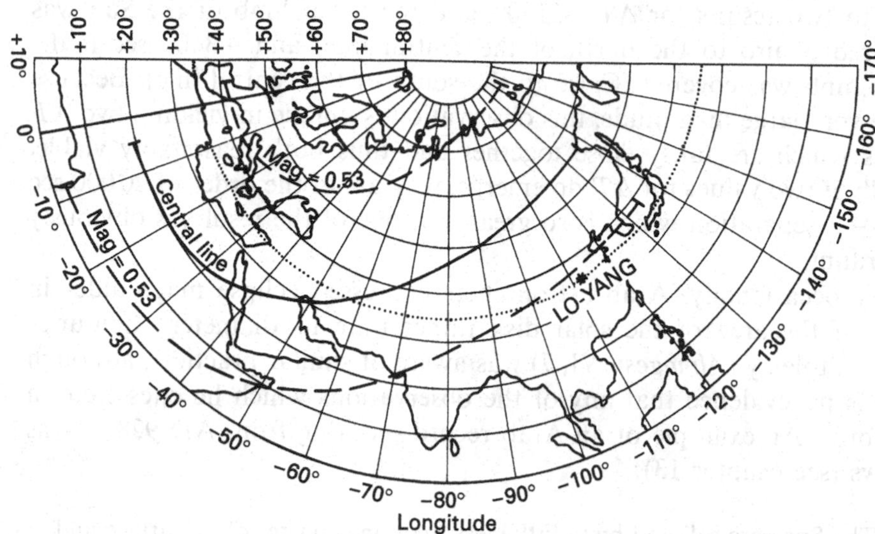

Fig. 3.15 Central line and two curves corresponding to magnitude 0.53 (each computed using $\Delta T = 0$) for the eclipse of AD 489 Mar 18. The Sun was estimated to be $\frac{8}{15}$ (i.e. 0.53) covered at the Chinese capital of Lo-yang.

AD 489 Mar 18. The record indicates that $\frac{8}{15}$ (i.e. 0.53) of the solar diameter was covered at maximum (see chapter 9). For this eclipse, the central line and the two curves corresponding to magnitude 0.53 (each computed using $\Delta T = 0$) are charted in figure 3.15. It can be seen that only the more southerly curve for magnitude 0.53 intersects the parallel of latitude for Lo-yang – and at a single longitude ($\Lambda' = -98.1$ deg). Hence using equation (3.4), the equivalent value of ΔT is approximately +3450 sec, although with a very large uncertainty. As in the case of central and near-central solar eclipses, I have preferred to use an iterative technique to solve for ΔT. This more convenient procedure selects those values of ΔT which yield the observed magnitude at the required location with the Sun above the horizon.

The Arab report of a very large partial eclipse in AD 1004 has been considered in section 3.8.2. Observers in Cairo expressed the magnitude in two different ways:

> ...what remained of it resembled the crescent Moon on the first day of the month. I estimated the magnitude of the eclipse as 11 digits...

These two estimates are in poor accord with one another. When the crescent Moon is about 24 hours old, only some two per cent of the lunar diameter is sunlit, which represents a phase considerably more than the estimated magnitude of 11 digits (0.92). However, the allusion to the crescent Moon may be purely descriptive. Accepting the latter value (0.92) leads to two results for ΔT: +150 sec if the lower limb of the Sun was obscured (Cairo to the north of the central line) and +3400 sec if the upper limb was covered (Cairo to the south of the central line). Because of the very large magnitude, this observation is unique in yielding two ΔT results which are fairly close together and thus both reasonably viable. Usually if two values for ΔT do emerge, they are of the order of 20 000 sec apart – a separation which is so great that one of the results is obviously discordant.

Very occasionally, Arab astronomers expressed eclipse magnitudes in terms of the area of the solar disk rather than its diameter. Centuries before, Ptolemy (*Almagest*, VI, 7) was aware of similar practices, although there is no evidence that any of the observations which he cites were in this form. An example of an Arab record – dating from AD 928 – is as follows (see chapter 13):

> The Sun rose eclipsed by (a little) less than one quarter of its surface and the eclipse continued to increase until one-quarter of it was eclipsed ...

Such observations are readily reduced to linear magnitude by simple geometry. In table 3.1, I have listed the area and linear equivalents for solar and lunar eclipses. For solar obscurations, I have assumed equal semi-diameters for the Moon and Sun, while for lunar obscurations I have adopted a typical umbral radius of 2.70 times that of the Moon. It will be noted that discrepancies (all of the same sign) are greatest for moderately small eclipses. An eclipse in which half of the area of the solar disk (but as much as much as 0.60 of the diameter) is covered by the Moon is represented diagrammatically in figure 3.16.

Ptolemy (*Almagest*, VI, 8) gives a similar table to table 3.1, all magnitudes being, however, expressed in digits. In general, there is close accord between his results and those given in table 3.1.

3.8.5 Observations that the Sun rose or set eclipsed

Several observations indicating that the Sun was eclipsed when it was on the horizon are to be found among the records of Babylonian, Chinese and Arab astronomers. In most instances, the fraction of the Sun which was obscured at these moments is not specified; it is merely implied that the Sun was visibly eclipsed when it rose or set. Observations of this kind set limits to ΔT since at least some part of the solar disk must be eclipsed when the Sun is on the appropriate horizon. For a recent photograph showing the partially eclipsed Sun on the point of setting, see figure 3.17.

If it can only be established from the record that the Sun reached the horizon between first and last contact, ΔT can only be fixed within very

Table 3.1 Conversion from area to linear magnitude for both solar and lunar eclipses.

	Solar	Lunar		Solar	Lunar
Area mag.	Lin. mag.	Lin. mag.	Area mag.	Lin. mag.	Lin. mag.
0.05	0.12	0.11	0.55	0.64	0.57
0.10	0.19	0.17	0.60	0.68	0.61
0.15	0.26	0.23	0.65	0.72	0.65
0.20	0.31	0.28	0.70	0.76	0.69
0.25	0.36	0.32	0.75	0.80	0.73
0.30	0.41	0.37	0.80	0.84	0.77
0.35	0.46	0.41	0.85	0.88	0.81
0.40	0.51	0.45	0.90	0.92	0.86
0.45	0.55	0.49	0.95	0.96	0.92
0.50	0.60	0.53	1.00	1.00	1.00

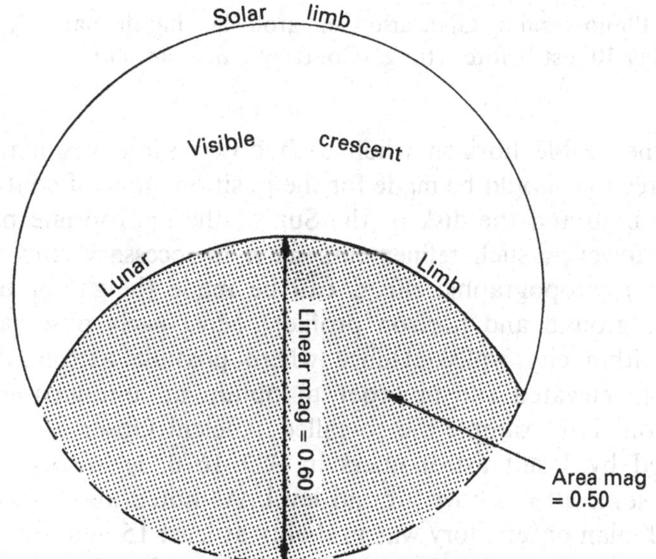

Fig. 3.16 Solar eclipse in which half of the area of the Sun's disk (but as much as much as 0.60 of the diameter) is covered by the Moon.

wide limits. However, it is sometimes possible to decide from the text if the phase was growing or diminishing at the time – thus roughly halving the indicated range of ΔT. Further refinements cannot usually be made.

In calculating the LT of sunrise or sunset under these circumstances, I have assumed that the *centre* (rather than the upper limb) of the Sun

Fig. 3.17 Photograph by Giancarlo Gengaroli showing the partially eclipsed Sun in 1994 May 10 just before setting. (Courtesy: *l'astronomia.*)

was on the visible horizon when each observation was made. Ideally, some correction should be made for the position angle of contact between the lunar limb and the disk of the Sun at the appropriate phase of the eclipse. However, such refinements seem unnecessary since only crude allowance for topographic effects can be made (height of the observer above the ground and horizon profile). Most early observatories were located within city walls. Hence where possible astronomers usually selected an elevated site in order to obtain an uninterrupted view of the horizon. For example the so-called 'Ancient Observatory' at Beijing, established by Jesuit astronomers in AD 1670, is located on a raised platform some 10 m above ground level. As noted above (section 3.8.3), the Babylonian observatory was probably at least 15 m above the ground. Adopting a mean observer height of some 10 or 15 m, leads to an assumed solar depression of roughly 0.65 deg under the above circumstances.

In the vicinity of Babylon and several other ancient cities which were equipped with observatories, the horizon is extremely level. However, at certain other important sites the surrounding countryside is quite hilly and careful allowance for horizon profile would be needed in order to deduce the LT of sunrise or sunset accurately. To give an arbitrary example, 200 m hills on the eastern or western horizon at a distance of 10 km would delay sunrise or advance sunset by at least 5 min. Such an amount would materially affect the derived ΔT limits – especially in medieval times when the value of this parameter was of the order of 1000 sec. In summary, if the necessary correction is likely to be significantly less than this amount, I have retained the observation but have made no allowance for horizon profile. Otherwise, I have rejected it. In chapter 14, where the various results obtained in this book are analysed, I have reconsidered the more critical ΔT limits.

A useful illustration is provided by the solar eclipse of BC 322 Sep 26. This was observed at Babylon – see chapter 5. The eclipse was recorded in the following words:

> Year 2 of (king) Philip, (month VI). The 28th, around 3 deg before sunset, solar eclipse ... It set eclipsed.

The recorded time before of only 3 deg (i.e. 12 minutes) before sunset leads to a discrete value for ΔT, as will be discussed in chapter 5. However, the mere fact that some part of the eclipse was visible before sunset can set a useful lower limit to ΔT. The calculated LT of sunset at Babylon on the day in question was 18.08 h, a UT of 15.01 h. It may thus be inferred that the UT of first contact was earlier than 15.01 h; otherwise no part of the eclipse would be visible before the Sun set. Computing iteratively, the TT of first contact is 18.74 h, leading to the result $\Delta T > +13\,400$ sec. Although the text implies that the eclipse began only a few minutes before sunset, there is always the possibility of a scribal error in the time interval. Hence to obtain an upper limit to ΔT, it seems best to assume only that last contact took place after the Sun reached the western horizon – i.e. that some part of the eclipse remained visible until sunset. In this case, the LT would be > 18.08 h, and the UT > 15.01 h. Comparing with the computed TT for this phase of 20.02 h gives $\Delta T < +18\,000$ sec. When the two separate limits are combined, it may be concluded that $+13\,400 < \Delta T < +18\,000$ sec, both of which represent fairly firm bounds.

Since estimates of the degree of obscuration of the Sun at sunrise or sunset are quite rare, these will be considered in section 3.9.3 along with the more frequent lunar observations of this kind.

3.8.6 Records which merely note the occurrence of a solar eclipse

The vast majority of ancient and medieval eclipse records mention no more than that an eclipse of the Sun or Moon occurred on a specified date. Lunar reports of this kind are clearly valueless in the determination of ΔT. In principle, solar observations can be used to determine ΔT, but there are several problems inherent in the analysis of such data. If it is assumed that a bare record of a solar eclipse can imply any magnitude (from scarcely visible to central), the range in ΔT obtained from a single observation is huge – some 40 000 sec. Even in ancient times – when in any case such data are less frequent – results of this sort tell us very little. Additionally, experience shows that because of the non-random distribution of population centres the ΔT values obtained in this way may well seem to imply large-scale variations in the LOD which prove to be wholly spurious. Apart from a brief discussion in the present section, such observations will not be considered in this book. They are included here mainly because Newton (1970 and 1972b) attached much importance to them.

The difficulties associated with the analysis of laconic records of solar eclipses are perhaps best illustrated by consideration of the almost contemporaneous total solar eclipses of AD 1178 Sep 13 and 1187 Sep 4. In the interval of only nine years between them, ΔT is unlikely to have varied by more than a few tens of seconds. As a result of an extensive search through published compilations of European chronicles, Ginzel (1884a) was able to assemble 14 separate records of the first event and as many as 27 for the second. No progress in uncovering further records of these two eclipses has since been made. Most of the accounts which Ginzel compiled were from local annals (e.g. of monasteries) and hence the place of observation may be confidently established. However, a few other reports were contained in chronicles which drew on information from a wide area. In almost every case, it is merely reported that an eclipse of the Sun occurred on a specified day without giving any descriptive details. For instance, the following brief report is from Italy in AD 1178:

> 1178 ... On the Ides of September (Sep 13), the Sun was obscured [*sol obscuratus est*].
>
> [*Chronicon Fossae Nuovae.*]

Figure 3.18 marks the locations of each observing site for the eclipse of AD 1178 and figure 3.19 for the eclipse of 1187. Both charts are drawn on a rectangular grid in order to best demonstrate the effect of altering ΔT. The computed tracks of totality on each date, based on the assumption of a constant length of day of 86 400 SI seconds ($\Delta T = 0$), are shown. On the seemingly plausible assumption that the individual observations should be equally weighted, it is possible by displacing the calculated central lines (i) in 1178 towards the east by about 20 deg and (ii) in 1187 towards the west by some 10 deg to derive what appear to be their most likely true positions. The implied values for ΔT are roughly +5000 sec in 1178 and −2500 sec in 1187. These results are at considerable variance with one another – and also with those obtained from roughly contemporaneous Chinese and Arab timed measurements, which indicate a value for ΔT at this date of around +1000 sec (see chapters 9 and 13).

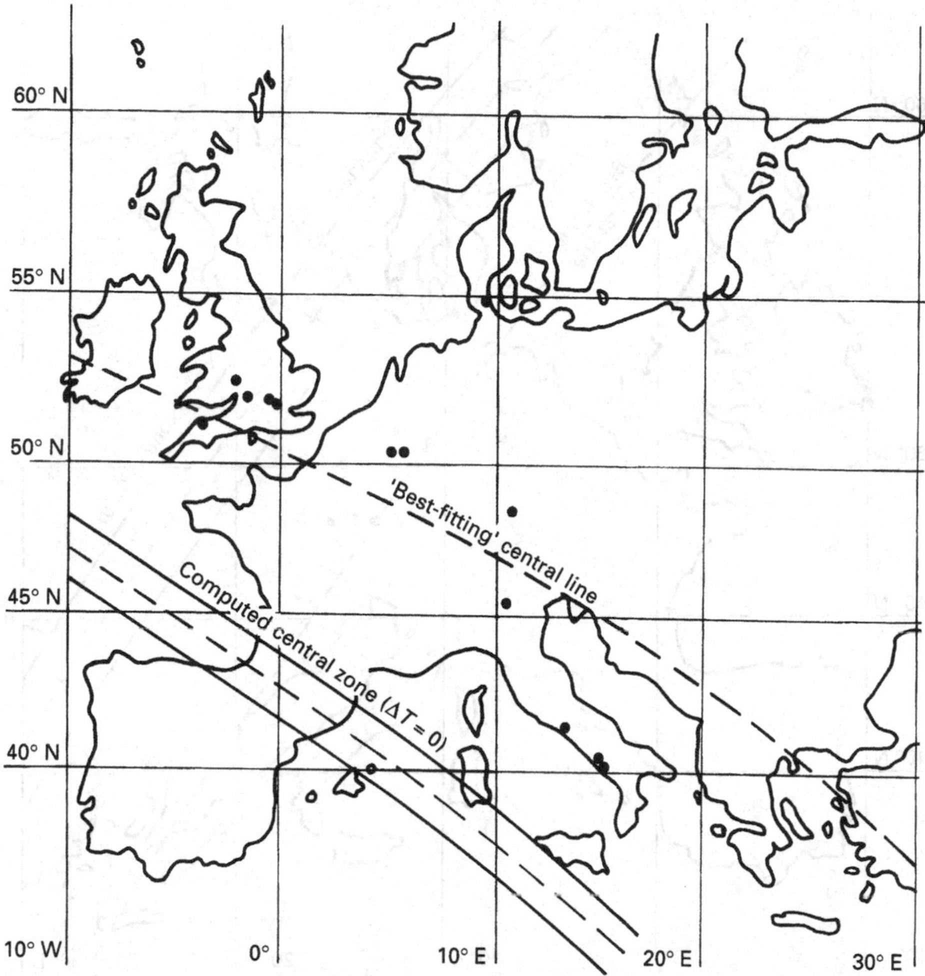

Fig. 3.18 Locations of the various European sites where the solar eclipse of AD 1178 Sep 13 was observed. Equally weighting the observations, the 'best-fitting' central line lies some 20 deg to the east of the computed central line.

Rather than assume huge variations in ΔT in such a short time, it is much more reasonable to conclude that analysis of this sort is very much influenced by 'population bias' – the non-random distribution of population centres (or at least those places from which preserved eclipse records are accessible) in medieval Europe. A similar illustration, involving the eclipse of AD 1039, is given by Muller and Stephenson (1975). For independent comments, see Lambeck (1980), p. 314.

In summary, the use of *careful* descriptions of eclipses – whether qualitative or quantitative records – is much to be preferred.

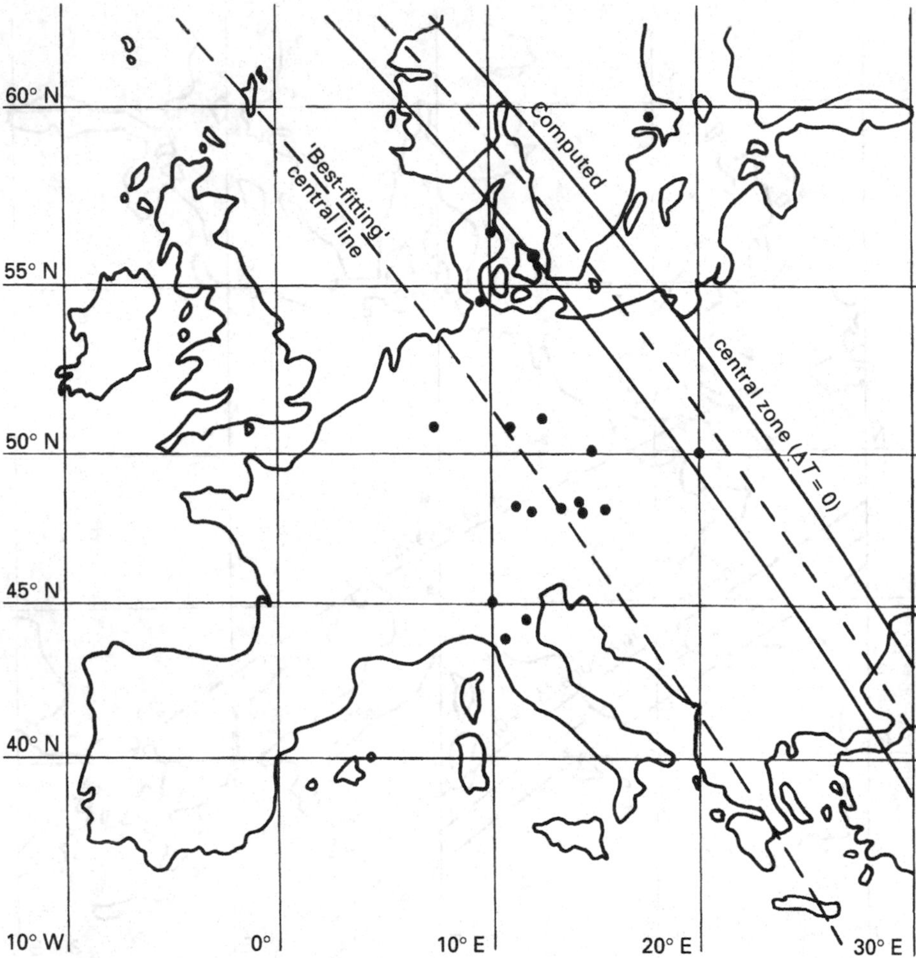

Fig. 3.19 Locations of the various European sites where the solar eclipse of AD 1187 Sep 4 was observed. Equally weighting the observations, the 'best-fitting' central line lies some 10 deg to the west of the computed central line.

3.9 Lunar eclipse observations and their reduction to ΔT

As noted earlier, lunar eclipses have more restricted usage than their solar counterparts in the study of the Earth's past rotation. However, the fact that relatively more timings and rising and setting observations of lunar eclipses have survived does much to redress this balance.

Before I discuss the various ways in which lunar eclipse records can be used to determine ΔT, some comments on the adopted dating scheme are necessary. Throughout this book I have assigned a double date on the Julian calendar – e.g. Jun 13/14 – for all eclipses of the Moon. This is partly because these events, which can last for several hours, often begin before local midnight and end after it. Furthermore, since most observing sites are much to the east of the Greenwich meridian, the LT tends to be several hours ahead of the UT. An additional reason for this choice is that several cultures began the new day at times other than midnight. For instance, the Babylonian day began at sunset (i.e. roughly 6 hours before the start of the civil day) and this practice was later followed by Muslims. On the other hand, it seems that Chinese astronomers did not change the calendar date until sunrise (some 6 hours after the start of the civil day). A double date thus helps to avoid confusion.

3.9.1 Timed lunar eclipse contacts

Ancient and medieval timings of partial lunar eclipses generally relate to three phases: first contact, maximal phase (i.e. mid-eclipse) and last contact. In the case of total obscurations, four separate times are usually recorded; these are the two external contacts and also the beginning and end of totality (second and third contact). Although most early astronomers determined times directly (e.g. with the aid of a clepsydra), Arab observers were in the habit of measuring altitudes of the Moon or a suitably placed bright star instead.

In analysing all observations of the umbral contacts for lunar eclipses, I have applied the standard increment of two per cent to the radius of the Earth's shadow to approximately allow for the effect of the atmosphere. There is also some evidence that an observer using the unaided eye may mistake the deep penumbral shadow for the umbra itself. For example, at the very beginning of the seventeenth century (a little before the introduction of the telescope), the Belgian astronomer Wendelin criticised a contemporary for failing to make this distinction (Pingre, 1901, p. 20). However, recent investigation of recorded lunar eclipse *durations* (from first to last contact) by unaided-eye observers shows that the mean ratio of the measured and computed values is very close to unity, although there is significant scatter among individual results. A survey of such measurements by Stephenson and Fatoohi (1994b) revealed that the appropriate mean ratios for Babylonian (after 300 BC), Chinese and early seventeenth century European astronomers were respectively 1.02 ± 0.03, 0.99 ± 0.04 and 1.01 ± 0.01. Hence, when a series of contact timings is used to determine ΔT, penumbral effects should be negligible.

Once a lunar eclipse timing is reduced to UT, computation of the corresponding TT leads to a result for ΔT by direct subtraction. There is no need for an iterative solution (unlike in the case of a solar eclipse) since at any given moment the local circumstances of a lunar eclipse are virtually identical over the entire hemisphere of the Earth's surface from which the Moon is visible at the time.

The total lunar eclipse of BC 226 Aug 1/2 was one of the many carefully timed by Babylonian astronomers. The record contains the following details (see chapter 5):

> Night of the 14th ... At 52 deg after sunset, when α Cyg culminated, lunar eclipse; when it began on the east side, in 17 deg (of) night time it covered it completely; 10 deg night time maximal phase; when it began to clear, it cleared in 15 deg night time from south to north ... 42 deg onset, maximal phase and clearing ... .
>
> [Trans. Sachs and Hunger (1989, p. 141).]

The reference to the culmination of the star α Cyg provides an independent check on the time of first contact – in addition to the sunset measurement. It will be noted that the reported duration of 42 deg is indeed the sum of the durations of the individual phases. The four intervals after sunset correspond to 3.47, 4.60, 5.27 and 6.27 h respectively. Since sunset occurred at a LT of 18.94 h, the corresponding LTs for the contacts are 22.41, 23.54, 0.21 h and 1.21 h. Although the Babylonian date (commencing at sunset) of all four contacts was Aug 2, the eclipse began on a civil date (starting at local midnight) of Aug 1 and ended on Aug 2. The various results are shown diagrammatically in figure 3.20. Converting the LTs to UT (all on Aug 1) and comparing with the computed TTs (all on Aug 2), we have the following results (to the nearest 50 sec):

- First contact: UT = 19.51 h, TT = 0.38 h, ΔT = +17 550 sec.
- Second contact: UT = 20.63 h, TT = 1.50 h, ΔT = +17 550 sec.
- Third contact: UT = 21.30 h, TT = 2.57 h, ΔT = +18 950 sec.
- Last contact: UT = 22.30 h, TT = 3.68 h, ΔT = +19 350 sec.

During the eclipse, α Cyg (RA 19.43 h) was 11.00 h to the east of the Sun. Hence the LT of culmination (identifying first contact) was 12.00 + 11.00 = 23.00 h, a UT of 20.09 h on Aug 1. This is as much as 0.68 h after the corresponding sunset measurement and leads to a result for ΔT of +15 450 sec. The relation between the UT, TT and ΔT values for each of the five measurements is illustrated in figure 3.21. All of the ΔT results are in fair (but by no means excellent) accord with one another. Fortunately there are many further observations from around this period.

To give an additional example, astronomers in Baghdad (ϕ = +33.33 deg, Λ=−44.42 deg) determined the time of first and second contact at the eclipse of AD 854 Feb 16/17 by measuring star altitudes (see chapter 13). The altitude of α Tau at first contact was found to be 45.5 deg in the east, while that of α CMi at the beginning of totality was between 22 and 23 deg in the east. At the date in question, the celestial co-ordinates (RA

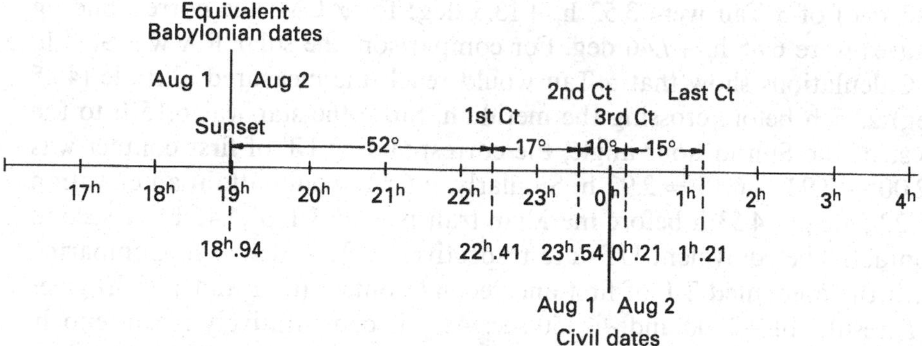

Fig. 3.20 LT measurements made by Babylonian astronomers for the lunar eclipse of BC 226 Aug 1/2.

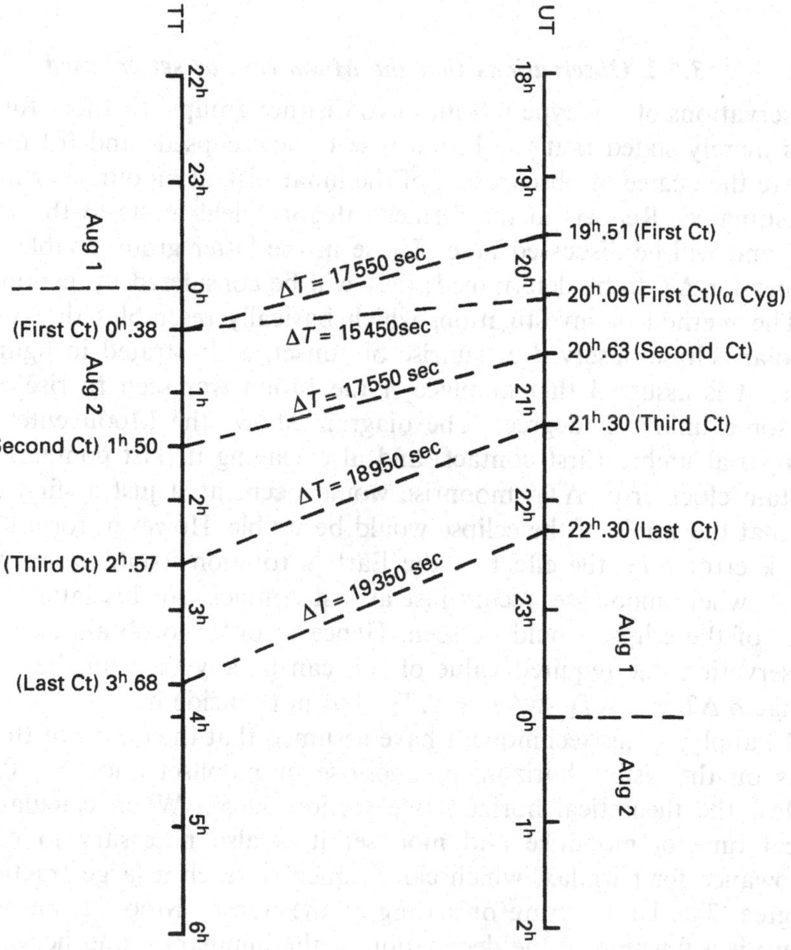

TT

UT

22h 23h 0h 1h 2h 3h 4h 5h 6h

18h 19h 20h 21h 22h 23h 0h 1h 2h

Aug 1 Aug 2

Aug 1 Aug 2

(First Ct) 0h.38

(Second Ct) 1h.50

(Third Ct) 2h.57

(Last Ct) 3h.68

19h.51 (First Ct)

20h.09 (First Ct)(α Cyg)

20h.63 (Second Ct)

21h.30 (Third Ct)

22h.30 (Last Ct)

ΔT = 17 550 sec

ΔT = 15 450sec

ΔT = 17 550 sec

ΔT = 18950 sec

ΔT = 19 350 sec

Fig. 3.21 Relation between the UT, TT and equivalent ΔT values for each of the five measurements made at the lunar eclipse of 226 BC.

and dec) of α Tau were 3.52 h, +13.5 deg; for α CMi the corresponding figures were 6.65 h, +7.46 deg. For comparison, the solar RA was 9.67 h.

Calculations show that α Tau would reach the measured altitude (45.5 deg) 2.93 h before crossing the meridian. Since the star was 6.15 h to the west of the Sun in hour angle, the corresponding LT of first contact was $12.00 - 2.93 - 6.15 = 2.92$ h. Similarly, α CMi would attain an elevation of 22.5 deg at 4.53 h before meridian transit – an LT of 4.45 h for second contact. The equivalent UTs are respectively 0.01 and 1.53 h. Comparing with the computed TT of first and second contact (0.71 and 1.91 h) gives ΔT results of +2500 and +1350 sec. At this comparatively recent epoch, ΔT was fairly small. Hence the discrepancy between the individual values (1150 sec), although rather less than that noted among the Babylonian observations in 226 BC is relatively much more significant.

3.9.2 *Observations that the Moon rose or set eclipsed*

Observations of this type fall into two distinct groups: (i) those for which it is merely stated that the Moon rose or set eclipsed; and (ii) instances where the degree of obscuration of the lunar disk at moonrise or moonset is estimated. Reports in the former category yield limits to the value of ΔT and will be discussed here. Those in the latter group enable specific values for ΔT to be determined; they will be considered in section 3.9.3.

The method of investigation, which basically resembles that used for a solar eclipse observed at sunrise or sunset, is illustrated in figure 3.22. Here it is assumed that at place *A* the Moon was seen to rise eclipsed to some unknown degree. The diagram shows the Moon entering the terrestrial umbra (first contact) and also leaving it (last contact). For a certain clock error ΔT_1, moonrise would occur at *A* just at first contact so that the whole of the eclipse would be visible. However, for a different clock error ΔT_2, the effect of the Earth's rotation would be to move *A* to *A*′ where moonrise occurs just at last contact. In this latter case, no part of the eclipse would be seen. Hence in order to obtain accord with observation, the required value of ΔT can be anywhere in the indicated range $\delta\, \Delta T$ (i.e. $\Delta T_1 < \Delta T < \Delta T_2$), but not outside it.

In applying this technique, I have assumed that the *centre* of the Moon was on the visible horizon at moonrise or moonset and thus 0.65 deg below the theoretical horizon (see section 3.8.5). When calculating the local time of moonrise and moonset it is also necessary to make an allowance for parallax, which can frequently reach a large fraction of a degree. The LT of rising or setting of the Sun or Moon at any selected place is a function of the declination of the luminary – and hence of ΔT. However, on account of the very slow change in the *solar* declination (never more than 1 arcmin in an hour), in computing the local time of

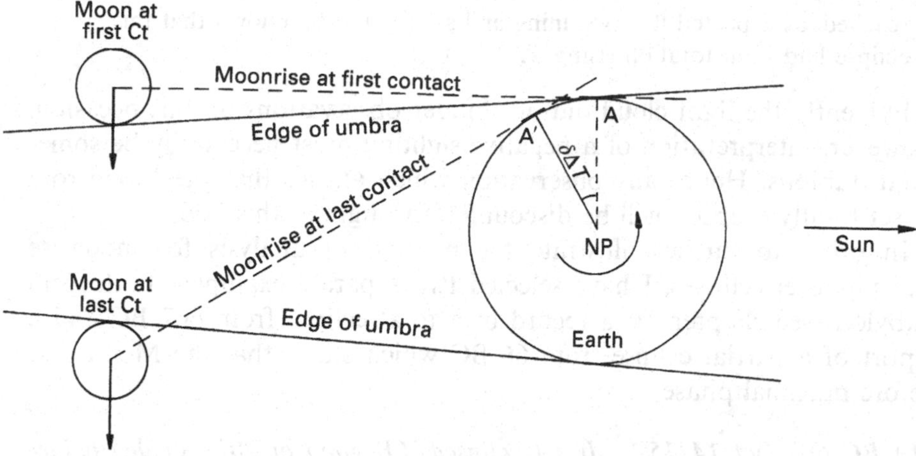

Fig. 3.22 Method of investigation for an eclipse in which the Moon was observed to rise eclipsed to some unknown degree. A discrete range of ΔT (i.e. $\delta\Delta T$) is indicated.

sunrise or sunset, it is quite unnecessary to make any allowance for ΔT, corrections amounting to less than 0.01 h. This very minor variation has been ignored in previous discussion. On the other hand, since the Moon moves so much more rapidly than the Sun (by a factor of approximately 13), it is necessary to deduce the LT of moonrise or moonset iteratively using successive approximations to the final choice of ΔT. The response to changes in ΔT is particularly significant for spring moonrise and autumn moonset.

If an eclipse was generally total, yet the record states that the Moon rose or set whilst partially covered, it is usually possible to decide from the text whether the Moon reached the horizon before or after the occurrence of the total phase. Otherwise, two discrete ranges of ΔT are obtained, one corresponding to moonrise or moonset between first and second contact and the other between third and last contact. When the totally eclipsed Moon is on the horizon, it is probably always invisible to the unaided eye. In principle, if the sky was clear, experienced observers, who would be well aware that the eclipsed Moon was accurately in opposition to the Sun, would probably be able to determine whether any trace of the lunar partial phase was discernible when the Sun reached the opposite horizon. This is illustrated by a Chinese account from Lin-an ($\phi = +30.25$ deg, $\Lambda = -120.17$ deg) of the total lunar eclipse of AD 1168 Mar 25/26 (see also chapter 9):

> On this evening, when the Moon rose there was light cloud. Until the fall of darkness it could not be seen that the Moon had been totally eclipsed. When the 3rd mark of the initial half of the hour *hsu* was

reached, as expected it was shining and so it could be known that this eclipse had been total on rising ...

Evidently the light cloud did not hinder observations on this occasion. However, interpretation of a negative sighting must necessarily be somewhat dubious. Hence any observation which alleges that the Moon rose or set totally eclipsed will be discounted throughout this book.

In order to further illustrate the method of analysis for moonrise and moonset eclipses, I have selected two separate examples, both from Babylon (see chapter 7): a record of a total eclipse from 667 BC and a report of a partial eclipse from 66 BC which states that the Moon rose before maximal phase.

(1) BC 667 Oct 14/15 '...It set eclipsed. (Began) at 20[+x] deg before sunrise'.
This record is found on a tablet which lists many reports of lunar eclipses. Although the eclipse was actually total, the text is so badly damaged that it can only be inferred that the Moon set between first and last contact.

(i) For first contact before moonset, LT of moonset = 6.15 h. Hence LT of contact < 6.15 h, UT < 3.02 h. Comparing the UT with the computed TT of 6.19 h, one gets $\Delta T > 11\,400$ sec.

(ii) For last contact after moonset, LT of moonset = 6.31 h. Hence LT of contact > 6.31 h, UT > 3.20 h. Computed TT of 9.73 h, thus $\Delta T < 23\,500$ sec.

When these two ΔT limits are combined, $11\,400 < \Delta T < 23\,500$ sec.

It will be noted that the computed LTs of moonset differ by as much as 0.16 h, mainly on account of the large ΔT range.

(2) BC 66 Dec 28/29 '...When the Moon rose, 2 fingers on the south side [were eclipsed]. In 9 deg night, over a third of the disk [was eclipsed]; 8 deg duration of maximal phase, until it began to become bright ...'
It is evident that the Moon rose between first contact and maximum.

(i) For first contact before moonrise, LT of moonrise = 16.87 h. Hence LT of contact < 16.87 h, UT < 14.00 h. Computed TT = 16.64 h; thus $\Delta T > 9500$ sec.

(ii) For maximum phase after moonrise, LT of moonrise = 16.90 h. Hence LT of contact > 16.90 h, UT > 14.14 h. Computed TT = 17.75 h; thus $\Delta T < 13\,350$ sec.

Combining these limits, one obtains $9500 < \Delta T < 13\,350$ sec.

Here the two calculated LTs of moonrise are not significantly different, mainly because the ΔT range is relatively small.

3.9.3 Estimates of the degree of obscuration of the Moon at moonrise or moonset

Several careful estimates of the fraction of the Moon which was enveloped in the Earth's shadow when it rose or set are preserved. As for both solar and lunar eclipse magnitudes generally, Babylonian and Arab estimates of this ratio are usually expressed to the nearest twelfth of the disk and Chinese estimates to the nearest tenth or fifteenth. Since the phase of a central lunar eclipse changes by such an amount in only about 5 min, it may be inferred that an observation of this type might compete fairly well with a carefully timed contact.

When the Moon is rising or setting, atmospheric refraction is maximal. Nevertheless, distortion of the lunar limb is not necessarily critical. Several good quality photographs which have been published in various journals etc. show the eclipsed *Sun* on the horizon – see figure 3.17 – but few seem available for the eclipsed Moon in a similar situation. Photographs such as these lead to the inference that on most occasions a reasonable estimate of the proportion of the disk in shadow would be possible, whether for the Sun or the Moon. In any case, if the limb of the luminary were very distorted, an observer in the ancient or medieval world might well feel that such an exercise was pointless. As will become apparent in subsequent chapters, individual results for ΔT obtained from these observations prove to be remarkably self-consistent.

If an estimate of the extent to which the Moon was covered at moonrise or moonset is recorded, it would seem best to suppose that the whole of the disk was just clear of the horizon when the observation was made. Allowing for horizontal refraction and assumed observer height (10–15 m), this corresponds to a depression for the Moon's centre of about 0.4 deg below the theoretical horizon. Using an iterative solution for ΔT, the calculated phase at moonrise or moonset may be varied until it equals the observed fraction. Each estimate of the proportion of the Moon covered at its rising or setting normally leads to a single value for ΔT (unless it is uncertain whether the phase was increasing or declining at the time).

The profile of the horizon is less important for this type of observation since specific limits to ΔT are not set. Horizon irregularities merely increase the scatter in the ΔT results which are obtained. Unless very significant, they can thus be ignored.

As noted above (section 3.8.4) Stephenson and Fatoohi (1994a) investigated the accuracy with which the unaided eye can determine the peak magnitude of a lunar eclipse. They analysed many lunar magnitude estimates from the pre-telescopic period originating from Babylon, China, the Arab lands and Europe. In each case there was nothing in the various records to suggest that the observer was prevented from viewing the Moon around maximal phase – either by cloud or the Moon being below the horizon at the time.

Comparison between observation and computation for these reference data showed that the median error in estimating magnitude (as a proportion of the lunar diameter) was 0.08. More significantly, it was also demonstrated that the various observers systematically overestimated the magnitudes of small eclipses (less than half covered) by up to about 0.15 (in units of the lunar diameter). For larger obscurations much the reverse was true. The effect of the deep penumbral shadow may well lead to overestimates of magnitude for small eclipses but physiological effects may well be important at all phases.

The results of this investigation are shown in figure 3.23, which is taken from the paper by Stephenson and Fatoohi (1994a). In this diagram, the observational error in estimating magnitude (in the sense computed–observed) is plotted as a function of the computed magnitude. A least squares straight line fit (for want of a better solution) to the data emphasises the remarkable skewness of this distribution, the gradient being as much as 0.19. Relative to this line, the median error of observation is only 0.05.

Each estimate of the proportion of the Moon covered at its rising or setting normally leads to a single value for ΔT (unless it is uncertain whether the phase was increasing or declining at the time).

As an example, I have selected the eclipse of AD 438 Jun 23/24, which was observed in the Chinese capital of Chien-k'ang (ϕ = +32.03 deg, $\Lambda = -118.78$ deg). The record reads as follows (see chapter 9):

> When the Moon first rose, it was already eclipsed. Its brightness had already regained 1/4 (of itself)...

The proportion of the lunar diameter covered at moonrise was thus estimated as 0.75. The eclipse was actually total, but it seems clear from the text that the phase was already declining when the Moon first came up. Making a minor adjustment of +0.03 to the estimated fraction of 0.75 to allow for systematic errors (as derived from figure 3.23) gives a result of 0.78.

Chien-k'ang (on the site of the modern city of Nanjing) was situated in a fairly level plain beside the Yangtze river. Using an iterative procedure, as in section 3.9.2, the computed LT of moonrise (assuming a depression below the theoretical horizon of 0.4 deg) is 19.18 h, a UT of 11.20 h. Since the computed TT when 0.78 of the lunar diameter was obscured after maximum is 12.82 h, the equivalent value of ΔT is +5800 sec.

As already noted, it is rare to find an estimate of the proportion of the *Sun* eclipsed when on the horizon. Investigation of the few observations of this type follows very much the above pattern. Although no allowance need

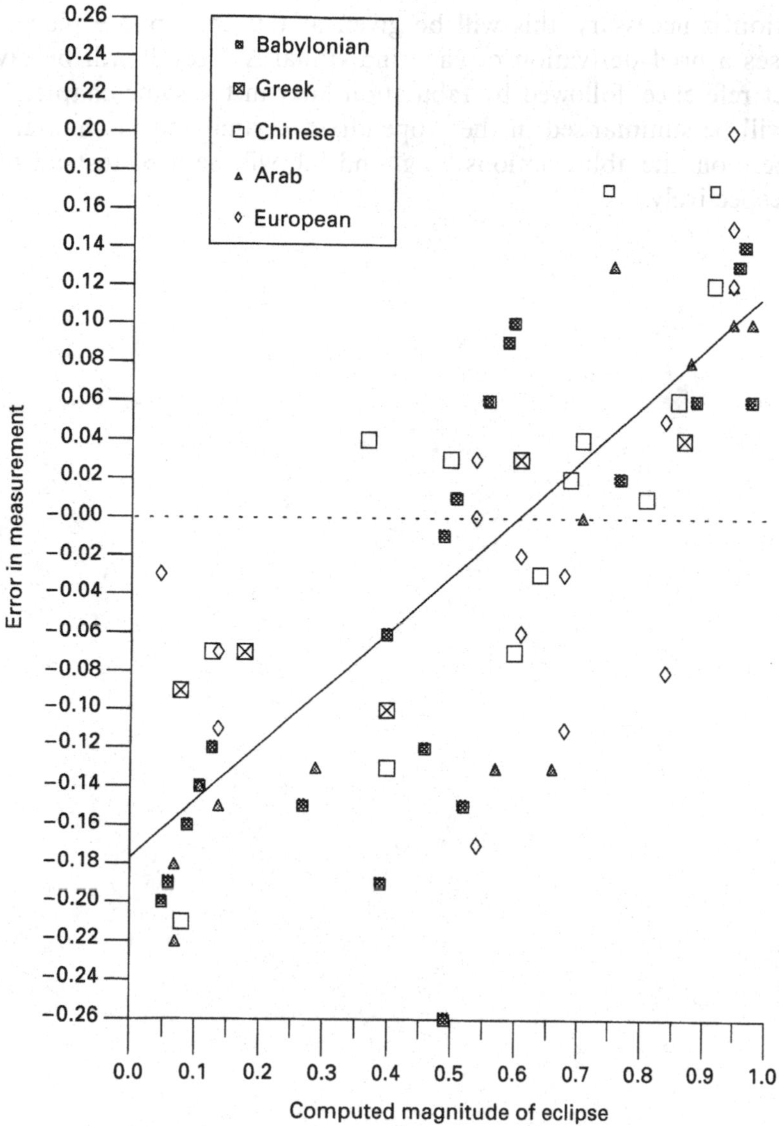

Fig. 3.23 Results of an investigation of recorded lunar eclipse magnitudes (Stephenson and Fatoohi, 1994a).

be made for ΔT in computing the LT of sunrise or sunset, an iterative solution is necessary in order to match the calculated phase with observation.

3.10 Conclusion

The various techniques developed in this chapter will be used to analyse the observations discussed in chapters 4 to 13. When it is felt that further explanation is necessary, this will be given at the appropriate place. In most cases a brief derivation of each individual ΔT result will be given for direct reference, followed by tabulation later in the same chapter. All results will be summarised in the Appendix and analysed in chapter 14. (From here on, the abbreviations long. and lat. will be used instead of Λ and ϕ respectively.)

4

Babylonian and Assyrian records of eclipses

4.1 Introduction

More celestial observations are preserved from Babylon than from any other contemporary civilisation. Yet until about a century ago, when large numbers of clay tablets devoted to astronomy began to be unearthed at the site of Babylon, little was known about the achievements of the skywatchers of this once great city. What could be established was mainly based on ancient Greek texts and the Old Testament. Both the Prophet Isaiah (e.g. 47:13) and the ancient Greek writer Strabo (*Geography*, XVI, 1.6) stress the Babylonian preoccupation with astrology. As noted in chapter 3, the ancient Greek historian Diodorus Siculus (*Library of History*, II, 9) implies that the lofty ziggurat – built during the reign of Nebuchadrezzar II (604–563 BC) was used as an observatory. Figure 4.1 shows a schematic view of Babylon in the days of Nebuchadrezzar II – as visualised by Herbert Anger (Unger, 1931, figure 7).

Among writers of the ancient Greek and Roman world whose works are still extant, only the great Alexandrian astronomer Claudius Ptolemy (*c.* AD 150) hints at the true scale on which celestial observation was practised at Babylon. In his *Mathematike Syntaxis* (Mathematical Systematic Treatise) – which later became known as the *Almagest* – Ptolemy specifically mentions sets of Babylonian eclipse observations to which he had access. Examples of his comments are as follows:

(i) First, the three ancient eclipses which are selected from those observed in Babylon...

[*Almagest*, IV, 6; trans. Toomer (1984, p. 191).]

(ii) First then, to correct the actual mean motion in latitude, we looked for [pairs of] lunar eclipses (among those securely recorded) separated by as great an interval as possible...

[*Almagest*, IV, 9; trans. Toomer (1984, p. 206).]

Fig. 4.1 A schematic view of Babylon in the days of Nebuchadrezzar II – as visualised by Herbert Anger (Unger, 1931).

> (iii) He (Hipparchus) says that these three eclipses which he adduces are from the series brought over from Babylon and were observed there...
>
> [*Almagest*, IV, 11; trans. Toomer (1984, p. 211).]

It is regrettable that Ptolemy actually cites no more than ten Babylonian records of *lunar* eclipses from the apparently large number available to him. Furthermore, no observations of *solar* eclipses from Babylon are preserved in the *Almagest*, although there are a few references to other celestial phenomena reported from this site, such as conjunctions of planets with stars. These various observations range in date from 721 to 229 BC, lunar eclipses covering the period from 721 to 382 BC.

The discovery of vast numbers of astronomical cuneiform texts at the site of Babylon during the 1870s and 1880s was eventually to revolutionise knowledge of Babylonian astronomy. These texts, which are in the form of inscribed clay tablets, range in date from about 730 BC to AD 75. Many of the eclipse records which they contain have proved so important in studies of the Earth's past rotation that the investigation of this material will occupy fully three chapters of the present book (5, 6 and 7). In particular, the earlier observations must have formed the ultimate source of the material used by Hipparchus and Ptolemy. Despite the existence of this huge archive, the ten Babylonian lunar eclipse records cited in the *Almagest* form an interesting set of data and deserve to be analysed in their own right. These observations will be investigated in the present chapter, followed by a discussion of the historical background to the astronomical cuneiform texts. At the close of this chapter, the eclipse records found on the Assyrian cuneiform tablets of the eighth and seventh centuries BC will be briefly considered. However, it should be stressed that these records are mainly in the form of astrological texts and – unlike the Babylonian material – contain little quantitative information.

4.2 The Babylonian lunar eclipse records in the *Almagest*

The very earliest astronomical records quoted in the *Almagest* are exclusively of Babylonian origin. (For a discussion of the Greek eclipse observations which Ptolemy cites, see chapter 10.) Ptolemy evidently did not have access to material much older than 721 BC – the earliest Babylonian eclipse record which he discusses. He states (III, 7) that beginning with the reign of King Nabonassar (correct name: Nabu-nasir) of Babylon (747–733 BC) 'the ancient observations are, on the whole, preserved down to our own time'. This is apparently why Ptolemy chose the era of Nabonassar (747 BC) for numbering years. There is a Hellenistic tradition that King Nabonassar (like Emperor Ch'in Shih-huang in China – see chapter 8) destroyed the records of his predecessors. His alleged purpose was to ensure 'that the reckoning of the Chaldean kings might start with himself' (Brinkman, 1968). Although this tradition may lack firm grounds, it is a fact that Babylonian history and chronology are extremely weak in the two centuries immediately prior to Nabonassar's reign. Even the names of several kings who ruled during this interval are unknown (cf. Oates, 1979, p. 201). However, beginning with Nabonassar, Babylonian chronology is securely established.

It seems very likely that Ptolemy did not compile the list of Babylonian

Table 4.1 Egyptian month-names as found in the Almagest.

Number	Name	Number	Name
I	Thoth	VII	Phamenoth
II	Phaophi	VIII	Pharmouthi
III	Athyr	IX	Pachon
IV	Choiak	X	Payni
V	Tybi	XI	Epiphi
VI	Mechir	XII	Mesore

observations himself. The evidence points instead to Ptolemy's great predecessor Hipparchus of Rhodes (*c.* 150 BC). As noted above, Ptolemy specifically mentions a series of eclipse observations which had been 'brought over from Babylon' and investigated by Hipparchus. Toomer (1988) is of the opinion that the entire Babylonian record available to Ptolemy was compiled by Hipparchus, further suggesting that Hipparchus 'arranged them in a form suitable for use by Greek astronomers'. It is clear from reading the accounts of individual eclipses quoted by Ptolemy (see section 4.3 below) that he did not receive them at first hand. Sadly, apart from the material preserved in the *Almagest*, all trace of Hipparchus' compilation has long been lost.

No ancient manuscripts of the *Almagest* now exist; the earliest copies date from the ninth century AD (Toomer, 1984, pp. 2–4). However, as Toomer points out, there is in general very close accord between the text of individual manuscripts.

For the very earliest eclipse observations (721 and 720 BC) which he cites, Ptolemy numbers the year from the accession of Mardokempad (Marduk-apla-iddin), who was ruler of Babylon at the time. However, all later years are counted from the era of Nabonassar. Although the eclipse dates would originally be expressed in terms of the Babylonian luni-solar calendar (see section 4.7), Ptolemy invariably specifies the month in terms of the Egyptian calendar. Numbering days from a fixed epoch was considerably simplified on this latter system. Each Egyptian year contained 12 equal months of 30 days followed by 5 *epagomenal* ('additional') days. This fixed year of 365 days was not adjusted to the solar year by intercalation so that the first day of the year gradually retrograded through the seasons, making a complete circuit in 1460 years (the Sothic cycle). Greek transliterations of the Egyptian month-names – as used by Ptolemy – are listed in table 4.1.

In quoting days of the month for lunar eclipses, Ptolemy was in the habit of using double dates – e.g. Thoth 18/19. This is because although the civil day began at sunset in Babylon, according to Egyptian convention it commenced at the following sunrise (Toomer, 1984, p. 12). As noted in chapter 3, I have found it convenient for similar reasons to use double dates on the *Julian* calendar for all eclipses of the Moon.

Comparison with the many lunar records found on the astronomical tablets recovered from Babylon (see chapter 6) makes it clear that the eclipse times quoted by Ptolemy are not in original form (i.e. using time-degrees) but have been modified to correspond to the Greek method (equinoctial or seasonal hours). Presumably Hipparchus was responsible for these reductions. It is a pity that the original measurements are not preserved; it is likely that some loss of accuracy would occur when the times were reduced to the Greek system. However, no attempt at restoration is possible. Only in a single case (523 BC) is there a parallel inscription on an extant cuneiform tablet (see chapter 6), and even this is problematical.

Despite the assertion by Diodorus Siculus (*Library of History*, II, 9) that the astronomers of Babylon observed from the ziggurat, there is no confirmatory evidence in other ancient sources. In any case, the tower is reported by Strabo (*Geography*, XVI, 1.5) to have been damaged by Xerxes I after a rebellion in 482 BC, while around 330 BC Alexander the Great is said to have demolished it (Strabo, *loc. cit.*). Alexander's plans to rebuild the ziggurat were never carried out. A recent aerial photograph of the site is shown in figure 4.2.

The geographic co-ordinates of Babylon are: lat. = +32.55 deg, long. = −44.42 deg. Since the plain in which the site of the city is located is so flat, there is no need to make any allowance for horizon profile in deducing the time of sunrise or sunset, etc. Photographs of the horizon taken from the ruins of the city show that it is remarkably level in all directions (e.g. Ravn, 1942, plates II, IXb and Xa).

4.3 Investigation of the Babylonian lunar eclipses cited by Ptolemy

The ten Babylonian observations of lunar eclipses which Ptolemy discusses in his *Almagest* are all cited as textbook examples to illustrate the derivation of certain lunar parameters. For instance, several observations (from 721, 720, 383 and 382 BC) were used by Ptolemy to deduce the principal lunar anomaly, including correcting errors of calculation made by Hipparchus. Further data (from 720, 502 and 491 BC) were applied to the investigation of the Moon's mean motion in latitude, while observations in 621 and 524 BC yielded a good result ($31\frac{1}{3}$ arcmin) for the Moon's apparent diameter. These various techniques are discussed by Neugebauer (1975, pp. 81 ff., 104 ff., etc.).

Fig. 4.2 A recent aerial photograph of the site of the Babylonian ziggurat.

Each of the Babylonian lunar eclipse observations cited by Ptolemy is investigated below. In every case, I have given the computed Julian date and eclipse magnitude, followed by a translation of the text quoted from Toomer (1984). In some cases it is not clear whether times are expressed in equinoctial hours (*horai isemerinai*) or seasonal hours (*horai kairikai*). Where necessary I have appended further comments by Ptolemy if these help clarify the main text. The Julian dates reduced by Toomer from the historical dates are given in square brackets using negative integers for years. It will be noted that when compared with the computed Julian dates, these are invariably accurate. As mentioned in chapter 1, negative years differ by unity from their BC equivalents.

In each case, Ptolemy only quotes the time of one phase of an eclipse. This is usually first contact, but on four occasions (in 720, 523, 502 and 490 BC) he implies mid-eclipse (i.e. maximal phase) instead. Fotheringham (1915), following Nevill (1906b), inferred that the times of all but the first of these latter events related to first contact. However, since Ptolemy is our most direct source, there seems little alternative to accepting his interpretation. For a valuable historical discussion, see Britton (1985).

(1) BC 721 Mar 19/20 (mag. = 1.53)

> First the three ancient eclipses which are selected from those observed in Babylon. The first is recorded as occurring in the first year of Mardokempad, Thoth [month I] 29/30 in the Egyptian calendar [−720 Mar 19/20]. The eclipse began, it [the report] says, well over an hour after moonrise and was total. Now since the Sun was near the end of Pisces, and [therefore] the night was about 12 equinoctial hours long, the beginning of the eclipse occurred, clearly, $4\frac{1}{2}$ equinoctial hours before midnight.
>
> [*Almagest*, IV, 6; trans. Toomer (1984, p. 191).]

Ptolemy evidently interpreted the phrase 'well over an hour after moonrise' as implying an interval of $1\frac{1}{2}$ hours; this will be assumed here. On the supposition that sunset and moonrise occurred simultaneously, Ptolemy inferred that the eclipse began $1\frac{1}{2}$ hours after sunset. However, computation shows that the Moon would actually rise a significant time (about 0.2 hours) before sunset so that it seems preferable to specifically use moonrise as the reference moment. Whether the interval between moonrise and first contact was expressed in equinoctial or seasonal hours is unimportant since the Sun was so close to the (vernal) equinox; 1 seasonal hour would be equal to 1.01 h.

RESULTS

First contact on Mar 19 at $1\frac{1}{2}$ hours after moonrise. LT of moonrise = 17.73 h, hence LT of first contact = 19.23 h, UT = 16.45 h. Computed TT = 22.46 h, thus ΔT = 21 650 sec.

(2) BC 720 Mar 8/9 (mag. = 0.12)

> ...The second eclipse is recorded as occurring in the second year of Mardokempad, Thoth [month I] 18/19 in the Egyptian calendar [−719 Mar 8/9]. The [maximum] obscuration was 3 digits from the south exactly at midnight. So since mid-eclipse was exactly at midnight at Babylon ...
>
> [*Almagest*, IV, 6; trans. Toomer (1984, p. 191).]

Essentially the same record is to be found later in the *Almagest* (IV, 9) – Toomer, p. 208.

Results

Maximal phase on Mar 8 at midnight. LT of maximum = 24.00 h on Mar 8, hence UT = 21.28 h. Computed TT = 2.56 h on Mar 9, thus ΔT = 19 000 sec.

(3) BC 720 Sep 1/2 (mag. = 0.50)

> ...The third eclipse is recorded as occurring in the (same) second year of Mardokempad, Phamenoth [month VII] 15/16 in the Egyptian calendar [−719 Sep 1/2]. The eclipse began, it says, after moonrise, and the [maximum] obscuration was more than half [the disk] from the north ...The beginning of the eclipse was about 5 equinoctial hours before midnight (since it began after moonrise) ...
>
> [*Almagest*, IV, 6; trans. Toomer (1984, p. 192).]

Ptolemy's rough inference that the eclipse began 'about 5 equinoctial hours before midnight' will be ignored since this is not based on measurement. However, the observation that commencement did not occur until *after* moonrise sets a limit to the value for ΔT.

Results

First contact on Sep 1 after moonrise. LT of moonrise = 18.52 h, hence LT of first contact > 18.52 h, UT > 15.58 h. Computed TT = 21.52 h, thus ΔT < 21 400 sec.

(NB for values of ΔT > 21 400 sec, the Moon would rise already eclipsed.)

(4) BC 621 Apr 21/22 (mag. = 0.15)

> In the fifth year of Nabopolassar, which is the 127th year from Nabonassar, Athyr [month III] 27/28 in the Egyptian calendar [−620 Apr 21/22], at the end of the 11th hour in Babylon, the Moon began to be eclipsed; the maximum obscuration was $\frac{1}{4}$ of the diameter from the south. Now since the beginning of the eclipse occurred 5 seasonal hours after midnight...
>
> [*Almagest*, V, 14; trans. Toomer (1984, p. 253).]

A time 5 seasonal hours after midnight corresponds to the end of the 11th (seasonal) hour. Hence the measured time will be taken as exactly 11 seasonal hours after sunset.

Results

First contact on Apr 22 at 5 seasonal hours after midnight. LT of sunset = 18.48 h, hence length of night = 11.04 h. 1 seasonal hour = 0.92 h, thus LT of first contact = 4.60 h, UT = 1.60 h. Computed TT = 6.54 h, thus ΔT = 17 800 sec.

(5) BC 523 Jul 16/17 (mag. = 0.53)

> ... Again in the seventh year of Kambyses, which is the 225th year from Nabonassar, Phamenoth [month VII] 17/18 in the Egyptian calendar [−522 Jul 16/17], 1 [equinoctial] hour before midnight at Babylon, the Moon was eclipsed half its diameter from the north. Thus the eclipse occurred about 1 $\frac{5}{6}$ equinoctial hours before midnight at Alexandria.
>
> [*Almagest*, V, 14; trans. Toomer (1984, p. 253).]

It seems evident from the text that the time of maximal phase is recorded. Since Ptolemy adopted a longitude difference between Alexandria and Babylon of $\frac{5}{6}$ of an equinoctial hour [IV, 6], it is apparent that the time of mid-eclipse in Babylon was measured in equinoctial hours.

The extant cuneiform record of this same eclipse seems to contain predicted rather than observed details (see chapter 6).

Results

Maximal phase on Jul 16 at 1 equinoctial hour before midnight. LT of mid-eclipse = 23.00 h, UT = 20.05 h. Computed TT = 1.40 h on Jul 17, thus ΔT = 19 250 sec.

(6) BC 502 Nov 19/20 (mag. = 0.19)

> ... The second, which Hipparchus too used, occurred in the twentieth year of that Darius who succeeded Kambyses, Epiphi [month XI] 28/29 in the Egyptian calendar [−501 Nov 19/20], when $6\frac{1}{3}$ equinoctial hours of the night had passed; at this eclipse the Moon was, again, obscured from the south $\frac{1}{4}$ of its diameter. The middle of the eclipse was $\frac{2}{5}$ of an equinoctial hour before midnight in Babylon (for the length of half the night was about 6 $\frac{3}{4}$ equinoctial hours on that date) ...
>
> [*Almagest*, IV, 9; trans. Toomer (1984, p. 208).]

It is evident from the times given by Ptolemy in the last sentence of the above quotation that he assumed that the phase which was observed 'when $6\frac{1}{3}$ equinoctial hours of the night had passed' was maximal eclipse.

RESULTS

Maximal phase on Nov 19 at $6\frac{1}{3}$ equinoctial hours after sunset. LT of sunset = 17.26 h, hence mid-eclipse = 23.59 h, UT = 20.46 h. Computed TT = 1.72 h on Nov 20, thus ΔT = 18 950 sec.

(7) BC 491 Apr 25/26 (mag. = 0.09)

> The first eclipse we used is the one in Babylon in the thirty-first year of Darius I, Tybi [month V] 3/4 in the Egyptian calendar [–490 Apr 25/26], at the middle of the sixth hour [of night]. It is reported that at this eclipse the Moon was obscured 2 digits from the south ... For the time of mid-eclipse was $\frac{1}{2}$-hour before midnight at Babylon, and [hence] $1\frac{1}{3}$ equinoctial hours before midnight at Alexandria ...
>
> [*Almagest*, IV, 9; trans. Toomer (1984, pp. 206-7).]

Use of an ordinal numeral to fix the time (i.e. 'at the middle of the sixth hour') suggests seasonal hours; this was Ptolemy's normal practice. However, since the Sun was close to the equinox, the choice of units is unimportant: half a seasonal hour before midnight would be equal to 0.45 equinoctial hours. A local time of 23.55 h will be assumed.

RESULTS

Maximal phase on Apr 25 at 1/2 seasonal hour before midnight. LT of mid-eclipse = 23.55 h, hence UT = 20.53 h. Computed TT = 0.37 h on Apr 26, thus ΔT = 13 800 sec.

(8) BC 383 Dec 22/23 (mag. = 0.21)

> He (Hipparchus) says that these three eclipses which he adduces are from the series brought over from Babylon, and were observed there; that the first occurred in the archonship of Phanostratos at Athens in the month Poseidon, a small section of the Moon's disk was eclipsed from the summer rising-point [i.e. the north-east] when half an hour of night was remaining. He adds that it was still eclipsed when it set. Now this moment is in the 366th year from Nabonassar, in the Egyptian calendar (as Hipparchus himself says) Thoth 26/27 [–382 Dec 22/23], $5\frac{1}{2}$ seasonal hours after midnight (since half an hour of night was remaining).
>
> [*Almagest*, IV, 11; trans. Toomer (1984, pp. 211–2).]

Further remarks by Ptolemy indicate that he understood the moment when 'a small section of the Moon's disk was eclipsed' as meaning first contact. His statement that this occurred '$5\frac{1}{2}$ seasonal hours of night after midnight since half an hour of night was remaining' implies that the time before the end of night (i.e. sunrise) was also expressed in seasonal hours. The fact that the Moon set eclipsed provides independent limits to the value for ΔT. In order to determine these, it seems best to ignore the recorded times (in case of possible scribal error) and simply assume that the Moon set at some time between first and last contact.

NB Phanostratos was *archon* (chief magistrate) at Athens in the year 383/2 BC. (For a list of the Athenian archons, each of whom held office for one year, see Bickerman, 1980, pp. 138–9.)

RESULTS

(i) First contact on Dec 23 at $\frac{1}{2}$ seasonal hour before sunrise. LT of sunrise = 7.00 h, hence length of night = 14.00 h. 1 seasonal hour = 1.17 h, thus LT of first contact = 6.41 h, UT = 3.50 h. Computed TT = 8.43 h, thus $\Delta T = 17\,750$ sec.

(ii) Moon set eclipsed. (a) First contact: LT of moonset = 6.98 h on Dec 23, hence for Moon to set eclipsed after first contact, LT < 6.98 h, UT < 4.07 h. Computed TT = 8.43 h, thus $\Delta T > 15\,700$ sec. (b) Last contact: LT of moonset = 7.05 h, hence for Moon to set eclipsed before last contact, LT > 7.05 h, UT > 4.14 h. Computed TT = 10.10 h, thus $\Delta T < 21\,450$ sec. Combining these limits, one obtains $15\,700 < \Delta T < 21\,450$ sec. (NB the LT of moonset has been computed iteratively for each contact, as explained in chapter 3.)

(9) BC 382 Jun 18/19 (mag. = 0.49)

> ...He (Hipparchus) says that the next eclipse occurred in the archonship of Phanostratos at Athens in the month Skirophorion, Phamenoth 24/25 in the Egyptian calendar, and that the Moon was eclipsed from the summer rising-point [i.e. the north-east] when the first hour [of night] was well advanced. This moment is in the 366th year from Nabonassar, Phamenoth [month VII] 24/25 [–381 Jun 18/19], about $5\frac{1}{2}$ seasonal hours before midnight.
>
> [*Almagest*, IV, 11; trans. Toomer (1984, pp. 212).]

Since Ptolemy inferred that the eclipse began 'about $5\frac{1}{2}$ seasonal hours before midnight', he presumably understood that 'the first hour [of night]' was also expressed in the same units. As mentioned above, use of ordinal numerals for times quoted in seasonal hours was his customary style. Ptolemy thus interprets the clause 'when the first hour [of night] was well advanced' to mean half an hour after sunset, and this will be assumed.

In addition to the timed observation, which yields an estimate of the value of ΔT, the fact the eclipse began soon after sunset may set a useful limit to ΔT.

RESULTS

(i) First contact on Jun 18 at $\frac{1}{2}$ seasonal hour after sunset. LT of sunset = 19.15 h, hence duration of night = 9.70 h, 1 seasonal hour = 0.81 h.

LT of first contact = 19.56 h, UT = 16.51 h. Computed TT = 21.06 h, thus ΔT = 16 400 sec.

(ii) First contact after sunset. LT of contact > 19.15 h, UT > 16.10 h. Computed TT = 21.06 h, thus ΔT < 17 850 sec.

NB for values of ΔT > 17 850 sec the eclipse would have already begun *before* sunset (the Moon would rise several minutes before sunset).

(10) BC 382 Dec 12/13 (mag. = 1.48)

> ...He (Hipparchus) says that the third eclipse occurred in the archonship of Euandros at Athens, in the month Poseidon I, Thoth 16/17 in the Egyptian calendar, and that (the Moon) was totally eclipsed, beginning from the summer rising-point [i.e. the north-east], after four hours [of night] had passed. This moment is in the 367th year from Nabonassar, Thoth [month I] 16/17 [−381 Dec 12/13], about $2\frac{1}{2}$ hours before midnight. Now when the Sun is about two-thirds through Sagittarius, one hour of night at Babylon is about 18 time-degrees. So $2\frac{1}{2}$ seasonal hours produce 3 equinoctial hours. Therefore the beginning of the eclipse was 9 equinoctial hours after noon on the 16th ...
>
> [*Almagest*, IV, 11; trans. Toomer (1984, pp. 213).]

As Toomer (p. 213n) remarks, the statement that the eclipse began 'after four hours of night' is incompatible with Ptolemy's interpretation that it commenced 'about $2\frac{1}{2}$ hours before midnight'. Ptolemy consistently assumes $2\frac{1}{2}$ hours before midnight in his subsequent argument and it seems only reasonable to follow him. From Toomer, inspection of the various extant manuscripts of the *Almagest* does not clarify this issue since all give identical readings here.

NB Euandros was *archon* at Athens in the year 382/1 BC.

RESULTS

First contact on Dec 12 at $2\frac{1}{2}$ seasonal hours before midnight. LT of sunset = 17.03 h, hence length of night = 13.94 h, 1 seasonal hour = 1.16 h. LT of first contact = 21.10 h, UT = 18.10 h. Computed TT = 22.55 h, thus ΔT = 16 000 sec.

In table 4.2 are summarised the ΔT results obtained from each individual observation discussed above. Years are given in terms of the Julian calendar – using a negative integer rather than BC (see chapter 1). It should be noted that −719a refers to the first eclipse in the year −719 (i.e. Mar 8/9), etc. In this table, Ct stands for contact and M for mid-eclipse.

The various values for ΔT listed in the above table are plotted in figure 4.3. It can be seen from this diagram that with only a single exception (−490), the results are fairly self-consistent, bearing in mind that – as

Table 4.2 ΔT results from Babylonian lunar eclipse observations recorded in the Almagest.

Year	Ct	ΔT (s)
−720	1	21 650
−719a	M	19 000
−719b	1	<21 400
−620	1	17 800
−522	M	19 250
−501	M	18 950
−490	M	13 800
−382	1	17 750
−382	1	>15 700
−382	4	<21 450
−381a	1	16 400
−381a	1	<17 850
−381b	1	16 000

reported in the *Almagest* – the measured times are not expressed in their original form and are further only quoted to the nearest half hour or so. In particular, the observations in −382 and −381 set rather narrow limits to ΔT ($15\,700 < \Delta T < 17\,850$ sec) at this specific epoch.

4.4 Historical background to the astronomical cuneiform texts

The history of Babylon from the time of Nabonassar onwards encompasses several distinct historical periods (for a detailed account, see Oates, 1979). Soon after the death of this king (733 BC) there followed a century of Assyrian domination. However, during the rule of Nabopolassar (Nabu-apla-usur) from 625 to 605 BC, Assyria was eliminated, its capital of Nineveh being destroyed in 612 BC. Babylon reached its zenith in the reign of Nabopolassar's son Nebuchadrezzar II (Nabu-kudurra-usur), when it was the political centre of an extensive empire. Over the following centuries the city suffered a gradual decline. It was captured by Cyrus the Great in 538 BC, becoming then part of the large Persian empire. After more than two centuries of Persian rule, Babylon was annexed by Alexander the Great in 330 BC. Alexander planned to restore some of the glory of Babylon by making the city his eastern capital. However, his untimely death there (in the palace of Nebuchadrezzar II) in 323 BC frustrated these plans. Around 275 BC, Antiochus I ordered much of the population of Babylon to move to the newly founded metropolis

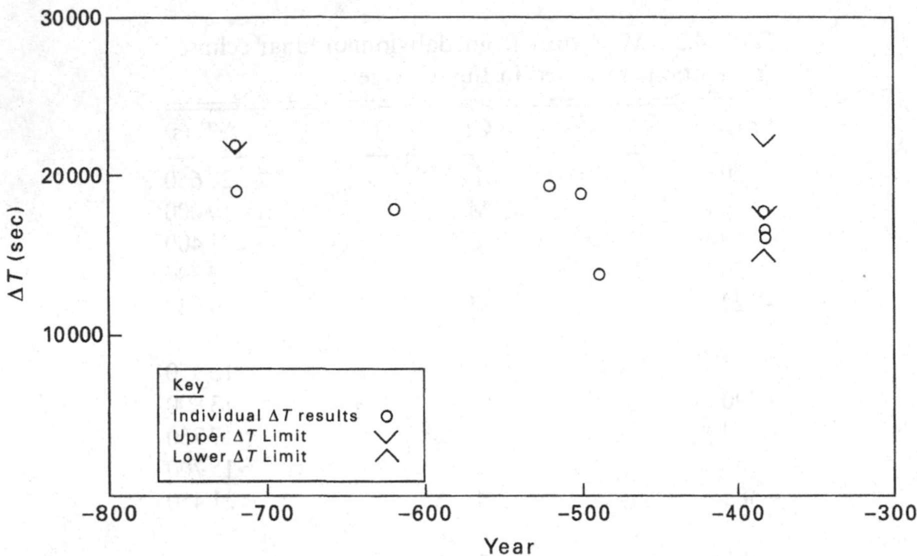

Fig. 4.3 ΔT values and limits obtained from the Babylonian eclipse observations in Ptolemy's *Almagest*.

of Seleucia, some 100 km to the north. This was a blow from which Babylon never recovered, although both Antiochus I and his immediate successors continued to attach special importance to the city. Hellenistic rule at Babylon continued until 122 BC, when Mithridates II established Parthian domination there.

Both Diodorus Siculus (II, 9) and Strabo (XVI, 1.5) remark that in their time (late first century BC) the greater part of Babylon was deserted and given over to agriculture. About AD 24, merchants from Palmyra established a trading colony there, but half a century later this was moved to a site near Seleucia (Oates, 1979, p. 96). The last we hear of Babylon in the ancient world was in AD 116 when the Roman emperor Trajan wintered there. He found 'nothing but mounds and stones and ruins' (Dio, LXVII). Not until relatively recent times was the city rediscovered.

Unlike other great ancient cities such as Nineveh, Babylon was never finally destroyed by invaders; it gradually died out. Because Ptolemy wrote several decades after Babylon had ceased to be inhabited, he could have had no direct contact with the astronomers of that city. This would explain why he had to rely on Hipparchus for the Babylonian observations which he used.

After Trajan's visit in AD 116, there is no historical reference to Babylon for more than a thousand years. Budge (1925, pp. 58 ff.) has given a fascinating account of the rediscovery of the city, commencing with Benjamin de Tudela – a twelfth century Jewish rabbi who was a native of

Spain. De Tudela journeyed over much of the known world of the time, visiting the ruins at Babil (as the site was then called) around AD 1173. We have no definite knowledge of any other traveller reaching Babylon until 1620, when Pietro della Valle collected a number of inscribed bricks from the site and took them back to his native Italy. Little further was heard about Babylon for more than a hundred years. However, during the latter half of the eighteenth century a number of Western explorers reached the city. These men expressed concern at the extensive pillaging of the site by workmen from nearby towns and villages for bricks to re-use in new constructions. Around 1795, the French zoologist Antoine Olivier found the ruins in utter confusion, further remarking that many neighbouring towns such as Hillah had been built with bricks from Babylon. About the same time, Abbé Beauchamp was told by a builder from Hillah that sometimes the workmen uncovered inscribed cylinders, etc. These were left among the rubble; only the bricks were salvaged.

Digging for bricks was still continuing a century later when the first archaeological excavations at Babil began under Sir Austen Layard (in 1851). However, soon afterwards the inhabitants of the area became aware of the pecuniary value of inscriptions, even when the items were badly damaged. During the 1870s and early 1880s, numerous clay tablets found their way to antique dealers in Baghdad. Between 1876 and 1882, the British Museum purchased virtually all of the available texts from these merchants via the London dealers Spartali and Shemtob. It was among this vast collection that the Late Babylonian astronomical texts, now numbering some 2000 tablets, came to light. At the time, no other academic institution was concerned to purchase such material. A very few texts of a similar nature were acquired by other museums such as the Staatliche Museen, Berlin, and the Morgan Library Collection, New Haven. However, the British Museum holds at least 98 per cent of the extant Babylonian tablets devoted to astronomy.

Obviously there is no archaeological context whatever for the Late Babylonian astronomical texts which were purchased from Baghdad. This situation was partly remedied as the result of planned excavations at the site of Babylon during the 1880s. These operations were undertaken by Hormuzd Rassam on behalf of the British Museum (Rassam, 1897; Reade, 1986). Some fifty astronomical texts (as well as many inscriptions of a non-astronomical nature) came to light as the result of his work. Unfortunately, Rassam did not keep a detailed record of his excavations so that the precise location in the vast ruins of Babylon where he made his discoveries is unknown. It appears that no other astronomical tablets have been unearthed from Babylon since the end of last century, although in recent years Iraqi archaeologists have undertaken extensive excavations there.

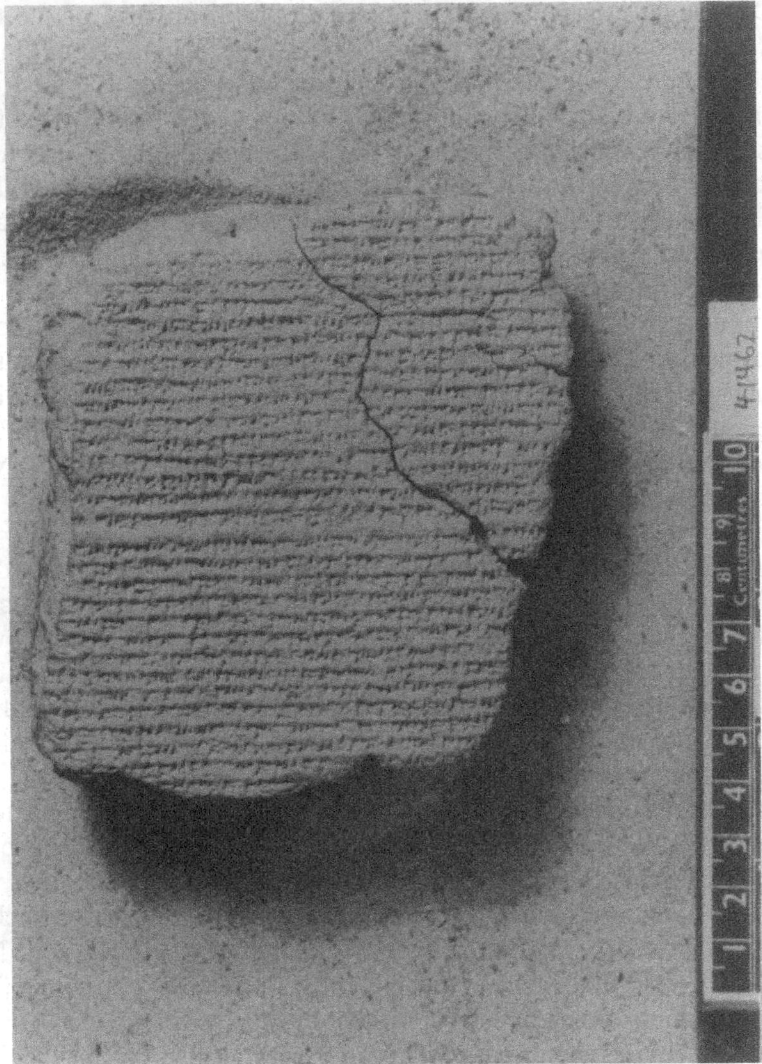

Fig. 4.4 (*a*) For caption see facing page.

4.5 Decipherment of the cuneiform texts

Today, approximately 2000 astronomical texts from Babylon have been carefully catalogued and many of these are currently undergoing extensive investigation. These texts are in the form of clay tablets inscribed with *cuneiform* writing. Individual signs (which represent syllables) were made up of triangular wedges, each formed by dipping a pointed reed into moist clay. The clay was afterwards baked or sun-dried. The name cuneiform is derived from the Latin for wedge – i.e. *cuneus*. This name owes its

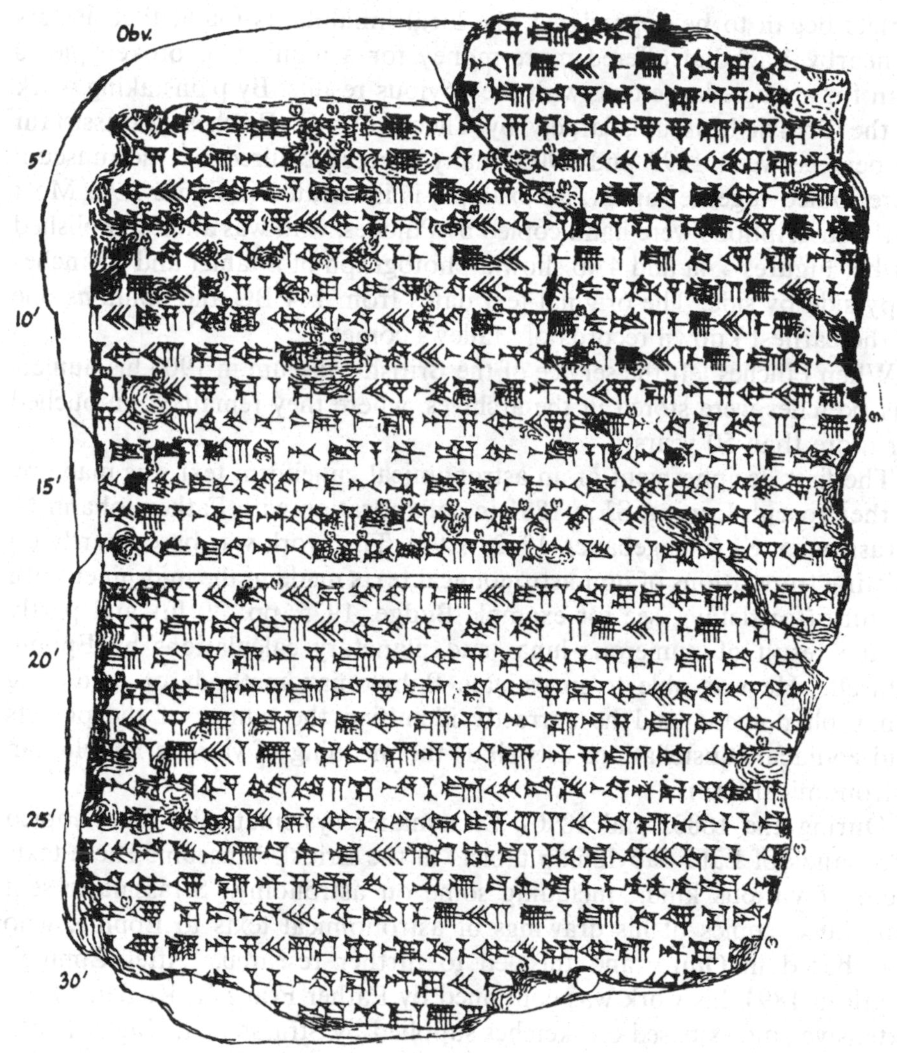

Fig. 4.4 Photograph (a) of an astronomical diary dating from 164 BC together with a drawing (b) of the same text by T. G. Pinches. (Courtesy: British Museum.)

origin to the Dutch explorer Engelbert Kampfer, who in 1686 copied a Babylonian inscription. He called the written characters *cunatae*.

Cuneiform was written from left to right. After one side of a tablet was fully inscribed, it was turned over in a vertical direction and the inscription was continued – starting at the top left corner of the reverse side.

When the astronomical texts were first dug up at the site of Babylon, many were badly damaged – either accidentally or deliberately – and most now consist of mere fragments. As the pecuniary value of the tablets began to be appreciated, the local inhabitants found that dealers in nearby Baghdad offered more money for a number of broken pieces than for a single large text, with the obvious results. By painstaking work at the British Museum, largely by Theophilus G. Pinches (an assistant keeper) between 1895 and 1900, many pieces acquired by the museum were joined together again; an example is illustrated in figure 3.9a. Most of the inscriptions were hand copied by Pinches, who was an accomplished scribe. Figures 4.4a and 4.4b show a photograph of a tablet and a Pinches copy side by side. The original text dates from 164 BC and contains one of the earliest known records of Halley's comet.

When Pinches left the service of the British Museum in 1900 his numerous sketches were stored in the archives, where they remained untouched for more than 50 years.

The first decipherment of an astronomical cuneiform text was made by Father Joseph Epping, SJ in 1881 in collaboration with Father Johann N. Strassmaier, SJ (Neugebauer, 1952, p. 98). This work was based partly on existing translations of non-astronomical texts made in the mid-nineteenth century (for details, see for example Budge, 1925, pp. 39 ff.) and partly as the result of numerous lunar and planetary calculations by Epping himself. He was able to investigate Babylonian methods of predicting lunar phenomena and he correctly identified the names of the planets and zodiacal constellations as well as the meaning of various Babylonian astronomical terms.

During the 1880s and 1890s, Strassmaier systematically hand-copied thousands of Late Babylonian tablets in the British Museum. These texts were of various kinds, including some on astronomy. Strassmaier sent annotated copies of his drawings of astronomical texts to Epping, who was based in Quito, and this led to further research. After Epping's death in 1894, his work was continued by Father Franz X. Kugler, SJ. By extensive studies based on sketches supplied by Strassmaier, Kugler (1907, 1909–24, 1913–14) built up a detailed picture of Babylonian mathematical astronomy. This work was continued by Otto Neugebauer (e.g. 1955) and Father Johann Schaumberger, SJ.

Until 1948, relatively little interest had been shown in Babylonian observational astronomy because few relevant texts were known. However, in that year Abraham Sachs (1948) attempted the first classification of observational and other non-mathematical cuneiform texts based on the then known sample of only about 20 tablets of this kind. In 1953, when Sachs spent a year at the British Museum, he became aware of the full extent both of the texts themselves and also of the sketches made by Pinches. Not long afterwards, Sachs and Schaumberger (1955) published many of the drawings made by Pinches (and also some by Strassmaier) but without translations. In their book entitled *Late Babylonian Astronomical and Related Texts*, Sachs and Schaumberger cited both the British Museum (BM) number and their own reference number for each tablet, the latter usually known as LBAT. Where appropriate, both of these designations will be used below.

When possible Sachs and Schaumberger assigned a date for each inscription. Some dates were still preserved on the tablets, but they derived many further dates by retrospective calculation of the various planetary and lunar observations which they contained. Today, some 1500 texts devoted to observation and other aspects of non-mathematical astronomy have been catalogued. However, less than half of these have been dated; numerous small fragments – some no more than 3 cm across – remain undatable.

Judging from the content of the extant astronomical texts, it may be conservatively estimated that less than ten per cent of the original material has so far been found. Whether the remainder was destroyed over the centuries when the site of Babylon was being pillaged – or whether further astronomical tablets still lie buried, awaiting future discovery – is an unanswered question. Nevertheless, cataloguing, dating, transliteration and translation of the extant material has proved an enormous task, which is still unfinished. Sachs spent some 30 years on this project and since his death in 1983 Hermann Hunger of the University of Vienna has continued his work. Photographs, transliterations and translations of many texts have already been published (Sachs and Hunger, 1988, 1989, 1996). However, it will be several years before completion is in sight.

4.6 Origin and classification of the Late Babylonian astronomical texts

In principle, the astronomical texts which the British Museum acquired from antique dealers in Baghdad could either have originated from Babylon or one of the other ancient sites in Southern Iraq such as Borsippa or Sippar. That they in fact came from Babylon can be established from the following features: (i) the very obvious continuity of style and content between these and the tablets excavated at Babylon by Rassam; (ii) the frequent mention of events occurring in the city of Babylon in the inscriptions (other cities are seldom alluded to); (iii) the specifically Babylonian deities (e.g. Bel and Beltija) to which the introductory invocations are addressed; and (iv) the character of the few personal names which occur in the colophons. For details, see for example Sachs (1948, 1974), and also the many translations published by Sachs and Hunger (1988, 1989, 1996). There is, in fact, little evidence that any of the Late Babylonian texts on non-mathematical astronomy in the British Museum originated anywhere but Babylon itself.

The period covered by the tablets is towards the end of the active life of Babylon, and hence they have become known as the Late Babylonian astronomical texts. Most dated inscriptions originate from between 400 and 50 BC, but several texts contain older material – one eclipse prediction dating from 731 BC (first year of Ukin-zer). Huber (1973, p. 1) estimates that the information on one damaged text must have originally extended as far back as some time between 750 and 740 BC It thus seems quite likely that the records could have extended as far back as the reign of Nabonassar, as Ptolemy asserted.

The most recent surviving astronomical texts, dating from AD 61/62 and 74/75, are almanacs prepared from then contemporary observations (Sachs, 1976). The latter tablet, now in Dropsie College, Philadelphia, is also the latest datable text which is written in cuneiform.

The astronomical texts reveal that through centuries of both greatness and decline, celestial observations continued at Babylon on a regular basis with little change of pattern. Just when astronomical activity ceased cannot be established with certainty, but it must have been at some time between AD 75 and 116 (when Trajan visited the deserted ruins).

In all, the Late Babylonian astronomical texts may be divided into five main groups: diaries, tables of specific phenomena, 'goal-year texts', almanacs and texts devoted to mathematical astronomy. Observational reports can be found on tablets in the first three categories. The main motive for stargazing in Babylon was astrological. In essence, the astronomers utilised their observations to produce almanacs which in turn formed the basis of astrological prognostications. It is noteworthy that the casting of horoscopes originated in Babylon around 400 BC (Sachs, 1967). Celestial phenomena which were regularly noted include conjunctions of the Moon with planets and selected stars, conjunctions of planets with one another with these same stars, the time of moonrise and moonset (measured relative to sunrise or sunset) at both new and full Moon, eclipses, the heliacal rising or setting of planets and Sirius, and equinoxes and solstices.

The Babylonian name for the Sun was *samas* and for the Moon was *sin*. An eclipse was identified by the term AN-KU_{10}. (Many different signs were pronounced as KU.)

The following detailed account of the partial lunar eclipse of Apr 10/11 in 80 BC – recorded on both the obverse and reverse of a British Museum tablet devoted to this event (BM 33562A; LBAT 1445) – gives some indication of the variety of information which the Babylonian astronomers noted:

> Year 168 (Arsacid), that is year 232 (SE), Arsaces, king of kings, which is in the time of King Orodes (I). Month I, night of the 13th, moonrise to sunset 5,50 (= 5 $\frac{5}{6}$) deg, cloudy (?), measured (?). 5 deg before μ Her

culminated, lunar eclipse, beginning on the south-east side. In 20 deg of night it made 6 fingers. 7 deg duration of maximal phase, until it began to become bright. In 13 deg from south-east to north-west, 4 fingers lacking to brightness, it set [...]. During this eclipse, north wind (?) went. During this eclipse, Jupiter, Saturn and Mars stood there. When becoming bright, Saturn set. The other planets did not stand there. (The Moon was) $\frac{2}{3}$ cubit behind α Lib, 6 fingers toward the south beneath. (Began) at 40 deg before sunrise. On the 13th, moonset to sunrise 1 deg, mist, measured.

[Trans. Huber (1973 pp. 75-76) – with minor emendations by the present author.]

In the above text, the year is given in terms of both the Arsacid era (247 BC) and the Seleucid era (SE: 311 BC). The eclipse was only partial; when the whole of the Moon (or, rarely, the Sun) was obscured the term TIL ('total') was used. The various units (deg, cubits, fingers) used will be discussed below (section 4.10).

The names of the five planets were normally cited in the following order: MUL-BABBAR (Jupiter), *dele-bat* (Venus), GU_4-UD (Mercury), GENNA (Saturn) and AN (Mars). During the Seleucid period, this sequence was strictly followed.

Although there was considerable interest in cyclical events – e.g. lunar and planetary movements and eclipses – little concern was shown for celestial bodies appearing at irregular intervals such as comets and meteors. Brief records of Halley's comet have been found on tablets dating from 164 and 87 BC (Stephenson *et al.*, 1985; Stephenson and Walker, 1985).

Of the various types of astronomical text, the diaries represent the most original material. They contain a day to day (and night to night) account of celestial and meteorological observations, each text typically covering a period of six or seven months. Historical events are also frequently noted – e.g. the accessions and deaths of kings and major events in Babylon – as well as changes in the level of the adjacent River Euphrates and prices of various commodities such as dates and wool. Virtually all of the datable diaries which were written prior to 164 BC have been published by Sachs and Hunger (1988, 1989, 1996).

The earliest surviving diaries originate from 652 and 568 BC, the latter being fairly well preserved. Sachs (1974) gives an interesting bar diagram showing the years from which fragments of diaries are preserved; this is reproduced in figure 4.5. Despite recent research by Hunger on additional diaries, this diagram still closely represents the present temporal distribution of texts.

During the Hellenistic period (fourth century BC onwards) extensive use was made of the material in past diaries by the Babylonian astronomers. It would thus appear that copies of diaries extending over several centuries were kept on file. Using these primary sources, the astronomers were

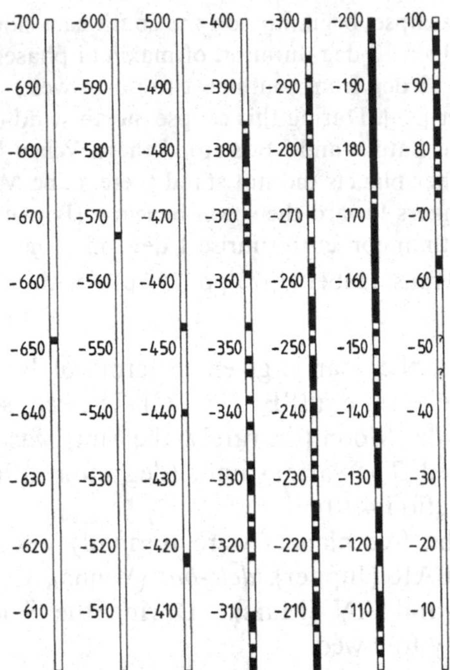

Fig. 4.5 Bar diagram showing the years from which fragments of diaries are preserved (after Sachs, 1974). Each bar represents a fraction of a year covered.

able to compile: (i) tables listing occurrences of a particular phenomenon (e.g. eclipses or planetary data) and (ii) 'goal-year texts'. The tables often extended over several centuries, eclipses usually being listed at 18-year intervals (i.e. the saros period). For example, Huber (1973, p. 1) gives the following description of one such series of texts – of which the tablets numbered BM 32238, 45640, 35115, 35789 and 32234 (LBAT 1414, 1415, 1416, 1417 and 1419) are surviving fragments:

> A big, tightly organised compilation, originally covering 24 saros cycles, that is 24 × 38 = 912 eclipse possibilities (i.e. including both observations and predictions) or 24 × 18 years. Extending over several tablets, it must have started not earlier than –749 Sep 1 and not later than –739 Mar 20, and must have ended between –316 Dec 13 and –307 Jul 9.

Some texts cite every eclipse observation and prediction over a selected short interval – e.g. BM 38462 (LBAT 1420) covers the period from 604 BC to 576 BC. However, as in the case of the eclipse of 80 BC cited above, other tablets are entirely devoted to a single eclipse.

Goal-year texts represented an intermediate stage in the preparation of almanacs. On these tablets were assembled observations from past years which were to form the basis of predictions for a selected future year (the goal-year). Observations would include eclipses 18 years before the required year, Venus data from 8 years previously and so on. Very few translations of goal-year texts have so far been published. For sample transliterations and translations, dealing only with planetary observations, see Stephenson and Walker (1985, pp. 34–36).

In the study of Babylonian *lunar* eclipse records, the tables of specific phenomena – namely those devoted to lunar obscurations – provide a valuable supplement to the material in the diaries since so many diaries are missing. In particular, several tables of this kind contain observations made between about 700 and 400 BC, a period from which very few diaries are preserved. By chance, no tables of solar eclipses are extant and there are no known records of these events from before 370 BC. In later centuries, goal-year texts provide a further source of both lunar and solar eclipse observations. However, at all periods, surviving lunar eclipse reports are much more frequent than their solar counterparts.

The compilations of Sachs and Hunger (1988, 1989, 1996) and also Huber (an unpublished manuscript, produced in 1973 and freely circulated) have provided transliterations and translations of most of the available eclipse records. Sachs and Hunger's publications are so far exclusively devoted to the diaries. Huber's manuscript is specifically concerned with eclipses, which he extracted from diaries, eclipse tables and goal-year texts. His work contains many more eclipse observations than the publications of Sachs and Hunger. However, he relied mainly on the texts for which drawings were published by Sachs and Schaumberger (1955) so that his list, although invaluable, contains several important omissions.

4.7 The Babylonian calendar

In common with many ancient calendars, the Babylonian calendar was luni-solar. The year began with the month Nisannu around the time of the spring equinox. Until the late fourth century BC, years were counted from the accession of each ruler, but from 311 BC, the start of the reign of Seleucus I (known as the Seleucid era – customarily abbreviated to SE), a continuous system of numbering was adopted. This remained in use at Babylon for the rest of the period covered by the astronomical texts. The precise date of the Seleucid era adopted in Babylon was 1 Nisannu (i.e. Apr 3) in 311 BC However, on the Macedonian calendar it was some 6 months earlier (Dios 1, equivalent to Oct 8 in 312 BC). Even today some Christian churches in Egypt and Iraq still number years in terms of the Seleucid era; for medieval examples see chapters 11 and 12. In Babylonian texts the Arsacid era is occasionally used. Of Iranian origin, the Arsacid dynasty was established in 247 BC

Most years contained 12 lunar months, each of either 29 or 30 days,

Table 4.3 Babylonian month-names and their abbreviations as found in the astronomical cuneiform texts.

Number	Name	Abbrev	Number	Name	Abbrev
I	Nisannu	bar	VII	Tashritu	du
II	Ayaru	gu_4	VIII	Arahsamnu	apin
III	Simanu	sig	IX	Kislimu	gan
IV	Du'uzu	su	X	Tebetu	ab
V	Abu	izi	XI	Shabatu	ziz
VI	Ululu	kin	XII	Addaru	se

making a total of about 354 days. Names of the Babylonian months and their one-syllable abbreviations as found in the astronomical texts are listed in table 4.3. In translations, it is customary to give Roman numerals rather than full month-names.

In order to keep the calendar roughly in step with the seasons, every two or three years an extra month was added. This intercalary month was placed immediately after either the 6th or 12th month. (There is an interesting parallel here with modern leap-seconds, which are always inserted on Jun 30 or Dec 31.) At first, intercalation in Babylon was somewhat haphazard, but around 400 BC a regular scheme based on the Metonic cycle was adopted. Intercalary months are denoted by VI_2 or XII_2.

At Babylon (as for several other early civilisations), each day began at sunset – normally a well-defined moment. Towards the close of the 29th day of a month, a watch was kept for the young crescent Moon low in the western sky just after sunset. If the crescent was sighted, the new month was regarded as having just begun; otherwise an extra day was added to the month. When cloud prevailed, calculation superseded observation to determine whether or not the crescent should have been visible. Such calculations seem to have been fairly successful, although the rules are not known in detail. It has not been ascertained whether, in view of the more advanced astronomy of later centuries, calculation ever completely replaced observation for fixing the start of each month.

From an extensive study of the Babylonian calendar, involving detailed investigation of intercalary months and also the visibility of the crescent Moon, Parker and Dubberstein (1956) were able to produce comprehensive tables for the conversion of Babylonian dates to the Julian calendar over the entire period from 626 BC to AD 75. These tables are highly reliable, errors seldom exceeding a single day; they have proved very valuable in the present investigation.

4.8 Techniques used in dating observational tablets

On a well-preserved astronomical diary, the date is typically given in the first line of the obverse and sometimes in the last line of the reverse. For example a diary from 247 BC, which now consists of four joined pieces (BM 32889 + 32967 + 41614 + 41618; LBAT 271, 273 and 272) begins and ends as follows (Sachs and Hunger (1989, pp. 59–65)):

> [First line] Diary for the year 65 [Seleucid], king Antiochus. Month I...
> [Last line] Diary from month I to month VI, year 65, [king] Antiochus.

This covers the period from Apr 15 to Oct 8 in 247 BC. For most diaries, the date is broken away and here it is necessary to resort to computations based on the planetary and lunar information which the diary contains. As noted above, much pioneering work in dating damaged texts was done using this method by Sachs and Schaumberger (1955). In recent years, I have employed similar techniques to date a number of further diaries based on transliterations and translations supplied to me by Hermann Hunger. The following brief comments are based on my experiences.

Often it is necessary to begin with a date range of several hundred years, as derived from historical and linguistic considerations. Calculation of the planetary phenomena is best undertaken first and this may indicate only a very few potentially viable dates, often widely spaced. Afterwards using the lunar observations it may prove possible to fix a unique date. Generally speaking, if a single date cannot be fairly quickly derived in this way no amount of calculation will probably help; the text evidently contains too little information to allow a unique solution. For examples of the techniques used, see Stephenson and Walker (1985).

Goal-year texts usually present few dating problems since in addition to the goal-year each individual eclipse entry carries a separate date which is 18 years prior to the reference year. Tablets containing lists of lunar eclipses at 18-year intervals can be fairly easily dated – even if a text is badly damaged; one or two fully preserved entries or the mention of the accession or death of a ruler can provide the key to the whole sequence. If a tablet devoted to a single eclipse is only partially preserved, the problems of dating are more serious. However, if the position of the eclipsed Moon relative to a reference star is reported – as in a text from 80 BC (see section 4.7) – it may well prove possible to derive a unique date by comparing observation with computation. For this purpose a preliminary value for ΔT may be utilised since the motion of the Moon relative to the background stars is fairly slow.

It is important to state here that none of the Babylonian eclipse observations investigated below have been dated by comparing the observed time-interval between sunset or sunrise and first contact with the interval computed using a provisional value for ΔT. Such a technique would merely be an example of the 'identification game' (cf. Newton, 1970, p. xiv, etc.). If an independent date cannot be derived, the observation will not be considered.

4.9 Units of time

The principal unit of time adopted by the Babylonian astronomers was the US. It is usually asserted that one US corresponded to four minutes of time, so that there were precisely 15 of these units in an hour. A larger measure, the *beru*, was equal to 30 US and was thus equivalent to two hours. (For these definitions, see for example Neugebauer (1955, vol. I, p. 39). The US was subdivided into 60 GAR (sometimes read as NINDA). Since the US was the time-interval for the celestial sphere to turn through 1 deg, an appropriate translation of the term is 'time-degree', although it is customary to abbreviate this to 'degree'. Use of these same units also spread to Greece, where they were known as *chronoi isemerinoi* ('equatorial times'), also frequently rendered as 'time-degrees'.

It should be noted that the Babylonians employed separate units for measuring the angular separation between two celestial bodies; these were the cubit (KUS), equivalent to about 2 or $2\frac{1}{2}$ deg and the finger (SI), one cubit being divided into 24 fingers (Neugebauer, 1955, vol. I, p. 39). As discussed in chapter 3, fingers were also used in expressing eclipse magnitude, but here the definition is quite separate, corresponding to one-twelfth of the diameter of the luminary obscured.

Eclipse contacts were systematically timed in terms of US relative to sunrise or sunset, depending on which was nearer. Probably a water clock was used for this purpose (Neugebauer, 1947) although little is known regarding the practicalities of time-keeping in Babylon. Eclipses were never timed to better than the nearest US, but the time of moonrise or moonset relative to sunrise or sunset was often expressed to the nearest 10 GAR.

In response to a query by Dr C. B. F. Walker of the British Museum, a recent investigation by Stephenson and Fatoohi (1994b) based on the recorded durations of lunar eclipses by Babylonian astronomers demonstrated that the US was indeed of fixed duration, and equal to $\frac{1}{15}$ hour. It was concluded that down the centuries covered by the Late Babylonian astronomical texts there is no evidence of changes in the definition of the unit or of any seasonal variation. Lunar eclipses were made the subject of this investigation since (unlike in the case of solar obscurations) the duration of any phase is independent of ΔT. Hence it is a relatively simple matter to compare a measured interval with its computed equivalent. Stephenson and Fatoohi investigated a series of 60 lunar eclipse durations, timed to the nearest degree. These covered the period from about 560 to 50 BC Although a few rather earlier measurements are preserved on the

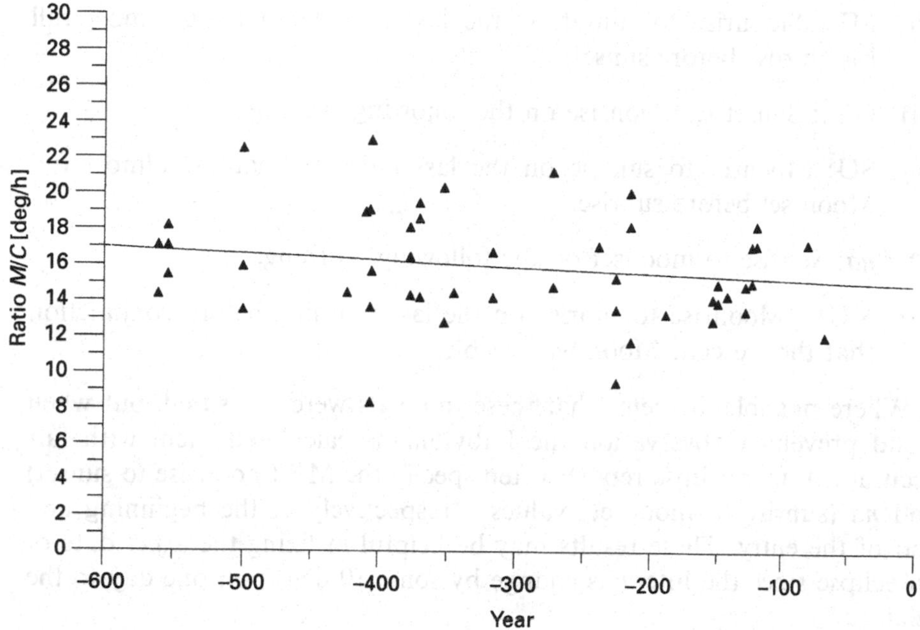

Fig. 4.6 Constancy of the US during the Late Babylonian period. (Stephenson and Fatoohi, 1994b.)

Babylonian texts, they are only rounded to the nearest 10 degrees. From the selected set of data, Stephenson and Fatoohi derived a mean result of 1 hour = 15.6 ± 0.4 US, which effectively includes the accepted equivalent of 15.0 US. It is noteworthy that the more recent observations (after 300 BC) yielded a result of 15.3 ± 0.3 US.

A graph showing the individual results obtained from each observation for the number of US in an hour shows negligible variation down the centuries – see figure 4.6. In particular, there is no evidence that the definition of the unit was ever revised during the period covered by the diagram. (NB in this diagram, M/C stands for measured interval divided by computed interval.)

4.10 'Lunar sixes'

Each lunar month the Babylonian astronomers determined six time-intervals, two around new Moon and four around full Moon. These intervals, appropriately termed by Sachs 'lunar sixes', were quoted to the nearest degree. They are as follows:

(i) *na*: Sunset to moonset when the young crescent Moon was first sighted; this began the month.

(ii) ME: Moonrise to sunset on the last evening that the almost full Moon rose before sunset.

(iii) GE_6: Sunset to moonrise on the following evening.

(iv) SU: Moonset to sunrise on the last morning that the almost full Moon set before sunrise.

(v) *na*: Sunrise to moonset on the following morning.

(vi) KUR: Moonrise to sunrise on the last morning before conjunction that the crescent Moon was visible.

Where possible, it seems that these intervals were measured, but when cloud prevented observation the Babylonians calculated them with fair accuracy. Lunar eclipse reports often specify the ME (moonrise to sunset) and *na* (sunrise to moonset) values – respectively at the beginning and end of the entry. These results may be helpful in fixing the *exact* date of an eclipse since the intervals change by some 10 deg from one day to the next.

4.11 Observation and prediction of eclipses

The various Babylonian non-mathematical texts (diaries, goal-year texts and tables) contain both observations and (unsuccessful) predictions of eclipses. If an eclipse was successfully predicted – meaning that it was observable – this fact is never stated; the texts contain scarcely any comparisons between theory and observation. Although the main emphasis in this chapter will be on observed eclipses, it is necessary to comment also on the basic form in which predictions are recorded and to discuss how such accounts differ from observational reports.

4.11.1 Observations

Descriptions of observed eclipses usually contain the more detailed information. In addition to the time of first contact (expressed relative to sunrise or sunset) the following details are often given: approximate position angle of first and last contact; time interval between first contact and greatest phase; maximum degree of obscuration of the Sun; time interval between greatest phase and last contact; whether the Sun or Moon rose or set whilst eclipsed; wind direction; and which planets were above the horizon during the eclipse. Ordinarily, the term AN-KU *samas* or AN-KU *sin* indicates a predicted eclipse, whereas the opposite order *samas* AN-KU or *sin* AN-KU denotes an observed event. However, there are occasional exceptions to this rule, perhaps resulting from scribal errors.

The report of the solar eclipse of Mar 14 in 190 BC is particularly detailed and contains most of the above features. The main description is found on a tablet specifically devoted to the event (BM 33812, LBAT 1438) which was excavated by Rassam. A badly damaged diary (BM 32951, no LBAT number) duplicates some of this information (Sachs and Hunger, 1989, pp. 314–15). A translation of the first of these inscriptions is as follows:

> Year 121 (SE), King An(tiochus), month XII, 29 solar eclipse beginning on the north-west side. In 15 deg day [...] over a third of the disk was eclipsed. When it began to become bright, in 15 deg day from north-west to east it became bright. 30 deg total duration. [During this eclipse], east (wind) went. During this eclipse [...], Venus, Mercury and Saturn [stood there]. Towards the end of becoming bright, Mars rose (?) The other planets did not stand there. (Began) at 30 deg (= 1) *beru* after sunrise.
>
> [Trans. Sachs and Hunger 1989, p. 315).]

The reference to the year on the Seleucid era establishes that this event occurred during the reign of King Antiochus III. There is no suggestion in the above text that any of the planets mentioned were actually seen during the eclipse. Since the phase was only partial, the loss of daylight would be small. The Babylonian astronomers were in the habit of calculating which planets, and also whether the bright star Sirius, would be above the horizon during an eclipse. All the remaining data – including, of course the wind direction – are apparently based on observation.

When converted to the Julian calendar, the above date (BC 190 Mar 14) proves to be in exact accord with that of a tabular solar eclipse (e.g. as listed by von Oppolzer, 1887). Such accuracy is characteristic of all Late Babylonian observations of both solar and lunar eclipses for which a date is well preserved. On this occasion, sunrise seems to have been clearly defined. However, when heavy cloud prevailed around sunrise or sunset, a substantial error in fixing the time of this moment (and hence the local time of the start of the eclipse) might have been incurred.

Comparison between measured lunar eclipse durations expressed to the nearest degree and their computed intervals shows that timing errors may be significant. Employing a mean fit to the ΔT results obtained from these data, Stephenson and Fatoohi (1993) showed that such discrepancies averaged 8 deg for intervals up to about 6 hours. Evidently the timing devices (e.g. water clocks) of the period were crudely designed and presumably did not incorporate a constant head device. Although these deficiencies set a limit to the accuracy with which ΔT can be obtained from individual Babylonian timed observations, the substantial number of data available is a major compensating factor.

4.11.2 Predictions

Reports of eclipse predictions give little more than the date and the expected time of first contact together with a comment whether the eclipse was 'watched for but not seen' (*ki* PAP NU IGI) or merely 'passed by' (*sa* DIB). As a general rule, if the former expression was used, the eclipse was anticipated to occur when the luminary was above the horizon; in the case of an eclipse which 'passed by' the converse was expected. Examples are as follows:

(1) BC 731 Apr 9

1,50 Year 1 Ukin-zer, month I, [lunar eclipse] which passes (*sa* DIB). (Began) at 1,0 (i.e. 60) deg after sunrise.

[BM 35789 (= LBAT 1414): Obv. col. I; trans. Huber (1973, p. 2).]

Ukin-zer (Nabu-mukin-zeri) reigned from 731 to 729 BC This record is found on a list of lunar eclipses at 18-year intervals, so that the term for lunar eclipse is not included in the entry. The number 1,50 in the first line is unexplained. This eclipse was correctly predicted for the hours of daylight at Babylon and thus would be expected to be invisible there. (NB I have given only a single Julian date in this instance – i.e. Apr 9 – since the eclipse was anticipated during daylight.)

(2) BC 247 Sep 7

[D]iary for the year 65 (SE), king Antiochus... [month V]. The 28th, 74 deg after sunrise, solar eclipse (at) 5 months' distance; when I watched I did not see it (*ki* PAP NU IGI).

[BM 32889 + 32967 + 41614 + 41618 (= LBAT 271 + 272 + 273, etc.), Rev. line 4; trans. Sachs and Hunger (1989, p. 63).]

This event took place during the reign of Antiochus II. Calculations show that the eclipse, although correctedly expected in the daytime, would only be visible far to the north of Babylon.

With only a single exception (when cloud prevented observation in 125 BC), no explanation is given in the preserved inscriptions for the failure of an anticipated eclipse to become visible. However, for such an early epoch, the Babylonian astronomers were remarkably successful in their predictions. Investigation of the extant records of solar obscurations reveals that virtually all of their expected eclipses would be visible from some part of the terrestrial surface – often remote from Babylon. For lunar eclipses, the success rate was similar, although about one-third of their predicted eclipses were actually only penumbral and thus would very likely pass unnoticed, even if the Moon was above the horizon at the time. Of course the astronomers themselves could only measure success by events that were observable at Babylon, representing only a small fraction of the quota for the Earth as a whole. Nevertheless, they would know on which dates in a year to keep watch for possible eclipses, even though they could never be sure that their predictions would prove reliable.

Predictions were based on the angular distance of the Moon from either node at conjunction with the Sun (Aaboe *et al.*, 1991). The Babylonian astronomers were well aware that eclipses of the same type – whether observable or not – tended to recur at intervals of five or six lunar months on account of the motion of the Sun from one node to the other during this time. However, it is not certain just when the Babylonians first understood the cause of eclipses. Surviving inscriptions do not mention how eclipses were produced, but the empirical rules of prediction did not require a physical explanation. Knowledge of the cause of eclipses may well have passed to the Babylonians during the Hellenistic period when there was much interaction between Greek and Babylonian astronomy.

It will be noted that for the observed solar eclipse of 190 BC the time of first contact is cited at the end of the account. This practice was very common for both solar and lunar eclipses. That the quoted time in such cases is indeed based on observation (rather than a mere repetition of the originally predicted time) would seem evident from the fact that the reported *durations* of the various phases were undoubtedly measured. Durations are *never* mentioned in entries which are concerned only with (unsuccessful) predictions. However, further discussion seems desirable.

In order to resolve this issue, I have compared (Stephenson, 1997) the mean error in the recorded times of first contact for both observed and predicted eclipses listed by Sachs and Hunger (1988, 1989, 1996) and Huber (1973). As mentioned above, predictions were identified by the use of either of the terms 'passed by' or 'watched for but not seen' in the records. In this investigation, I used the preliminary ΔT formula derived from a mixture of early observations by Stephenson and Morrison (1984). This may be expressed as follows:

$$\Delta T = +1360 + 320t + 44.3t^2, \tag{4.1}$$

where t is measured in Julian centuries from 1800.0. This equation is adequate for most historical purposes, although refinements will be made in chapter 14 (see also Stephenson and Morrison, 1995). When a predicted lunar eclipse was only penumbral, I have made computations for the moment when the umbral shadow made its closest approach to the Earth, since there was no true first contact. In the case of a solar eclipse where the lunar shadow passed fully to the north or the south of Babylon, I computed for the moment when the shadow made its closest approach to this site.

In the above survey, about 50 observations of lunar and solar eclipses were analysed, giving a result for the mean error in the time of first contact of 8 deg (see also Stephenson and Fatoohi, 1993). By comparison, investigation of records of solar eclipse predictions (26 in number) yielded a mean error as large as 60 deg, while for predicted lunar obscurations (28 in number) the corresponding result was nearly 40 deg. In both sets of predictions, the scatter proved to be very large. The implications are clear: whenever possible, the Babylonian astronomers replaced predicted times of first contact by observation. This conclusion will be assumed in all subsequent ΔT investigations in chapters 5 and 6.

Since observed eclipses were timed relative to sunrise or sunset, it would seem that when a solar obscuration was expected, the Babylonian astronomers would start timing as soon as the Sun rose and continue either until an indentation in the solar limb was observed or until sunset if no eclipse materialised. In the case of a lunar eclipse, timing would commence at sunset. When an observable solar eclipse occurred significantly after noon, the time of onset was measured relative to sunset rather than sunrise, the converse being the case for a lunar eclipse beginning after midnight.

4.12 Assyrian records of solar and lunar eclipses

Many allusions to both solar and lunar eclipses are found among the divination reports to Assyrian kings around 700 BC. These reports, inscribed in cuneiform on clay tablets, originate from the royal archives at Nineveh (lat. = 36.40 deg, long. = −43.13 deg). This city, the last capital of Assyria, was destroyed by a coalition of Babylonians, Medes and Scythians in 612 BC. Most of the divination texts recovered from the ruins are now in the British Museum. The Assyrian rulers employed specialists in divination to record and interpret unusual phenomena of various kinds. Celestial omens included eclipses, conjunctions of the Moon with planets and stars, planetary movements, haloes, meteors and comets.

The astrological reports, which cover the period from 709 to 649 BC, were first investigated by Thompson (1900). A full transliteration, translation and commentary has recently been published by Hunger (1992). Many of the allusions to eclipses (identified by the term AN.KI) relate to attempts at prediction. As most eclipse omens were unfavourable, a successful prediction might give an opportunity to take action against whatever danger was expected. An interesting example of a full Moon when a lunar eclipse was *not* anticipated is as follows. This is recorded by the Babylonian astrologer Tab-silli-Marduk.

> '[In *Iy*]*yar* (month II)...the night of the 14th day, is the [da]y of the watch (to be held), and there will be no eclipse. I guarantee it seven times,

an eclipse will not take place. I am writing a definitive word to the king.' From Tab-silli-Marduk, nephew of Bel-nasir.

[Trans. Hunger (1992, p. 251).]

Few actual observations are noted in the astrological texts, although prognostications such as the following give such detailed astronomical circumstances as to imply real observations:

'If the Sun at its rising is like a crescent and wears a crown like the Moon: the king will capture his enemy's land; evil will leave the land, and (the land) will experience good...' [From Ras]il the older, servant of the king.

[Trans. Hunger (1992, p. 220).]

The above eclipse has been identified by Parpola (1983) as that of BC 669 May 27. Rasil was a Babylonian scribe so that the place of observation may be presumed as somewhere in Babylonia.

Among clearly expressed observations among the Assyrian astrological reports, the most detailed relates to the solar eclipse of BC 657 Apr 15 (Parpola, 1983). This event is recorded on a badly damaged British Museum tablet:

'On the 28th day, at $2\frac{1}{2}$ double hou[rs of the day...] in the west [...] it also cover[ed...] 2 fingers towards [...] it made [an eclipse], the east wind [...] the north wind ble[w. This is its interpretation]...' From [Akkullanu].

[Trans. Hunger (1992, p. 63).]

Akkullanu, the writer of a number of surviving texts, was an Assyrian scribe. Since the time is only quoted to the nearest hour and the place of observation is somewhat uncertain, the record is only of minimal value for the determination of ΔT. However, this and observations such as the following are of more general interest:

'The eclipse of the Moon which took place in Marchesvan (month VIII) began [in the east]. That is bad for Subartu. What [is wrong]? After it, Jupiter ent[ered] the Moon three times. What is being done to (make) its evil pass? ...' From the lamentation priest Bel-suma-iskun.

[Trans. Hunger (1992, p. 263).]

Bel-suma-iskun was a Babylonian scribe. According to Parpola (1983), the text dates from 675 BC.

By far the most well known observation of a solar eclipse from Assyria occurred in 763 BC. This is recorded not on an astrological text but in the Assyrian Chronicle:

(Eponym of) Bur-Saggile of Guzana. Revolt in the citadel; in (the month) Siwan, the Sun had an eclipse (*samas attalu*).

[Assyrian Chronicle; trans. Millard (1994, p. 58).]

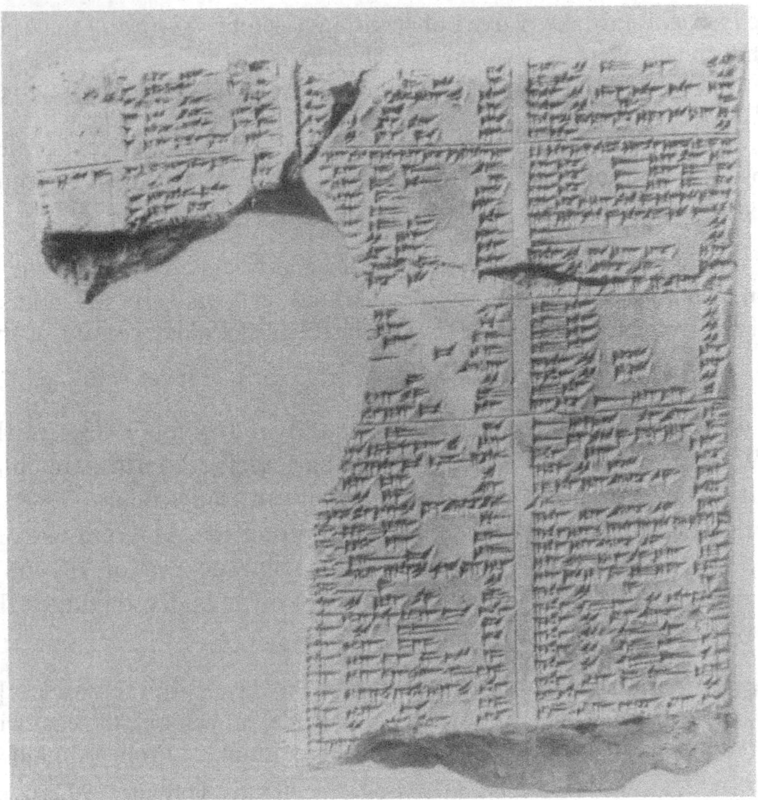

Fig. 4.7 A copy of the Assyrian Chronicle containing a record of the solar eclipse of 763 BC. (Courtesy: British Museum.)

Several damaged copies of the Assyrian Chronicle are preserved in the British Museum. A photograph of one of these tablets, recording the eclipse, is shown in figure 4.7.

The Assyrian Chronicle records very few natural events. It is mainly a list of the annual *limmu*, senior officials after whom the year was named. The practice of appointing *limmu* began as early as the eighteenth or nineteenth centuries BC In translations, the Greek equivalent of the term *limmu* – i.e. *eponym* – is usually preferred. This in turn has found its way into English and means 'a person who gives his name to something'. During later centuries in Assyria, the eponym was a provincial governor; however, several kings also held this office. The Assyrian system of annual eponyms was probably the origin of the Greek practice of appointing archons (see section 4.3) and the Roman system of consuls, each of whom held office for one year.

A complete list of Assyrian eponyms is preserved from 910 to 646 BC (Millard, 1994, pp. 55–62). Since several kings of Assyria mentioned in the list were also rulers of Babylon, it is possible to deduce the year when Bur-Saggile was eponym from Ptolemy's *Canon of Kings* (see Toomer, 1984, p. 11). This year corresponds approximately to BC 763. The month Siwan, the third month of the year, was equivalent to May–June. Reference to the charts of Oppolzer (1887) shows that between 777 and 745 BC only two eclipses could have been large in Assyria: BC 765 Feb 10 and 763 Jun 15. Of these, Feb 10 is far too early for Siwan and hence the date of the eclipse must be BC 763 Jun 15. This date was accepted by Fotheringham (1920b).

As no other eclipse is mentioned in the Assyrian Chronicle, Fotheringham supposed that it must have been total somewhere in Assyria. However, this suggestion is unfounded; the record gives no information regarding magnitude, although the eclipse was presumably very striking. It may well have been seen at the Assyrian capital of Ashur (lat. = 35.48 deg, long. = −43.23 deg), but the report could have come from some provincial location instead.

4.13 Conclusion

Although the Assyrian eclipse records are of undoubted historical interest, they are of negligible significance in the study of the Earth's past rotation. Fortunately, the roughly contemporaneous Babylonian observations are of great value in the determination of changes in the length of the day. Indeed, without them little would have been known concerning the terrestrial rotation in the BC period. This material forms the subject of the next three chapters.

5

Investigation of Babylonian observations of solar eclipses

5.1 Introduction

In this chapter, all accessible reports of solar eclipses preserved on the Late Babylonian astronomical texts which are of value in studying the Earth's past rotation will be investigated. Compared with the substantial number of observations which must have originally been made, very few Babylonian records of solar eclipses are now extant. Reasons for this are briefly discussed in chapter 4. Reliable observations only range in date from 369 to 136 BC – less than half the period covered by lunar eclipse sightings.

The solar eclipses which form the subject of the present chapter will be divided into four categories: (i) the only known example of a total obscuration of the Sun observed at Babylon (dating from 136 BC); (ii) measurements of the local times of a variety of eclipse contacts; (iii) estimates of solar eclipse magnitude at maximum phase; and (iv) instances when the Sun rose or set whilst obscured. Several of these observations have already been discussed in chapter 3, where they were used to illustrate various analytical techniques. However, they will now be considered in greater detail. At the close of this chapter, a *possible* allusion to a total solar eclipse recorded in the *Religious Chronicle* of Babylon will be considered. This event probably took place around 1000 BC – much earlier than the period covered by the Late Babylonian texts.

All eclipse records on the Late Babylonian texts where the date is doubtful will be rejected, as will questionable readings of damaged signs. In some instances, the sign representing the time of first contact is missing or damaged. Hence even if the durations of other phases are preserved intact, the observations are of no value for determining ΔT. As a rule, only those measurements which seem capable of yielding useful results for ΔT are discussed below. In each case, I have cited the Julian date of the eclipse as derived from modern astronomical computations. When the recorded date is fully preserved, the equivalent on the Julian calendar is almost invariably identical to the computed date. In the case of damaged texts, I have given brief comments on the restoration of the original date. For further details, see Sachs and Schaumberger (1955), Sachs and Hunger (1988, 1989, 1996) or Huber (1973). Because of the frequent references to the three volumes published by Sachs and Hunger, these works will be referred to as SH I, SH II and SH III throughout this chapter and also in chapters 6 and 7.

Most translations are taken from either SH I, SH II and SH III or Huber (1973). When these two separate works translate the same texts, there are often stylistic differences between the individual renderings. However, these are slight. I have avoided any attempts to standardise the various translations to a set pattern, preferring when possible to quote the *published* versions in SH I, SH II and SH III. This is by no means a criticism of Huber's excellent manuscript.

5.2 The total solar eclipse of BC 136 Apr 15

Although this is the only total solar eclipse recorded in the Late Babylonian astronomical texts, fortunately the preserved descriptions are quite detailed. The phenomenon is described in two separate British Museum texts: (i) BM 34034 – an unusually well-preserved goal-year text – and (ii) BM 45745 – a small remnant of a diary. The date of the eclipse recorded in the goal year text is fully specified as the 29th day of the intercalary 12th month of SE 175; this corresponds exactly to BC 136 Apr 15, so that there can be no doubt about the identification. The year of the diary (SE 175) is also preserved, although the month is broken away. Photographs of the appropriate sides of the two tablets are shown in figures 3.8a and 3.8b. Translations of the relevant entries are as follows:

(i) SE 175, month XII_2. The 29th, solar eclipse. When it began on the south-west side, in 18 deg daytime in the morning it became entirely total (TIL *ma* TIL *ti gar* AN). (It began) at 24 deg after sunrise.

[BM 34034 (= LBAT 1285), Rev. 24–28; trans Hunger (1995).]

A similar translation is given by Huber (1973), pp. 93–94. The term TIL (rendered in the above account as 'total') is commonly used in the Late Babylonian astronomical texts to identify complete eclipses of the Moon (see chapter 3); the repetition of TIL here can perhaps best be rendered 'entirely total'. This same goal-year text also contains a record of the partial lunar eclipse of BC 136 Apr 1 (see chapter 6).

(ii) SE 175, [king] Arsaces, [month XII_2]. The 29th, at 24 deg after sunrise, solar eclipse; when it began on the south and west side, [...] [Ven]us, Mercury and the Normal stars were visible; Jupiter

> and Mars, which were in their period of invisibility, were visible in its eclipse [...] it threw off (the shadow) from west and south to north and east; 35 deg onset, maximal phase and clearing; in its eclipse, the north wind which was set [to the west? side blew ...].
>
> [BM 45745 (= LBAT 429), Rev. 13′–15′; trans. SH III, p. 185.]

A comparable translation by Abraham Sachs was published by Clark and Stephenson (1977, p. 15). Since so much of the original tablet is missing, there are several lacunae in the text; these are indicated here (as elsewhere) by [...]. When the top line (or lines) of a text is missing (as in the above example), it is customary to number lines from the first preserved line using a prime (e.g. 13′).

In the central zone, the computed magnitude of this eclipse was 1.05. Even without the direct evidence of totality from the goal-year text, it would be clear from the diary entry that the eclipse was fully complete at Babylon. As well as stars, as many as four planets were sighted (including Jupiter and Mars, which were too close to the Sun to be seen under normal circumstances). The 'Normal Stars' (MUL-SID) mentioned in the text were a series of 31 zodiacal stars which were often used to indicate the position of the Moon and planets (see chapter 6). The term as used in this context presumably means no more than the stars in general, implying that several stars had become visible.

The computed elongations of the five bright planets from the Sun and also their stellar magnitudes on this occasion were as follows: Mercury (20 deg E, mag. = +1.3), Venus (31 deg E, mag. = −4.1), Mars (29 deg W, mag. = +1.5), Jupiter (1.5 deg E, mag. = −1.5) and Saturn (166 deg W, mag = +0.4). This last planet would probably be below the western horizon during the eclipse, which would explain why it was not mentioned. Both Mercury and Mars were rather faint and hence would be somewhat difficult to detect. Jupiter, was extremely close to the Sun – only 1.2 deg from the solar limb. Near the central line, totality would last for about 3 min 25 sec. Although no estimate of the duration of totality is given, it probably must have lasted at least a minute to enable the various planets to be distinguished and identified.

The altitude and azimuth for maximal phase of the Sun, four observed planets and brighter stars in the vicinity of the Sun, are shown in figure 5.1. These positions are computed for the measured time of totality at Babylon (i.e. 42 deg after sunset or a local time of 8.76 h – see section 5.3).

RESULTS

For totality at Babylon, $11\,210 < \Delta T < 12\,140$ sec.

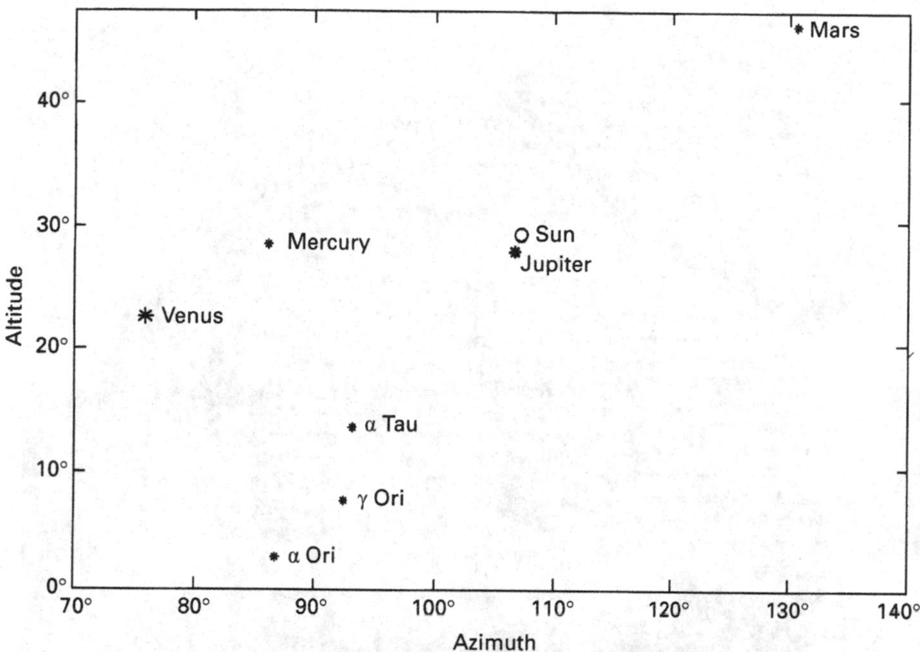

Fig. 5.1 Visibility of the planets Mercury, Venus, Mars and Jupiter and also bright stars during the total solar eclipse of 136 BC at Babylon. NB Saturn would be below the horizon.

If a duration of at least one minute is assumed, the above ΔT interval reduces by only 40 sec (to $11\,230 < \Delta T < 12120$ sec), a negligible difference at this early period.

The measured local times of the various phases lead to independent results for ΔT; these will be discussed in section 5.3.

5.3 Timed observations of solar eclipses

At the head of each entry below (after the Julian date), I have given – for reference only – the computed degree of obscuration of the Sun at maximum phase; this is based on a provisional ΔT formula (equation (4.1)).

(1) BC 322 Sep 26 (computed mag. at Babylon = 0.17)

> Year 2 of (king) Philip...(month VI)...The 28th, around 3 deg before sunset, solar eclipse (*ana* SU *samas* AN-KU$_{10}$) [...] gusty west and north wind. It set eclipsed.
>
> [BM 34093 (= LBAT 212), Rev. 23′; trans. SH I, p. 227; see also Huber, p. 83.]

Fig. 5.2 A photograph of the reverse of British Museum Tablet 34093 (an astronomical diary) containing a record of the sunset eclipse of 322 BC as seen from Babylon.

A photograph of the reverse of this tablet – containing the above eclipse record – is shown in figure 5.2.

Although the sign for month VI is missing in this diary, the symbol for the following month VII is clearly preserved a few lines later (Rev. 26′). Philip Arrhidaeus was the successor to Alexander the Great. The term *ana* SU *samas* means sunset, so that an extra *samas* would be expected before AN-KU$_{10}$ in order to identify the type of eclipse. Perhaps this is merely an example of haplography, as suggested by Huber.

This observation was used by Fotheringham (1935) in an investigation of the lunar and solar accelerations on UT. However, he used an incorrect value (4 deg) for the interval between the start of the eclipse and sunset. The observation that the Sun set eclipsed will be discussed below (section 5.5).

RESULTS

First contact on Sep 26 at 0.20 h before sunset. LT of sunset = 18.08 h, hence measured LT of first contact = 17.88 h, UT = 14.81 h. Computed TT = 18.75 h, thus ΔT = 14 150 sec.

(2) BC 281 Jan 30 (computed mag. at Babylon = 0.21)

> Year 30 (SE), kings Seleucus and Antiochus...(month X)...The 29th, [solar] eclipse; when the Sun came out (i.e. rose), 2 fingers were obscured on the south side; at 6 deg daytime, [...] 20 deg daytime onset and clearing; during its eclipse, the north wind which was slanted to the [...] blew; during its eclipse, Mars, Venus...
>
> [BM 41660 (= LBAT 232), Rev. 9–10; trans SH I, p. 313.]

In the late summer of the 30th year of his reign (equivalent to 282/1 BC), Seleucus I died and his son Antiochus I ascended the throne; this is why two kings are mentioned in the above diary entry. Although the month is missing, it can readily be restored from the frequent lunar and planetary data. As noted in chapter 4, the Babylonians frequently calculated which planets would be above the horizon during both solar and lunar obscurations. The planets would not, of course, be actually visible at most solar eclipses.

The observation that the Sun rose eclipsed will be discussed below (section 5.5). It should be noted that the term E-a, which SH translate as 'came out', is the normal term for sunrise. Presumably the text implies that maximum phase occurred 6 deg after this moment, following which the Sun began to clear. However, the relevant part of the entry is only partially preserved. Hence only the measurement of last contact (20 deg after sunrise) will be considered here.

RESULTS

Last contact on Jan 30 at 1.33 h after sunrise. LT of sunrise = 6.76 h, hence measured LT of last contact = 8.09 h, UT = 5.42 h. Computed TT = 9.02 h, thus ΔT = 12 950 sec.

(3) BC 254 Jan 31 (computed mag. at Babylon = 0.26)

> [SE 57]...month X...The 28th, 56 deg before sunset, solar eclipse; when it began, in 12 deg daytime [...] when it began [to cl]ear, it cleared from south to north in 11 deg daytime; 23 deg onset and clearing. During its eclipse blew the west wind which was slanted to the north.
>
> [BM 34278 (= LBAT 596) + BM 35418 (= LBAT 258), Rev. 11, 12; trans. SH II, p. 29.]

Only the month and day of the eclipse are preserved on this diary, which consists of two joined fragments, but the year can readily be established from the many planetary observations which are preserved (SH II, p. 30). The text indicates that first contact occurred 56 deg before sunset, maximal phase 44 deg before sunset and end 33 deg before sunset. Although the reference to maximum is damaged, it is clear from the various recorded durations that this phase is referred to.

RESULTS

(i) First contact on Jan 31 at 3.73 h before sunset. LT of sunset = 17.26 h, hence measured LT of first contact = 13.53 h, UT = 10.86 h. Computed TT = 14.04, thus ΔT = 11 450 sec.

(ii) Maximal phase at 2.93 h before sunset, hence measured LT = 14.33 h, UT = 11.66 h. Computed TT = 14.86 h, thus ΔT = 11 500 sec.

(iii) Last contact at 2.20 h before sunset hence measured LT = 15.06 h, UT = 12.39 h. Computed TT = 15.63 h, thus ΔT = 11 650 sec.

These three results for ΔT are in remarkably close accord with one another.

(4) BC 249 May 4 (computed mag. at Babylon = 0.80)

Observations of this eclipse are recorded on two separate tablets.

(i) (SE 63, month I)... The 28th, clouds were in the sky; 90 deg daytime [...solar eclipse ...] [...] onset and clearing; during its eclipse [...].

[BM 45723 (= LBAT 267), Obv 8′; trans. SH II, p. 51; see also Huber, p. 84.]

BM 45723 is an astronomical diary. Only a small piece remains, without reference to either year or month, but the few planetary and lunar observations are sufficient to enable a date to be identified (SH II, p. 53).

(ii) (SE 63, month I), 28 solar eclipse; over two-thirds of the disk it made. (It began) at 90 deg after sunrise.

[BM 32154 (= LBAT 1216) + BM 32408 (= LBAT 1217), Obv. 3; trans. Huber, p. 84.]

The joined fragments BM 32154 + BM 32408 are part of a goal-year text dating from SE 81 (231–230 BC). The eclipse occurred 18 years before the goal-year. Unfortunately only the local time of *first* contact – i.e. 90 deg after sunset – is preserved.

RESULTS

First contact on May 4 at 6.00 h after sunrise. LT of sunrise = 5.28 h, hence measured LT = 11.28 h, UT = 8.22 h. Computed TT = 12.07 h, thus ΔT = 13 850 sec.

(5) BC 195 Jun 6 (computed mag. at Babylon = 0.19)

Year 117 (SE), month II 28, solar eclipse, beginning on the south side. 4 fingers it made. (Began) at 2 *beru* (= 60 deg) after sunrise.

[BM 34048 (= LBAT 1249), Rev. 3′–8′; trans. Huber, p. 86.]

This tablet is a goal-year text dating from SE 135 (177–176 BC). The eclipse, whose date is fully specified on the tablet, occurred 18 years before the goal-year.

RESULTS

First contact on Jun 6 at 4.00 h after sunrise. LT of sunrise = 4.91 h, hence measured LT = 8.91 h, UT = 5.83 h. Computed TT = ?? ΔT = ??

NB no value of ΔT can satisfy this observation; the computed UT of first contact (i.e. TT − ΔT) cannot be earlier than 6.03 h, otherwise the calculated eclipse magnitude reduces to zero at Babylon. Evidently either the original measurement was faulty or a scribe made an error in reporting the time. An upper limit to the value of ΔT is set by the fact that the eclipse was visible at all (mag. > 0) at Babylon; this leads to the result $\Delta T < 15\,350$ sec. The actual magnitude estimate (4 fingers) will be considered in section 5.4.

(6) BC 190 Mar 14 (computed mag. at Babylon = 0.77)

> Year 121 (SE), king An[tiochus] month XII 29 solar eclipse beginning on the north-west side. In 15 deg day [...] over a third of the disk was eclipsed. When it began to become bright, in 15 deg of day from north-west to east it became bright. 30 deg total duration. [During this eclipse] east (wind) went. During this eclipse [...] Venus, Mercury and Saturn [stood there]. Toward the end of becoming bright, Mars rose (?) The other planets did not stand there. (Began) at 30 deg (= 1) *beru* after sunrise.
>
> [BM 33812 (= LBAT 1438), Obv. and Rev.; trans. Huber, p. 87, with amendment by C. B. F. Walker.]

This clearly dated tablet from the reign of Antiochus III is devoted specifically to this eclipse. The text is remarkably well preserved. Huber questioned the reading of 30 deg but direct consultation of the original tablet by Dr Christopher Walker of the British Museum (personal communication) shows that this figure is clear and is confirmed by further expressing the interval as 1 *beru.*

It will be noted that the computed magnitude at Babylon (0.77) is much greater than the recorded figure. For an investigation of the estimate of magnitude, see section 5.4. As discussed in chapter 10, this same eclipse may have been observed to be total in the Hellespont strait (Dardanelles) and partial at Alexandria. However, in neither case is the time of day reported.

RESULTS

(i) First contact on Mar 14 at 2.00 h after sunrise. LT of sunrise = 6.10 h, hence measured LT of first contact = 8.10 h, UT = 5.34 h. Computed TT = 8.92 h, thus ΔT = 12 900 sec.

(ii) Maximum phase at 3.00 h after sunrise, hence measured LT of maximum = 9.10 h, UT = 6.34 h. Computed TT = 10.07 h, thus ΔT = 13 400 sec.

(iii) Last contact at 4.00 h after sunrise, hence measured LT of maximum = 10.10 h, UT = 7.34 h. Computed TT = 11.27 h, thus ΔT = 14 150 sec.

(7) BC 170 Jul 28 (computed mag. at Babylon = 0.44)

> Year 142 (SE) of the kings An(tiochus) and An(tiochus) his son, month IV 28 solar eclipse, beginning on the north-west side. In 12 deg day it made one third of the disk. (Began) at 20 deg day before sunset.
>
> [BM 35387 (= LBAT 1263), Rev. 3′–7′; trans. Huber, p. 90.]

At the time, Antiochus IV reigned jointly with his son (Sachs and Wiseman, 1954). This tablet is a goal-year text dating from SE 160 (152–151 BC). The eclipse, whose date is fully preserved, occurred 18 years before the goal-year.

RESULTS

(i) First contact on Jul 28 at 1.33 h before sunset. LT of sunset = 18.98 h, hence measured LT of first contact = 17.65 h, UT = 14.74 h. Computed TT = 18.16 h, thus ΔT = 12 300 sec.

(ii) Maximal phase at 0.53 h before sunset, hence measured LT of maximum = 18.45 h, UT = 15.54 h. Computed TT = 18.93 h, thus ΔT = 12 200 sec.

(8) BC 136 Apr 15 (computed mag. at Babylon = 1.05)

> SE 175 month XII_2. The 29th, at 24 deg after sunrise, solar eclipse; when it began on the south-west side, in 18 deg day towards noon it became entirely total. (Time interval of) 35 deg for onset, maximal phase and clearing.
>
> [Composite description based on translations by Hunger (1995) and SH III, p. 185 of BM 34034 (= LBAT 429), Rev. 24–28 and BM 45745 (= LBAT 1285), Rev. 13′–15′.]

The observation of totality has already been discussed above (section 5.2).

RESULTS

(i) First contact on Apr 15 at 1.60 h after sunrise. LT of sunrise = 5.56 h, hence measured LT of first contact = 7.16 h, UT = 4.19 h. Computed TT = 7.69 h, thus ΔT = 12 600 sec.

(ii) Total at 2.80 h after sunrise, hence measured LT of maximum = 8.36 h, UT = 5.39 h. Computed TT = 8.76 h, thus ΔT = 12 100 sec.

Table 5.1 ΔT results from Late Babylonian timings of solar eclipses.

Year[a]	Ct[b]	SR/SS[c]	Int (deg)[d]	ΔT (sec)[e]
−321	1	SS	−3	14150
−280	4	SR	+20	12950
−253	1	SS	−56	11450
−253	M	SS	−44	11500
−253	4	SS	−33	11650
−248	1	SR	+90	13850
−189	1	SR	+30	12900
−189	M	SR	+45	13400
−189	4	SR	+50	14150
−169	1	SS	−20	12300
−169	M	SS	−8	12200
−135	1	SR	+24	12600
−135	2	SR	+42	12100
−135	4	SR	+59	12250

[a] The year of each eclipse on the Julian calendar, using negative integers.
[b] The contact, etc., whose time was measured (M = maximal phase).
[c] Whether the eclipse was timed relative to sunrise (SR) or sunset (SS).
[d] The measured interval in degress (positive if after sunrise, negative if before sunset).
[e] The calculated value of ΔT (in sec) which satisfies the observation.

(iii) Last contact at 3.93 h after sunrise, hence measured LT of end = 9.49 h, UT = 6.52 h. Computed TT = 9.93 h, thus ΔT = 12 250 sec.

Here is another set of very self-consistent values for ΔT. These results closely correspond to the limits set by the observation of totality at the same eclipse – i.e. 11 210 < ΔT < 12 140 sec (see section 5.2)

The results obtained in this section from timed observations are listed in table 5.1; the ΔT results in this table are in very good accord with one another.

5.4 Estimates of solar eclipse magnitude

Six preserved Babylonian records of solar eclipses contain estimates of the maximum degree of obscuration of the Sun: BC 369, 249, 195, 190, 170 and 166. Timings of all but the first and last of these eclipses have been discussed earlier (section 5.3) so that for these only the essential details will be repeated below. However, full accounts of the eclipses of 369 and 166 BC will be given.

(1) BC 369 Apr 11

> [Artaxerxes II, year 35, month XII, end... solar eclipse...]. In 6 deg daytime $\frac{1}{3}$ of the disk was covered ... [...] At that time Jupiter was in Leo [... to] the 7th day of month I of year 36, the river level [...] 2 cubits 8 fingers.
>
> [BM 37097 + 37211 (no LBAT number), Rev. 4′–7′; trans. SH I, p. 125.]

The date of this small diary fragment may be derived from the reference to year 36 on line 7′. This mentions a measurement of the river level at Babylon from some day of the current month to the 7th day of month I in year 36. Only the 36th year of either Artaxerxes I (i.e. 430–429 BC) or Artaxerxes II (370–369 BC) – both of whom had unusually long reigns – or 36 SE (277–276 BC) can be intended. In their commentary (SH I, p. 125), Sachs and Hunger state that:

> ... it appears that the reverse (of the tablet) deals with the last month of a year 35, because a measurement of the river level from some day of the current month 'to the 7th of month I of year 36' is mentioned.

There were solar eclipses in the last month of both the 35th year of Artaxerxes II (i.e. BC 369 Apr 11) and SE 35 (i.e. BC 276 Apr 3). However, no similar event took place after the 9th month in the 35th year of Artaxerxes I (BC 429 Jan 17). Hence only the eclipses of BC 369 and 276 would seem viable (see also SH II, p. 125). The location of Jupiter in Leo (a sign of the zodiac covering the longitude range 120–150 deg) fits only the earlier year. At the eclipse of BC 369 the longitude of Jupiter was 136 deg, whereas at the eclipse of 276 BC it was 82 deg. It should be pointed out that when the eclipse of BC 429 occurred, the longitude of the planet was 120 deg, but as noted above, this eclipse occurred at the wrong time of year.

RESULTS

To attain the observed magnitude of 0.33 in 369 BC, ΔT = ?? sec.

NB no value of ΔT can produce a maximum visible obscuration as small as the observed figure of 0.33 unless greatest phase occurred a little before sunrise. If this were the case, the eclipse would already be declining when the Sun rose. However, the record seems clear that the eclipse was already visible for 6 deg before maximal phase was reached. For a value of ΔT (16 500 sec) which produces greatest phase 6 deg (i.e. 0.4 h) after sunrise, the magnitude would be 0.39. Although the Babylonian estimate of '$\frac{1}{3}$ of the disk' would thus be quite reasonable, a result for ΔT cannot be derived.

(2) BC 249 May 4

Two observations of this eclipse are recorded on separate tablets – a diary and a goal-year text (see section 5.3) – but only the latter inscription contains an estimate of magnitude.

> (SE 63, month I), 28 solar eclipse; over two-thirds of the disk it made. (Began) at 90 deg after sunrise.
>
> [BM 32154 (= LBAT 1216) + BM 32408 (= LBAT 1217), Obv. 3; trans. Huber, p. 84.]

It will be assumed that the magnitude estimate of 'over two-thirds' means somewhere between $\frac{2}{3}$ (8 fingers) and $\frac{3}{4}$ (9 fingers) – i.e. approximately $8\frac{1}{2}$ twelfths or 0.71.

RESULTS

To attain the observed magnitude of 0.71, $\Delta T = 14\,650$ sec (only one possible solution).

(3) BC 195 Jun 6

> Year 117 (SE), month II 28 solar eclipse, beginning on the south side. 4 fingers it made. (Began) at 2 *beru* (= 60 deg) after sunrise.
>
> [BM 34048 (= LBAT 1249), Rev. 3′–8′; trans. Huber, p. 86.]

RESULTS

To attain the observed magnitude of 0.33, with the lower limb of the Sun covered, $\Delta T = 10\,250$ sec.

(4) BC 190 Mar 14

> Year 121 (SE)... month XII 29 solar eclipse beginning on the north-west side. In 15 deg day [...] over a third of the disk was eclipsed. When it began to become bright... (Began) at 30 deg (= 1) *beru* after sunrise.
>
> [BM 33812 (= LBAT 1438), Obv. and Rev.; trans. Huber, pp. 87–88.]

It will be assumed that the magnitude estimate of 'over a third' means somewhere between $\frac{1}{3}$ (4 fingers) and 5 fingers – i.e. some $4\frac{1}{2}$ twelfths or 0.38.

RESULTS

To attain the observed magnitude of 0.38 with the upper limb of the Sun covered, $\Delta T = 6850$ sec.

(5) BC 170 Jul 28

> Year 142 (SE)... month IV 28 solar eclipse beginning on the north-west side. In 12 deg day it made one third of the disk. (Began) at 20 deg day before sunset.
>
> [BM 35387 (= LBAT 1263), Rev. 3′–7′; trans. Huber, p. 90.]

RESULTS

To attain the observed magnitude of 0.33 with the upper limb of the Sun covered, ΔT = 35 850 sec.

NB this result is highly discordant, as comparison with the fairly self-consistent ΔT values in table 5.1 shows. Furthermore with the derived figure for ΔT of 35 850 sec, the computed LT would be around 10.00 h instead of the recorded time of about 8 deg (i.e. 20–12 deg) before sunset (18.45 h). Even if it is supposed that the *lower* limb of the Sun was obscured, no value of ΔT can produce a maximum visible obscuration quite so small as the observed figure unless the eclipse reached its peak some time after sunset. This is not in accord with observation. It would appear that the magnitude estimate is unreliable – presumably representing a scribal error.

(6) BC 166 May 17

> [Y]ear 146, [king] Antiochus [month I, end... solar eclipse]. In 13 deg of day more than one-third of the disk...
>
> [BM 32844 (= LBAT 376), Obv. A13; trans. SH II, p. 486.]

This entry is towards the end of the first month covered by a diary for the year 146 SE (166/165 BC). At the time, Antiochus IV was sole ruler. Although the month is missing, assumption of month I is readily confirmed from the various lunar observations noted in the text.

As above, it will be assumed that the magnitude estimate of 'over a third' means somewhere between $\frac{1}{3}$ (4 fingers) and 5 fingers – i.e. $4\frac{1}{2}$ twelfths or 0.38.

RESULTS

To attain the observed magnitude of 0.38, either ΔT = 15 250 sec or −6850 sec. Only the first value for ΔT need be considered; the alternative is quite discordant.

The rather mediocre set of results obtained in this section is summarised in table 5.2. This table gives the year (expressed as a negative integer), observed magnitude and ΔT result for each observation.

Table 5.2 ΔT results from Babylonian estimates of solar eclipse magnitude.

Year	Est. mag.	ΔT (sec)
−368	1/3	??
−248	>2/3	14650
−194	4/12	10250
−189	>1/3	6850
−169	1/3	??
−165	>1/3	15250

5.5 Solar eclipses occurring near sunrise or sunset

There are only two known Babylonian records in this category (322 and 281 BC). In the latter instance, the degree of obscuration of the Sun is specified at sunrise; this estimate will also be considered. A further observation (241 BC) is also worth investigating. Although computation based on a preliminary value of ΔT (equation (4.1)) indicates that the eclipse occurred near sunset, the text itself mentions the clearance of the Sun. Pre-selection of this kind in no way introduces bias; it merely pinpoints potentially viable observations.

Timings of both the eclipses of 322 and 281 BC have already been discussed in section 5.3. Hence only partial quotations, omitting irrelevant details, will be given here. However, the record from 241 BC will be translated in full. Following earlier practice, computed magnitudes are given purely for reference.

(1) BC 322 Sep 26 (computed mag. at Babylon = 0.17)

Year 2 of (king) Philip, (month VI). The 28th, around 3 deg before sunset, solar eclipse [...]... It set eclipsed.

[BM 34093 (= LBAT 212), Rev. 23′; trans. SH I, pp. 226–227; see also Huber, p. 83.]

As discussed in some detail in chapter 3, for the present purpose the most viable inference is that the Sun set at some time between first and last contact. Although the '3 deg' of the text suggests that sunset occurred long before maximal phase, there is always a slight possibility that this could represent a scribal error for a much larger number.

RESULTS

(i) First contact on Sep 26 before sunset. LT of sunset = 18.08 h, hence LT of first contact < 18.08 h, UT < 15.01 h. Computed TT = 18.74 h, thus $\Delta T > 13\,400$ sec.

(ii) Last contact after sunset. LT of contact > 18.08 h, UT > 15.01 h. Computed TT = 20.02 h, thus $\Delta T < 18\,000$ sec.

Combining the above limits, one obtains $13\,400 < \Delta T < 18\,000$ sec.

(2) BC 281 Jan 30 (computed mag. at Babylon = 0.21)

> Year 30 (SE)...(month X). The 29th [solar] eclipse; when the Sun came out (i.e. rose), 2 fingers were obscured on the south side; at 6 deg daytime, [...]...20 deg day onset and clearing ...
>
> [BM 41660 (= LBAT 232), Rev. 9–10; trans SH I, p. 313.]

One-sixth of the solar diameter was estimated to be obscured at sunrise, after which the phase apparently increased and then decreased. Hence it may reasonably be concluded that the Sun rose between first contact and maximum phase.

RESULTS

(i) First contact on Jan 30 before sunrise. LT of sunrise = 6.78 h, hence LT of first contact < 6.78 h, UT < 4.11 h. Computed TT = 7.53 h, thus $\Delta T > 12\,350$ sec.

(ii) Maximum phase after sunrise. LT of maximum > 6.78 h, UT > 4.11 h. Computed TT = 8.13 h, thus $\Delta T < 14\,450$ sec.

Combining the above limits, one obtains $12\,350 < \Delta T < 14\,450$ sec.

(iii) Fraction of Sun covered at rising. For any value of ΔT in the above interval, the computed eclipse magnitude ranges from 0.17 to 0.22. These figures are probably too close to the estimated degree of obscuration at sunrise (2 fingers or 0.17) for a useful independent value of ΔT to be deduced from this observation.

(3) BC 241 Nov 28 (computed mag. at Babylon = 0.38)

> [SE 70, month VIII, end... Solar eclipse ...] cleared; 30 deg onset and clearing; during the eclipse [...].
>
> [Rm 720 + 732 + BM 41522 (= LBAT 278, 277, 883), Obv. 3′; trans. SH II, p. 79.]

This is a fairly well preserved diary consisting of several joining fragments, two of which were uncovered by Rassam at the site of Babylon. Although both the year and month have been broken off, the planetary and lunar observations enable the date to be unambiguously derived (SH

Table 5.3 ΔT limits from Babylonian observations of solar eclipses.

		ΔT range (sec)	
Year	Observation	LL (sec)	UL (sec)
−321	sunset	13 400	18 000
−280	sunrise	12 350	14 450
−240	sunset	12 800	—
−194	mag.>0	—	15 350
−135	total	11 210	12 140

II, p. 83). The time of first contact for the solar eclipse relative to sunrise or sunset is, unfortunately, wanting.

The value of ΔT computed according to equation (4.1) indicates that the eclipse ended close to sunset. However, since clearing is mentioned (twice), it must have been seen to end before sunset. In this instance, it is only possible to determine one useful (lower) limit to ΔT.

RESULTS

Last contact on Nov 28 before sunset. LT of sunset = 17.11 h, hence LT of contact < 17.11 h, UT < 14.01 h. Computed TT = 17.57 h, thus ΔT > 12 800 sec.

In table 5.3 are assembled the results obtained from the three observations analysed in this section together with the ΔT limits derived from the total eclipse of 136 BC (section 5.2) and the very small eclipse of 195 BC (section 5.3). Individual columns of this table contain the year (negative), type of observation, lower limit (LL) to ΔT, and upper limit (UL) to ΔT. In the year −194 (i.e. 195 BC), the fact that the eclipse was visible from Babylon at all provides an upper limit only to ΔT.

5.6 A possible allusion to a total solar eclipse in the Babylonian *Religious Chronicle*

The *Religious Chronicle* of Babylon, a fragmentary copy of which is preserved in the British Museum (tablet BM 35968), makes reference to darkness by day accompanied by 'fire in the sky'. This has been interpreted as a large eclipse of the Sun. Consecutive extracts from the chronicle may be translated as follows:

> On the sixteenth day of the month Ab, in the seventh year, two deer entered Babylon and were killed.

On the twenty-sixth day of the month Sivan, in the seventh year, day turned to night and (there was) a fire (*isatu*) in the sky.

In the month Elul, in the eleventh year, water flowed within the wall of the lower forecourt

[*Religious Chronicle*, ii 12–15; trans. Grayson (1975, p. 135).]

NB as pointed out by Grayson, the scribe put the events in Ab (the 5th month) and Sivan (the 3rd month) in the wrong order.

The unknown king in whose reign the above events occurred was on the throne for at least 17 years. The chronicle is mainly concerned with portents observed at Babylon, although there are occasional references to other places. Fotheringham (1920b) noted that as a rule no more than two or three portents are recorded in any one year and he thus inferred that the celestial phenomenon described 'was a rare one, was observed at Babylon, and may reasonably have been a total eclipse of the Sun'. He identified the 'fire in the sky' (rendered in his source as 'fire in the midst of heaven') as the solar corona. On the presumption that the event occurred during the eleventh century BC, he calculated that the eclipse of BC 1063 Jul 31, was 'the only one which can have been total at Babylon within that century'.

However, this identification is far from secure. For instance:

(i) The name of the king is not known.

(ii) July is too late for the month Sivan, the third month of the Babylonian year (which began in March–April).

(iii) Since the start of each month was fixed by visibility of the crescent Moon, a solar eclipse could not occur as early as the 26th day. Thus in the Late Babylonian texts, solar eclipses are invariably recorded on the 28th or 29th days (see the various examples quoted earlier in this chapter).

(iv) Many of the portents recorded in the chronicle are quite mundane, as can be seen in the above extract. They closely resemble the omens found in the Late Babylonian texts – translations of which are given by SH I, SH II and SH III. Thus there is no need to assume that the phenomenon causing darkness was unusually rare.

(v) It is by no means certain that the occurrence described was indeed an eclipse. It could have been a severe storm, in which case *isatu* would mean lightning (Grayson, 1975, 135n; see also Brinkman, 1968, p. 68n). The Babylonians at this period should have been very familiar with eclipses but in the above description there is nothing which could be identified as a technical term for an event of this

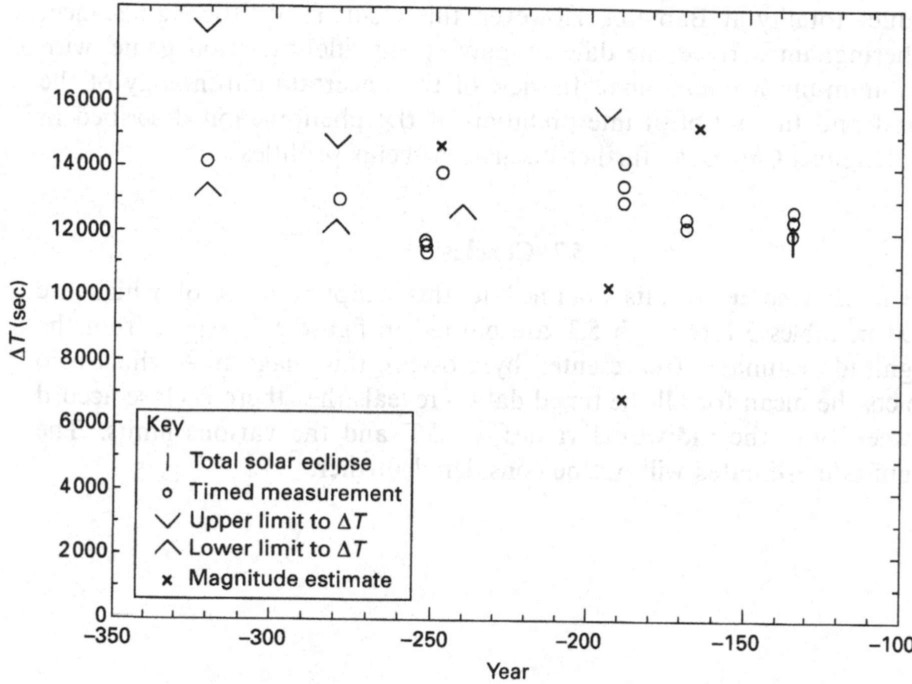

Fig. 5.3 Diagram showing the ΔT values and limits obtained from Late Babylonian observations of solar eclipses.

kind – such as AN-KU_{10} found in the Late Babylonian texts. Even if the event was an unusually spectacular obscuration of the Sun, it would begin and end like any other eclipse.

(vi) Explanation of the 'fire in the sky' as the corona seems very unlikely. The corona is, in fact, scarcely ever alluded to in ancient and medieval literature, even when the effects of a total eclipse are vividly described (see especially chapter 11).

Rowton (1946) favoured a solar eclipse at sunset and proposed BC 1012 May 9. However, according to Wiseman (1965), this was in the *15th* year of a king (namely Simbar-sipak), rather than the 7th. It should be stressed that the chronology of this period is still poorly established. Although calculations using a value for ΔT derived from equation (4.1) (i.e. 28 000 sec) indicate that the eclipse of 1012 BC would be fairly large at sunset in Babylon, no figure for ΔT can lead to a visible magnitude greater than 0.96 there. A significantly greater phase would be needed to effectively turn the day into night.

Similar calculations based on equation (4.1) ($\Delta T = 29\,000$ sec) indicate that the eclipse of 1063 BC would reach a magnitude of 0.85 at Babylon in the early morning (6.0 h). Decreasing ΔT to about 25 000 sec would produce totality at Babylon. However, this result is of little significance; Fotheringham derived the date by playing the 'identification game' with the minimum of constraints. In view of the uncertain chronology of the period and the doubtful interpretation of the phenomenon described in the *Religious Chronicle*, further discussion seems profitless.

5.7 Conclusion

The miscellaneous results obtained in this chapter, most of which are listed in tables 5.1 through 5.3, are plotted in figure 5.3. Apart from the magnitude estimates (represented by crosses), this diagram – which also depicts the mean for all the timed data – reveals that there is close accord between both the individual values for ΔT and the various limits. The magnitude estimates will not be considered further.

6
Timed Babylonian lunar eclipses

6.1 Introduction

Because so many observations of lunar eclipses are preserved on the Late Babylonian astronomical texts, it is necessary to devote two chapters of this book to their discussion. This is a reflection of the great importance of the lunar eclipse records from Babylon in the study of long-term changes in the length of the day. The present chapter is restricted to timed contacts, while in chapter 7 a variety of untimed observations of eclipses of the Moon will be considered.

Most Babylonian lunar eclipse timings are expressed relative to sunrise or sunset, although very occasionally they are referred to moonrise or moonset instead. I have divided these measurements into four categories: (i) those for which only a single contact timing is preserved; (ii) timings of two separate contacts (mainly partial eclipses); (iii) three or four contact timings (total eclipses only); and (iv) timings of eclipse maxima (partial eclipses only). Observations in these categories will be discussed in turn in sections 6.2 through 6.5. In addition, a number of eclipses were also timed in relation to the culmination of certain reference stars (see section 6.6).

Times of the very earliest eclipses (whether observed or predicted) were nearly always quoted with low precision. Between 731 BC (the earliest surviving record, representing a prediction) and just before 600 BC times were almost exclusively expressed to the nearest 10 deg. The only known exception to this rule is an observation dating from 666 BC; here it is said that the Moon commenced to be obscured only 3 deg after sunset. Evidently 10 deg was the maximum precision that the Babylonian astronomers felt able to achieve under normal circumstances at this archaic period. However, two significant improvements in accuracy can be traced. Between about 600 and 560 BC (a period which roughly coincides with the reign of Nebuchadrezzar II), rounding of times to the nearest 5 deg was the norm, but after 560 BC virtually all measurements were estimated to the nearest degree. It may be presumed that solar eclipse timings

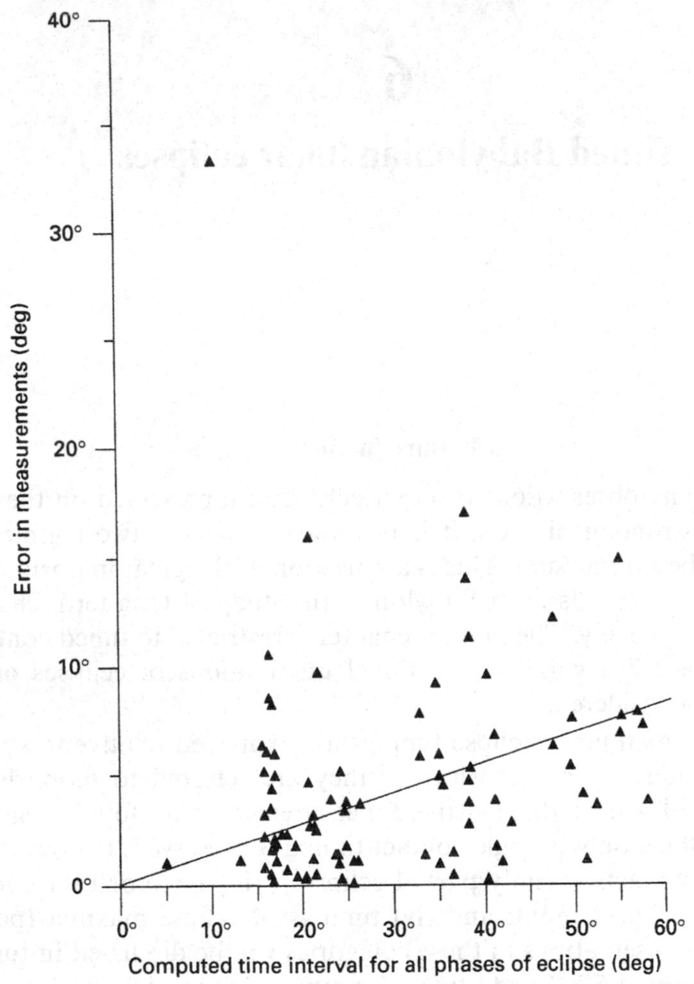

Fig. 6.1 Errors in measured lunar eclipse durations (Stephenson and Fatoohi, 1993).

originally followed much the same pattern but – as noted in chapter 5 – no observations of this kind have survived from before 370 BC.

Despite the marked decrease in rounding errors around 560 BC, during later centuries progress in measuring time-intervals appears to have been relatively slow. This is apparent from figure 6.1, which plots the errors in measuring lunar eclipse durations made by the Late Babylonian astronomers from 560 to 50 BC (Stephenson and Fatoohi, 1993). It thus seems a pity that observers at earlier periods were not more optimistic when rounding their measurements.

As is true of several solar eclipse records on the astronomical texts, the date of a lunar obscuration is often broken away. Techniques of restoration have already been discussed (chapter 4). When considering an observation which is reported on either a diary, goal-year text, fragmentary lunar eclipse table, or single tablet devoted to the event, I have given specific comments with regard to the date of that record. However, a substantial number of observations (as well as predictions) are reported on a series of extensive lunar eclipse tables; these typically cite events over many years. Some preliminary remarks on these tables seem necessary here in order to avoid needless repetition when discussing the observations themselves. The following comments are largely based on the scholarship of Huber (1973).

(i) All of the surviving observations (and predictions) of lunar eclipses from earliest times (731 BC) to 609 BC – as well as many later observations down to 317 BC – are recorded on a series of five British Museum tablets. Their reference numbers are: BM 32238 (= LBAT 1414), BM 45640 + 35115 + 35789 (= LBAT 1415 + 1416 + 1417: three joining pieces) and BM 32234 (= LBAT 1419). This major compilation, which lists eclipses at 18-year intervals, originally covered 24 saros cycles or 432 years and extended from some time between 749 and 740 to between 317 and 308 BC. The extant remnant listed as BM 32238 cites eclipses from 731 to 659 BC (obverse) and from 389 to 317 BC (reverse). Tablets BM 45640 + 35115 + 35789 contain data from 703 to 632 BC (obverse) and from 415 to 360 BC (reverse), while BM 32234 extends from 609 to 537 BC (obverse) and from 519 to 447 BC (reverse). Many names of rulers are preserved on these tablets: e.g. Nabu mukin-zeri (who reigned from 731 to 726 BC), Bel-ibni (702–699 BC), Samas-sum-ukin (667–647 BC), Kandalanu (647–625 BC), Nebuchadrezzar II (604–562 BC), Xerxes I (485–465 BC) and Philip (323–316 BC). From the well-defined chronological sequence on this series of texts, virtually all eclipse dates can be confidently restored.

(ii) BM 38462 (= LBAT 1420) reports lunar eclipses for almost every year from the beginning of the reign of Nebuchadrezzar II (604/3 BC) to his 29th year (576/5 BC). The damaged (but still recognisable) name of Nebuchadrezzar is given on the first line of the tablet.

(iii) Between them, BM 34787 (= LBAT 1426) and BM 34684 (= LBAT 1427) cover the period from the 22nd year of Artaxerxes I (443/2 BC) to the 18th year of his successor Darius II (406/5 BC). Eclipses are mentioned in most years. Darius II became king during months IX/X of the 41st year of Artaxerxes I. His accession ('Ochos, whose name is Darius') is noted in column II of tablet BM 34787.

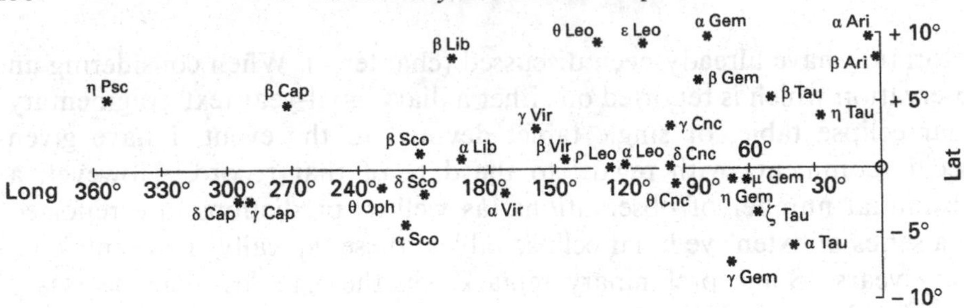

Fig. 6.2 Co-ordinates of the Babylonian 'Normal Stars' at 164 BC.

Further justification for the dates of lunar eclipses mentioned on any of the above tablets will not usually be given below.

In later texts (from the fifth century BC onwards), the location of the eclipsed Moon relative to the nearest 'Normal Star' (MUL-SID) is frequently given. The Normal Stars were a series of 32 stars in the zodiacal belt. Conjunctions between the Moon and planets with these stars are mentioned very frequently on the Late Babylonian astronomical texts and calculations of the circumstances of these events have readily revealed their identity. For a full list of names and identifications (in terms of Bayer Greek letters) see Stephenson and Walker (1985), p. 16 and Sachs and Hunger (1988), pp. 17–19. The celestial co-ordinates of the Normal Stars for the epoch 164 BC (the date of the earliest recorded sighting of Halley's comet in Babylonian history) are shown in figure 6.2. Spacing of these stars was quite irregular and in particular there was a significant gap between about 290 and 350 deg longitude; the reason for this gap is unknown.

Many examples of eclipse observations containing allusions to Normal Stars will be found in this chapter. For instance, at the eclipse of BC 317 Dec 13/14 the Moon was said to be '$1\frac{1}{2}$ cubits in front of β Gem', while at the eclipse of BC 80 Apr 10/11 the Moon was '$\frac{2}{3}$ cubit behind α Lib'. As mentioned in chapter 4, 1 cubit was roughly equivalent to 2 or $2\frac{1}{2}$ deg. Estimates of the position of the eclipsed Moon in relation to a Normal Star can sometimes provide valuable aids to verifying dates (see section 6.3).

The terms 'in front of' (*ina* IGI) and 'behind' (*ar*) – as used in the above examples – respectively imply locations 'to the west of' and 'to the east of' a celestial body – following the diurnal motion of the celestial sphere. Similarly the expressions 'above' (*e*) and 'below' (SIG) are also frequently used in the Late Babylonian astronomical texts. It has yet to be established whether an ecliptic frame of reference or some other system was used in these cases. In making calculations, I have somewhat arbitrarily assumed an ecliptic framework.

For each individual eclipse investigated below, the computed magnitude is given in parentheses for reference, following the calculated Julian date. Where only one contact measurement is extant (as in section 6.2) I have rejected the observation if the magnitude is very small (less than about 0.10). Individual contacts for small partial eclipses are probably difficult to define with the unaided eye since the Moon then enters the Earth's umbral shadow obliquely. Under these circumstances, the boundary between the umbra and the deep penumbral shadow is likely to be poorly defined. However, where both first and last contact are preserved for an eclipse of small magnitude, I have utilised the pair of observations to derive a single mean result for ΔT (section 6.3).

6.2 Lunar eclipses for which only a single contact measurement relative to sunrise or sunset is preserved

In this section those measurements which are rounded to the nearest degree will be considered first (section 6.2.1) followed by the older observations of lower precision (section 6.2.2). A continuous sequence of numbering entries will nevertheless be adopted. In some instances, dates have been derived by Sachs and Schaumberger (1955). This reference will be abbreviated to SS in the remainder of this chapter. As mentioned in chapter 5, the publications by Sachs and Hunger (1988, 1989 and 1996) are cited as SH I, SH II and SH III.

For partial eclipses, a duration of maximal phase is often recorded – e.g. 5 deg in 143 BC. This is not a real effect and merely arises from the limited resolution of the human eye. A detailed discussion is given in section 6.5.

6.2.1 Measurements expressed to the nearest degree

(1) BC 666 Apr 10/11 (mag. = 0.48)

[Samas-sum-ukin], year 2, month I, beginning(?) on [south-east(?)]. (Began) at 3 deg after sunset.

[BM 45640 (= LBAT 1415), Obv. III, bottom; trans. Huber, p. 8.]

NB direct consultation of the original tablet by Dr Christopher Walker (personal communication) shows that the 3 deg (questioned by Huber) is clear.

RESULTS

First contact on Apr 10 at 0.20 h after sunset. LT of sunset = 18.29 h, hence LT of first contact = 18.49 h, UT = 15.57 h. Computed TT = 21.42 h, thus ΔT = 21 050 sec.

(2) BC 562 Mar 2/3 (mag. = 1.77)

> [Nebuchadrezzar] year 42, month XII 15, 1,30 (= 90) deg after sunset [...] 25 deg duration of maximal phase. In 18 deg it became bright. West (wind) went. 2 cubits below γ Vir eclipsed.
>
> [BM 41536 (= LBAT 1421), col. II, 5′–8′; trans. Huber, pp. 22–23.]

This small tablet is part of a lunar eclipse table. It now records only two eclipses: in the 6th and 12th months of the 42nd year of an unspecified king. Such a long reign could only refer to either Nebuchadrezzar II (42nd year = BC 563/2) or Artaxerxes II (BC 363/2). However, the terminology is early, and in any case there were only eclipses in the 3rd and 9th months of the appropriate year of Artaxerxes II. Hence the former year must be intended. Since the interval between first and second contact in the above text is not preserved, only the timing of the start of the eclipse can be utilised. Calculation shows that when the eclipse occurred, the Moon would be about 3 deg to the south of γ Vir, in close accord with the record.

RESULTS

First contact on Mar 2 at 6.00 h after sunset. LT of sunset = 17.66 h, hence LT of first contact = 23.66 h, UT = 20.97 h. Computed TT = 1.41 h (Mar 3), thus ΔT = 15 950 sec.

(3) BC 537 Oct 16/17 (mag. = 1.50)

> [Cyrus, year 2], month VII [...it made] $\frac{2}{3}$ of the disk towards totality. Not (yet) total, it set eclipsed... 5 deg in front of the Pleiades eclipsed. (Began) at 14 deg before sunrise.
>
> [BM 32234 (= LBAT 1419), Obv. V, middle; trans. Huber, p. 25.]

In this text, the term US is used both as an angular unit and as a measure of time. Before dawn on the morning of Oct 17, the Moon would be about 8 deg to the west of η Tau, rather more than – but still in fair agreement with – the estimated figure.

RESULTS

First contact on Oct 17 at 0.93 h before sunrise. LT of sunrise = 6.24 h, hence LT of first contact = 5.30 h, UT = 2.17 h. Computed TT = 7.39 h, thus ΔT = 18 800 sec.

(4) BC 483 Nov 18/19 (mag. = 1.47)

> [Xerxes, year 3], month VIII 13. Beginning on the south side, maximal phase not observed, it set eclipsed. During the eclipse [Venus] stood (there), the other planets did not stand there. (Began) at 10 deg before sunrise.
>
> [BM 32234 (= LBAT 1419), Rev. III; trans. Huber, pp. 27–28.]

The death of Xerxes 18 years later (465 BC) is recorded in the next entry of this tablet, following a note of a lunar eclipse in that same year (BC 465 Jun 5/6). Unfortunately, the account of this latter eclipse is too badly damaged to provide usable information on ΔT.

RESULTS

First contact on Nov 19 at 0.67 h before sunrise. LT of sunrise = 6.74 h, hence LT of first contact = 6.07 h, UT = 2.93 h. Computed TT = 7.74 h, thus ΔT = 17 300 sec.

(5) BC 421 Feb 2/3 (mag. = 1.64)

> [Darius II], year 2, month XI, 15. (At) 19 deg after sunset, beginning on the east. In 14[+x deg...] [...cubits] in front of ε Leo eclipsed...
>
> [BM 34787 (= LBAT 1426), col. II, 11′–12′; trans. Huber, p. 34.]

The accession of Darius II (423 BC) is noted in line 4′ of this same column. Since the interval between first and second contact is damaged, only the timing of the start of the eclipse can be utilised. In the early evening of Feb 2, the Moon would be more than 20 deg to the *east* of ε Leo, but only about 1.3 deg to the west of θ Leo (one of the adjacent Normal Stars). Possibly the star has been misidentified in the text.

RESULTS

First contact on Feb 2 at 1.27 h after sunset. LT of sunset = 17.26 h, hence LT of first contact = 18.53 h, UT = 15.86 h. Computed TT = 20.17 h, thus ΔT = 15500 sec.

(6) BC 406 Oct 9/10 (mag. = 0.97)

> [Darius II, year 18, month VII...] (At) 14 deg before sunrise, beginning on the east [...] Mercury and [...].
>
> [BM 34684 (= LBAT 1427), Rev. 9–10; trans. Huber, p. 41.]

RESULTS

First contact on Oct 10 at 0.93 h before sunrise. LT of sunrise = 6.12 h, hence LT of first contact = 5.19 h, UT = 2.07 h. Computed TT = 6.66 h, thus ΔT = 16 500 sec.

(7) BC 371 May 17/18 (mag. = 0.78)

> [Artaxerxes II, year 34, month II...eclipsed]. During the eclipse, Saturn stood above α Sco. (Began) at 1,6 (= 66) deg after sunset.
>
> [BM 32238 (= LBAT 1414), Rev. II, top; trans. Huber, pp. 45–46.]

On the above date, Saturn would be almost in conjunction with α Sco and about 7 deg to the north of it – in satisfactory agreement with the record.

RESULTS

First contact on May 17 at 4.40 h after sunset. LT of sunset = 18.88 h, hence LT of first contact = 23.28 h, UT = 20.19 h. Computed TT = 23.73 h, thus ΔT = 12 750 sec.

(8) BC 367 Aug 29/30 (mag. = 1.35)

[Artaxerxes II, year 38] month V, night of the 13th, moonrise to sunset: 9 deg. When 56 deg night were left to sunrise, lunar eclipse; on the south-east side [...] cubits behind the 'rear container' (*qup-pu ar*) of Aquarius it was eclipsed; the north wind blew. During its eclipse Mars (?) [...].

[BM 35184 (= LBAT 186), Rev. 5–6; trans. SH I, p. 135; see also Huber, p. 48.]

On this astronomical diary, the date is given at the lower edge of the tablet as: '[...] to the end of month VI, year 38 of king Arses, who is ca[lled] king Artaxerxes'. The star referred to as the 'rear container' of Aquarius has not been identified.

RESULTS

First contact on Aug 30 at 3.73 h before sunrise. LT of sunrise = 5.44 h, hence LT of first contact = 1.71 h, UT = 22.76 h (Aug 29). Computed TT = 4.19 h (Aug 30), thus ΔT = 19 550 sec.

(9) BC 317 Jun 18/19 (mag. = 0.37)

[Philip, year 7, month III...] ...4(?) [cubits] in front of β Cap eclipsed. (Began) at 10 deg after sunset.

[BM 32238 (= LBAT 1414), Rev. IV + V, top; trans. Huber, p. 50.]

The date (year 7 of Philip) is given at the end of the next entry, which records the eclipse of BC 317 Dec 13/14 (see section 6.4). Soon after sunset on Jun 18, the Moon would be about 10 deg to the west of β Cap, in close accord with observation.

RESULTS

First contact on Jun 18 at 0.67 h after sunset. LT of sunset = 19.15 h, hence LT of first contact = 19.82 h, UT = 16.78 h. Computed TT = 21.11 h, thus ΔT = 15 550 sec.

(10) BC 308 Jul 8/9 (mag. = 0.99)

[SE 4, month IV, night of the 13th...] 10 deg before sunrise, lunar eclipse [...] behind the 'rear horn of the goat-fish' (β Cap). The 13th, sunrise to moonset: 50′ ($\frac{5}{6}$ deg). Gusty wind.

[BM 40122 (= LBAT 218), 5′–6′; trans. SH I, p. 243.]

Although this tablet is only a small fragment of an astronomical diary, it contains a variety of planetary and lunar observations. From this information, the date may be confidently derived by calculation (see SH I, p. 243). With regard to the eclipse report, before dawn on the morning of Jul 9 the Moon would be about 8 deg to the east of (or 'behind') β Cap. The computed interval between sunrise and moonset on this occasion is 0.3 deg, in adequate accord with the recorded $\frac{5}{6}$ deg.

RESULTS

First contact on Jul 10 at 0.67 h before sunrise. LT of sunrise = 4.86 h, hence LT of first contact = 4.19 h, UT = 1.22 h. Computed TT = 5.14 h, thus ΔT = 14 100 sec.

(11) BC 240 Nov 2/3 (mag. = 1.40)

> [SE 72], month VIII 14, at 3 deg before sunrise, beginning on the east side. It set eclipsed. Month XII_2 (? or: cloudy).
>
> [BM 32236) (= LBAT 1218), Rev. 3–5; trans. Huber, p. 55.]

This tablet is a goal-year text for SE 90 (222–221 BC) - (SS, p. xxv). As is typical of such texts, the eclipse observation was made 18 years before this goal-year. The last two signs in the entry (SE? DIR) most probably imply that there was an intercalary 12th month in the year of the eclipse; this is confirmed for SE 72 (BC 240/239) by the tables of Parker and Dubberstein (1956). However, the sign DIR can also mean 'cloudy'.

RESULTS

First contact on Nov 3 at 0.20 h before sunrise. LT of sunrise = 6.54 h, hence LT of first contact = 6.34 h, UT = 3.17 h. Computed TT = 7.12 h, thus ΔT = 14 200 sec.

(12) BC 212 Apr 29/30 (mag. = 0.62)

> Year 100 (SE), king Antiochus (III). Month I 13 moonrise to sunset: 9 deg, measured. Lunar eclipse, beginning on the south side. Around(?) maximal phase cloudy, not observed. It set eclipsed. (Began) at 20 deg before sunrise.
>
> [BM 32222 (= LBAT 1237), Rev. 48–53; trans. Huber, p. 56.]

This accurately dated observation is reported on a goal-year text for SE 118 (194–193 BC) - (SS, p. xxv). The computed interval between moonrise and sunset on Apr 29 is 8.1 deg.

RESULTS

First contact on Apr 30 at 1.33 h before sunrise. LT of sunrise = 5.34 h, hence LT of first contact = 4.01 h, UT = 0.97 h. Computed TT = 4.25 h, thus ΔT = 11 800 sec.

(13) BC 212 Oct 24/25 (mag. = 0.94)

Year 100 (SE), king Antiochus (III). Month VII 15 moonrise to sunset: 3 deg, measured. Lunar eclipse, beginning on the north-east side. 2 fingers were lacking to totality. (Began) at 28 (or 27?) deg after sunset.

[BM 32222 (= LBAT 1237), Rev. 56–59; trans. Huber, p. 57.]

This record immediately follows the entry just cited (i.e. BC 212 Apr 30). Inspection of the original tablet by Walker (personal communication) indicates that despite slight damage, the reading 'after sunset' (i.e. GE_6 *gin*) is quite certain. However, Walker notes that the text is not clear whether the interval after sunset is 28 or 27 deg, although he prefers the former. In calculations, I have thus assumed a mean of 27.5 deg. The computed interval between moonrise and sunset on Oct 24 is 2.3 deg.

RESULTS

First contact on Oct 24 at 1.83 h after sunset. LT of sunset = 17.60 h, hence LT of first contact = 19.43 h, UT = 16.27 h. Computed TT = 22.20 h, thus ΔT = 21 350 sec.

(14) BC 194 Nov 4/5 (mag. = 0.92)

The observations of this eclipse are preserved in more than one text:

(i) [SE 118, month VIII.] Night of the 14th moonrise to sunset: 6 deg, measured (despite) clouds [...lunar eclipse...when ...culmina]ted; when it began on the east side, in 12 deg night [it made...] it entered a cloud. In its eclipse, Jupiter, Venus, Mars and Sirius stood there; the rem[ainder of the planets did not stand there...] [...] it was eclipsed 1 cubit [...] α Tau; (began) at 12 deg before sunrise.

[BM 35331 (= LBAT 324), Obv. 17′–19′; trans. SH II, p. 279; see also Huber, p. 58.]

Although the date of this astronomical diary is broken away, it has been restored using the numerous planetary and lunar observations which are cited (SH II, 285). When the eclipse occurred, the Moon would be rather less than 2 deg to the west of α Tau, in close accord with the record. The computed interval between moonrise and sunset on Nov 4 is 5.0 deg.

(ii) SE 118, month VIII 14, when 'the 4 of (the Lion's) breast' (α, γ, η, ζ Leo) culminated, (lunar eclipse), beginning on the east side. In 12 deg of night it made two-thirds of the disc [...] 1,52 (?)...

[BM 34236 (= LBAT 1436), Obv. 4–6; trans. Huber, p. 58.]

This lunar eclipse text contains only two records: a prediction for the 13th day of month II and the above observation.

RESULTS

First contact on Nov 5 at 0.80 h before sunrise. LT of sunrise = 6.57 h, hence LT of first contact = 5.77 h, UT = 2.59 h. Computed TT = 6.38 h, thus ΔT = 13 650 sec.

(15) BC 160 Jan 26/27 (mag. = 1.45)

> [SE 151...] king Demetrius... [month IX] when Gemini culminated, [lunar eclipse...] the north? wind blew; in its eclipse clouds were in the sky [...] α Leo it was eclipsed; (began) at 48 deg after sunset.
>
> [BM 46003 (= LBAT 385), Obv. 3′–5′; trans. SH III, p. 41.]

The date of this astronomical diary has been established by calculation using the numerous planetary and lunar observations which it contains (SH III, p. 41). The eclipse actually took place in the 2nd year of king Demetrius I.

RESULTS

First contact on Jan 26 at 3.20 h after sunset. LT of sunset = 17.21 h, hence LT of first contact = 20.41 h, UT = 17.73 h. Computed TT = 21.64 h, thus ΔT = 14 050 sec.

(16) BC 143 Feb 17/18 (mag. = 0.88)

> [SE 168], month XI 13 moonrise to sunset: 3 deg, measured. 5 deg after β Aur culminated, lunar eclipse, beginning on the south-east side. In 20 deg night it made 9 fingers. 5 deg duration of maximal phase. (Began) at 7 deg after sunset.
>
> [BM 35787 (= LBAT 1278), Rev. 3′–8′; trans. Huber, p. 68.]

Although the year in which this eclipse occurred is no longer preserved, the event is reported in a goal-year text for SE 186 (126–125 BC). The computed interval between moonrise and sunset on Feb 17 is 2.3 deg. A photograph of BM 35787 and the joining fragment BM 34658 is shown in figures 6.3 (obverse) and 6.4 (reverse).

RESULTS

First contact on Feb 18 at 0.47 h after sunset. LT of sunset = 17.51 h, hence LT of first contact = 17.98 h, UT = 15.31 h. Computed TT = 18.80 h, thus ΔT = 12 550 sec.

(17) BC 134 Mar 9/10 (mag. = 0.26)

> SE 177, month XII [...] when α Lyr culminated, lunar eclipse; when it began on the east side, it covered 2 fingers in 9 deg of night [...] [behind] γ Vir it became eclipsed; (began) at 9 deg before sunrise. The 13th, sunrise to moonset: 1 deg, measured (despite) clouds and mist; the north wind blew.
>
> [BM 34669 (= LBAT 433), Rev. 16′–17′; trans. SH III, p. 197.]

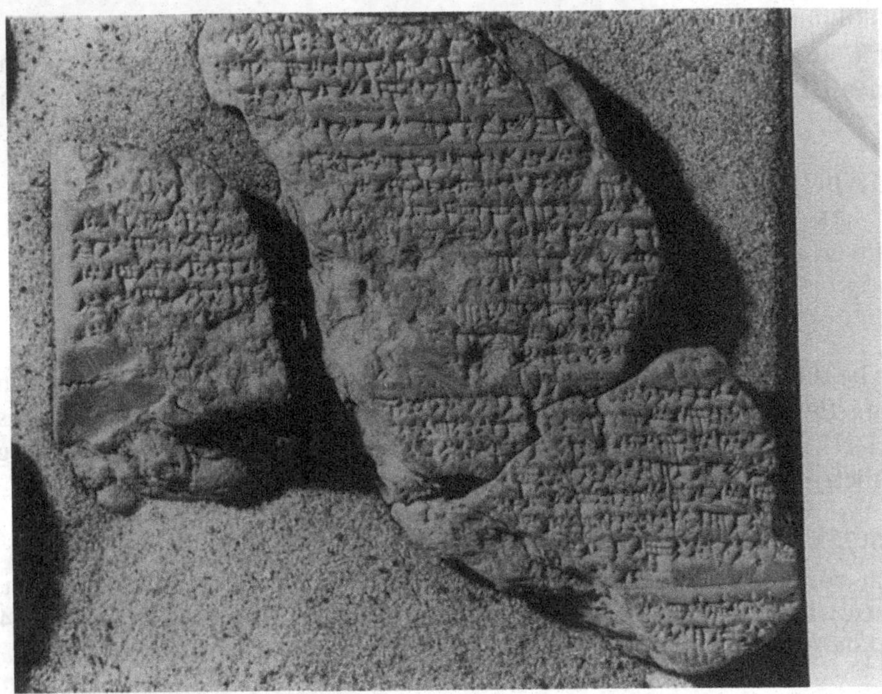

Fig. 6.3 Photograph of BM 35787 and the joining fragment BM 34658 – obverse. (Courtesy: British Museum.)

This tablet is an astronomical diary. Its year (177 SE), during the reign of King Arsaces, is preserved in two places. The computed interval between sunrise and moonset on Mar 10 is 0.3 deg.

RESULTS

First contact on Mar 10 at 0.60 h before sunrise. LT of sunrise = 6.16 h, hence LT of first contact = 5.56 h, UT = 2.81 h. Computed TT = 5.87 h, thus ΔT = 10 950 sec.

(18) BC 109 May 1/2 (mag. = 0.52)

SE 203, Arsaces, king of kings [month I]. Night of the 15th, moonrise to sunset: 3 deg, measured (despite) mist. 8 deg after sunset, lunar eclipse on the south side [...] when it began, it made 6 fingers in 8 deg of night [...] maximal phase and clearing; its eclipse was red, its redness was red brown, the east wind blew; in its eclipse, Jupiter, Venus, Saturn [and Mars stood there ...].

[BM 40622 (= LBAT 469), Obv. 7′–8′ + BM 45646 (+ LBAT 466), Obv 2′; trans SH III, p. 355.]

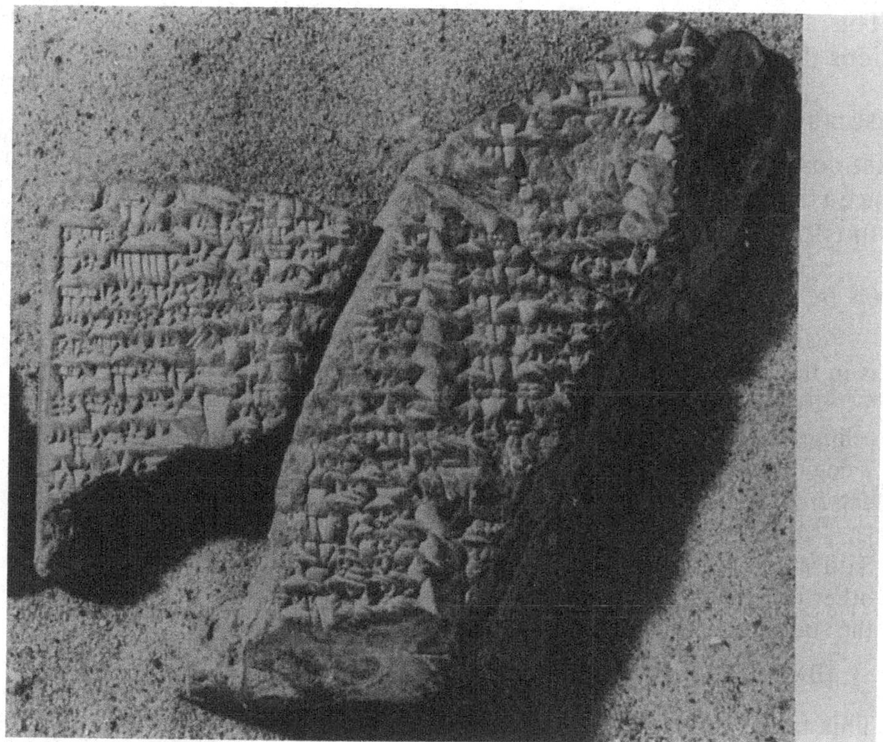

Fig. 6.4 Photograph of BM 35787 and the joining fragment BM 34658 – reverse. (Courtesy: British Museum.)

The computed interval between moonrise and sunset on May 1 is 2.5 deg.

The red colour of the Moon was presumably due to a local meteorological phenomenon (e.g. a dust storm). Unless an eclipse is almost total, reddening of the Moon caused by dispersion of sunlight in the Earth's atmosphere before it reaches the lunar surface is relatively insignificant.

RESULTS

First contact on May 1 at 0.53 h after sunset. LT of sunset = 18.69 h, hence LT of first contact = 19.22 h, UT = 16.17 h. Computed TT = 19.53 h, thus ΔT = 12 050 sec.

(19) BC 96 Aug 3/4 (mag. = 0.71)

[SE 216, month V...] when it began to clear, in 20 deg night [it cleared] from south to north and we[st ...] [...the remainder of the planet]s did not stand there; in Aquarius it became eclipsed; (began) at 57 deg after sunset. The 13th [...]

[BM 45847 (= LBAT 492), 7′–8′; trans. SH III, p. 425; see also Huber, p. 73.]

This astronomical diary has been dated by planetary and lunar calculations (SH III, p. 425).

RESULTS

First contact on Aug 3 at 3.80 h after sunset. LT of sunset = 18.91 h, hence LT of first contact = 22.71 h, UT = 19.81 h. Computed TT = 23.47 h, thus ΔT = 13 150 sec.

(20) BC 80 Apr 10/11 (mag. = 0.60)

> Year 168 (Arsacid), that is year 232 (SE), Arsaces, king of kings, which is in the time of king Orodes (I). Month I 13 moonrise to sunset: 5,50 (= $5\frac{5}{6}$) deg, cloudy(?), measured(?). 5 deg before μ Her culminated, lunar eclipse, beginning on the south-east side. In 20 deg night it made 6 fingers. 7 deg duration of maximal phase, until it began to become bright. In 13 deg from south-east to north-west, 4 fingers lacking to brightness, it set [...] During this eclipse, north wind (?) went. During this eclipse, Jupiter, Saturn and Mars stood there. When becoming bright, Saturn set. The other planets did not stand there. $\frac{2}{3}$ cubits behind α Lib, 6 fingers towards the south beneath. (Began) at 40 deg before sunrise.
>
> [BM 33562A (= LBAT 1445), Obv. and Rev.; trans. Huber, pp. 75–76.]

This tablet is devoted to this eclipse. Years are reckoned both from the Arsacid era (247 BC) and the Seleucid era. Orodes I, a Parthian king, began to reign in 80 BC. When the eclipse occurred, the Moon would be about 0.6 deg to the east of α Lib and about 2 deg to the south of it – in fair (but not good) accord with the record. The computed interval between moonrise and sunset on Apr 10 is 7.4 deg.

RESULTS

First contact on Apr 11 at 2.67 h before sunrise. LT of sunrise = 5.62 h, hence LT of first contact = 2.95 h, UT = 0.00 h. Computed TT = 3.19 h, thus ΔT = 11 450 sec.

(21) BC 80 Oct 5/6 (mag. = 0.39)

> [...] year 232 (SE) [king Arsa]ces [...] which is in the time of king Orodes (I) and Is [...his wife] [...month VII...] (Began) at 30 deg after sunset.
>
> [BM 42073 (= LBAT 1446), Obv. and Rev.; trans. Huber, p. 76.]

This badly damaged tablet is concerned specifically with this particular eclipse. There were only two lunar eclipses in the year 232 SE. The earlier of these two events, which occurred in month I, has already been discussed in the previous entry. In view of the differing details, the present record must relate to the later eclipse.

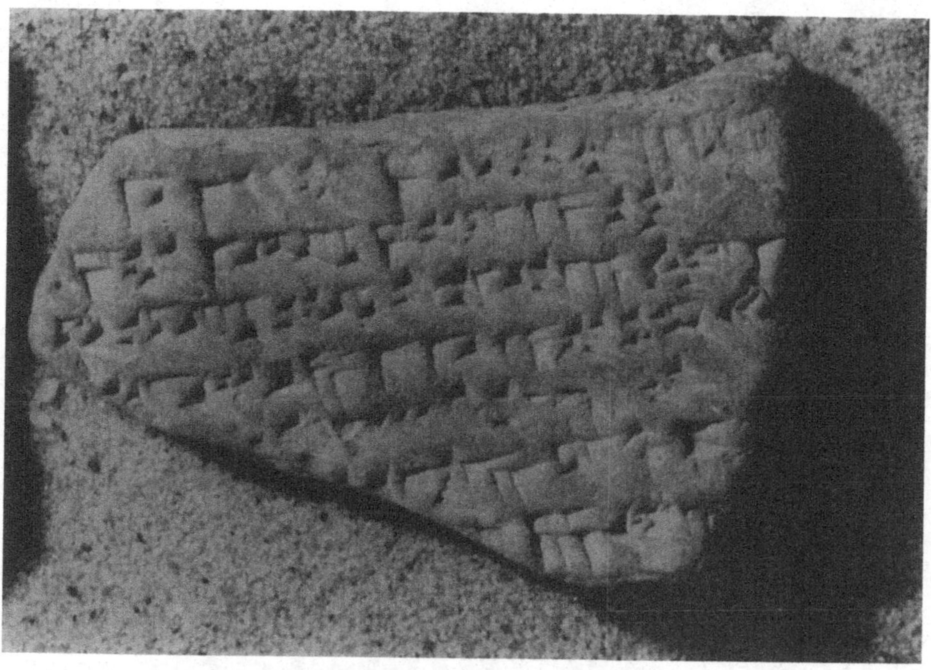

Fig. 6.5 Photograph of BM 41565 – obverse. (Courtesy: British Museum.)

RESULTS

First contact on Oct 5 at 2.00 h after sunset. LT of sunset = 17.90 h, hence LT of first contact = 19.90 h, UT = 16.78 h. Computed TT = 20.13 h, thus ΔT – 12 050 sec.

(22) BC 67 Jan 19/20 (mag. = 0.81)

> Year 180 (Arsacid), that is year 244 (SE), king Arsaces (IV) and Pir'ustana, his wife, the lady, month X 15. Moonrise to sunset: 1 deg, mist, measured. As the Moon rose, two-thirds of the disk on the north-east side were eclipsed. 6 deg night duration of maximal phase, until it began to become bright. In 16 deg night from south-east to north-west it became bright. 23 deg total duration. During this eclipse the sky was overcast. During this eclipse, north (wind) went. During this eclipse, Venus, Saturn and Sirius stood there, the other planets did not stand there. 1 1/2 cubits in front of α Leo. (Began) at 16 deg before sunset.
>
> [BM 45628 (= LBAT 1448) + Dupl. BM 41565 (= LBAT 1447), Obv. and Rev; trans. Huber, pp. 77–78.]

These two tablets are duplicates of one another. Each records only this particular eclipse. A photograph of the surviving fragment of BM 41565 (obverse) is shown in figure 6.5. It is evident that the recorded time of first contact (i.e. 16 deg before sunset) represents a prediction. When the

Moon rose it was said to be already largely obscured. The eclipse appears to have ended 22 or 23 deg after moonrise. The discrepancy of 1 deg in the arithmetic (i.e. 6 + 16 = 22 rather than 23) is presumably due to a scribal error. A measured time-interval of 22.5 deg (1.50 h) after moonrise will be assumed for last contact. When the eclipse occurred, the Moon was 2 deg to the west of α Leo, which accords well with the text. The computed interval between moonrise and sunset on Jan 19 is 0.7 deg.

RESULTS

Last contact on Jan 19 at 1.50 h after moonrise. LT of moonrise = 17.10 h, hence LT of last contact = 18.60 h, UT = 15.89 h. Computed TT = 18.71 h, thus ΔT = 10 150 sec.

The various results obtained above are summarised in table 6.1.

6.2.2 Other less accurate measurements (to the nearest 5 or 10 degrees)

(23) BC 695 Apr 30/May 1 (mag. = 0.74)

[Assur-nadin-sum], year 5, month II 1[0+x... not] complete, it set eclipsed. (Began) at 30 deg before sunrise.

[BM 32238 (=LBAT 1414), Obv. III; trans. Huber, p. 4.]

RESULTS

First contact on May 1 at 2.00 h before sunrise. LT of sunrise = 5.40 h, hence LT of first contact = 3.40 h, UT = 0.36 h. Computed TT = 5.01 h, thus ΔT = 16 750 sec.

(24) BC 686 Apr 21/22 (mag. = 0.55)

[Se]nnacherib [year 3], month II(?)... 5] months [...] beginning on the [north-] east side. $\frac{2}{3}$ of the disk [...] west (wind) went. (Began at) 1,40 (= 100) deg after sunset. Month VI_2.

[BM 35789 (= LBAT 1417), Obv. I + BM 35115 (= LBAT 1416), Obv. I, top; trans. Huber, p. 5.]

Sennacherib, who became ruler of Assyria in 704 BC, assumed the ancient title of 'King of Sumer and Akkad' in 688 BC following his capture of the city of Babylon (Oates, 1986, pp. 119–20). The VI_2 at the end of the above entry identifies the intercalary month in that year; this occurred several months after the eclipse.

RESULTS

First contact on Apr 22 at 6.67 h after sunset. LT of sunset = 18.46 h, hence LT of first contact = 1.13 h, UT = 22.14 h (Apr 21). Computed TT = 4.40 h (Apr 22), thus ΔT = 22 500 sec.

Table 6.1 ΔT results from Babylonian timings of lunar eclipses estimated to the nearest degree (single contacts only).

Year[a]	Ct	SR/SS[b]	Int (deg)[c]	ΔT (sec)[d]
−665	1	SS	+3	21 050
−561	1	SS	+90	15 950
−536	1	SR	−14	18 800
−482	1	SR	−10	17 300
−420	1	SS	+19	15 500
−405	1	SR	−14	16 500
−370	1	SS	+66	12 750
−366	1	SR	−56	19 550
−316	1	SS	+10	15 550
−307	1	SR	−10	14 100
−239	1	SR	−3	14 200
−211a	1	SR	−20	11 800
−211b	1	SS	+27.5	21 350
−193	1	SR	−12	13 650
−159	1	SS	+48	14 050
−142	1	SS	+7	12 550
−133	1	SR	−9	10 950
−108	1	SS	+8	12 050
−95	1	SS	+57	13 150
−79a	1	SR	−40	11 450
−79b	1	SS	+30	12 050
−66	4	MR	+22.5	10 150

[a] The year of the eclipse on the Julian calendar, using negative integers.
[b] Whether the eclipse was timed relative to sunrise (SR), sunset (SS), or moonrise (MR) – the last instance only in the year −66.
[c] The measured interval in degrees (positive if after sunset or moonrise, negative if before sunrise).
[d] The value of ΔT (in seconds) which satisfies the observation.
NB −211a refers to the first eclipse in the year −211 (Apr 29/30), etc.

(25) BC 685 Oct 3/4 (mag. = 1.83)

[Sennacherib, year 4, month VI(?)...] 6(?) deg [...] middle, Aries [...] south. (Began) at 20 deg after sunset.

[BM 45640 (= LBAT 1415), Obv. II, top; trans. Huber, p. 6.]

RESULTS

First contact on Oct 3 at 1.33 h after sunset. LT of sunset = 18.01 h, hence LT of first contact = 19.34 h, UT = 16.27 h. Computed TT = 21.55 h, thus ΔT = 19 000 sec.

(26) BC 667 Oct 14/15 (mag. = 1.86)

> [Samas-sum-ukin, year 1, month VII(?)...] It set eclipsed. (Began) at 20[+x] deg before sunrise.
>
> [BM 45640 (= LBAT 1415), Obv. III, top; trans. Huber, pp. 7–8.]

The immediately following entry on the same tablet (Obv. III, bottom) records an eclipse in month I of the second year of the same ruler – see entry (1) above. Hence, although the month is not specified in the text under discussion, it should be either 5 or 6 months previously and thus in month VII or VIII. Only the eclipse of BC 667 Oct 14/15 can fit this date.

According to a personal communication from Walker, who has examined the original tablet on my behalf, the 20 in 20[+x] is clear and almost certainly not a damaged 30. He is in doubt whether the sign between the 20 and GE_6 *ana* ZALAG ('before sunrise') represents a numeral or not. I have taken 25 deg as a fair average.

RESULTS

First contact on Oct 15 at 1.67 h before sunrise. LT of sunrise = 6.18 h, hence LT of first contact = 4.51 h, UT = 1.39 h. Computed TT = 6.19 h, thus ΔT = 17 250 sec.

NB if the text had actually read 20 deg before sunrise, the corresponding value of ΔT would have been 16 050 sec.

(27) BC 601 Apr 10/11 (mag. = 0.84)

> [Nebuchadrezzar II, year 4] month I, 13, middle watch, 3 *beru*, 5 US (= 95 deg) after sunset [...] north-east side, 3 (fingers) lacked to totality [...] on north it became bright. North (wind) went.
>
> [BM 38462 (= LBAT 1420), Obv. I, 8–10; trans. Huber, p. 13.]

The three watches into which the night was divided – USAN ('first part of the night'), $MURUB_4$ ('middle part') and ZALAG ('last part') – are only mentioned in fairly early texts.

RESULTS

First contact on Apr 11 at 6.33 h after sunset. LT of sunset = 18.31 h, hence LT of first contact = 0.64 h, UT = 21.71 h (Apr 10). Computed TT = 2.72 h (Apr 11), thus ΔT = 18 000 sec.

(28) BC 599 Feb 19/20 (mag. = 0.75)

[Nebuchadrezzar II, year 5], month VI_2 passed. [Month XI] (after) 5 months, $3\frac{1}{2}$ *beru* (= 105 deg) after sunset [...] $\frac{2}{3}$ covered, on the south covered [...during] this eclipse [...].

[BM 38462 (= LBAT 1420), Obv. I, 12–15; trans. Huber, p. 13.]

The true lunar month (XI) can be readily restored since it was said to be 5 months after a predicted eclipse in the intercalary 6th month.

RESULTS

First contact on Feb 20 at 7.00 h after sunset. LT of sunset = 17.49 h, hence LT of first contact = 0.48 h, UT = 21.83 h (Feb 19). Computed TT = 1.83 h (Feb 20), thus ΔT = 14 400 sec.

(29) BC 587 Jan 7/8 (mag. = 1.80)

[Nebuchadrezzar II, year] 17, month IV (eclipse) passed. [Month X] 13, morning watch(?) 1 *beru* 5 U[S] (= 35 deg) [before sunrise...] it set totally [eclipsed].

[BM 38462 (= LBAT 1420), Obv. II, 16–18; trans. Huber, p. 17.]

Although the interval to which the measurement of 35 deg refers is broken away, it can confidently be restored as 'before sunrise'. The characteristic style of BM 38462 is to cite the time interval measured after sunset *or* before sunrise near the start of a record. Since the Moon set eclipsed, it is clear that the 35 deg cannot have been measured after sunset.

RESULTS

First contact on Jan 8 at 2.33 h before sunrise. LT of sunrise = 6.97 h, hence LT of first contact = 4.64 h, UT = 1.86 h. Computed TT = 7.10 h, thus ΔT = 18 850 sec.

(30) BC 580 Aug 14/15 (mag. = 1.82)

[Nebuchadrezzar II] year 25, month V, $1\frac{1}{2}$ *beru* (= 45 deg) after sunset.

[BM 38462 (= LBAT 1420), Rev. I, 3′; trans. Huber, p. 18.]

This very brief record is intact. Evidently the scribe who compiled the eclipse table could obtain no information on the times of the later phases. There is no reference to the eclipse 'passing by', which suggests that it was observed.

RESULTS

First contact on Aug 14 at 3.00 h after sunset. LT of sunset = 18.81 h, hence LT of first contact = 21.81 h, UT = 18.89 h. Computed TT = 0.19 h (Aug 15), thus ΔT = 19 050 sec.

(31) BC 573 Apr 1/2 (mag. = 1.73)

[Nebuchadrezzar II, year 31, month XII...] South (wind) went (?) 1 (?) cubit in front of Libra eclipsed. Saturn in Capricorn... Mars 2 cubits in front of α Sco. (Began) at 1,30 (= 90) deg after sunset.

[BM 32234 (= LBAT 1419), Obv. III, middle; trans. Huber, p. 21.]

The immediately following entry in the same column of this text specifically mentions Nebuchadrezzar in recording an eclipse which 'passed by' in month VI of his 32nd year (i.e. BC 573 Sep 26). On the night of Apr 1/2 in BC 573, Mars was about 5 deg (or roughly 2 cubits) distant from α Sco but it was located to the *north* of this star rather than west of it (as implied in the text).

RESULTS

First contact on Apr 2 at 6.00 h after sunset. LT of sunset = 18.16 h, hence LT of first contact = 0.16 h, UT = 21.28 h (Apr 1). Computed TT = 2.30 h (Apr 2), thus ΔT = 18 050 sec.

(32) BC 523 Jul 16/17 (mag. = 0.54)

Year 7 (Kambyses), month IV, night 14, $1\frac{2}{3}$ *beru* (= 50 deg) after sunset, the Moon makes a total eclipse, (but) a little is left over; north (wind) went.

[BM 33066 (= LBAT 1477), Rev. 19–20; trans. Huber, p. 25.]

Both this and the following entry are recorded on the same tablet from the reign of king Kambyses II (529–522 BC). The terminology of the entry ('the Moon makes a total eclipse, (but) a little is left over') is most unusual. The eclipse described above is also reported in the *Almagest* (see chapter 4), but with differing details. There it is stated that first contact took place 1 hour before midnight and that the Moon was eclipsed by half its diameter. As sunset occurred at 19.11 h, the local time of first contact as recorded on the cuneiform tablet would actually correspond to 1.56 h before midnight, while an almost total eclipse is (incorrectly) implied. Regardless of which measurement of time is the more accurate, the *Almagest* provides a sound estimate of magnitude.

As the recorded magnitude on the cuneiform text is considerably greater than the true value, Huber was of the opinion that this record represents a prediction, although the text mentions wind and the phrase *sin* AN KU normally relates to an observation. He further remarked that the obverse of the same tablet certainly contains predictions. Because of these uncertainties, the record will not be analysed for ΔT. Similar remarks apply to the immediately following report on BM 33066 (entry 33 below).

Table 6.2 ΔT results from Babylonian timings of lunar eclipses estimated to the nearest 5 or 10 deg (all single contacts).

Year	Ct	SR/SS	Int. (deg)	ΔT (sec)
−694	1	SR	−30	16750
−685	1	SS	+100	22500
−684	1	SS	+20	19000
−666	1	SR	−25	17250
−600	1	SS	+95	18000
−598	1	SS	+105	14400
−586	1	SR	−35	18850
−579	1	SS	+45	19050
−572	1	SS	+90	18050

(33) BC 522 Jan 9/10 (mag. = 1.85)

Year 7 (Kambyses), month X, night 14, $2\frac{1}{2}$ *beru* (= 75 deg) to sunrise are left over, the Moon makes a total eclipse. South and north, clouded, went.

[BM 33066 (= LBAT 1477), Rev. 21–22; trans. Huber, pp. 25–26.]

Totality is correctly described in this brief account, but for the reasons given in the previous entry the record will not be considered further.

Table 6.2 gives a summary of the results obtained in this section. The form of this table is similar to that of table 6.1.

The ΔT values listed in tables 6.1 and 6.2 are plotted in figure 6.6. This diagram provides clear evidence of a gradual decline in ΔT down the centuries, and with a single exception (the high value in −211), the results form a remarkably self-consistent set.

6.3 Eclipses for which two contact measurements relative to sunrise or sunset are extant

Most of the observations discussed in this section are of partial lunar eclipses, the times of both first and last contact being preserved. However, also considered here are two total eclipses in 189 and 120 BC. On the former occasion the Moon set whilst eclipsed and only the first two contacts were timed. Although the eclipse of 120 BC is computed to have been just total (magnitude 1.02), it was actually described as incomplete at maximal phase. Two observations, dated by calculation as BC 239 Apr 28/29 (SH II, p. 85) and BC 154 Mar 21/22 (Huber, pp. 66–67) have been rejected. In each case, the eclipse is reported on a tablet specifically

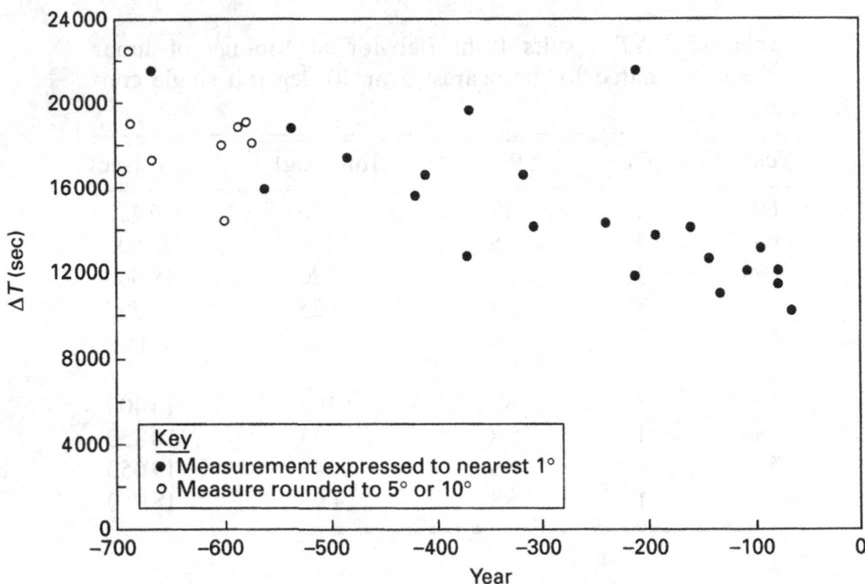

Fig. 6.6 ΔT values obtained from timed lunar eclipses – single contact only.

devoted to that event but the date is broken away. Although the derived dates are probably sound, there is very little independent information in either text to assist dating; without pre-supposing a value for ΔT, alternative identifications cannot be ruled out.

I have included in this investigation two eclipses of very small magnitude (close to 0.10). These date from 397 and 163 BC. Timings of both first and last contact are available for each event. As noted above (section 6.1), definition of the individual contacts seems likely to have been poor. This may explain why the duration of the eclipse of 397 BC – estimated by the Babylonian astronomers as 27 deg – contrasts with the computed figure of only 16 deg. Yet in 163 BC there was remarkably close agreement between observation (20 deg) and theory (21 deg) for this parameter. It is perhaps best to use only the mean result for both contacts in each case. This will be identified by the letter m in the summary table (6.3) at the end of this section.

(1) BC 424 Sep 28/29 (mag. = 0.93)

[Artaxerxes I, year 41], month VI 14, 50 deg after sunset, beginning on the north-east. After 22 deg, 2 fingers lacked to totality. 5 deg duration of maximal phase. In 23 deg towards [...] 50 deg total duration. The sky was overcast... 3 cubits below α + β Ari.

[BM 34787 (= LBAT 1426), col. II, 2′–3′; trans. Huber, pp. 32–33.]

Direct consultation of the original tablet by Walker (personal communication) shows that the 50 deg (questioned by Huber) is very clear. When the eclipse occurred, the Moon was about 10 deg to the south of α and β Ari, in fair agreement with the record.

RESULTS

(i) First contact on Sep 28 at 3.33 h after sunset. LT of sunset = 18.07 h, hence LT of contact = 21.40 h, UT = 18.33 h. Computed TT = 22.98 h, thus ΔT = 16 700 sec.

(ii) Last contact on Sep 29 at 6.67 h after sunset. LT of contact = 0.74 h, UT = 21.66 h (Sep 28). Computed TT = 2.40 h (Sep 29), thus ΔT = 17 050 sec.

(2) BC 408 Oct 31/Nov 1 (mag. = 0.18)

> [Darius II, year 16], month VIII 15, 15 deg after sunset, beginning on the south-east. A quarter of the disk was eclipsed [...] 27 deg total duration... 1 cubit in front of Taurus eclipsed [...] and Mars stood there [...].
>
> [BM 34684 (= LBAT 1427), Obv. 10′–13′; trans. Huber, pp. 38–39.]

When the eclipse occurred, the Moon was about 2.5 deg to the west of α Tau (the principal star of Taurus). This is in good accord with observation.

RESULTS

(i) First contact on Oct 31 at 1.00 h after sunset. LT of sunset = 17.51 h, hence LT of contact = 18.51 h, UT = 15.34 h. Computed TT = 19.57 h, thus ΔT = 15 250 sec.

(ii) Last contact at 2.80 h after sunset. LT of contact = 20.31 h, UT = 17.14 h. Computed TT = 21.11 h, thus ΔT = 14 300 sec.

(3) BC 397 Apr 5/6 (mag. = 0.09)

> [Artaxerxes II, year 7], month XII_2, 14. Beginning on the south side, a quarter of the disk covered. It became bright towards the west. 27 deg total duration. The sky was overcast, south (wind) went. (Began) at 48 deg after sunset.
>
> [BM 35115 (= LBAT 1416), Rev. II, middle; trans. Huber, p. 41.]

RESULTS

(i) First contact on Apr 5 at 3.20 h after sunset. LT of sunset = 18.25 h, hence LT of contact = 21.45 h, UT = 18.54 h. Computed TT = 22.84 h, thus ΔT = 15 450 sec.

(ii) Last contact at 5.00 h after sunset. LT of contact = 23.25 h, UT = 20.34 h. Computed TT = 23.92 h, thus ΔT = 12 850 sec.

Since the magnitude was so small, only the mean ΔT value of 14 150 sec will be retained in further analysis.

(4) BC 189 Feb 16/17 (mag. = 1.28)

> [Year 122 (SE)] month XI 14. Moonrise to sunset 9 deg, mist, measured. Lunar eclipse beginning on the east side. After 16 deg of night totally covered. It set eclipsed. (Began) at 34 deg before sunrise.
>
> [BM 34519 (= LBAT 1251), Rev. 23–26; trans. Huber, p. 61.]

This tablet is a goal-year text for SE 140 (172–171 BC). Following the usual practice on this type of tablet, the eclipse observation was made 18 years prior to this goal-year. The computed interval between moonrise and sunset on Feb 16 (8.2 deg) is close to the recorded figure.

RESULTS

(i) First contact on Feb 17 at 2.27 h before sunrise. LT of sunrise = 6.51 h, hence LT of contact = 4.24 h, UT = 1.58 h. Computed TT = 4.57 h, thus ΔT = 10 750 sec.

(ii) Second contact at 1.20 h before sunrise. LT of contact = 5.31 h, UT = 2.65 h. Computed TT = 5.78 h, thus ΔT = 11 250 sec.

(5) BC 163 Mar 30/31 (mag. = 0.12)

The observations of this eclipse are preserved in three separate texts. Each tablet is damaged but the surviving accounts supplement one another.

> (i) [Year 148 (SE) king Antiochus IV/V, month XII_2 15...], cloudy, measured. During the whole night cirrus clouds (?). 3 deg after υ Boo culminated, [...] From south to west it became bright. 20 deg total duration. During this eclipse the sky was overcast.... During this eclipse north wind... [...] before sunrise. On the 15th sunrise to moonset 5,30 (= $5\frac{1}{2}$) deg.
>
> [Composite translation of BM 41628 (= LBAT 378), Rev. 12′–13′ and BM 41462 (= LBAT 380), Rev. 20′–21′; trans. Huber, p. 64.]

Both BM 41628 and BM 41462 are astronomical diaries. Their dates, which prove to be identical, have been independently derived by calculation using the planetary and lunar observations which they contain (SS, p. xvi). They thus represent separate copies of the same diary. The computed interval between sunrise and moonset on Mar 31 is 4.1 deg.

> (ii) Year 14[8 (SE) king Antiochus IV/V], month XII_2 [15]. 3 deg after υ Boo culminated, lunar eclipse, beginning on the south side. In 10 deg of night it made 3 fingers. (Began) at 85 deg before sunrise.
>
> [BM 34037 (= LBAT 1264), Rev. 3′–9′; trans. Huber, pp. 63–64.]

This tablet is a goal-year text for SE 167 (145–144 BC). The eclipse observation was made 18 years before this goal-year.

NB Antiochus IV died a few months before the eclipse, in the same year (month IX of SE 148). This is why Huber names both Antiochus IV and his son Antiochus V in restoring the text.

RESULTS

(i) First contact on Mar 31 at 5.67 h before sunrise. LT of sunrise = 5.81 h, hence LT of contact = 0.14 h, UT = 21.27 h (Mar 30). Computed TT = 23.94 h (Mar 30), thus ΔT = 9600 sec.

(ii) Last contact at 4.33 h before sunrise. LT of contact = 1.48 h, UT = 22.61 h (Mar 30). Computed TT = 1.32 h (Mar 31), thus ΔT = 9750 sec.

Since the magnitude was so small, only the mean ΔT value of 9700 sec will be retained in subsequent analysis.

(6) BC 129 Nov 4/5 (mag. = 0.63)

> Year 120 (Arsacid), that is year [183] (SE), king Arsaces, month VIII 13. Moonrise to sunset 4,30 (= $4\frac{1}{2}$) deg, measured [...] after (α, β) Gem culminated, lunar eclipse beginning on the [north-east] side. [In 18 deg] night [it made ... fingers] 6 deg duration of maximal phase [...] In 16 deg from north(?) [...]. 40 deg total duration [...] distant; in the beginning of a halo [...]. During this eclipse, Sirius [stood] there. The other planets did not [stand there]. 2 cubits above α Tau [...]. (Began) at 55 deg before sunrise. On the 13th, sunrise to moonset 3,30 (= $3\frac{1}{2}$) deg, cloudy, not observed.
>
> [BM 33982 (= LBAT 1441), Obv. and Rev.; trans. Huber, pp. 70–71.]

This tablet is devoted to the record of a single eclipse. Huber notes an error of one year in the date relative to the Arsacid era. He proposes year 119 rather than 120. There was no eclipse in month VIII of the latter year, and although one did occur in month VII – i.e. BC 128 Oct 25 – the calculated circumstances do not accord with observation. For instance, the computed duration of maximal phase (totality) would be fully 25 deg; this is much greater than the 6 deg of the text. Further, using a preliminary value for ΔT of 11 900 sec (from equation (4.1)) – a figure supported by other roughly contemporary observations discussed in this chapter – this eclipse would occur around noon at Babylon. It would thus be entirely invisible from this site. At greatest phase, the Moon would be about 11 deg to the west of α Tau (rather than 2 cubits *above* this star), while the reported intervals between moonrise and sunset and sunrise to moonset are considerably at variance with theory.

On the contrary, the eclipse of BC 129 Nov 4/5 occurred in the correct lunar month. At the time, the Moon would be about 1 deg to the west of α Tau and some 6 deg to the north of this star – in good accord with observation. Further, at the measured time of 55 deg before sunrise (LT = 2.92 h), all of the planets (but not the star Sirius) would indeed be below the horizon. The computed interval between moonrise and sunset on Nov 4 was 6.3 deg, while that between sunrise and moonset on Nov 5 was 2.9 deg. Both of these are in fairly good agreement with the record. For these various reasons, the date proposed by Huber would seem to be satisfactory.

Results

(i) First contact on Nov 5 at 3.67 h before sunrise: LT of sunrise = 6.58 h, hence LT of contact = 2.92 h, UT = 23.74 h (Nov 4). Computed TT = 3.04 h (Nov 5), thus ΔT = 11 850 sec.

(ii) Last contact at 1.00 h before sunrise. LT of contact = 5.58 h, UT = 2.41 h. Computed TT = 5.91 h, thus ΔT = 12 600 sec.

(7) BC 120 Jun 1/2 (mag. = 1.02)

> Year 128 (Arsacid), that is year 192 (SE), king Arsaces. Month II 14. Moonrise to sunset 6,30 (= $6\frac{1}{2}$) deg, measured. 5 deg after α Her culminated, lunar eclipse, beginning on the north-east side. After 24 deg, 1 UD lacked to totality. 6 deg duration of maximal phase, until it began to become bright. In 24 deg, it became bright from north-east to south-west. 54(?) deg total duration...(Began) at 66 deg after sunset.
>
> [BM 45845 (= LBAT 1442), Obv. and Rev.; trans. Huber, pp. 71–72.]

This tablet records no more than the above eclipse. Although the eclipse is calculated to have been just total, the record implies that a portion of the Moon ('1 UD': probably a small fraction) remained unobscured. The interval from start to finish is damaged but this can readily be restored from the sum of the durations of the individual phases (i.e. 54 deg). The computed interval between moonrise and sunset on Jun 1 is 5.5 deg.

Results

(i) First contact on Jun 1 at 4.40 h after sunset. LT of sunset = 19.06 h, hence LT of contact = 23.46 h, UT = 20.38 h. Computed TT = 23.86 h, thus ΔT = 12 500 sec.

(ii) Last contact on Jun 2 at 8.00 h after sunset. LT of contact = 3.06 h, UT = 23.98 h (Jun 1). Computed TT = 3.01 h (Jun 2), thus ΔT = 10 900 sec.

The results derived in this section are summarised in table 6.3.

Table 6.3 ΔT results from Babylonian timings of lunar eclipses for which two contacts were observed.

Year[a]	Ct[b]	SR/SS[c]	Int (deg)[d]	ΔT (sec)[e]
−423	1	SS	+50	16700
−423	4	SS	+100	17050
−407	1	SS	+15	15250
−407	4	SS	+42	14300
−396	m	SS	+61.5	14150
−188	1	SR	−34	10750
−188	2	SR	−18	11250
−162	m	SR	−75	9700
−128	1	SR	−55	11850
−128	4	SR	−15	12600
−119	1	SS	+66	12500
−119	4	SS	+120	10900

[a] The year of the eclipse on the Julian calendar, using negative integers.
[b] The contact; for the very small eclipses of −396 and −162, mid-eclipse (m) is used.
[c] Whether the eclipse was timed relative to sunrise (SR) or sunset (SS).
[d] The measured interval in degrees (positive if after sunset or moonrise, negative if before sunrise).
[e] The value of ΔT (in seconds) which satisfies the observation.

6.4 Three or four timings relative to sunrise or sunset preserved (total eclipses only)

(1) BC 555 Oct 6/7 (mag. = 1.53)

[Nabunaid, year 1], month VII 13. In 17 deg (from the) east side total. 28 deg duration of the maximal phase. In 20 deg from east to west it became bright. This eclipse was clouded (?). Behind α and β Ari eclipsed. Towards becoming bright west (wind) went. (Began) at 55 deg before sunrise.

[BM 32234 (= LBAT 1419), Obv. IV, middle; trans. Huber, p. 23.]

Although the sum of the durations of the individual phases is 65 deg, this eclipse is said to have begun at 55 deg before sunrise. There is clearly a textual error somewhere; otherwise last contact would occur long after moonset. Walker has kindly verified the reading of the various numbers in the text. It will be seen that the measured interval between first and second contact (17 deg) is in fair accord with that between third and fourth contact (20 deg); these figures also agree well with the computed interval of 17 deg. In addition, the recorded duration of totality (28 deg) is close to the calculated figure of 25 deg. Hence it seems likely that the timing of first contact is incorrect. Huber (p. 23) preferred to read 'after sunset' (GE_6 *gin*) for 'before sunrise' (GE_6 *ana* ZALAG). This is not unreasonable, but it is also plausible that the scribe made a numerical error in recording the interval between first contact and sunrise. It thus seems best to reject this record.

(2) BC 501 Nov 7/8 (mag. = 1.47)

> [Darius I, year 21], month VIII. In 15 deg (from) the east total. 25 deg duration of the maximal phase. In 25 deg from east to west it became bright. Clouded (?) Towards becoming bright, south (wind) went. (Began) at 1,17 (= 77) deg after sunset.
>
> [BM 32234 (= LBAT 1419), Rev. II; trans. Huber, p. 27.]

Direct consultation of the original tablet by Walker (personal communication) shows that the first '1' in '1,17 deg' (questioned by Huber) is fairly clear.

RESULTS

(i) First contact on Nov 7 at 5.13 h after sunset. LT of sunset = 17.41 h, hence LT of contact = 22.55 h, UT = 19.39 h. Computed TT = 23.54 h, thus ΔT = 14 950 sec.

(ii) Second contact at 6.13 h after sunset. LT of contact = 23.55 h, UT = 20.39 h. Computed TT = 0.65 h (Nov 8), thus ΔT = 15 300 sec.

(iii) Third contact on Nov 8 at 7.80 h after sunset. LT of contact = 1.22 h, UT = 22.06 h (Nov 7). Computed TT = 2.22 h (Nov 8), thus ΔT = 14 950 sec.

(iv) Fourth contact at 9.47 h after sunset. LT of contact = 2.88 h, UT = 23.72 h (Nov 7). Computed TT = 3.33 h (Nov 8), thus ΔT = 13 000 sec.

(3) BC 407 Oct 20/21 (mag. = 1.38)

> [Darius II, year 17], month VII 14, 48 deg before sunrise, beginning in the east. In '21' deg total, 12(?) deg [duration of maximal phase]. In 15 deg not [complete it set(?)...].
>
> [BM 34684 (= LBAT 1427), Rev. 4–6; trans. Huber, pp. 39–40.]

Examination of the original tablet by Walker (personal communication) indicates that the '21' (questioned by Huber) is clear, but the subsequent interval – '12'(?) – is indeed doubtful. However, Huber's interpretation that the Moon set at 15 deg after the end of totality is probably correct so that the '12' may be restored (i.e. 21 + 12 + 15 = 48). This interval almost certainly relates to the duration of maximal phase, as inferred by Huber.

RESULTS

(i) First contact on Oct 21 at 3.20 h before sunrise. LT of sunrise = 6.31 h, hence LT of contact = 3.11 h, UT = 23.97 h (Oct 20). Computed TT = 4.12 h (Oct 21), thus ΔT = 14 950 sec.

(ii) Second contact at 1.80 h before sunrise. LT of contact = 4.51 h, UT = 1.37 h. Computed TT = 5.23 h, thus ΔT = 13 850 sec.

(iii) Third contact at 1.00 h before sunrise. LT of contact = 5.31 h, UT = 2.17 h. Computed TT = 6.68 h, thus ΔT = 16 200 sec.

(4) BC 378 Apr 6/7 (mag. = 1.32)

> [Artaxerxes II, year 26], month XII_2 15, beginning on the south side. [After] 15 deg total, 21 deg duration of maximal phase. When it began to become bright from the east, [in] 19 deg toward the west it became bright. 45 deg (error for 55 deg ?) total duration. 14 fingers in front of α Lib eclipsed...(Began) at 37 deg [after sunset *or* before sunrise].
>
> [BM 45640 (= LBAT 1415), Rev. III, middle; trans. Huber, p. 45.]

Huber was of the opinion that the sign for 15 deg (i.e the onset of totality) was indistinct. However, inspection of the tablet by Walker (personal communication) reveals that the sign is clear. The sum of the durations of the individual phases is 55 deg rather than the stated 45; the latter thus probably represents a scribal error. Whether the eclipse began 37 deg before sunrise or after sunset is not preserved. However, if this interval was measured relative to sunrise, the end would occur long after moonset and thus be unobservable. Hence it may reasonably be inferred that the time of first contact was measured after sunset. This conclusion is confirmed by calculation of the position of the Moon in relation to α Lib.

RESULTS

(i) First contact on Apr 6 at 2.47 h after sunset. LT of sunset = 18.26 h, hence LT of contact = 20.73 h, UT = 17.82 h. Computed TT = 22.27 h, thus ΔT = 16 000 sec.

(ii) Second contact at 3.47 h after sunset. LT of contact = 21.73 h, UT = 18.82 h. Computed TT = 23.32 h, thus ΔT = 16 200 sec.

(iii) Third contact at 4.87 h after sunset. LT of contact = 23.13 h, UT = 20.22 h. Computed TT = 0.60 h (Apr 7), thus ΔT = 15 750 sec.

(iv) Last contact on Apr 7 at 6.13 h after sunset, hence LT of contact = 0.40 h, UT = 21.49 h (Apr 6). Computed TT = 1.65 h (Apr 7), thus ΔT = 14 950 sec.

(5) BC 353 Nov 21/22 (mag. = 1.35)

[Artaxerxes III, year 6], month VIII 14, beginning on the south-east side. After 23 deg total. 18 deg duration of maximal phase. After 6 deg of night, a quarter of the disk had become bright (?) and it set eclipsed. The eclipse was clouded (?). $1\frac{1}{2}$ cubits behind β Tau eclipsed. During the eclipse, Saturn stood there; the other planets did not stand there... (Began) at 47 deg before sunrise.

[BM 32238 (= LBAT 1414), Rev. III, bottom, 1–9; trans. Huber, pp. 49–50.]

When the eclipse occurred, the Moon would be 4.5 deg to the east of β Tau, which agrees well with the $1\frac{1}{2}$ cubits of the text. Since the Moon set eclipsed, only the times of the first three contacts could be measured by the Babylonian astronomers.

Results

(i) First contact on Nov 22 at 3.13 h before sunrise. LT of of sunrise = 6.79 h, hence LT of contact = 3.66 h, UT = 0.53 h. Computed TT = 4.96 h, thus ΔT = 15950 sec.

(ii) Second contact at 1.60 h before sunrise. LT of contact = 5.19 h, UT = 2.06 h. Computed TT = 6.10 h, thus ΔT = 14550 sec.

(iii) Third contact at 0.40 h before sunrise. LT of contact = 6.39 h, UT = 3.26 h. Computed TT = 7.50 h, thus ΔT = 15250 sec.

(6) BC 317 Dec 13/14 (mag. = 1.34)

[Philip Arrhidaeus, year 7], month IX 15, beginning on the south-east side. After 19(?) deg total. 5 deg duration of maximal phase. In 16 deg on the north-east side it became bright. 40 deg total duration... This eclipse was clouded (?). $1\frac{1}{2}$ cubits in front of β Gem eclipsed. (Began) at 44 deg after sunset. Month IX, year(?) 7 of Philip(?), (the following year is) year 2 of Antigonus, son of [...].

[BM 32238 (= LBAT 1414), Rev. V, bottom, 1–10; trans. Huber, p. 51.]

When the eclipse occurred, the Moon was 3 deg to the west of β Gem, in fair agreement with the record. Although the reading of the interval between first and second contact as 19 deg is doubtful, it may be restored from the recorded duration of 40 deg. It should be noted that the recorded duration of totality is extremely short; for comparison, the calculated figure is about 21 deg. However, the text is quite legible at this point. The results given below are derived on the assumption that the individual measurements are correct.

NB an independent cuneiform text confirms that Antigonus began to rule in 316 BC. However, he is never given the title of king in contemporary documents (Parker and Dubberstein, 1956, p. 20).

Results

(i) First contact on Dec 13 at 2.93 h after sunset. LT of sunset = 17.02 h, hence LT of contact = 19.96 h, UT = 16.98 h. Computed TT = 21.58 h, thus ΔT = 16 550 sec.

(ii) Second contact at 4.20 h after sunset. LT of contact = 21.22 h, UT = 18.25 h. Computed TT = 22.72 h, thus ΔT = 16 100 sec.

(iii) Third contact at 4.53 h after sunset. LT of contact = 21.55 h, UT = 18.58 h. Computed TT = 0.12 h (Dec 14), thus ΔT = 19 950 sec.

(iv) Last contact at 5.60 h after sunset. LT of contact = 22.62 h, UT = 19.65 h. Computed TT = 1.26 h (Dec 14), thus ΔT = 20 200 sec.

Because of the serious error in the duration of maximum phase, the last two values for ΔT are in poor accord with the first two of the same set. However, there do not seem to be sufficient grounds for rejecting these results.

(7) BC 226 Aug 1/2 (mag. = 1.21)

[SE 86], month IV, night of the 14th, moonrise to sunset: 4 deg, measured (despite) mist; at 52 deg after sunset, when α Cyg culminated, lunar eclipse; when it began on the east side, In 17 deg night time it covered it completely; 10 deg night time maximal phase; when it began to clear, it cleared in 15 deg night time from south to north... 42 deg onset, maximal phase and clearing; its eclipse was red(?); (in) its eclipse, a gusty north wind blew; (in) its eclipse all of the planets did not stand there; 5 cubits behind δ Cap it became eclipsed.

[BM 33655 (no LBAT number), Rev. 3–8; trans. SH II, p. 141.]

This tablet, which was discovered at the site of Babylon by Rassam, is an astronomical diary. Its date has been derived by calculation using the numerous planetary and lunar observations which are preserved (SH II, p. 138). When the eclipse occurred, the Moon would be 11 deg to the east of δ Cap, which agrees well with the 5 cubits of the record. The computed interval between moonrise and sunset on Aug 1 is 3.5 deg.

Results

(i) First contact on Aug 1 at 3.47 h after sunset. LT of sunset = 18.94 h, hence LT of contact = 22.41 h, UT = 19.51 h. Computed TT = 0.38 h (Aug 2), thus ΔT = 17 550 sec.

(ii) Second contact at 4.60 h after sunset. LT of contact = 23.54 h, UT = 20.63 h. Computed TT = 1.50 h (Aug 2), thus ΔT = 17 550 sec.

(iii) Third contact on Aug 2 at 5.27 h after sunset. LT of contact = 0.21 h, UT = 21.29 h (Aug 1). Computed TT = 2.57 h, thus ΔT = 18 950 sec.

(iv) Last contact at 6.27 h after sunset. LT of contact = 1.21 h, UT = 22.30 h (Aug 1). Computed TT = 3.68 h, thus ΔT = 19 350 sec.

(8) BC 215 Dec 25/26 (mag. = 1.37)

Year 97 (SE), month IX, night of the 1[3th (?)...] measured; the bright star of the Old Man (α Per) stood in culmination, lunar eclipse; on the east side when it began, in 21 deg of night, all of it became covered; 16 deg of night totality; when it began to clear, it cleared in 19 deg of night from east and north to the west(?); 56 deg onset, totality [and clear]ing; (began) at one-half *beru* (i.e. 15 deg) after sunset. [...] eclipse; in its eclipse, Sirius [stood there; Sa]turn set; Mars came out (i.e. rose); the remainder of the planets did not stand there. In its eclipse, the north wind which was set to the west side blew; it was cold. 2 cubits in front of the front stars of Cancer (θ and γ Cnc) it became eclipsed. The 13th, sunrise to moonset: 9 deg 10′ (= $9\frac{1}{6}$ deg), measured; the north wind blew, the cold became severe.

[BM 36402 (= LBAT 294) + BM 36865 (no LBAT number), Obv. 1–9 and Rev. 1–5; trans. SH II, p. 157.]

This remarkably detailed description is recorded on an astronomical diary, consisting of two joining fragments. When the eclipse occurred, the Moon would be about 5 deg to the west of θ and γ Cnc, in close accord with the text. The computed interval between sunrise and moonset on Dec 26 is 8.2 deg.

RESULTS

(i) First contact on Dec 25 at 1.00 h after sunset. LT of sunset = 16.99 h, hence LT of contact = 17.99 h, UT = 15.10 h. Computed TT = 20.04 h, thus ΔT = 17 750 sec.

(ii) Second contact at 2.40 h after sunset. LT of contact = 19.39 h, UT = 16.50 h. Computed TT = 21.09 h, thus ΔT = 16 500 sec.

(iii) Third contact at 3.47 h after sunset. LT of contact = 20.46 h, UT = 17.57 h. Computed TT = 22.47 h, thus ΔT = 17 650 sec.

(iv) Last contact at 4.73 h after sunset. LT of contact = 21.72 h, UT = 18.83 h. Computed TT = 23.52 h, thus ΔT = 16 850 sec.

In table 6.4 are summarised the various results obtained in this section. The form of this table is similar to that of table 6.3

The results from tables 6.3 and 6.4 are plotted in figure 6.7. In this diagram, the general trend towards decreasing ΔT values in more recent centuries is obscured by the anomalously high results derived from both the eclipses of −225 and −214.

6.5 Eclipse maxima timed relative to sunrise or sunset (partial eclipses only)

As a rule, the Babylonian astronomers do not seem to have determined the moment when a partial lunar eclipse was at its height. Instead they distinguished a discrete period around mid-eclipse during which no change

Table 6.4 ΔT results from Babylonian timings of lunar eclipses for which three or four contacts were observed.

Year	Ct	SR/SS	Int. (deg)	ΔT (sec)
−500	1	SS	+77	14950
−500	2	SS	+92	15300
−500	3	SS	+117	14950
−500	4	SS	+142	13000
−406	1	SR	−48	14950
−406	2	SR	−27	13850
−406	3	SR	−12	16200
−377	1	SS	+37	16000
−377	2	SS	+52	16200
−377	3	SS	+73	15750
−377	4	SS	+92	14950
−352	1	SR	−47	15950
−352	2	SR	−24	14550
−352	3	SR	−6	15250
−316	1	SS	+47	16550
−316	2	SS	+63	16100
−316	3	SS	+68	19950
−316	4	SS	+84	20200
−225	1	SS	+52	17550
−225	2	SS	+69	17550
−225	3	SS	+79	18950
−225	4	SS	+94	19350
−214	1	SS	+15	17750
−214	2	SS	+36	16500
−214	3	SS	+52	17650
−214	4	SS	+71	16850

in the degree of obscuration of the Moon was noticeable. This typically lasted around 6 deg, after which a gradual decline began to be observed. In sections 6.2 and 6.3 above, several examples have already been cited in passing – e.g. BC 143 Feb 17/18, when it was recorded that:

...In 20 deg night it made 9 fingers. 5 deg duration of maximal phase.

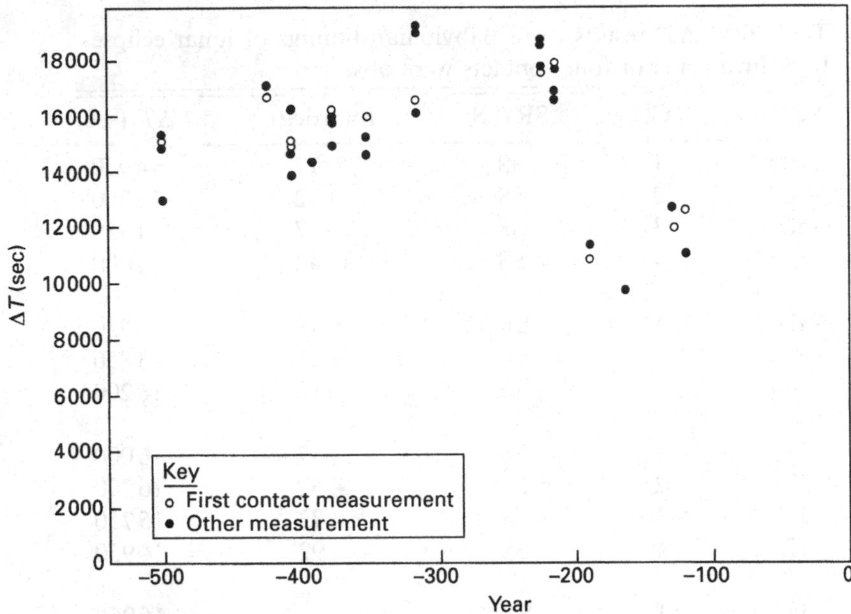

Fig. 6.7 ΔT values obtained from timed lunar eclipses – two, three and four contact measurements.

No other ancient or medieval civilisation seems to have assigned an actual duration to the greatest phase of a partial eclipse. Although purely artificial, and arising from the limited acuity of the unaided eye, this implies considerable care in observation. Computation of the change in the degree of obscuration of the Moon over an interval of 3 deg (half of the typical duration) yields a result close to two per cent of the lunar diameter for all but the smallest eclipses. This is illustrated in figure 6.8, which depicts an eclipse of central magnitude 0.75 at 3 deg (12 min) before maximum, at greatest phase itself, and at 3 deg after maximum. In angular measure, the change corresponds to only about 0.5 arcmin. Interestingly, the Babylonian astronomers never seem to have reported a duration of maximal phase for a solar eclipse (as is clear from the records quoted in chapter 5). Probably the great brilliance of the Sun prevented the necessary close scrutiny. For further discussion, see Stephenson and Fatoohi (1993).

Assuming that greatest phase occurred near the mid-point of the interval during which no change in the lunar obscuration was noticed, such records may enable a reasonable estimate of the moment of mid-eclipse to be inferred directly, as in the examples discussed below. For all but the last of these observations, timings of first and/or last contact have already been considered in sections 6.2 and 6.3. Hence in these cases only those

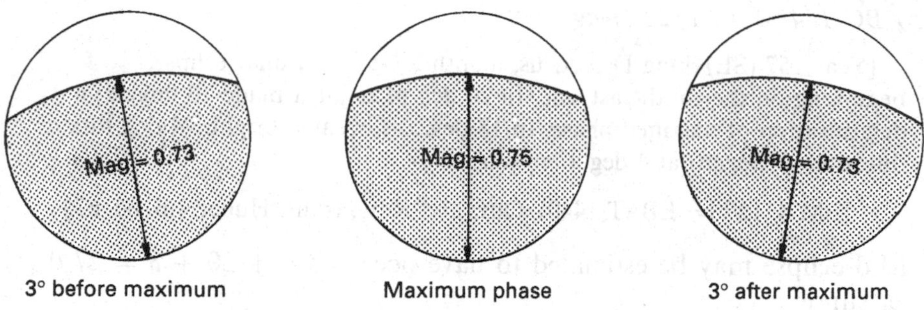

Fig. 6.8 Change in magnitude of a partial lunar eclipse around maximal phase.

portions of the records which need to be repeated will be cited here. However, the remaining account – from 66 BC – will be quoted in full.

(1) BC 424 Sep 28/29 (mag. = 0.93)

> [Artaxerxes I, year 41], month VI 14, 50 deg after sunset, beginning on the north-east. After 22 deg, 2 fingers lacked to totality. 5 deg duration of maximal phase. In 23 deg towards [...] 50 deg total duration...
>
> [BM 34787 (= LBAT 1426), col. II, 2′–3′; trans. Huber, pp. 32–33.]

Halving the estimated duration of greatest phase (i.e. 5 deg), mid-eclipse may be assumed to have occurred 50 + 22 + 2.5 = 74.5 deg after sunset.

RESULTS

Maximum on Sep 28 at 4.97 h after sunset. LT of sunset = 18.07 h, hence LT of maximum = 23.04 h, UT = 19.96 h. Computed TT = 0.69 h (Sep 29), thus ΔT = 17 000 sec.

(2) BC 239 Apr 28/29 (mag. = 0.41)

> [SE 73], month I, night of the 13th... At 80 deg after sunset, lunar eclipse; it began on the south and east side, in 15 deg night it made a little over $\frac{2}{3}$(?) of the disk; 10 deg of night maximal phase. When it began to clear, in 15 deg of night it cleared from the east to the west; 40 deg onset, maximal phase and clearing...
>
> [BM 55511 (no LBAT number); Obv. and Rev.; trans. SH II, pp. 84–85.]

On this occasion, an unusually long duration of maximal phase is recorded. As in the previous example, it will be assumed that mid-eclipse occurred 80 + 15 + 5 = 100 deg after sunset.

RESULTS

Maximum on Apr 29 at 6.67 h after sunset. LT of sunset = 18.63 h, hence LT of maximum = 1.30 h, UT = 22.26 h (Apr 28). Computed TT = 0.56 h (Apr 29), thus ΔT = 8250 sec.

(3) BC 154 Mar 21/22 (mag. = 0.85)

> [Year 157 (SE), king Demetrius, month XII] 15... Lunar eclipse, beginning on the south-east side. In 20 deg of night it made 10 fingers; 6 deg duration of maximal phase. In 18 deg... it became bright. 44 deg total duration... (Began) at 4 deg after sunset.
>
> [BM 41129 (= LBAT 1440), Obv. and Rev.; trans. Huber, pp. 66–67.]

Mid-eclipse may be estimated to have occurred 4 + 20 + 3 = 27 deg after sunset.

RESULTS

Maximum on Mar 21 at 1.80 h after sunset. LT of sunset = 18.02 h, hence LT of maximum = 19.82 h, UT = 17.01 h. Computed TT = 20.67 h, thus ΔT = 13 150 sec.

(4) BC 143 Feb 17/18 (mag. = 0.88)

> [SE 168], month XI 13... lunar eclipse... In 20 deg of night it made 9 fingers. 5 deg duration of maximal phase. (Began) at 7 deg after sunset.
>
> [BM 35787 (= LBAT 1278), Rev. 3′–8′; trans. Huber, p. 68.]

A time for mid-eclipse of 7 + 20 + 2.5 = 29.5 deg after sunset will be adopted.

RESULTS

Maximum on Feb 17 at 1.97 h after sunset. LT of sunset = 17.51 h, hence LT of maximum = 19.48 h, UT = 16.81 h. Computed TT = 20.40 h, thus ΔT = 12 900 sec.

(5) BC 129 Nov 4/5 (mag. = 0.63)

> Year 120 (Arsacid)... month VIII, night of the 13th... lunar eclipse... [In 18 deg] of night [it made ... fingers] 6 deg duration of maximal phase [...] In 16 deg ... 40 deg total duration... (Began) at 55 deg before sunrise.
>
> [BM 33982 (= LBAT 1441), Obv. and Rev.; trans. Huber, pp. 70–71.]

Mid-eclipse will be assumed to have occurred 55 − 18 − 3 = 34 deg before sunrise.

RESULTS

Maximum on Nov 5 at 2.27 h before sunrise. LT of sunrise = 6.58 h, hence LT of maximum = 4.31 h, UT = 1.13 h. Computed TT = 4.48 h, thus ΔT = 12 050 sec.

(6) BC 120 Jun 1/2 (mag. = 1.02)

> Year 128 (Arsacid), that is year 192 (SE)...month II 14...lunar eclipse...After 24 deg, 1 UD lacked to totality. 6 deg duration of maximal phase, until it began to become bright. In 24 deg, it became bright...54(?) deg total duration...(Began) at 66 deg after sunset.
>
> [BM 45845 (= LBAT 1442), Obv. and Rev.; trans. Huber, pp. 71–72.]

It will be inferred that mid-eclipse occurred 66 + 24 + 3 = 93 deg after sunset.

Results

Maximum on Jun 2 at 6.20 h after sunset. LT of sunset = 19.06 h, hence LT of maximum = 1.26 h, UT = 22.18 h (Jun 1). Computed TT = 1.44 h, thus ΔT = 11 700 sec.

(7) BC 80 Apr 10/11 (mag. = 0.60)

> Year 168 (Arsacid), that is year 232 (SE)...month I 13...lunar eclipse...In 20 deg night it made 6 fingers. 7 deg of night duration of maximal phase, until it began to become bright. In 13 deg...4 fingers lacking to brightness, it set...(Began) at 40 deg before sunrise.
>
> [BM 33562A (= LBAT 1445), Obv. and Rev.; trans. Huber, pp. 75–76.]

Mid-eclipse will be assumed to have occurred 40 – 20 – 3.5 = 16.5 deg before sunrise.

Results

Maximum on Apr 11 at 1.10 h after sunset. LT of sunrise = 5.62 h, hence LT of maximum = 4.52 h, UT = 1.57 h. Computed TT = 4.58 h, thus ΔT = 10 800 sec.

(8) BC 67 Jan 19/20 (mag. = 0.81)

> Year 180 (Arsacid), that is year 244 (SE)...month X, 15...As the Moon rose, two-thirds of the disk...were eclipsed. 6 deg night duration of maximal phase, until it began to become bright. In 16 deg night ...it became bright. 23 deg total duration...(Began) at 16 deg before sunset.
>
> [BM 45628 (= LBAT 1448) + Dupl. BM 41565 (= LBAT 1447 0, Obv. and Rev.; trans. Huber, pp. 77–78.]

Since the Moon rose so close to mid-eclipse, it would probably be difficult for the observers to define this phase. Hence the record is best rejected.

(9) BC 66 Dec 28/29 (mag. = 0.34)

Year 182 (Arsacid), that is year 246 (SE), Arsaces, king of kings, and [Pir'ustana], Istar, his wife(?), the lady. Month IX 14 [...] When the Moon rose, 2 fingers on the south side [were eclipsed]. In 9 deg night, over a third of the disk [was eclipsed]; 8 deg duration of maximal phase, until it began to become bright. in 11[+x deg from...] to south-west it became bright. 22(?) [total duration]. During this eclipse the sky was overcast... During this eclipse, north (wind) [...] During this eclipse, a halo enci[rcled(?)...] Jupiter and Saturn stood there [...] rose. The other p[lanets did not stand there]. $4\frac{1}{2}$ cubits behind β Gem [...]. (Began) at 6 deg before [sunset].

[BM 32845 (= LBAT 1450), Obv. and Rev.; trans. Huber, pp. 79–80.]

This tablet is devoted specifically to this eclipse. Its date is fully preserved. The text has not been considered previously since first contact was unobservable (estimated to occur 6 deg before sunset) while the time of last contact is only partially preserved. Although the magnitude was small, it is clear that the Moon rose well before maximal phase. Halving the estimated duration of greatest phase (i.e. 8 deg), mid-eclipse may be assumed to have occurred 9 + 4 = 13 deg after moonrise.

RESULTS

Maximum on Dec 28 at 0.87 h after moonrise. LT of moonrise = 16.85 h, hence LT of maximum = 17.72 h, UT = 14.85 h. Computed TT = 17.75 h, thus ΔT = 10450 sec.

In table 6.5 are summarised the results obtained in this section. This table follows much the same form as the previous tables in this chapter. Measurements were all made relative to sunrise (SR) or sunset (SS) with the exception of the eclipse of −65, for which the time was determined relative to moonrise (MR).

6.6 First contacts timed relative to culmination of stars

By the third century BC, in addition to the traditional sunrise and sunset measurements, the Babylonian astronomers began to time the beginning of lunar eclipses relative to the culmination of certain stars or small star groups. (Several examples have already been cited above in other contexts.) There were 26 of these *ziqpu* ('culminating') stars, occupying a rather broad zone which lay significantly to the north of the celestial equator, but was roughly parallel to it. According to the detailed study by Schaumberger (1952), declinations of individual *ziqpu* stars ranged from about +20 to +40 deg. Spacings in RA between adjacent reference objects varied from about 5 to 30 deg.

Just when this method of timing eclipses was introduced is not known, but the earliest surviving example dates from 226 BC. Over the next 150

Table 6.5 ΔT results from Babylonian timings of lunar eclipse maxima.

Year	SR/SS	Int. (deg)	ΔT (sec)
−423	SS	+74.5	17 000
−238	SS	+100	8 250
−153	SS	+27	13 150
−142	SS	+29.5	12 900
−128	SR	−34	12 050
−119	SS	+93	11 700
−79a	SR	−16.5	10 800
−65	MR	+13	10 450

years or so (down to 80 BC), some 15 observations are extant. Probably this represents only a small fraction of the original material. In some cases it is simply asserted that a particular eclipse began 'when' a certain *ziqpu* star was on the meridian. However, often a time interval 'before' (*ina* IGI) or after (*ar*) culmination is specified. Such intervals are typically of the order of 5 to 10 deg. I have rejected those records for which the interval of time between the culmination of the selected *ziqpu* star and the start of the eclipse is missing or damaged.

I have disregarded three lunar eclipses of very small magnitude since first contact would probably be poorly defined. The eclipse of BC 188 Aug 1/2 was said to begin when β Cyg culminated (BM 42053, Obv. 1; Huber, p. 61). This was in fact a penumbral eclipse, although the computed magnitude came very close to zero. Interestingly, this represents the only known ancient or medieval report of a penumbral eclipse. The eclipse of BC 185 Nov 24/5, which reached a calculated magnitude of only 0.03, was stated to have begun when the star group consisting of π, ο, ξ and ν Cas culminated (BM 35330, Obv. 7′; SH II, p. 355; Huber, p. 62). Finally, the eclipse of BC 163 Mar 30/31, which began 3 deg after υ Boo culminated, attained a magnitude of only 0.12.

The calculated magnitudes of all other eclipses which were timed in relation to star culminations exceeded 0.25. Each of the observations for which a reliable date can be deduced will be considered below.

It is unfortunate that no *Late Babylonian* lists of *ziqpu* stars exists. The most recent list probably dates from the eighth century BC and is found on tablet AO 6478 (Antiquite orientale) in the Louvre. This was published with a detailed commentary by Schaumberger (1952). An earlier list is contained on MUL.APIN ('Plough Star'), several partial copies of which are preserved in the British Museum (Hunger and Pingree, 1989).

AO 6478 gives an estimate of the difference in RA (i.e. difference in the local time of culmination) between each of the 26 *ziqpu* stars – usually to the nearest 5 US. However, no values for declination are preserved. Identities of the brighter stars (such as α Boo, α Cor, α Her, α Lyr, α Cyg, α Per, α Aur, and α and β Gem) appear reasonably well established, but some of the fainter ones seem decidely questionable. In the discussion of individual eclipse records below, I have accepted the results of Schaumberger (1952), but with some hesitation in certain cases. In general, Schaumberger's identifications have been accepted by both Huber (1973) and SH.

For most of these observations, contact timings relative to sunrise or sunset have already been considered in sections 6.2 to 6.4. In such instances, only those parts of the record which need to be repeated will be cited below; otherwise the full text will be quoted.

(1) BC 226 Aug 1/2 (mag. = 1.21)

> [SE 86], month IV, night of the 14th... At 52 deg after sunset, when α Cyg culminated, lunar eclipse; when it began on the east side....
>
> [BM 33655 (no LBAT number), Rev. 3–8; trans. SH II, p. 141.]

Results

RA of star = 19.43 h, RA of Sun = 8.44 h; hence LT of first contact = 12.00 + 11.00 h = 23.00 h, UT = 20.09 h. Computed TT = 0.38 h, thus ΔT = 15 450 sec.

(2) BC 215 Dec 25/26 (mag. = 1.37)

> Year 97 (SE), month IX, night of the 1[3th(?)...] measured; the bright star of the Old Man (α Per) stood in culmination, lunar eclipse; when it began on the east side...
>
> [BM 36402 + BM 36865 (= LBAT 294), Obv. 1–9 and Rev. 1–5; trans. SH II, p. 157.]

Results

RA of star = 1.07 h, RA of Sun = 18.07 h; hence LT of first contact = 12.00 + 7.00 h = 19.00 h, UT = 16.11 h. Computed TT = 20.04 h, thus ΔT = 14 150 sec.

(3) BC 194 Nov 4/5 (mag. = 0.92)

> [SE 118], month VIII 14. When 'the 4 of (the Lion's) breast' (α, γ, η, ζ Leo) culminated, (lunar eclipse) beginning on the [north-]east side....
>
> [BM 34236 (= LBAT 1436), Obv. 4–6; trans. Huber, p. 58.]

RESULTS

Mean RA of star group = 8.12 h, RA of Sun = 14.45 h; hence LT of first contact = 12.00 + 17.67 h = 5.67 h, UT = 2.49 h. Computed TT = 6.38 h, thus ΔT = 14 000 sec.

(4) BC 150 Jul 2/3 (mag. = 1.11)

> [SE 162] king Alexander... king Demetrius... [month III...] moonrise to sunset: [x]+6? deg, measured (despite mist); the north wind blew. When (the point) 4 deg in front of 15 Lac culminated, lunar ec[lipse ...] when it began, in 20 deg night it was completely covered; 12 deg of night maximal phase; it set eclipsed. 32 deg onset and [maximal phase ...] in its eclipse, Jupiter and Saturn stood there [...] the remainder of the planets did not stand there; $2\frac{1}{2}$ cubits below β Cap, [having passed a little to the east. [it became eclipsed ...]
>
> [BM 34632 (= LBAT 400), Obv. 5′–8′; trans. SH III, p. 85]

This tablet is an astronomical diary. Its date has been established from the numerous planetary and lunar observations which it contains and the references to both Alexander and Demetrius (SH III, p. 87). No previous reference has been made to this eclipse since the measurement of first contact relative to sunrise or sunset is not preserved. Accordingly, the full record is translated above. It is most unusual for a text to record the *sum* of the individual phases for an eclipse whose visibility was eventually interrupted by moonset. When the eclipse occurred, the Moon was 6 deg to the south of β Cap, i.e. roughly $2\frac{1}{2}$ cubits.

An alternative rendering of the statement referring to the *ziqpu* star would be: '4 deg before 15 Lac culminated, lunar eclipse...'. The star 15 Lac is very faint (magnitude 4.9) and it may well be doubted whether its identification by Schaumberger is correct. Hence this observation will not be considered further.

(5) BC 143 Feb 17/18 (mag. = 0.88)

> [SE 168], month XI 13, moonrise to sunset 3 deg measured. 5 deg after β Aur culminated, lunar eclipse beginning... In 20 deg of night it made 9 fingers....
>
> [BM 35787 (= LBAT 1278), Rev. 3′–8′; trans. Huber, p. 68.]

RESULTS

RA of star = 3.47 h, RA of Sun = 21.89 h; hence LT of first contact = 12.00 + 5.59 + 0.33 = 17.92 h, UT = 15.25 h. Computed TT = 18.80 h, thus ΔT = 12 750 sec.

Table 6.6 ΔT results from Babylonian timings of first contacts for lunar eclipses measured relative to meridian transits of *ziqpu* stars.

Year	Int. (deg)	ΔT (sec)
−225	0	15450
−214	0	14150
−193	0	14000
−142	+5	12750
−135	0	11800
−119	+5	11950
−79a	−5	11300

(8) BC 136 Mar 31/Apr 1 (mag. = 0.73)

Year 175 (SE), month XII/2 15 moonrise to sunset 17,40 (= $17\frac{2}{3}$) deg, cloudy, not observed. When α Cor culminated, lunar eclipse, beginning on the south-east side. In 18(?) deg of night it made 7 fingers. At 1(?) *beru* before sunrise.

[BM 34034 (= LBAT 1285), Rev. 17–23; trans. Huber, p. 69.]

This tablet is a goal-year text for the year SE 194 (118–117 BC). A total solar eclipse recorded on the same tablet (date equivalent to BC 136 Apr 15) was considered in chapter 5. Both events occurred 18 years before the goal-year. The lunar eclipse was not considered earlier in this chapter since the measurement of first contact relative to sunrise is damaged; although not questioned by Huber, Walker (personal communication) doubts the reading, which the scribe has written over an erasure. The full record is translated above. Fortunately, the reference to the culminating star is intact; this implies that the eclipse began when α Cor was on the meridian.

The recorded interval between moonrise and sunset ($17\frac{2}{3}$ deg) is much too large for the evening before a lunar eclipse; the computed equivalent is 6.2 deg. Perhaps the text should have read 7,40 rather than 17,40.

RESULTS

RA of star = 14.08 h, RA of Sun = 0.47 h; hence LT of first contact = 12.00 + 13.61 = 1.61 h, UT = 22.73 h. Computed TT = 2.00 h, thus ΔT = 11 800 sec.

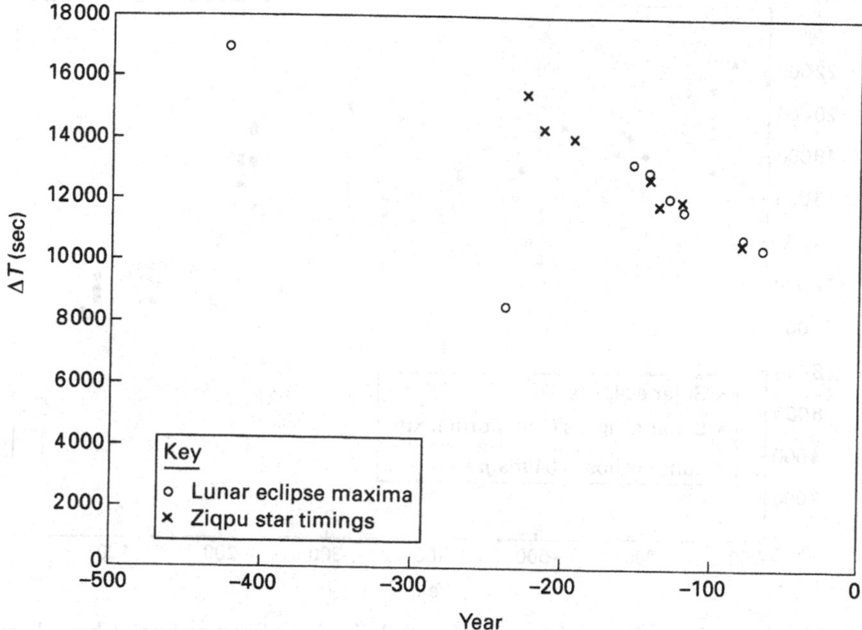

Fig. 6.9 ΔT values obtained from timed lunar eclipses – maximal phase and *ziqpu* star observations.

(7) BC 120 Jun 1/2 (mag. = 1.02)

Year 128 (Arsacid), that is year 192 (SE)... Month II 14 moonrise to sunset 6,30 (= $6\frac{1}{2}$ deg) measured. 5 deg after α Her culminated, lunar eclipse, beginning on the north-east side...

[BM 45845 (= LBAT 1442), Obv. and Rev.; trans. Huber, pp. 71–72.]

RESULTS

RA of star = 15.65 h, RA of Sun = 4.35 h; hence LT of first contact = 12.00 + 11.30 + 0.33 = 23.63 h, UT = 20.53 h. Computed TT = 23.86 h, thus ΔT = 11 950 sec.

(8) BC 80 Apr 10/11 (mag. = 0.60)

Year 168 (Arsacid), that is year 232 (SE)... Month I 13, moonrise to sunset 5,50 (= $5\frac{5}{6}$ deg), cloudy(?), measured(?)... 5 deg before μ Her culminated, lunar eclipse, beginning on the south-east side...

[BM 33562A (= LBAT 1445), Obv. and Rev.; Huber, pp. 75–76.]

Although the '5' of '5 deg before μ Her culminated' is slightly damaged, it is fairly certainly '5' (the only alternative is '4'; Walker, personal communication).

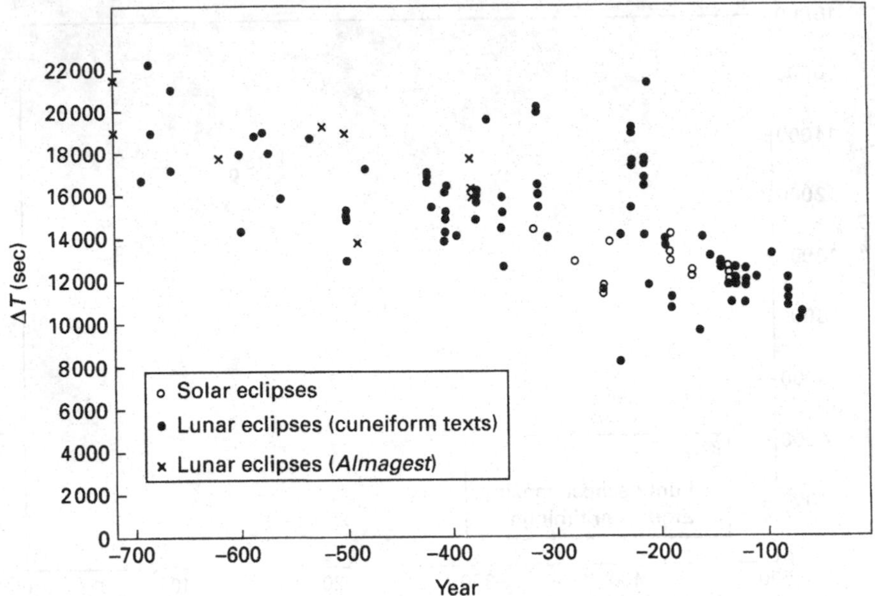

Fig. 6.10 ΔT values obtained from timed Babylonian observations of both lunar and solar eclipses.

RESULTS

RA of star = 16.43 h, RA of Sun = 1.10 h; hence LT of first contact = 12.00 + 15.33–0.33 = 3.00 h, UT = 0.05 h. Computed TT = 3.19 h, thus ΔT = 11 300 sec.

The various results obtained in this section are summarised in table 6.6. In figure 6.9 are plotted the ΔT results listed in both tables 6.5 and 6.6. With the obvious exception of –238, the results follow much the same pattern as in figure 6.6.

6.7 Conclusion

Figure 6.10 displays all of the ΔT results obtained in this chapter along with values obtained from eclipse timings shown in figure 4.3 (Babylonian lunar eclipse observations recorded in the *Almagest*) and figure 5.3 (solar eclipses cited on the Late Babylonian astronomical texts). These various data represent over 95 per cent of the carefully timed observations of eclipses preserved from the *ancient* world (up to about AD 400). In general, there is close accord between the ΔT values obtained from such a diverse assortment of measurements. However, since many of the time-intervals measured by the Babylonian astronomers amounted to several hours, there is a distinct possibility of significant clock drift; this may well

Table 6.7 ΔT results from Babylonian lunar and solar eclipse timings estimated to the nearest degree (interval measured < 25 deg).

Year	Type	SR/SS	Int. (deg)	ΔT (sec)
−665	Moon	SS	+3	21 050
−536	"	SR	−14	18 800
−482	"	SR	−10	17 300
−420	"	SS	+19	15 500
−407	"	SS	+15	15 250
−406	Moon	SR	−12	16 200
−405	"	SR	−14	16 500
−352	"	SR	−24	14 550
−352	"	SR	−6	15 250
−321	Sun	SS	−3	14 100
−316	Moon	SS	+10	15 550
−307	"	SR	−10	14 100
−280	Sun	SR	+20	12 950
−239	Moon	SR	−3	14 200
−214	"	SS	+15	17 750
−211	Moon	SR	−20	11 800
−193	"	SR	−12	13 650
−188	"	SR	−18	11 250
−169	Sun	SS	−20	12 300
−142	Moon	SS	+7	12 550
−135	Sun	SR	+24	12 600
−133	Moon	SR	−9	10 950
−128	"	SR	−15	12 600
−108	"	SS	+8	12 050
−66	Moon	MR	+22.5	10 150

account for much of the scatter in figure 6.10. On the other hand, some intervals were very short (in some cases as small as 3 deg), and in such cases clock drift should have been minimal. I have somewhat arbitrarily selected 25 deg (1.67 hours) as a dividing line between 'short' and 'long' intervals.

Table 6.7 lists the ΔT values obtained from measurements of both lunar and solar eclipse times where the recorded interval is less than 25 deg. (NB results from times rounded to the nearest 5 or 10 deg, observations

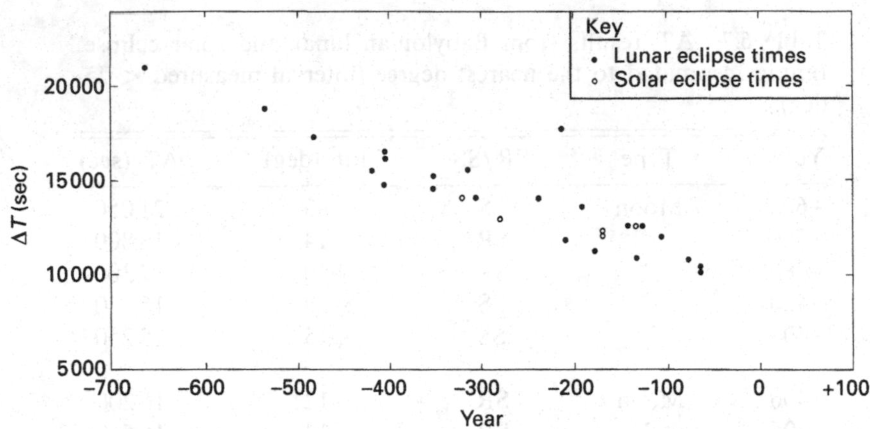

Fig. 6.11 ΔT values obtained from timed Babylonian observations of lunar and solar eclipses for which the measured interval was less than 25 deg.

of mid-eclipse and all *ziqpu* star measurements are not included.) These various values are plotted in figure 6.11. With only a very few exceptions, the ΔT results derived from these data are remarkably self-consistent, and in the final analysis (chapter 14) they will be assigned double weight. All other ΔT values as displayed in figure 6.10 will be given only unit weight.

7

Untimed Babylonian observations of lunar eclipses: horizon phenomena

7.1 Introduction

In the previous chapter, a variety of Babylonian timings of lunar eclipse contacts were analysed. Several of these records also noted that the Moon rose (*ki* E-*a*) or set (SU) whilst eclipsed. Additionally, some damaged texts do not contain useful measurements of time but nevertheless affirm that the Moon was eclipsed at its rising or setting. Such fairly straightforward observations (which enable limits on the value of ΔT to be deduced), would require no instrumental aid. In the following pages these various observations will be investigated together with a few rather more careful reports which give an estimate of the fraction of the Moon covered at moonrise or moonset.

Since the eclipsed Moon is in direct opposition to the Sun, it invariably rises close to sunset or sets near sunrise. It thus usually reaches the horizon when the sky is quite bright – often when the Sun is above the opposite horizon. However, as noted in chapter 4, despite these seemingly unfavourable conditions the Babylonians systematically measured the time of moonrise (relative to sunset) and moonset (in relation to sunrise) around full Moon with considerable care. Many examples of this practice are found in one of the earliest surviving astronomical diaries – dating from 568 BC, the 37th year of Nebuchadrezzar II (see SH I, pp. 47 ff.) – and it may extend back much further in time.

Because of this regular observing routine, if the sky was clear around the middle of a lunar month the Moon would be systematically scrutinised at its rising and setting. Hence any small indentation on the lunar limb would probably be noticed at these moments, whether or not an eclipse was expected at this time. The fact that Babylonian and also Chinese astronomers (see chapter 9) were in the habit of estimating the proportion of the Moon which was eclipsed when it rose or set suggests that they were able to discern the Moon fairly clearly when on the horizon.

Three Babylonian records – from 587, 171 and 134 BC – assert that the

Moon rose or set totally eclipsed. For example, the report of the event whose date corresponds to BC 171 Aug 23/24 reads as follows:

> [SE 141, Month V...] When the Moon came out (i.e rose), it was completely covered; when it began to clear, [...the remainder of the pla]nets did not stand there; it was eclipsed in the end of Aquarius. (Began) at 42 deg before sunset. The 15th [...].
>
> [BM 45654 (= LBAT 365), Rev. 12′–13′; trans. SH II, p. 451.]

As noted in chapter 3, it seems unlikely that the lunar disk would be visible under these conditions. However, because of the monthly routine practised by the astronomers of Babylon, if the Moon were to rise totally obscured, it should have been detected soon after emersion. Equally, if the Moon set totally eclipsed, its failure to reappear after immersion should have been noticed. Nevertheless, observations of this kind, based on indirect evidence, seem likely to be of reduced reliability, especially if cloud or mist prevailed at the time. Accordingly, reports that the Moon rose or set totally eclipsed will not be considered here.

The observations discussed in this chapter are divided into two categories:

(i) Moon rising or setting eclipsed (section 7.2);

(ii) Estimates of the proportion of the Moon covered at its rising or setting (section 7.3).

If the timing aspects of a certain text have already been discussed in chapter 6, only the necessary details regarding rising or setting will usually be reproduced below. Such entries will be marked with an asterisk [*]. However, if a record has not yet been considered, a full translation together with comments on dating (where appropriate) will be given. As in the previous chapter, all translations are either by Huber or SH.

7.2 Moon rising or setting eclipsed

Since many texts are badly damaged, often it can only be inferred that the Moon rose or set visibly eclipsed – i.e. between first and last contact. Unless an observation indicates the contrary, these circumstances will normally be assumed without comment. Since the local time of moonrise and moonset is a function of ΔT, these times have been computed iteratively, as discussed in chapters 3 and 4.

(1) BC 702 Mar 19/20 (mag. = 0.22)

> Bel-ibni, year 1 [month I ... beginning] on the south side [...it] set eclipsed [... before sunri]se.
>
> [BM 45640 (= LBAT 1415), Obv. I, bottom; trans. Huber, p. 4.]

The eclipse of BC 702 Mar 19/20 was the first to occur in the short reign of Bel-ibni (702–699 BC). Its record represents the oldest extant observation on the cuneiform texts which has been reliably dated. However, this is fully twenty years after the earliest Babylonian eclipse reported by Ptolemy (see chapter 4).

Results

(i) First contact on Mar 20 before moonset. LT of moonset = 6.17 h. Hence LT of contact < 6.17 h, UT < 3.39 h. Computed TT = 8.49 h, thus $\Delta T > 18\,350$ sec.

(ii) Last contact after moonset. LT of moonset = 6.21 h. Hence LT of contact > 6.21 h, UT > 3.43 h. Computed TT = 10.37 h, thus $\Delta T < 25\,000$ sec.

Combining these limits yields $18\,350 < \Delta T < 25\,000$ sec.

(2) BC 695 Apr 30/May 1 (mag. = 0.74) [*]

(Assur-nadin-sum), year 5, month II, 1[0+x, not] complete, it set eclipsed. (Began) at 30 deg before sunrise.

[BM 32238 (= LBAT 1414), Obv. III; trans. Huber, p. 4.)

This eclipse was only partial. It is not clear whether the statement that it was '(not) complete' means that the Moon set before the eclipse had attained its maximum or before the disk had fully cleared. Hence it will only be assumed the Moon set between first and last contact.

Results

(i) First contact on May 1 before moonset. LT of moonset = 5.36 h. Hence LT of contact < 5.36 h, UT < 2.32 h. Computed TT = 5.01 h, thus $\Delta T > 9700$ sec.

(ii) Last contact after moonset. LT of moonset = 5.43 h. Hence LT of contact > 5.43 h, UT > 2.38 h. Computed TT = 7.90 h, thus $\Delta T < 19\,850$ sec.

Combining these limits yields $9700 < \Delta T < 19\,850$ sec.

(3) BC 667 Oct 14/15 (mag. = 1.86) [*]

[Samas-sum-ukin, year 1]... It set eclipsed. (Began) at 20[+x] deg before sunrise.

[BM 45640 (= LBAT 1415), Obv. III, top; trans. Huber, p. 7.)

Although this eclipse was total, so little information is preserved that it can only be inferred that the Moon set between first and last contact.

RESULTS

(i) First contact on Oct 15 before moonset. LT of moonset = 6.15 h. Hence LT of contact < 6.15 h, UT < 3.02 h. Computed TT = 6.19 h, thus $\Delta T > 11\,400$ sec.

(ii) Last contact after moonset. LT of moonset = 6.31 h. Hence LT of contact > 6.31 h, UT > 3.20 h. Computed TT = 9.73 h, thus $\Delta T < 23500$ sec.

Combining these limits yields $11\,400 < \Delta T < 23\,500$ sec.

(4) BC 591 Mar 22/23 (mag. = 1.84)

> [Nebuchadrezzar] year 13, month VI [eclipse passed]. [Month XII 1]4()?, it rose eclipsed... 1 *beru* 10 deg [duration?] ...
>
> [BM 38462 (= LBAT 1420), Obv. II, 13′–15′; trans. Huber, p. 15.]

As pointed out in chapter 4, the Babylonian astronomers knew that lunar eclipses occurred at intervals of five or six months. Hence although the month in which the observed eclipse reported in the above text is missing, it may be concluded that it would take place five or six months after the predicted event in month VI. The date can thus be restored with fair confidence. The time-interval of '1 *beru* 10 deg' may well indicate the measured duration but this is only conjecture.

RESULTS

(i) First contact on Mar 22 before moonrise. LT of moonrise = 17.87 h. Hence LT of contact < 17.87 h, UT < 15.07 h. Computed TT = 18.56 h, thus $\Delta T > 12\,550$ sec.

(ii) Last contact after moonrise. LT of moonrise = 18.02 h. Hence LT of contact > 18.02 h, UT > 15.22 h. Computed TT = 22.09 h, thus $\Delta T < 24\,700$ sec.

Combining these limits yields $12550 < \Delta T < 24\,700$ sec.

(5) BC 588 Jan 18/19 (mag. = 0.54)

> [Nebuchadrezzar] year 16, month IV [eclipse passed]. [Month X], 14, morning watch(?), $\frac{2}{3}$ (?) *beru* night [before sunrise]. By its $\frac{1}{3}$ (??) covered, set eclipsed.
>
> [BM 38462 (= LBAT 1420), Obv. II, 13ff.; trans. Huber, p. 17.]

Both the time interval and degree of obscuration of the Moon are damaged so that it may be only reliably inferred that the Moon set at some time between first and last contact.

RESULTS

(i) First contact on Jan 19 before moonset. LT of moonset = 6.88 h. Hence LT of contact < 6.88 h, UT < 4.16 h. Computed TT = 7.22 h, thus $\Delta T > 11\,000$ sec.

(ii) Last contact after moonset. LT of moonset = 6.95 h. Hence LT of contact > 6.95 h, UT > 4.24 h. Computed TT = 10.03 h, thus $\Delta T <$ 20 850 sec.

Combining these limits yields $11\,000 < \Delta T < 20\,850$ sec.

(6) BC 577 Jun 13/14 (mag. = 0.99)

> [Nebuchadrezzar] year 27, month III 14? [...north] east side [...] it set eclipsed.
>
> [BM 38462 (= LBAT 1420), Rev. I, 7′–9′; trans. Huber, p. 19.]

RESULTS

(i) First contact on Jun 14 before moonset. LT of moonset = 4.85 h. Hence LT of contact < 4.85 h, UT < 1.77 h. Computed TT = 4.06 h, thus $\Delta T > 8250$ sec.

(ii) Last contact after moonset. LT of moonset = 4.96 h. Hence LT of contact > 4.96 h, UT > 1.89 h. Computed TT = 7.53 h, thus $\Delta T <$ 20 300 sec.

Combining these limits yields $8250 < \Delta T < 20\,300$ sec.

(7) BC 563 Sep 5/6 (mag. = 0.35)

> [Nebuchadrezzar] year 42, month VI 14. It rose eclipsed [...] and became bright. 6 US to become bright. (Began) at 35 deg [before sunset].
>
> [BM 41536 (= LBAT 1421), II, 2′ff.; trans. Huber, p. 22.]

The time of beginning is based on prediction rather than observation. Quite possibly the eclipse ended 6 deg after moonrise but since the text is damaged this inference may be incorrect. It will only be assumed that the Moon rose between first and last contact.

RESULTS

(i) First contact on Sep 5 before moonrise. LT of moonrise = 18.46 h. Hence LT of contact < 18.46 h, UT < 15.49 h. Computed TT = 19.12 h, thus $\Delta T > 13\,050$ sec.

(ii) Last contact after moonrise. LT of moonrise = 18.50 h. Hence LT of contact > 18.50 h, UT > 15.54 h. Computed TT = 21.26 h, thus ΔT < 20 600 sec.

Combining these limits yields $13\,050 < \Delta T < 20\,600$ sec.

(8) BC 537 Oct 16/17 (mag. = 1.50) [*]

[Cyrus, year 2] month VII 10[+x...it made] $\frac{2}{3}$ of the disk towards totality. Not (yet) total, it set eclipsed ... Began at 14 deg before sunrise.

[BM 32234 (= LBAT 1419), Obv. V, middle; trans. Huber, p. 25.]

Although the observers did not witness totality, apparently they were able to judge from the almost central entry of the Moon into the shadow that the eclipse would ultimately be complete. The Moon evidently set between first and second contact.

RESULTS

(i) First contact on Oct 17 before moonset. LT of moonset = 6.24 h. Hence LT of contact < 6.24 h, UT < 3.10 h. Computed TT = 7.39 h, thus $\Delta T > 15\,450$ sec.

(ii) Second contact after moonset. LT of moonset = 6.27 h. Hence LT of contact > 6.27 h, UT > 3.13 h. Computed TT = 8.50 h, thus $\Delta T < 19\,300$ sec.

Combining these limits yields $15\,450 < \Delta T < 19\,300$ sec.

(9) BC 483 Nov 18/19 (mag. = 1.47) [*]

[Xerxes, year 3] month VIII 13. Beginning on the south side. Maximal phase not observed; it set eclipsed...(Began) at 10 deg before sunrise.

[BM 32234 (= LBAT 1419), Rev. III; trans. Huber, p. 27.]

As in the previous example, the observers were aware that the eclipse had not reached its height by the time the Moon set. It is thus clear that moonset took place between first and second contact – an interpretation in accord with the very short time-interval between the start of the eclipse and sunrise.

RESULTS

(i) First contact on Nov 19 before moonset. LT of moonset = 6.73 h. Hence LT of contact < 6.73 h, UT < 3.59 h. Computed TT = 7.74 h, thus $\Delta T > 14\,900$ sec.

(ii) Second contact after moonset. LT of moonset = 6.78 h. Hence LT of contact > 6.78 h, UT > 3.64 h. Computed TT = 8.84 h, thus $\Delta T < 18\,700$ sec.

Combining these limits yields $14\,900 < \Delta T < 18\,700$ sec.

(10) BC 442 Mar 24/25 (mag. = 1.04)

[Artaxerxes (I), year 22...] it set eclipsed [...] $\frac{2}{3}$ (?) it made and then the Sun rose (?) 5 *us*(?) behind α Vir (?).

[BM 34787 (= LBAT 1426), col. I, top; trans. Huber, p. 30.]

Since the text is damaged, it seems best to assume no more than that the Moon set at some time between first and last contact. When the eclipse began, the Moon was about 7 deg to the east of α Vir, in fair accord with observation.

RESULTS

(i) First contact on Mar 25 before moonset. LT of moonset = 5.98 h. Hence LT of contact < 5.98 h, UT < 3.16 h. Computed TT = 6.87 h, thus $\Delta T > 13\,350$ sec.

(ii) Last contact after moonset. LT of moonset = 6.07 h. Hence LT of contact > 6.07 h, UT > 3.25 h. Computed TT = 10.10 h, thus $\Delta T <$ 24 650 sec.

Combining these limits yields $13\,350 < \Delta T < 24\,650$ sec.

(11) BC 353 Nov 21/22 (mag. = 1.35) [*]

> [Artaxerxes III, year 6] Month VIII 14. Beginning on the south-east side. After 23 deg total. 18 deg duration of maximal phase. After 6 deg night, one quarter of the disc had become bright (?) (and) it set eclipsed...
>
> [BM 32238 (= LBAT 1414), Rev. III, bottom; trans. Huber, p. 49.]

It is clear that the Moon set between third and last contact.

RESULTS

(i) Third contact on Nov 22 before moonset. LT of moonset = 6.89 h. Hence LT of contact < 6.89 h, UT < 3.76 h. Computed TT = 7.50 h, thus $\Delta T > 13\,450$ sec.

(ii) Last contact after moonset. LT of moonset = 6.94 h. Hence LT of contact > 6.94 h, UT > 3.81 h. Computed TT = 8.64 h, thus $\Delta T <$ 17 350 sec.

Combining these limits yields $13450 < \Delta T < 17\,350$ sec.

(12) BC 248 Oct 2/3 (mag. = 1.01)

> [SE 64, month VII...] Night of the 14th, [...] a little cloudburst, overcast; it set eclipsed; (during) its eclipse, Venus, Mercury and Ma[rs? stood there...].
>
> [BM 45949 (= LBAT 268), A3–A4; trans. SH II, p. 55; see also Huber, p. 55.]

This tablet is an astronomical diary. Although its date has been broken away, this has been derived by calculation using the various planetary and lunar observations which are preserved (Sachs and Schaumberger, 1955, p. xiv; SH II, p. 58).

RESULTS

(i) First contact on Oct 3 before moonset. LT of moonset = 6.01 h. Hence LT of contact < 6.01 h, UT < 2.90 h. Computed TT = 6.09 h, thus ΔT > 11 450 sec.

(ii) Last contact after moonset. LT of moonset = 6.15 h. Hence LT of contact > 6.15 h, UT > 3.06 h. Computed TT = 9.48 h, thus ΔT < 23 100 sec.

Combining these limits yields 11 450 < ΔT < 23 100 sec.

(13) BC 240 Nov 2/3 (mag. = 1.40) [*]

> [SE 72], Month VIII 14, at 3 deg before sunrise, beginning on the east side. It set eclipsed. Month XII_2 (? or cloudy).
>
> [BM 32286 (= LBAT 1218), Rev. 3–5; trans. Huber, pp. 55–56.]

Taken by itself, the extremely short duration of visibility is insufficient to prove that the Moon set before totality (because of the slight possibility of a scribal error in reporting this time). However, when combined with the lack of further description in the text (which is preserved intact) it may be reasonably inferred that the Moon reached the western horizon between first and second contact.

RESULTS

(i) First contact on Nov 3 before moonset. LT of moonset = 6.50 h. Hence LT of contact < 6.60 h, UT < 3.33 h. Computed TT = 7.12 h, thus ΔT > 13 650 sec.

(ii) Second contact after moonset. LT of moonset = 6.52 h. Hence LT of contact > 6.52 h, UT > 3.35 h. Computed TT = 8.16 h, thus ΔT < 17 300 sec.

Combining these limits, 13 650 < ΔT < 17 300 sec.

(14) BC 212 Apr 29/30 (mag. = 0.62) [*]

> Year 100 (SE)... month I 13... Lunar eclipse, beginning on the south side. Around(?) maximal phase cloudy, not observed. It set eclipsed. (Began) at 20 deg before sunrise...
>
> [BM 32222 (= LBAT 1237), Rev. 48–53; trans. Huber, p. 56.]

Apparently the sky was clear when the eclipse began and also when the Moon set. On account of the clouds around maximal phase, the observers may not have been able to judge whether the eclipse was on the decline when the Moon reached the horizon. It thus seems best to assume no more than that moonset occurred between first and last contact.

RESULTS

(i) First contact on Apr 30 before moonset. LT of moonset = 5.40 h. Hence LT of contact < 5.40 h, UT < 2.36 h. Computed TT = 4.25 h, thus $\Delta T > 6800$ sec.

(ii) Last contact after moonset. LT of moonset = 5.47 h. Hence LT of contact > 5.47 h, UT > 2.43 h. Computed TT = 7.03 h, thus $\Delta T <$ 16 550 sec.

Combining these limits yields $6800 < \Delta T < 16\,550$ sec.

(15) BC 190 Feb 27/28 (mag. = 1.05)

> SE 121 month XII 15 [...] lunar eclipse, beginning on the south-east(?) side. After 20(?) deg total [...] duration of maximal phase. When it began to get bright, from the east side to the west. Toward [...] thunder(?) [...] becoming bright not observed. Mars and Mercury in Capricorn. During the eclipse [...] eclipsed [...Began at ...] night before sunrise. It set eclipsed. On the 15th, sunrise to moonset 3,30 (i.e. $3\frac{1}{2}$) deg, cloudy(?) measured(?).
>
> [BM 33643 (= LBAT 1437); Obv. and Rev.; trans. Huber, pp. 59–60.]

This tablet, part of the surface of which is badly damaged, is devoted specifically to this eclipse. Although the date is fully preserved, the time of onset is missing. It is clear from the text – which mentions the brightening of the Moon after maximal phase – that moonset took place between emersion and last contact.

RESULTS

(i) Third contact on Feb 28 before moonset. LT of moonset = 6.42 h. Hence LT of contact < 6.42 h, UT < 3.72 h. Computed TT = 6.70 h, thus $\Delta T > 10\,700$ sec.

(ii) Last contact after moonset. LT of moonset = 6.44 h. Hence LT of contact > 6.44 h, UT > 3.75 h. Computed TT = 8.15 h, thus $\Delta T <$ 15 850 sec.

Combining these limits, $10\,700 < \Delta T < 15\,850$ sec.

(16) BC 190 Aug 23/24 (mag. = 1.02)

> SE 122, King An[tiochus III], month V 14. Sunset to moonrise 1 deg, cloudy, measured. When the Moon rose from a cloud(?), 2 fingers on the [west] side lacked to brightness. (Began) at 1 *beru* before sunset.
>
> [BM 34579 (= LBAT 1251), Rev. 14–18: trans. Huber, p. 60.]

This tablet is a goal-year text for SE 140 (172–171 BC). The eclipse occurred 18 years before the goal-year. Although it was apparently cloudy when the Moon rose, it was possible to measure the interval between sunset and moonrise. Hence the Moon must have been discernible at the time.

Normally only thin cloud prevails in Iraq in August. In describing the eclipse, the standard term for moonrise (*sin ki* E-*a*) is used. The statement that the eclipse began at 1 *beru* before sunset clearly represents a prediction of an unobservable event.

The lack of reference to maximum phase in the above text and the statement that 2 fingers 'lacked to brightness' (*ana* ZALAG *kat*) both suggest that the eclipse was already nearly over by moonrise. It will thus be assumed that the Moon rose at some time between emersion (although the eclipse was only marginally total) and last contact.

RESULTS

(i) Third contact on Aug 23 before moonrise. LT of moonrise = 18.63 h. Hence LT of contact < 18.63 h, UT < 15.70 h. Computed TT = 17.99 h, thus $\Delta T > 8250$ sec.

(ii) Last contact after moonrise. LT of moonrise = 18.68 h. Hence LT of contact > 18.68 h, UT > 15.75 h. Computed TT = 19.39 h, thus $\Delta T < 13\,100$ sec.

Combining these limits, $8250 < \Delta T < 13\,100$ sec.

(17) BC 189 Feb 16/17 (mag. = 1.28) [*]

> [SE 122] month XI 14...Lunar eclipse, beginning on the east side. After 16 deg night totally covered. It set eclipsed. (Began) at 34 deg before sunrise.
>
> [BM 34579 (= LBAT 1251), Rev. 23–26: trans. Huber, p. 61.]

This entry is recorded on the same goal-year text as the previous record, the eclipse also occurring 18 years before the goal-year. Recovery of the Moon after totality is not mentioned, suggesting that the Moon set totally eclipsed. However, this is not definitely implied and it seems best to conclude no more than that the Moon set between second and last contact.

RESULTS

(i) Second contact on Feb 17 before moonset. Local time of moonset = 6.53 h. Hence LT of contact < 6.53 h, UT < 3.86 h. Computed TT = 5.78 h, thus $\Delta T > 6900$ sec.

(ii) Last contact after moonset. LT of moonset = 6.60 h. Hence LT of contact > 6.60 h, UT > 3.93 h. Computed TT = 8.33 h, thus $\Delta T < 15\,800$ sec.

Combining these limits, $6900 < \Delta T < 15\,800$ sec.

(18) BC 150 Jul 2/3 (mag. = 1.11) [*]

> [SE 162 month III...] lunar ec[lipse...] when it began, in 20 deg night it was completely covered; 12 deg of night maximal phase; it set eclipsed. 32 deg onset and [maximal phase ...]
>
> [BM 34632 (= LBAT 400), 5′–8′; trans. SH III, p. 85.]

Although recovery of the Moon after totality is not mentioned, the text does not definitely imply that it set totally eclipsed; we can only conjecture what the lacuna at the end of the preserved text may have originally contained. As in the previous example, it seems best to assume no more than that the Moon set between second and last contact.

RESULTS

(i) Second contact on Jul 3 before moonset. Local time of moonset = 4.87 h. Hence LT of contact < 4.87 h, UT < 1.88 h. Computed TT = 4.21 h, thus $\Delta T > 8350$ sec.

(ii) Third contact after moonset. LT of moonset = 4.95 h. Hence LT of contact > 4.95 h, UT > 1.97 h. Computed TT = 6.43 h, thus $\Delta T < 16\,050$ sec.

Combining these limits yields $8350 < \Delta T < 16\,050$ sec.

(19) BC 99 Apr 11/12 (mag. = 1.82)

> Year 213 (SE), month I ... Night of the 14th, when the Moon came out (i.e. rose), it was eclipsed 3 fingers on the east side in the end of Libra, 2 cubits behind α Lib; in its eclipse, Jupiter in Gemini and Sirius stood there.
>
> [BM 140677 (no LBAT number), Obv. 4–6; trans. SH III, p. 407.]

This tablet is an astronomical diary. When the eclipse occurred, the Moon was in longitude 200 deg, and thus approaching the end of Libra (210 deg); it would be about 4 deg to the east of α Lib (in close accord with observation.) The statement that when the Moon rose it was partially obscured on the *east* side implies that the eclipse had just begun. (A lunar eclipse starts on the eastern edge of the disk and ends on the western limb, the opposite to a solar obscuration.) However, it is surprising that the text (which is intact) makes no reference to totality. Rather than attach too much weight to the recorded position angle of first contact – in case of possible scribal error – it would seem better to assume either that the Moon rose between first and second contact *or* between third and last contact.

RESULTS

(i) First contact on Apr 11 before moonrise. LT of moonrise = 18.26 h. Hence LT of contact < 18.26 h, UT < 15.30 h. Computed TT = 15.36 h, thus $\Delta T > 200$ sec.

(ii) Second contact after moonrise. LT of moonrise = 18.29 h. Hence LT of contact > 18.29 h, UT > 15.33 h. Computed TT = 16.43 h, thus $\Delta T < 3950$ sec.

(iii) Third contact before moonrise. LT of moonrise = 18.35 h. Hence LT of contact < 18.35 h, UT < 15.40 h. Computed TT = 18.21 h, thus $\Delta T > 10\,100$ sec.

(iv) Last contact after moonrise. LT of moonrise = 18.40 h. Hence LT of contact > 18.40 h, UT > 15.45 h. Computed TT = 19.29 h, thus $\Delta T < 13\,800$ sec.

Combining these limits yields either $200 < \Delta T < 3950$ sec or $10\,100 < \Delta T < 13\,800$ sec.

NB the second set of values is in much better accord with results obtained from other roughly contemporaneous observations and only this set will be retained.

(20) BC 80 Apr 10/11 (mag. = 0.60) [*]

> Year 168 (Arsacid), that is year 232 (SE)... month I 13 ... Lunar eclipse beginning on the south-east side. In 20 deg night it made 6 fingers. 7 deg duration of maximal phase until it began to become bright. In 13 deg from south-east to north-west, 4 fingers lacking to brightness, it set...
>
> [BM 33562A (= LBAT 1445), Obv. and Rev.; trans. Huber, pp. 75–76.]

It is evident that the Moon set between maximum and last contact.

RESULTS

(i) Maximum phase on Apr 11 before moonset. Local time of moonset = 5.63 h. Hence LT of contact < 5.63 h, UT < 2.68 h. Computed TT = 4.58 h, thus $\Delta T > 6850$ sec.

(ii) Last contact after moonset. LT of moonset = 5.66 h. Hence LT of maximum > 5.66 h, UT > 2.71 h. Computed TT = 5.97 h, thus $\Delta T < 11\,700$ sec.

Combining these limits yields $6850 < \Delta T < 11\,700$ sec.

(21) BC 67 Jan 19/20 (mag. = 0.81) [*]

> Year 180 (Arsacid), that is year 244 (SE)... month X 15. Moonrise to sunset 1 deg, mist, measured. As the Moon rose, two thirds of the disk on the north-east side were eclipsed. 6 deg night duration of maximal phase, until it began to become bright...
>
> [BM 45628 (= LBAT 1448) + Duplicate BM 41565 (= LBAT 1447), Obv. and Rev.; trans. Huber, pp. 77–78.]

The text implies that the Moon rose close to greatest phase. However, under these circumstances it would probably be difficult to judge whether the eclipse had already passed maximum by this time or whether this phase was still to come. Hence it seems best to conclude only that the Moon rose at some time between first and last contact.

RESULTS

(i) First contact on Jan 19 before moonrise. LT of moonrise = 17.05 h. Hence LT of contact < 17.05 h, UT < 14.33 h. Computed TT = 15.49 h, thus ΔT > 4150 sec.

(ii) Last contact after moonrise. LT of moonrise = 17.15 h. Hence LT of contact > 17.15 h, UT > 14.44 h. Computed TT = 18.71 h, thus ΔT < 15 350 sec.

Combining these limits yields 4150 < ΔT < 15 350 sec.

(22) BC 66 Dec 28/29 (mag. = 0.34) [*]

> Year 182 (Arsacid), that is year 246 (SE)... Month IX 14 [...] When the Moon rose, 2 fingers on the south side [were eclipsed]. In 9 deg night, over a third of the disk [was eclipsed]; 8 deg duration of maximal phase, until it began to become bright...
>
> [BM 32845 (= LBAT 1450), Obv. and Rev.; trans. Huber, pp. 79–80.]

It is apparent that the Moon rose between first contact and maximum.

RESULTS

(i) First contact on Dec 28 before moonrise. LT of moonrise = 16.87 h. Hence LT of contact < 16.87 h, UT < 14.00 h. Computed TT = 16.64 h, thus ΔT > 9500 sec.

(ii) Maximum phase after moonrise. LT of moonrise = 16.90 h. Hence LT of contact > 16.90 h, UT > 14.14 h. Computed TT = 17.75 h, thus ΔT < 13 350 sec.

Combining these limits yields 9500 < ΔT < 13 350 sec.

In addition to the above observations, Huber (pp. 17 and 22) concluded that the Moon rose or set eclipsed on BC 588 Jan 18/19 and BC 563 Sep 5/6. However, both texts are damaged at the appropriate point and Huber's inferences, although quite reasonable, cannot be reliably confirmed.

In the following example, the recorded time after sunset when the eclipse began is unusually small (only 3 deg). The mere fact that the eclipse was seen to commence after sunset provides a limit on ΔT. Only a single meaningful limit may be derived in this case (corresponding to first contact after sunset).

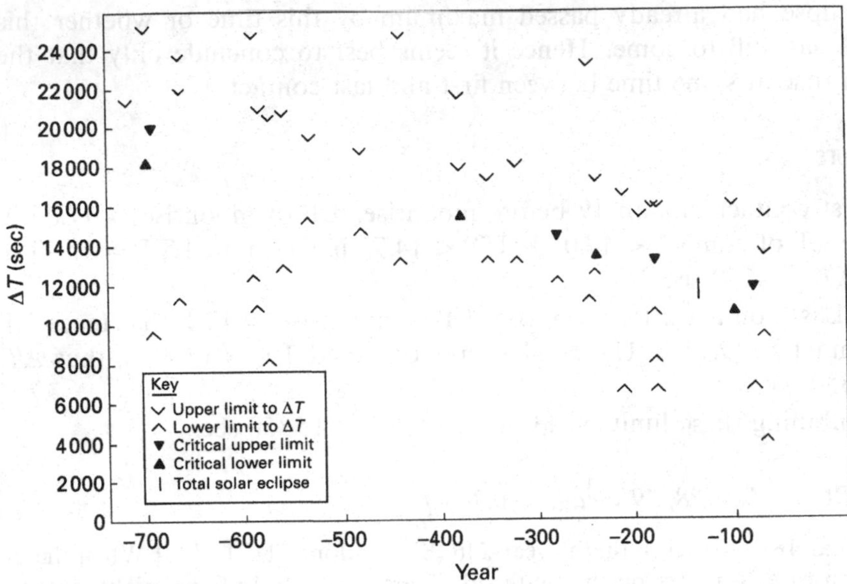

Fig. 7.1 ΔT limits obtained from rising and setting observations in chapters 4, 5 and 7, together with the total solar eclipse of 136 BC.

(23) BC 666 Apr 10/11 (mag. = 0.48) [*]

[Samas-sum-ukin] year 2, month I, beginning(?) on [...]. (Began) at 3 deg after sunset.

[BM 45640 (= LBAT 1415), Obv. III, bottom; trans. Huber, p. 8.]

Results

First contact on Apr 10 after sunset. LT of sunset = 18.29 h. Hence LT of contact > 18.29 h, UT > 15.37 h. Computed TT = 21.42 h, thus ΔT < 21 750 sec (only one limit).

NB on this occasion, the Moon would rise some 2 deg before sunset, so that it would be well clear of the horizon by the time the Sun set.

In table 7.1 are summarised the various results obtained in this section.

The results listed in table 7.1, along with the few ΔT ranges obtained from rising and setting observations in chapters 4 and 5 are shown in figure 7.1. (Also shown, for comparison, is the range derived from the total solar eclipse of 136 BC.) Although it is clear that many of the wider limits are redundant, several observations provide valuable bounds on ΔT at specific epochs. It should be noted that the limits set by a few observations between about −400 and −200 seem unusually critical. This question will be considered in chapter 14 when the full suite of data obtained throughout this book is analysed.

Table 7.1 ΔT limits from Babylonian observations where the Moon was seen to rise or set whilst eclipsed.

				ΔT Range (sec)	
Year[a]	MR/MS[b]	Ct[c]	Ct[c]	LL[d]	UL[d]
−701	MS	1	4	18 350	25 000
−694	MS	1	4	9 700	19 850
−666	MS	1	4	11 400	23 500
−665	MS	1	—	—	21 750
−590	MR	1	4	12 550	24 700
−587	MS	1	4	11 000	20 850
−576	MS	1	4	8 250	20 300
−562	MR	1	4	13 050	20 600
−536	MS	1	2	15 450	19 300
−482	MS	1	2	14 900	18 700
−441	MS	1	4	13 350	24 650
−352	MS	3	4	13 450	17 350
−247	MS	1	4	11 450	23 100
−239	MS	1	2	13 650	17 300
−211	MS	1	4	6 800	16 550
−189a	MS	3	4	10 700	15 850
−189b	MR	1	4	8 250	13 100
−188	MS	2	4	6 900	15 800
−149	MS	2	4	8 350	16 050
− 98	MR	3	4	10 100	13 800
− 79	MS	M	4	6 850	11 700
− 66	MR	1	4	4 150	15 350
− 65	MR	1	M	9 500	13 350

[a] Year (using negative integers rather than BC).
[b] Whether the eclipse was observed near moonrise (MR) or moonset (MS).
[c] The contacts between which the Moon rose or set.
[d] The range of ΔT indicated, from lower limit (LL) to upper limit (UL).
NB −189a and −189b refer to the first and second lunar eclipses in that year. The combined limits for that year are $10\,700 < \Delta T < 13\,100$ sec.

7.3 Estimates of the proportion of the Moon obscured at its rising or setting

In this section, I have discussed all reliably dated Babylonian observations for which the proportion of the Moon covered at moonrise or moonset is clearly stated. Texts in which the sign for the appropriate number of digits (or fraction) is damaged will be ignored. In each case a correction to the observed degree of obscuration derived from figure 3.23 (although scarcely significant) will be applied. Since all of the observational records have been translated in section 7.2, only summaries containing the relevant details will normally be cited here.

(1) BC 537 Oct 16/17 (mag. = 1.50)

> [Cyrus, year 2], month VII, 10[+x...it made] $\frac{2}{3}$ of the disk towards totality. Not (yet) total, it set eclipsed.... (Began) at 14 deg before sunrise.
>
> [BM 32234 (= LBAT 1419), Obv. V; trans. Huber, p. 25.]

It is clear that the estimate of phase was made before immersion. The proportion of the lunar diameter covered at moonset, adjusted according to the line of best fit in figure 3.23, will be taken as 0.70.

Results

LT of moonset on Oct 17 = 6.22 h, UT = 3.08 h. When 0.70 of lunar diameter obscured (phase increasing), TT = 8.14 h. Hence ΔT = 18 200 sec.

(2) BC 353 Nov 21/22 (mag. = 1.35)

> [Artaxerxes III, year 6] month VIII 14. Beginning on the south east side. After 23 deg total. 18 deg duration of maximal phase. After 6 deg night, one quarter of the disk had become bright (?) (and) it set eclipsed...
>
> [BM 32238 (= LBAT 1414), Rev. III, bottom; Huber, p. 49.]

NB from Huber, the translation of line 4 – 'one quarter of the disk had become bright' – is pure guesswork. However, this interpretation seems reasonable since the interval between emersion and moonset is roughly one-quarter of the interval from first contact to immersion. The proportion of the lunar diameter covered at moonset, adjusted as above, is 0.80.

Results

LT of moonset on Nov 22 = 6.87 h, UT = 3.74 h. When 0.80 of lunar diameter obscured (phase decreasing), TT = 7.75 h. Hence ΔT = 14 450 sec.

(3) BC 190 Aug 23/24 (mag. = 1.02)

SE 122...month V 14. Moonrise to sunset 1 deg, cloudy, measured. When the Moon rose from a cloud(?), 2 fingers on the [west] side lacked to brightness. (Began) at 1 *beru* before sunset.

[BM 34579 (= LBAT 1251), Rev. 14–18: trans. Huber, p. 60.]

As discussed in section 7.2, it seems most likely from the lack of reference to further phases and the terminology that the phase was decreasing at moonrise. The (adjusted) proportion of the lunar diameter covered at moonrise will be taken as 0.08.

RESULTS

LT of moonrise on Aug 23 = 18.70 h, UT = 15.77 h. When 0.08 of lunar diameter obscured (phase decreasing), TT = 19.31 h. Hence ΔT = 12 700 sec.

(4) BC 99 Apr 11/12 (mag. = 1.82)

Year 213 (SE), month I ... Night of the 14th, when the Moon came out (i.e. rose), it was eclipsed 3 fingers on the east side ...

[BM 140677 (no LBAT number), Obv. 4–6; trans. SH III, p. 407.]

As discussed in section 7.2, the statement that when the Moon rose it was partially obscured on the east side is at variance with the lack of reference to further stages in the text. Hence it will be assumed that the estimate of 3 fingers could refer to either an increasing or decreasing phase at moonrise. The (adjusted) proportion of the lunar diameter covered at moonrise will be taken as 0.18.

RESULTS

(i) LT of moonrise on Apr 11 = 18.29 h, UT = 15.33 h. When 0.18 of lunar diameter obscured at moonrise (phase increasing), TT = 15.54 h. Hence ΔT = 750 sec.

(ii) LT of moonrise = 18.42 h. When 0.18 of lunar diameter obscured (phase decreasing), TT = 19.09 h. Hence ΔT = 13 050 sec.

The former result is highly discordant. However, the latter figure is in close agreement with ΔT values derived from roughly contemporaneous observations and will thus be retained.

(5) BC 80 Apr 10/11 (mag. = 0.60)

Year 168 (Arsacid), that is year 232 (SE)...month I 13 ... Lunar eclipse beginning on the south-east side. In 20 deg night it made 6 fingers. 7 deg duration of maximal phase until it began to become bright. In 13 deg from south-east to north-west, 4 fingers lacking to brightness, it set...

[BM 33562A (= LBAT 1445), Obv. and Rev.; trans. Huber, pp. 75–76.]

The (adjusted) proportion of the lunar diameter covered at moonset will be taken as 0.28.

RESULTS

LT of moonset on Apr 11 = 5.62 h, UT = 2.67 h. When 0.28 of lunar diameter obscured (phase decreasing), TT = 5.55 h. Hence ΔT = 10 350 sec.

(6) BC 67 Jan 19/20 (mag. = 0.81)

> Year 180 (Arsacid), that is year 244 (SE)... month X 15... As the Moon rose, two thirds of the disk on the north-east side were eclipsed. 6 deg night duration of maximal phase, until it began to become bright...
>
> [BM 45628 (= LBAT 1448) + Duplicate BM 41565 (= LBAT 1447), Obv. and Rev.; trans. Huber, pp. 77–78.]

The (adjusted) proportion of the lunar diameter covered at moonrise will be taken as 0.70.

RESULTS

LT of moonrise on Jan 19 = 17.15 h, UT = 14.44 h. When 0.70 of lunar diameter obscured (phase increasing), TT = 17.61 h. Hence ΔT = 11 400 sec.

(7) BC 66 Dec 28/29 (mag. = 0.34)

> Year 182 (Arsacid), that is year 246 (SE)... Month IX 14 [...] When the Moon rose, 2 fingers on the south side [were eclipsed]. In 9 deg night, over a third of the disk [was eclipsed] 8 deg duration of maximal phase, until it began to become bright....
>
> [BM 32845 (= LBAT 1450), Obv. and Rev.; trans. Huber, pp. 79–80.]

The (adjusted) proportion of the lunar diameter covered at moonrise will be taken as 0.08.

RESULTS

LT of moonrise on Dec 28 = 16.90 h, UT = 14.14 h. When 0.08 of lunar diameter obscured (phase increasing), TT = 16.89 h. Hence ΔT = 9900 sec.

In table 7.2 are summarised the results obtained in this section. The following details are listed for each observation: year (negative); whether the eclipse was seen at moonrise (MR) or moonset (MS); the estimated fraction of the Moon obscured; and the resulting value for ΔT.

The few results in the above table are plotted in figure 7.2. It can be seen that these are in good accord with both one another and also the much larger quantity of data plotted in figure 6.10. Since they lead to discrete values of ΔT (rather than limits), they will be grouped with timed data in subsequent analysis.

Table 7.2 ΔT values from Babylonian observations for lunar eclipses in which the phase was estimated at moonrise or moonset.

Year	MR/MS	Est. mag.	ΔT (sec)
−536	MS	$\frac{2}{3}$	18 200
−352	MS	$\frac{3}{4}$	14 450
−189	MR	$\frac{1}{6}$	12 700
− 98	MR	$\frac{3}{12}$	13 050
− 79	MR	$\frac{4}{12}$	10 350
− 66	MR	$\frac{2}{3}$	11 400
− 65	MR	$\frac{2}{12}$	9 900

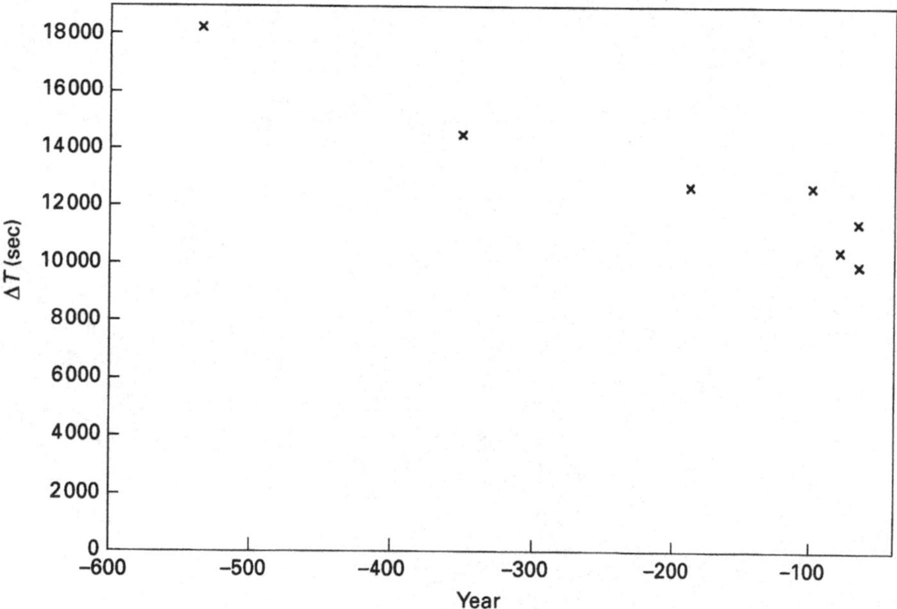

Fig. 7.2 ΔT values obtained from lunar eclipses in which the fraction of the Moon obscured at its rising or setting was estimated.

7.4 Conclusion

In the four chapters (4–7) of this book devoted to Babylonian data, a large number of valuable results and limits for ΔT in the period from 700 to 50 BC were deduced. The agreement between the ΔT values from timed measurements (figures 6.10 and 6.11) and untimed observations (figure 7.1 is remarkably good throughout the whole period covered by the data. Further results from ancient observations will be derived in chapters 8 (Chinese) and 10 (European), but there can be no doubt that the Babylonian data provide by far the most important source for information on ΔT before the medieval period.

8

Chinese and other East Asian observations of large solar eclipses

8.1 Introduction

More solar eclipses are recorded in the history of China than in the annals of any other civilisation. Not only were these events regarded as important astrological omens by the Chinese from an early period, but they also played a major role in the maintenance of the calendar.

Eclipses of the Sun have been systematically observed in China from at least the eighth century BC. The many hundreds of reports which are preserved since then are part of a huge corpus of accounts of celestial phenomena of various kinds (including eclipses of the Moon, lunar and planetary conjunctions, comets, novae and supernovae, meteors, sunspots and the aurora borealis). Most of these events were noted by official astronomers, who were employed by the ruler to keep a regular watch of the sky for ominous happenings. Nearly all of the original reports have long since been lost. Existing records – as found in the standard dynastic histories and other historical compendia – are usually no more than summaries of what may well have been detailed descriptions. These secondary sources are readily accessible in major libraries throughout the world, having been printed and reprinted many times. Block printing was discovered quite early in China (eighth century AD) and as a result older manuscripts have been phased out and are now relatively rare.

During the early centuries of its history, Chinese writing gradually evolved from simple pictograms to an advanced form of ideographic script. This script has remained essentially unchanged since the first century AD. Around 1950, the more complex characters were simplified in the People's Republic, but this revision has not found favour in Taiwan and Hong Kong, for example.

In my search for records of eclipses, I have made extensive use of the dynastic histories. A recent compilation of astronomical records in Chinese history entitled *Zhongguo Gudai Tianxiang Jilu Zongji* ('A Union Table of Ancient Chinese Records of Celestial Phenomena') has also proved extremely valuable. This compendium, published by Beijing Observatory (1988), is based on a systematic search of both national and local histories of China by a team of scholars.

Around the middle of the first millennium AD, Korean and Japanese astronomers independently began to make celestial observations, in each case following very much the traditional Chinese pattern. Numerous records of celestial phenomena are still preserved in the history of both countries, and these are also very largely found in printed works. Ancient and medieval Korean histories are invariably written in Classical Chinese and this is also true of most early histories of Japan. Although the Korean alphabet of 24 letters (known as *Han'gul*) was invented in the fifteenth century AD, it was not popular with the literati of the country until the present century. Medieval Japanese texts occasionally employ the *kana* syllabic signs in addition to Chinese characters; these were developed in the ninth century AD.

Among the accounts of solar eclipses from East Asia, more than fifty relate to instances where the Sun was said to be either totally or largely obscured. Observations of this type – most of which are untimed – will be the subject of the present chapter. After Chinese data have first been considered, Korean and Japanese material will then be discussed. However, it should be emphasised that as a rule the observations from China are superior to those from elsewhere in East Asia. Measurements of solar and lunar eclipse times, estimates of solar eclipse magnitudes and instances where the Sun or Moon was seen to rise or set eclipsed will be the subject of chapter 9.

Throughout this chapter (and also chapter 9), the Wade–Giles system of Chinese romanisation will be generally adopted; the only exceptions will be the personal names of current Chinese authors and the names of modern cities (for example Beijing) and provinces, for which the pinyin equivalent has become fairly standard. For Korean names the McCune–Reischauer method of romanisation is used, and for Japanese names the Hepburn system. Translations, except where otherwise indicated, are by the author. The help of Dr Kevin K. C. Yau (of Jet Propulsion Laboratory, Pasadena) with difficult texts is gratefully acknowledged.

A list of the principal dynasties in Chinese history – together with the intervening periods when the country was divided into two or more independent states – is given in table 8.1. This table also lists the inclusive dates for each era. The earliest dynasty whose existence can definitely be established is the Shang; numerous original inscriptions have survived from this period. According to tradition, the Shang was preceded by the Hsia dynasty. However, no contemporary Hsia documents are extant, and the very existence of this dynasty has been called into question (see, for example, Fitzgerald, 1976, pp. 26 ff.). The dates for the beginning and end

Table 8.1 Major Chinese dynasties and periods of partition.

Dynasty etc.	Date range
Shang	*c.* 1500–1050 BC
Chou	*c.* 1050–480 BC
Chan-kuo	*c.* 480–221 BC
Ch'in	221–206 BC
Former Han	206 BC–9 AD
Hsin	AD 9–23
Later Han	AD 23–220
San-kuo	220–265
Chin	265–420
Nan-pei	420–589
Sui	581–618
T'ang	618–907
Wu-tai	907–960
Sung	960–1279
Kin	1115–1234
Yuan	1271–1368
Ming	1368–1644
Ch'ing	1644–1911

of the Shang are uncertain, but most subsequent periods are accurately dated. As will be seen from table 8.1, there are occasional periods of overlap between dynasties. For example, although the Sui dynasty was founded in AD 581, it did not establish rule over the whole of China until 589.

8.2 Records of solar eclipses from the Shang dynasty

Only a very few allusions to eclipses survive from the Shang. Like other records of the period, these are inscribed on a variety of animal bones, etc. Since the beginning of the present century, vast numbers (some 150 000 in total) of Shang inscriptions have come to light as the result of excavations near the city of An-yang in north-east China. This location is generally regarded as the site of the Shang capital of Yin. The various texts which have so far been studied probably all range in date from *c.* 1350 BC to 1050 BC. At some time around the former epoch, Yin became the royal residence, while the most likely date for the end of the Shang dynasty is close to 1050 BC.

Shang inscriptions are largely in the form of oracular texts; it is clear that divination was practised on a major scale. The diviners of the period attempted to foretell the future by making cracks with a hot needle on the prepared surfaces of animal bones etc., such as ox shoulder blades and turtle plastrons (the relatively flat lower shell), and interpreting the form of these cracks (Keightley, 1978a; Chou, 1979). The results of their divination were inscribed on these same bones using a script which, although far from primitive, is much closer to picture writing than the present highly evolved system of ideographs (see figure 8.1). After use, the bones were buried to prevent defilement; this has led to their preservation in such large numbers.

The following text is fairly typical:

> The divination on day *chi-mao* (the 16th day of a 60-day cycle) was performed by Kuo. The King, after examining the crack forms, commented that it would rain on day *jen (-wu)*. On day *jen-wu* (the 19th day of the cycle) indeed it did rain.
>
> [Trans. Xu *et al.* (1989).]

Many other inscriptions of the period are concerned with similar mundane matters. Evidently the diviners were astute; they managed to achieve sufficient success in their predictions to ensure the continuation of this practice for several centuries. Since the inscriptions are largely devoted to various aspects of divination, there are few reports of historical events; astronomical observations are especially rare. In addition, most texts have suffered extensively with the passage of time and are now mere fragments. Despite these drawbacks, the names of nearly all of the Shang rulers have been discovered on the oracle bones, thus confirming the names in the traditional king list as found in later works.

The Shang calendar was basically lunar, but days were usually specified according to a 60-day cycle, examples of which are given in the above quotation. The origin of this cycle, which is independent of any astronomical parameters, is obscure. Nevertheless, the scheme continued in use – probably without any interruption – in the succeeding Chou dynasty and throughout later Chinese history; it still remains in operation today at the popular level. Ten 'celestial stems' (*t'ien-kan*) are combined consecutively with twelve 'earthly branches' (*ti-chih*) making 60 combinations in all. The ten *kan* are as follows: *chia, i, ping, ting, wu, chi, keng, hsin, jen, kuei*; the twelve *chih* are *tzu, ch'ou, yin, mao, ch'en, szu, wu, wei, shen, yu, hsu, hai.* In any particular cycle, the first day was *chia-tzu*, the second *i-ch'ou*, and so on until the 60th day *kuei-hai* was reached, after which a new cycle began. The full *kan-chih* cycle (as it is usually termed) is listed in table 8.2.

Fig. 8.1 Example of oracle bone inscription.

An eclipse was identified by the Shang people using the term *shih* ('to eat'); they evidently believed that some monster was devouring the luminary. This became the standard term to describe eclipses at all subsequent periods in Chinese history, long after the true explanation was understood (approximately first century AD).

Table 8.2 The Chinese sexagenary cycle.

	tzu	ch'ou	yin	mao	ch'en	szu	wu	wei	shen	yu	hsu	hai
chia	1		51		41		31		21		11	
i		2		52		42		32		22		12
ping	13		3		53		43		33		23	
ting		14		4		54		44		34		24
wu	25		15		5		55		45		35	
chi		26		16		6		56		46		36
keng	37		27		17		7		57		47	
hsin		38		28		18		8		58		48
jen	49		39		29		19		9		59	
kuei		50		40		30		20		10		60

Several references to solar and lunar eclipses have been identified on the Shang oracle bones (e.g. Pang *et al.*, 1988b; Xu *et al.*, 1989; Xu *et al.*, 1995). However, the records are often incomplete, while the meanings of many Shang ideographs are still poorly understood. In addition, eclipse dates are only partly preserved. Although the cyclical day is usually specified, in no case is the year cited – while only a single report gives the lunar month. Hence attempts to date a text which is believed to refer to an eclipse involves 'playing the identification game' (Newton, 1970, p. xiv).

Shang observations of lunar eclipses are briefly discussed in chapter 9. Among possible references to solar obscurations, only a single Shang text has ever been seriously proposed as referring to a *total* eclipse, but the interpretation involves a strong element of speculation. The phenomenon, recorded on a turtle plastron is stated to have occurred on the day *i-mao* (the 52nd day of the sexagenary cycle).

An eclipse identification was first proposed by Liu Chao-yang (1945). Recently, Pang *et al.* (1988b) revived this hypothesis. They gave a partial translation of the text as follows:

(i) Diviner Ko: ... day *i-mao* [cyclic day 52] to [next] dawn, fog. Three flames ate the Sun. Big stars [seen].

However, Prof. David N. Keightley, of the University of California, Berkeley – who has made an extensive study of Shang inscriptions – has suggested the following alternative translation of the full inscription in a personal communication to me:

(ii) On the next day *i-mao* it may not be sunny. The King read the cracks and said, 'There will be disaster but it will not rain'. On day *i-mao* at dawn it was foggy; when it came to the time of the ... meal, the day greatly cleared.

Although the term 'eclipse' is not mentioned in the text, Pang and his colleagues assumed that the phenomenon described was a total solar eclipse with a vivid corona. They accordingly computed the local circumstances of all solar obscurations which could have been visible at An-yang (lat. = 36.07 deg, long. =−108.88 deg) for several decades on each side of 1300 BC. No large eclipse proved to be visible at this location on any date whose cyclical day number was 52, but they identified a total obscuration of the Sun on a certain occasion when the day number was 53. The calculated Julian date of this event was BC 1302 Jun 5. Values of ΔT which produce totality at An-yang lead to a local time around 11 h.

Pang *et al.* did not attempt to justify their selection of day 53 (instead of the previous cyclical day). If the phenomenon had indeed occurred on day 53, a reference to this date would have been expected in the text since the oracle bones are careful to specify individual dates. As it happens, references to the corona – even in a clear sky – are extremely rare in the history of all ancient and medieval cultures (see also chapters 10 and 11). If the translation by Keightley is correct, the text resembles the weather prediction quoted near the beginning of this section. However, it is evident that scholars are in dispute over the rendering of several key characters in the inscription. Until these difficulties are resolved, the record would appear to be of negligible value in the study of long-term changes in the rate of rotation of the Earth.

8.3 The Chou dynasty and Warring States period (*c.* 1050–221 BC)

Around 1050 BC, the inhabitants of the Chou region in western China overthrew the Shang kingdom and established a new dynasty. Early Chou history is very fragmentary and there are no direct references to eclipses until the eighth century BC. Various attempts have been made to explain a supposed 'double-dawn' occurring around 900 BC as produced by a major eclipse at sunrise. This phenomenon is recorded in the 'Bamboo Annals' (*Chu-shu Chi-nien*), a chronicle recovered in AD 281 from the tomb of a prince who had died in 296 BC. Although the chronicle has long since been lost again, its content was partially reconstructed during the last (Ch'ing) dynasty from preserved quotations in a variety of writings (Keightley, 1978b). It might be mentioned here that before the invention of paper in the first century BC, bamboo was a common writing medium, a book being formed by tying together a number of strips, each of typical length about 50 cm. The record of interest, which like other entries in the Bamboo Annals is very brief, may be translated as follows:

> During the first year of King I (of western Chou), the day dawned twice in Cheng.

King I, the seventh Chou ruler, is believed to have reigned at some time between about 966 and 895 BC (Pang *et al.*, 1988a).

Cheng was located in central China, not far from present-day Loyang. (The approximate geographical co-ordinates of the site are: lat. = 34.50 deg, long. = −109.80 deg.) In principle, it would perhaps not be unreasonable to attribute the phenomenon described in the Bamboo Annals to a total eclipse occurring at sunrise. Rather similar phenomena were recorded during medieval times in England (AD 1230) and in Russia (1476) – see chapter 11. However, it should be stressed that in the Chou texts there is no direct reference to the cause of the phenomenon. If the Sun rose fully obscured, the closing stages of the event would resemble any other eclipse so that its true nature would be evident. It is clear from the descriptions given by the medieval European observers mentioned above that they were well aware that an eclipse was responsible for the darkness which they experienced. In contrast, the absence of the term *shih* in the Chinese account materially weakens an eclipse interpretation.

Several investigations on the eclipse theme have been published since the pioneering work of Liu Chao-yang (1944), but none can be regarded as successful. The most recent attempt is that of Pang *et al.* (1988a). From a computer check of all eclipses during the selected interval from 966 to 895 BC, these authors chose the sunrise eclipse of Apr 21 in 899 BC as the 'only possible match'. However, this eclipse was merely annular and even at those places where the central phase was witnessed fully ten per cent (by area) of the solar disk would remain unobscured. Hence the reduction in the level of illumination would be scarcely noticeable – less than on a typical overcast day.

As is apparent from the historical eclipse maps for East Asia produced by Stephenson and Houlden (1986), no eclipse in the preferred date range could possibly have been *total* at Cheng, even if large uncertainties in the value of ΔT at this epoch were supposed. It seems doubtful whether any alternative natural explanation (e.g. meteorological) can be found. Possibly the original account in the Bamboo Annals was merely drawing on legend rather than having a real, factual basis. For a recent critical re-appraisal of the evidence, see Stephenson (1992).

Commencing in the eighth century BC, we encounter the first reliable observations of solar eclipses in Chinese history. These are recorded in a Chou dynasty chronicle known as the *Ch'un-ch'iu* (Spring and Autumn Annals). This intriguing compilation, which is reputed to have been edited by Confucius, extends from 722 to 481 BC. As one of the accepted Confucian Classics, the text has been studied and commented on by generations of scholars. Between AD 837 and 841 the entire set of these Classics was engraved on series of stone tablets at the T'ang dynasty capital of Ch'ang-an (now Xi'an) – (Carter, 1925, p. 212). These stelae are

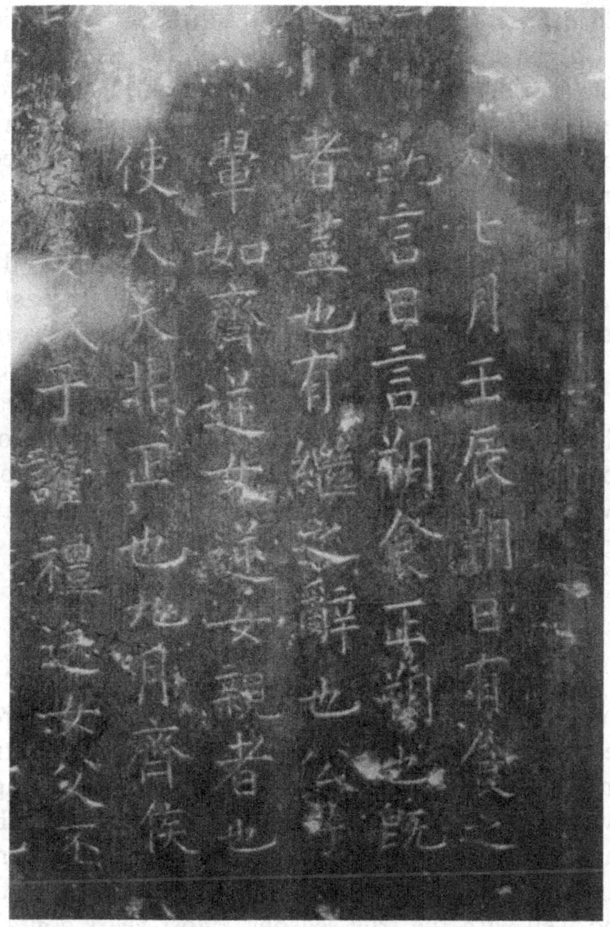

Fig. 8.2 Section of a ninth century stele in Xi'an Museum containing a record of the total solar eclipse of 709 BC. (Courtesy: Dr Liu Ciyuan.)

now preserved in the Beilin Museum (Museum of the Forest of Stelae) in Xi'an.

A photograph of a section of one of these stelae, containing the record of the total solar eclipse of 709 BC, is shown in figure 8.2.

The *Ch'un-ch'iu* has been translated into English by Legge (1872). The chronicle is largely concerned with the State of Lu (the birthplace of Confucius) and its interaction with the other semi-independent states into which China was divided at this period. Only a formal allegiance was paid to the Chou king by the rulers of individual states.

The only state chronicle of ancient China which is still intact – possibly as a result of the 'Burning of the Books' at the command of Emperor Ch'in Shih Huang-ti in 213 BC – the Spring and Autumn Annals records as many as 36 eclipses of the Sun. This series of observations, which commences with the event of Feb 22 in 720 BC, is the earliest from any part of the world. Only isolated eclipse sightings of more ancient date are preserved from elsewhere, notably an observation in Assyria in 763 BC (see chapter 5). Lately, a detailed discussion of the eclipses and other astronomical records in the *Ch'un-ch'iu* (including a few sightings of comets and meteors) has been published by Stephenson and Yau (1992a).

Despite the interest in *solar* eclipses at this early period, there are no allusions to *lunar* obscurations in the *Ch'un-ch'iu.* By at least the eighth century BC, eclipses of the Sun were regarded as major omens of ill, whereas their lunar counterparts were considered to be of minor significance. This is probably because lunar eclipses were already recognised to be periodic phenomena, and were therefore not considered worth reporting. Thus in one of the early poems in the *Shih-ching* (Book of Odes) we find the following remark concerning two successive eclipses that occurred during the eighth century BC:

> That this Moon is eclipsed is but an ordinary matter; but that this Sun is eclipsed – wherein lies the evil?
>
> [Trans. Karlgren (1950, p. 100).]

In the Spring and Autumn Annals, years are numbered from the accession of the current 'Duke' (*Kung*) of Lu. Within any particular year, entries are grouped in four conventionalised seasons (spring, summer, autumn and winter), each of duration three lunar months. The calendar was so regulated that the year began around – or rather earlier than – the time of the winter solstice. Hence the four divisions of the year do not correspond well with the true seasons. Most years contained twelve lunar months, but an occasional extra month was inserted to keep the calendar in step with the seasons. This intercalary month could be added at any time of year. There is no evidence that months ever began with the first sighting of the lunar crescent. Since solar eclipses recorded in the *Ch'un-ch'iu* invariably began on the first day of the lunar month, it may be supposed that they were used to regulate the calendar. How the start of a month was fixed during the long intervals between observable eclipses is not known. Use of a luni-solar calendar continued at all later stages in Chinese history.

Many dates in the chronicle are also expressed in terms of the sexagenary cycle and this is especially true of solar eclipses. Although intercalation may have been somewhat irregular at this early period, the use of the sexagenary cycle enables accurate conversion of dates to the Julian calendar with the minimum of assumptions. Nearly all dates, when so reduced, prove to be in *exact* accord with those of eclipses listed in modern tables. This is a remarkable result for such an ancient work and it gives strong support to the authenticity of the purely historical data recorded in the Classic.

The main focus of the *Ch'un-ch'iu* is on Lu affairs. Eclipse ceremonies held at the state capital of Ch'u-fu are described in the *Ch'un-ch'iu* on several occasions (BC 669 May 27, 664 Aug 28 and 612 Apr 28). It seems likely that all or nearly all of the eclipse observations in the chronicle were made at Ch'u-fu. This was situated on the site of the present city of the same name (pinyin equivalent Qifu) in north-east China. Already by the Chou dynasty, individual states are known to have had official astronomers, and the names of some of these are still preserved. These are recorded in the *Shih-chi* ('Historical Record': chaps. 6 and 27), a work compiled by the Grand Historian Szu-ma Ch'ien between 104 and 87 BC. In each case the record of the eclipse ceremonies at Ch'u-fu reads as follows:

> The Sun was eclipsed; drums were beaten and oxen were sacrificed at the temple.

Presumably, similar rites took place at other eclipses. Even in relatively recent centuries, it was the official custom to beat drums and gongs during an eclipse (in an attempt to rescue the Sun from the dragon which was imagined to be devouring it). Yet, already by the Ch'un-ch'iu period, eclipses seem to have been recognised as natural phenomena, at least among certain individuals. For example, the *Tso-chuan* (an enlargement of the *Ch'un-ch'iu* mainly composed around 300 BC) records the following extract from a speech made in 597 BC in support of a general who had recently suffered defeat and was in danger of execution:

> His defeat is like an eclipse of the Sun or Moon; does it harm the brightness (of these bodies)?

In view of these well-chosen words, the general was reinstated!

The style of the *Ch'un-ch'iu* is laconic and this applies to the eclipse records; in most cases it is merely reported that on a certain day the Sun was eclipsed. Although on three occasions (BC 709 Jul 17, 601 Sep 20 and 549 Jun 19) the Sun was said to be totally obscured, further details – e.g. darkness or the appearance of stars – are lacking. In each case the term *chi* is used to indicate totality. This character originally was a pictogram of a man turning his head away from a dish of food, indicating that he was replete; its astronomical meaning is thus fairly obvious. Throughout subsequent Chinese history this same expression continued in usage to describe major solar obscurations.

There are no references to large partial obscurations of the Sun in the *Ch'un-ch'iu* – e.g. using terms such as 'almost complete' or 'like a hook',

Table 8.3 Geographical co-ordinates of Chinese cities where the eclipse observations analysed in this book were made.

Place name	Modern name	Lat. (deg)	Long. (deg)
Ch'ang-an (Han)	Xi'an	34.35	−108.88
Ch'ang-an (T'ang)	Xi'an	34.27	−108.90
Chien-k'ang	Nanjing	32.03	−118.78
Ch'u-fu	Qifu	35.53	−117.02
Lin-an	Hangzhou	30.25	−120.17
Lo-yang	Luoyang	34.75	−112.47
Pien	Kaifeng	34.78	−114.33
P'ing-ch'eng	Datong	40.20	−113.20
Sung-chiang	Songjiang	31.00	−121.22
Ta-hsing Ch'eng	Xi'an	34.27	−108.90
Ta-tu	Beijing	39.92	−116.42

as found in later works from the Han dynasty onwards. Hence at this early period it is possible that the expression *chi* might have been loosely used to describe both total and nearly-total eclipses. However, even if the original reports had been fairly specific, loss of detail would almost certainly occur when they were included in the *Ch'un-ch'iu* in order to conform to its style. For these reasons, it might well be doubted whether the observers in 709, 601 and 549 BC actually witnessed the central phase or merely saw an eclipse in which the Sun was very largely covered.

At subsequent periods in Chinese history, records of totality are sometimes qualified by allusions to darkness or the visibility of stars by day. If so, an observation will be assigned a high rating (class 'A'). A brief account of a total eclipse which is supported by extra details in other sources will usually be included in category 'B', whereas one which merely notes the occurrence of totality without providing any further information will be assigned to the lowest category (class 'C'). In the case of the eclipse of 709 BC, a later text provides additional particulars. Hence the observation will be given a 'B' classification. However, the two records from 601 and 549 BC are not supported by any other sources and hence will be allocated to the lowest grade. The utility of this grading scheme will be reviewed at the close of this chapter.

The location of Ch'u-fu and those cities where later eclipse observations analysed in this and the following chapter were made are shown in figure 8.3; geographical co-ordinates (in degrees and decimals) are listed in table 8.3. On the map, the modern place names are used.

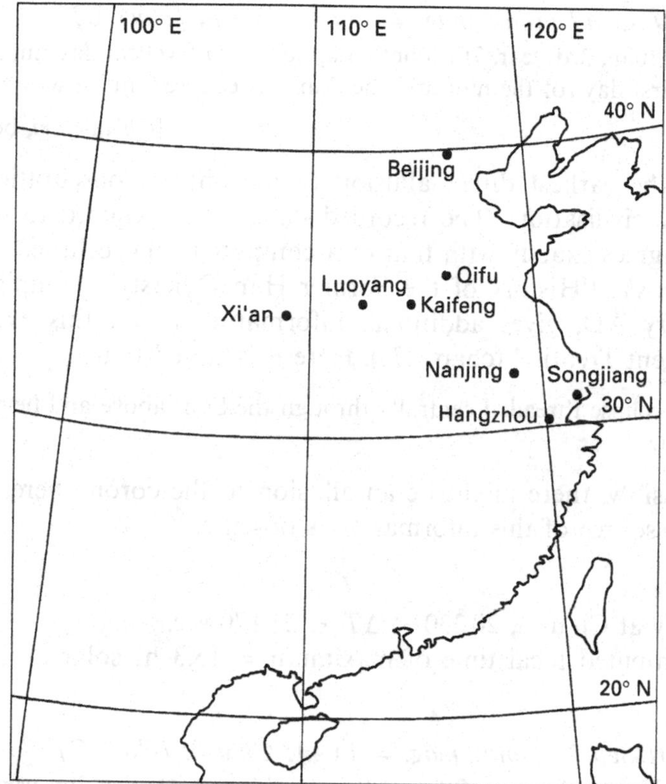

Fig. 8.3 Chart showing the locations of Chinese cities where eclipses were observed.

For each of the eclipses of 709, 601 and 549 BC, the following summary information is given below: calculated Julian date; type of eclipse (including computed central magnitude); presumed place of observation (namely Ch'u-fu); and classification. These details are followed by a full translation of the record – including the original date – together with brief additional remarks and also the computed range of ΔT (in seconds) necessary to render the eclipse total at Ch'u-fu. The calculated approximate local time and solar altitude at maximum phase, based on equation (4.1), are also given for reference; these latter details will not be used to interpret the records in any way. This same formula will also be used to calculate local times and altitudes in later sections of this chapter for all dates prior to AD 950. However, for all later eclipses, these peripheral details will be computed using a further ΔT equation derived by Stephenson and Morrison (1984). This is as follows:

$$\Delta T = 25.5t^2, \tag{8.1}$$

where t is in Julian centuries from the epoch AD 1800.

(1) BC 709 Jul 17 (total, mag. = 1.06): Ch'u-fu [class B]

Duke Huan, 3rd year, 7th month, day *jen-ch'en* [cyclical day number = 29], the first day (of the month). The Sun was eclipsed and it was total.

[*Ch'un-ch'iu*, book I.]

This is the earliest direct allusion to a complete obscuration of the Sun in any civilisation. The recorded date, when reduced to the Julian calendar, agrees exactly with that of a computed solar eclipse.

The *Han-shu* ('History of the Former Han Dynasty'), compiled in the first century AD, gives additional information about this event in its 'Five-Element Treatise' (chap. 27c). Here it is stated that:

> ...the eclipse threaded centrally through the Sun; above and below it was yellow.

Just possibly, there might be an allusion to the corona here. Unfortunately, the source of this information is obscure.

RESULTS

For totality at Ch'u-fu, $20\,230 < \Delta T < 21\,170$ sec.

(NB computed local time of maximum = 15.3 h; solar altitude = 45 deg.)

(2) BC 601 Sep 20 (total, mag. = 1.04): Ch'u-fu [class C]

Duke Hsuan, 8th year, 7th month, day *chia-tzu* [1]. The Sun was eclipsed and it was total.

[*Ch'un-ch'iu*, book VII.]

In order for the recorded and computed dates to agree, the 7th month of the text should be replaced by the 10th month. However, the written characters for these numerals are very similar. If this alteration is made, the reduced date is in precise accord with computation.

RESULTS

For totality at Ch'u-fu: $21\,140 < \Delta T < 21\,960$ sec.

(NB computed local time of maximum = 15.6 h; solar altitude = 31 deg.)

(3) BC 549 Jun 19 (total, mag. = 1.08): Ch'u-fu [class C]

Duke Hsiang, 24th year, 7th month, day *chia-tzu* [1], the first day of the Moon. The Sun was eclipsed and it was total.

[*Ch'un-ch'iu*, book IX.]

Here the recorded and computed dates are in exact agreement. Since the track of totality was unusually broad and ran nearly parallel to the equator as it crossed China, the range of ΔT which satisfies the observation is rather wide.

RESULTS

For totality at Ch'u-fu: $16150 < \Delta T < 21710$ sec.

(NB computed local time of maximum = 13.9 h; solar altitude = 63 deg.)

The above ΔT results, along with those derived from later Chinese (and other East Asian) solar eclipse observations, will be summarised at the end of this chapter (tables 8.8 and 8.9).

The *Chan-kuo* or Warring States period which followed the *Ch'un-ch'iu* era lasted until 221 BC. Only nine solar eclipses are recorded during this interval of 260 years. These are cited briefly in the *Piao* or Chronological tables of the *Shih-chi* (chap. 15). They are presumably taken from a lost chronicle of the State of Ch'in in western China, whose capital was Hsien-yang (close to the site of modern Xi'an). Years are numbered from the accession of each Ch'in ruler. The earliest eclipse date corresponds to 444/443 BC, and the latest to 249/248 BC.

On three occasions (444/443, 383/382 and 302/301 BC) the *Shih-chi* records that the daylight was darkened by an eclipse, while on the first of these dates it is stated that stars were also seen. However, in each case, no more than the year of occurrence is specified and there is no indication as to whether the phase was central or only partial. The earliest of these records may be translated as follows:

> Duke Li (of the Ch'in dynasty), 34th year. The Sun was eclipsed. It became dark in the daytime and stars were seen.
>
> [*Shih-chi*, chap. 15.]

On the Ch'in calendar, the year began with the 10th lunar month, and thus commenced either in late October or November on the Julian calendar. The eclipse may be identified as that of BC 444 Oct 24. However, as this was annular, only 93 per cent of the Sun being obscured in the central zone, the allusion to darkness must be exaggerated. Both Venus (magnitude −3.3; 20 deg east of the Sun) and Mercury (−0.5; 14 deg east) would be well placed for visibility. It might be mentioned that the eclipse is the earliest in any civilisation for which the visibility of stars is reliably reported. (A similar occurrence, dating from 431 BC, is described by the ancient Greek historian Thucydides – see chapter 10.)

The other two eclipses were reported in the third year of Duke Hsi-an (302/1 BC) and the sixth year of Duke Chao Hsiang. The corresponding Julian dates can be readily established as BC 382 Jul 3 and 300 Jul 26, respectively. Both of these eclipses were generally total. Since all three Chan-kuo observations make no direct reference to magnitude, they are probably valueless for determining ΔT and hence will not be considered further.

8.4 The Ch'in and Han dynasties (221 BC to AD 220)

Towards the end of the Chan-kuo, Ch'in became the dominant state and eventually, in 256 BC, the last Chou king abdicated. Not long afterwards (221 BC) the Ch'in ruler unified the whole country. He then adopted the title Shih Huang-ti ('First Sovereign Emperor'). One of the most important archaeological finds in recent decades – the so-called 'Terracotta Army' – was made in an annexe of his tomb near Xi'an. No eclipses are recorded during the brief Ch'in dynasty, which lasted until 206 BC. Many historical documents were probably destroyed when, following the collapse of Ch'in rule, the capital of Hsien-yang was stormed by rebel troops led by Hsiang Yu (Dubs, 1938, vol. I, p. 7). Fortunately, the two official histories of the subsequent Han dynasty, covering the period from 206 BC to AD 220, contain many accounts of eclipses. From the Han onwards, it became the practice for scholars to be commissioned to write a comprehensive history after the end of each dynasty, the compilers usually having free access to official records including those of the imperial astronomers.

The *Han-shu*, compiled by Pan Ku between AD 58 and 76, covers the Former Han dynasty together with the brief Hsin interregnum (AD 9 to 23). During the Former Han the capital was Ch'ang-an ('Everlasting Peace'), whose site lies a little to the north-west of the present city of Xi'an. Wang Mang usurped the throne in AD 9 but his reign proved extremely unpopular and when he was killed by Han loyalist forces in AD 23, Ch'ang-an was so ruined that the capital had to be transferred to Lo-yang – far to the east. Events of the second phase of the Han dynasty are reported in the *Hou-han-shu* ('History of the Later Han Dynasty'), which was composed by Fan Yeh towards the middle of the fifth century.

In common with most later dynastic histories, the *Han-shu* and *Hou-han-shu* consists of four main divisions. These are: *Pen-chi* ('Basic Annals'), *Piao* ('Chronological tables'), *Chih* ('Treatises') and *Lieh-chuan* ('Biographies'). In both Han official histories, solar eclipse records are assembled (along with certain other astronomical observations - as well as meteorological data) in special monographs known as *Wu-hsing Chih* or 'Treatises on the Five Elements'. (NB the five elements of Chinese philosophy were earth, metal, water, fire and wood.) In each treatise, much stress is placed on the omen-value of the various natural phenomena which are reported. There are many other treatises on a wide variety of topics: astrology, the calendar, music, rites, food and commodities, and offices, to name just a few (see Han, 1955, pp. 200–201).

The lengthy Five-Element monograph in the *Han-shu* forms chapter 27 of this work and eclipse observations are found towards the end of this treatise (section 27c). In the corresponding treatise of the *Hou-han-shu*, which covers several chapters, eclipse records are confined to chapter 28.

Table 8.4 Treatises of Chinese dynastic histories which contain series of solar eclipse records.

Name of history	Chapter	Title
Han-shu	27c	Wu-hsing Chih
Hou-han-shu	28	Wu-hsing Chih
Chin-shu	12	T'ien-wen Chih
Sung-shu	34	Wu-hsing Chih
Nan-ch'i-shu	12	T'ien-wen Chih
Wei-shu	105	T'ien-hsiang Chih
Sui-shu	21	T'ien-wen Chih
Chiu-t'ang-shu	36	T'ien-wen Chih
Hsin-t'ang-shu	32	T'ien-wen Chih
Chiu-wu-tai-shih	139	T'ien-wen Chih
Hsin-wu-tai-shih	59	Szu-t'ien K'ao
Sung-shih	52	T'ien-wen Chih
Kin-shih	20	T'ien-wen Chih
Yuan-shih	48	T'ien-wen Chih

References to lunar obscurations are still extremely rare at this period. Brief notices of solar eclipses also frequently occur in the *Pen-chi* ('Basic Annals'), the opening chapters of these same histories; these compilations describe the main events of each emperor's reign. Very occasionally, an eclipse may be alluded to in another part of an official history – for instance in the *Lieh-chuan* or biographical section. Incidentally, observations of celestial phenomena other than eclipses are mainly reported in the *T'ien-wen Chih* ('Treatises on Astrology') of the two Han histories. These form chapter 26 of the *Han-shu* and chapters 20–22 of the *Hou-han-shu*.

In most later histories, eclipse observations are cited in the Astrological Treatise rather than in the Five-Element section. This usually goes under the title *T'ien-wen Chih* but other variants occasionally occur. The chapter numbers in the appropriate histories of the various *chih* which contain series of solar eclipse observations are listed in table 8.4.

During the early decades of the Former Han, years were numbered from the accession of each ruler. However, before long the use of subdivisions of a reign – or reign periods – was established. This practice continued down to the end of the last dynasty (it also spread to Japan, but reign periods were never used in Korea). Dates within a given year were usually expressed in terms of the lunar month, cyclical day and day of the month – as in the *Ch'un-ch'iu* – but the year now began in the early spring, rather than near the winter solstice. Only minor changes were made to the calendar in succeeding dynasties. During the Former Han and the first few decades of the Later Han, most solar eclipses occurred on the last day of a lunar month. Afterwards the calendar was so adjusted that the first day became much more common.

In expressing dates of Former Han eclipses in terms of the Julian calendar, I have followed Dubs (1938–1955). However, in each case I have verified his reduction of cyclical day numbers using a computer program. I have effected date conversion in the Later Han – and all subsequent dynasties – with the aid of a specially designed computer program based on the extensive chronological tables of Hsueh and Ou-yang (1956). It should be emphasised that throughout Chinese history from the Han onwards, recorded dates of solar eclipses, when converted to the Julian calendar, usually agree precisely with the calculated dates of these phenomena.

In common with the numerous other celestial events (such as occultations of stars and planets by the Moon, planetary conjunctions, etc.) reported in the astrological treatises, the observations of eclipses are probably largely based on the records of the court astronomers. Because of the emphasis placed on astrology by the Chinese court, these officials were employed to maintain a systematic watch of the sky – both day and night – at the imperial observatory for any unusual phenomenon which might be regarded as a celestial omen. They were required to keep their business secret so that the detailed astronomical knowledge that they possessed could not be passed to the general public (see, for example, Needham, 1959, p. 193). Soon after the first Han emperor Kao Tzu ascended the throne in 202 BC, an observatory was built at his capital of Ch'ang-an (present-day Xi'an). This observatory remained active for more than two centuries – until Ch'ang-an was sacked when the reign of the usurper Wang Mang was forcibly brought to an end in AD 23. Following the re-establishment of Han rule, a new imperial observatory was founded at Lo-yang; the site of this can still be identified today.

Usually, it seems that the official histories take for granted that a reader would assume that an astronomical observation was made at the capital itself. Only very rarely does a text directly mention the capital – for example, to comment on the visibility of an eclipse at this city and its invisibility in the provinces, or vice versa. Thus, on a date corresponding to BC 16 Nov 1 it is stated in the Five-Element Treatise of the *Han-shu* that:

> Heaven caused the capital alone to know of it; the states in the four (quarters) did not see it.

In the whole of the *Han-shu*, only a single solar eclipse (BC 15 Mar 29) is stated to have been observed in the provinces and on this occasion the weather was cloudy in Ch'ang-an. Here, the *Wu-hsing Chih* notes that:

> The four quarters all saw it, but in the capital it was overcast.

Provincial observers seem to have been unusually active in the Later Han and as many as 17 accounts (nearly one quarter of the total) in the Five-Element Treatise of this work are said to have been reported from a provincial location. At no other period in Chinese history do we find a comparable situation.

A typical record of this kind in the *Hou-han-shu* treatise takes the following form:

> The Sun was eclipsed; it was not seen by the official astronomers but was communicated from ... province.

In each case the precise date is given and the name of the province is usually specified. However, there is nothing in these brief records to suggest that any of the eclipses were unusually large. In the absence of a reference to a provincial location, it seems reasonable to assume that an observation was made at the capital itself.

Many solar eclipse reports in both the *Han-shu* and *Hou-han-shu* are accompanied by astrological comments. Since at least the time of the philosopher Mencius (fourth century BC), the ruler was regarded as the Son of Heaven and it was thus believed that celestial portents could either (a) warn him of coming danger, or (b) level criticism at him for mis-rule. Thus following an obscuration of the Sun on a date corresponding to Nov 14 in AD 30, the Bearer of the Gilded Mace memorialised to the throne that this event was a warning against too frequent transfer of officials. After a further eclipse six months later (AD 31 May 10), the Grand Palace Grandee memorialised that this portent was a warning against the Emperor's unwillingness to accept candidates recommended by the high officials. (Both of these accounts are given in chapter 36 of the *Hou-han-shu*; see also Bielenstein, 1979.) Shrewd courtiers must often have taken advantage of the opportunities for criticising the ruler presented by the occurrence of eclipses and other celestial omens.

In both Han Five-Element Treatises the right ascension (RA) of the Sun at an eclipse is frequently recorded to the nearest degree (*tu*) in one of the 28 lunar lodges. (NB there were 365.25 Chinese degrees to a circle.) Often an astrological prognostication is based on the lunar lodge in which the Sun happened to be. These constellations were associated both with different aspects of Chinese administration and the regions occupied by the ancient states into which China had once been divided. Thus the total eclipse of AD 65 Dec 16 was said to have occurred while the Sun was 11 deg in *Nan-tou*. The *Hou-han-shu* (chap. 28) text continues:

> *Nan-tou* represents (the ancient state of) Wu. Kuang-ling as far as the constellations are concerned belongs to Wu. Two years later, Ching, King of Kuang-ling, was accused of plotting rebellion and committed suicide.

Kuang-ling lay much to the east of Lo-yang, not far from the site of present-day Nanjing. A delay of two years between the occurrence of a celestial portent and its acknowledged fulfilment is not unusual.

On two occasions (in 181 BC and AD 120 – see below), a major eclipse was regarded as presaging the death of a member of the Han imperial family. A passage in a later work, the *Chin-shu* ('History of the Chin Dynasty'), shows that the omen-value of an eclipse was believed to depend very much on its magnitude. Commenting on the solar eclipse of AD 360 Aug 28, the Astrological Treatise (chap. 12) of this history records the following remarks:

> Whenever an eclipse covers a small portion of the Sun the calamity it brings will be relatively small, but when it covers a large portion of the Sun the consequences will be much more serious.
>
> [Trans. Ho 1966, p. 159.]

Han and later records indiscriminately apply the same expression *chi* (i.e. 'total') to both total and annular eclipses. Thus both the eclipses of 198 and 181 BC are described in this way in the *Han-shu* treatise, but whereas the former event was, in fact, only annular (no more than 95 per cent of the Sun being covered by the Moon), the latter was fully total. Similar ambiguity occurs in several later dynasties, possibly because during a central annular eclipse the Sun is reduced to a complete ring. Not until as late as AD 1292 do we find a separate expression to describe an annular obscuration: 'the Sun was like a golden ring' (see section 8.7 below). However, even long after that date there was still no separate term in standard usage. It should be noted that accounts of large partial obscurations of the Sun in the Five-Element Treatises normally avoid the use of *chi*. Instead, the expression *chin* ('complete') – for example in the sense *pu-chin* (not complete) – is usually preferred.

Although the accounts of eclipses in the Han treatises are often paralleled by records in the annals of the same dynastic histories, these latter works normally omit almost all technical details. For instance, whereas the *Wu-hsing Chih* frequently specify the right ascension of the Sun during an eclipse (to the nearest degree in a particular lunar lodge), the *Pen-chi* are silent on such information. It is also noteworthy that the annals of both Han histories never describe partial eclipses; unless totality is alleged, no mention is made of magnitude. Thus, while the Five-Element Treatise in the *Hou-han-shu* records that in AD 120 the eclipse was 'almost complete; on the Earth it became like evening', the Annals (chap. 5) merely state that the Sun was totally obscured. Again, the eclipses of both 80 and 28 BC were said to be 'not complete, and like a hook' in the Five-Element Treatise of the *Han-shu* whereas the Annals simply record the total phase.

The lower reliability of accounts of celestial phenomena in the *Pen-chi* of both the *Han-shu* and *Hou-han-shu* is not confined to eclipses. It is also evident in the case of the various cometary records. For example, the report of the 12 BC apparition of Halley's comet in the Astrological Treatise of the *Han-shu* cites precise dates and carefully describes the motion of the comet through several constellations (see, for example, Stephenson and Yau, 1985). Such detailed information is doubtless based on the observations of the court astronomers. However, the corresponding entry in the Basic Annals (chap. 10) merely notes the month of visibility and only mentions one constellation through which the comet passed. Many similar illustrations could be cited and for these reasons notices of eclipses in the Han annals will not be considered further.

In the Five-Element Treatise of the *Han-shu*, three solar eclipses (198 BC, 181 BC and AD 2) were reported as total while as many as seven (188, 147, 89, 80, 35, 28 and 2 BC) were stated to be not complete. Only one eclipse of each kind (total in AD 65 and incomplete in AD 120) is noted in the corresponding treatise of the *Hou-han-shu*. It is unfortunate that all four records of totality in both the *Han-shu* and *Hou-han-shu* treatises are devoid of any supporting description; in each case it is merely asserted that an eclipse was 'total'. With the exception of the event of 181 BC, which is described in more detail in the *Shih-chi*, I have included each of these observations in the lowest category 'C'.

Where careful descriptions of partial eclipses are available, such as *jo-k'ou* ('like a hook'), I have ranked each observation in class 'A'. However, it is sometimes merely stated that an eclipse was 'almost complete' or 'not complete'. In the case of an eclipse which was total on the Earth's surface, the meaning of these terms is fairly obvious. Nevertheless, ambiguity might well occur in the case of an obscuration of the Sun which was generally annular. Since it is not known whether the term *chi* was consistently applied to central annular eclipses, in principle a description such as 'almost complete' could imply either the ring phase or a large partial eclipse. For these reasons, I have adopted a 'B' rating for generally total eclipses which – without any further details – were said to be either 'almost complete' or 'not complete' However, I have rejected all observations of annular eclipses which were described in this way.

Full translations of all 'A' and 'B' ranked accounts of large solar eclipses from both phases of the Han dynasty – as reported in the respective Five-Element Treatises – are given below, along with comments and results of calculation. The form of each entry follows that in section 8.2 above. For brevity, all records in category 'C' are merely summarised, along with the corresponding ΔT limits, in table 8.5. Throughout the Former Han, the place of observation will be taken as Ch'ang-an; in the Later Han, Lo-yang will be assumed. It should be noted that whereas an observation of totality yields an inclusive range of ΔT, a partial eclipse is best interpreted as excluding a set of ΔT values.

(1) BC 188 Jul 17 (total, mag. = 1.02): Ch'ang-an [class B]

> Emperor Hui, 7th year, 5th month, day *ting-mao* [4], the last day of the month. The Sun was eclipsed; it was almost complete. It was in the beginning of (the lunar lodge) *Ch'i-hsing.*
>
> [*Han-shu*, chap. 27c.]

The recorded RA of the Sun corresponds to 115 deg. This is in good agreement with the calculated figure of 112 deg.

RESULTS
For a partial eclipse at Ch'ang-an, either $\Delta T < 13\,830$ or $> 14\,140$ sec.
(NB computed local time of maximum = 15.2 h; solar altitude = 47 deg.)

(2) BC 181 Mar 4 (total, mag. = 1.04): Ch'ang-an [class B]

> Empress of Kao-tzu, 7th year, first month, day *chi-ch'ou* [26], the last day of the month. The Sun was eclipsed; it was total; it was 9 deg in (the lunar lodge) *Ying-shih,* which represents the interior of the Palace chambers. At that time the (Dowager) Empress of Kao-[tzu] was upset by it and said, 'This is on my account'. The next year it was fulfilled.
>
> [*Han-shu*, chap. 27c.]

The Empress Dowager died nearly 18 months afterwards, on 18 August in 180 BC The eclipse and the Empress' reaction are described in more detail in the Annals of the *Shih-chi,* a work composed some 150 years before the *Han-shu.* This is clearly based on an eyewitness report:

> On the day *chi-ch'ou,* the Sun was eclipsed and it became dark in the daytime. The Empress Dowager was upset by it and her heart was ill at ease. Turning to those around her she said, 'This is on my account'.
>
> [*Shih-chi*, chap. 9.]

Although in his *Shih-chi* Szu-ma Ch'ien notes the occurrence of many solar eclipses during the first century or so of the Former Han dynasty, only in this single instance does he give any descriptive details. His account suggests that the eclipse was extremely large at Ch'ang-an, but unfortunately there is no direct reference to the total phase. Conversely, although the *Han-shu* account states that the eclipse was total, it omits to give any qualifying information. Hence category 'B' seems most appropriate. The recorded RA of the Sun corresponds to 329 deg. This is in poor accord with the calculated figure of 342 deg.

RESULTS

For totality at Ch'ang-an, 11 800 sec < ΔT < 12 720 sec.

(NB computed local time of maximum = 14.7 h; solar altitude = 33 deg.)

(3) BC 147 Nov 10 (total, mag. = 1.02): Ch'ang-an [class B]

> Emperor Ching, 3rd year, 9th month, day *wu-hsi* [35], the last day of the month. The Sun was eclipsed; it was almost complete. It was 9 deg in (the lunar lodge) *Wei.*
>
> [*Han-shu*, chap. 27c.]

The recorded RA of the Sun corresponds to 228 deg. This is in excellent agreement with the calculated figure of 227 deg.

On this occasion, the track of totality did not extend further south than 36.2 deg N latitude at any point on the Earth's surface. Hence no value of ΔT will render the eclipse central at Ch'ang-an. In consequence, no limits to ΔT can be derived from the observation that the phase was partial there.

(NB computed local time of maximum = 11.1 h; solar altitude = 38 deg.)

(4) BC 89 Sep 29 (barely annular, mag. = 0.997): Ch'ang-an [class A]

> Cheng-ho reign period, 4th year, 8th month, day *hsin-yu* [58], the last day of the month. The Sun was eclipsed; it was not complete and like a hook. It was 2 deg in *K'ang.* At the hour of *fu* (= 15–17 h) the eclipse began from the north-west. Towards the hour of sunset it recovered.
>
> [*Han-shu*, chap. 27c.]

As this eclipse was only said to be partial and the zone of annularity was of negligible width, virtually any value of ΔT would be acceptable. The time of day when the eclipse began is too rough to provide any useful information regarding ΔT. However, the fact that the eclipse is said to have ended before sunset enables a limiting value for ΔT to be deduced. This observation will be considered in chapter 9 below. The recorded RA of the Sun is equivalent to 188 deg. This is in fair agreement with the calculated value of 183 deg.

(NB computed local time of maximum = 15.7 h; solar altitude = 27 deg.)

(5) BC 80 Sep 20 (total, mag. = 1.02): Ch'ang-an [class B]

> Yuan-feng reign period, first year, 7th month, day *chi-hai* [36], the last day of the month. The Sun was eclipsed; it was almost complete. It was 20 deg in (the lunar lodge) *Chang.*
>
> [*Han-shu*, chap. 27c.]

The recorded RA of the Sun corresponds to 143 deg. This is in very poor accord with the calculated figure of 175 deg.

RESULTS

For a partial eclipse at Ch'ang-an, either $\Delta T < 8200$ or > 8520 sec.

(NB computed local time of maximum = 13.2 h; solar altitude = 54 deg.)

(6) BC 35 Nov 1 (?) (annular, mag. = 0.985): Ch'ang-an [class B]

> Chien-chao reign period, 5th year, 6th month, day *jen-shen* [9], the last day of the month. The Sun was eclipsed; it was not complete and like a hook; then (*yin*) it set.
>
> [*Han-shu*, chap. 27c.]

The date in the above text, which is repeated in the Basic Annals (chap. 9), cannot be correct. It corresponds to Aug 23 in 34 BC, but no eclipse was visible on the Earth's surface on that date. Unfortunately, the RA of the Sun is not specified; this would have materially helped in establishing the correct date. From the eclipse maps of Stephenson and Houlden (1986), it can be seen that between 39 and 29 BC only a single eclipse could have been large in China: that of Nov 1 in 35 BC, which would indeed occur near sunset. This event, first identified by Dubs (1944, vol. II, p. 355), took place on the day *ting-ch'ou*, the last day of the ninth month in the fourth year of the Chien-chao reign period. Since the recorded and calculated dates are only some ten months apart, this is a viable alternative, but it is difficult to explain how such a rare combination of major calendrical errors occurred. Despite the careful description of the appearance of the Sun, I have therefore included the observation in category 'B'.

As the zone of annularity in 35 BC would not reach further north than about latitude 27 deg N anywhere on the Earth's surface, the eclipse can only have been partial at Ch'ang-an for any value of ΔT. Although this result is in keeping with observation, it is not possible to use the record to set limits to ΔT. The reference to the Sun setting largely eclipsed is more useful and this will be discussed in chapter 9.

(NB computed local time of maximum = 17.6 h; solar altitude = −3 deg.)

(7) BC 28 Jun 19 (total, mag. = 1.07): Ch'ang-an [class A]

> Ho-p'ing reign-period, first year, 4th month, day *chi-hai* [36], the last day of the month. The Sun was eclipsed; it was not complete and like a hook; it was 6 deg in *Tung-ching*... When the eclipse first began, it started from the south-west.
>
> [*Han-shu*, chap. 27c.]

The above entry is followed by a lengthy astrological commentary. This eclipse is also noted in the Biographical Section of the *Han-shu* in a chapter dealing with the maternal relatives of the emperors. The following account from this source implies that the magnitude was particularly large at Ch'ang-an:

> The Sun was eclipsed in *Tung-ching*. A moment later it was almost exhausted; it was not much different from a total eclipse.
>
> [*Han-shu*, chap. 97b.]

The solar RA reported in the Five-Element Treatise corresponds to 72 deg. This is in poor accord with the calculated figure of 84 deg.

Results

For a partial eclipse at Ch'ang-an, either $\Delta T < 8090$ sec or > 9530 sec.
(NB computed local time of maximum = 9.6 h; solar altitude = 57 deg.)

(8) BC 2 Feb 5 (barely total, mag. = 1.006): Ch'ang-an [class A]

> Yuan-shou reign period, first year, first month, day *hsin-ch'ou* [38], the first day of the month. The Sun was eclipsed; it was not complete and like a hook. It was 10 deg in *Ying-shih*.
>
> [*Han-shu*, chap. 27.]

Since this eclipse was said to be partial and the central zone was extremely narrow, only a very restricted range of ΔT (between about 2050 and 2150 sec) is excluded. The observation is thus only of minimal value. The solar RA corresponds to 332 deg – considerably more than the calculated value of 317 deg.
(NB computed local time of maximum phase = 7.8 h; solar altitude = 11 deg.)

(9) AD 120 Jan 18 (total, mag. = 1.03): Lo-yang [class A]

> Yuan-ch'i reign-period, 6th year, 12th month, day *wu-wu* [55], the first day of the month. The Sun was eclipsed. It was almost complete; on Earth it was like evening. It was 11 deg in *Hsu-nu*. The Female Ruler was upset by it; two years and three months later, Teng, the Empress Dowager, died.
>
> [*Hou-han-shu*, chap. 28.]

The Empress Dowager's death actually occurred 15 months later, on April 17 in AD 121 (inclusive counting of years is adopted in the text). An almost contemporary text, the *Fu-hou Ku-chin-chu* ('Lord Fu's Commentary on Things Old and New'), written *c.* AD 140, notes that 'the stars were all visible' but no other details are preserved. This is included along with the *Hou-han-shu* report. Other commentaries speak indirectly of the astrological consequences of a *total* eclipse but this may be merely a generalisation.

A relatively recent block-printed version of the text, together with the various commentaries (shown in smaller letters) is illustrated in figure 8.4.

In the *Hou-han-shu* account we have a good description of an eclipse which fell only marginally short of totality. Further, there is a particularly strong suggestion that the eclipse was indeed seen at the capital. Around this date the Five-Element Treatise is careful to specify when an eclipse was *not* seen at Lo-yang and in such instances the true place of observation (in the provinces) is usually given. This is the case for both the immediately preceding eclipse of AD 118 Sep 3 and the following one of AD 120 Aug 12. However, no such statement accompanies the AD 120 Jan 18 record.

The RA of the Sun is equivalent to 297 deg, very close to the calculated value of 299 deg.

RESULTS

For a partial eclipse at Lo-yang, either $\Delta T < 8150$ sec or > 8970 sec.

(NB computed local time of maximum = 13.7 h; solar altitude = 29 deg.)

In table 8.5 are given the following details for each of the remaining Han observations in categories 'B' and 'C': Julian date; type of eclipse (whether annular or total); computed magnitude in the central zone; summary of observation; lower and upper limit to ΔT (in seconds) necessary to satisfy the record.

Referring to the results in table 8.5, the following specific comments seem appropriate. In 198 BC, the fairly broad zone of annularity ran almost parallel to the equator; as a result, the calculated ΔT range to satisfy the observation is unusually wide. The belt of totality in AD 2 was of almost negligible width (some 5 km in the latitude of Ch'ang-an) and thus only an extremely narrow range in ΔT would satisfy the observation. The probability of totality actually being witnessed and reported – whether at Ch'ang-an or anywhere else in China – is very low. On this occasion, the most plausible assumption would seem to be that an account of an almost total eclipse from Ch'ang-an became corrupted into an assertion of totality by the time it found its way into the official history. The observation would thus appear to be valueless for determining ΔT.

8.5 From the San-kuo to the Sui dynasty (AD 220–617)

After the fall of the Han dynasty in AD 220, China was divided into three kingdoms or *San-kuo* (namely Shu Han, Wei and Wu). This period of partition lasted for only 45 years. In AD 265 the country was again unified, this time under the Chin dynasty, whose capital was initially Lo-yang – as in most of the Later Han. However, the northern half of the country was overrun by invaders half a century later and Lo-yang

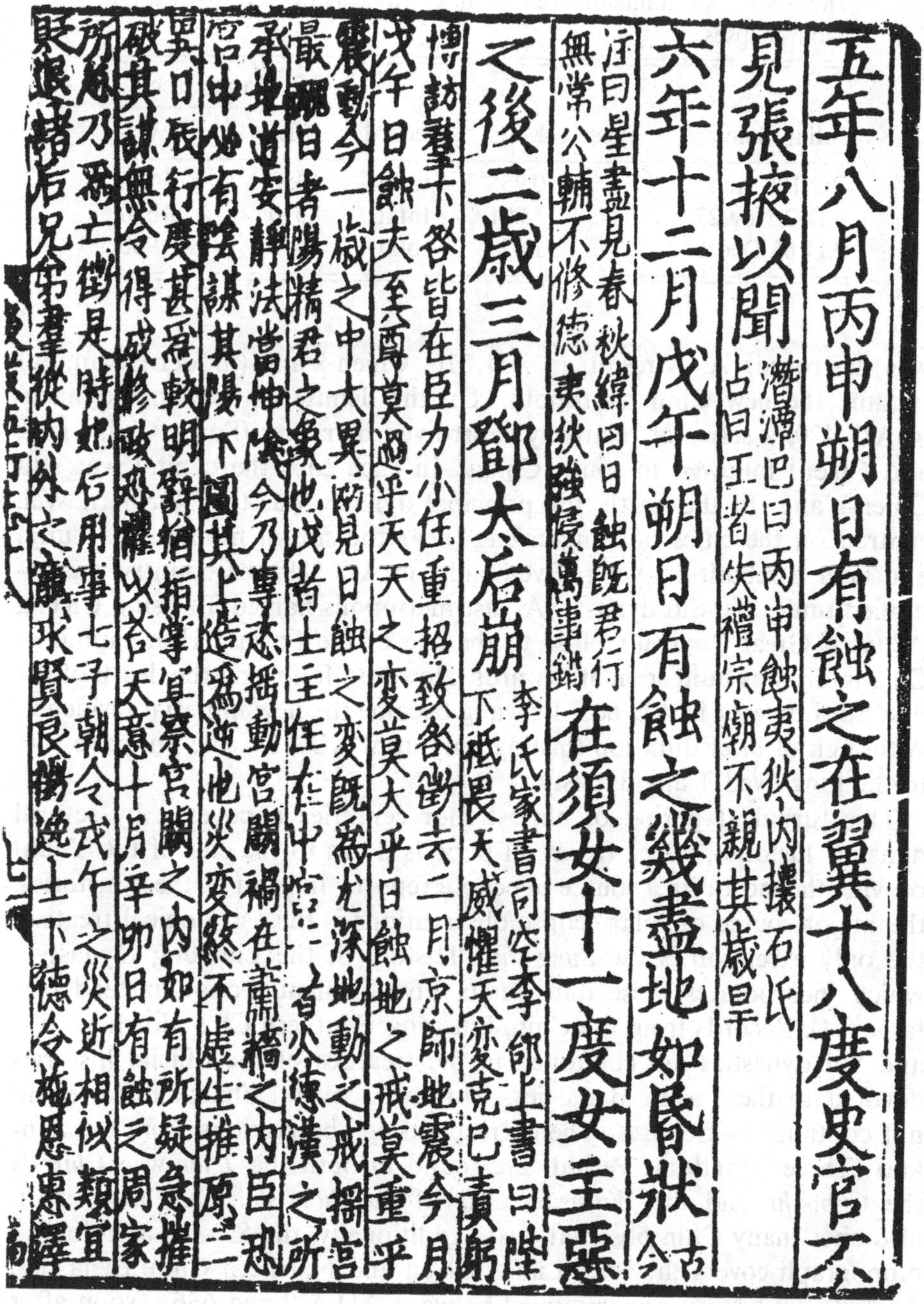

五年八月丙申朔日有蝕之在翼十八度史官不
見張掖以聞潛潭巴曰丙申蝕夷狄內攘石氏占曰王者失禮宗廟不親其歲旱
六年十二月戊午朔日有蝕之幾盡地如昏狀古今
注曰星盡見春秋緯曰日蝕既君行無常公輔不修德夷狄強侵萬事錯在須女十一度女主惡
之後二歲三月鄧太后崩李氏家書司空李郃上書曰陛下祗畏天威懼天變克己責躬
博訪群下咎皆在臣力小任重招致咎徵去二月京師地震今月
戊午日蝕夫至尊莫過乎天天之變莫大乎日蝕地之戒莫重乎
震動今一歲之中大異兩見日蝕之變既爲尤深地動之戒搖宮
最日者陽精君之象也戊者土主位在中宮午者火德漢之所
承地道安靜法當由陰今乃專恣搖動宮闕禍在蕭牆之內臣恐
宮中必有陰謀其陽下圖其上造爲逆也災變終不虛生推原二
異日辰行度甚爲較明對指事宜察宮闕之內如有所疑急摧
破其謀無令得成修政恐懼以荅天意十月辛卯日有蝕之周家
所忌乃爲亡徵是時也后用事七子朝令戊午之災近相似類宜
貶退諸后兄弟羣從內外之寵求賢良徵逸士下德令施恩惠澤

後漢五行志 七

Fig. 8.4 Nineteenth century block print containing AD 120 solar eclipse record.

Table 8.5 ΔT limits derived from 'C'-rated Han records of large solar eclipses.

Julian date	Type	Mag.	Record	ΔT Range (sec) LL	UL
BC 198 Aug 7	A	0.95	total	6350	13 110
AD 2 Nov 23	T	1.001	total	8570	8 590
AD 65 Dec 16	T	1.02	total	8450	8 900

was destroyed. As a result, in AD 316, Chien-k'ang (modern Nanjing) became the new Chin metropolis. On the demise of the Chin dynasty in AD 420, a series of relatively short-lived dynasties (Sung, Ch'i, Liang and Ch'en) followed in south China. In each case the capital was still Chien-k'ang. In the north, the principal dynasty was the Northern Wei, centred on the cities of P'ing-ch'eng (near the site of modern Da-t'ung) and later a rebuilt Lo-yang. Eventually, in AD 589, the country was re-unified under the Sui dynasty. A vast metropolis named Ta-hsing Ch'eng ('City of Great Revival'), close to the site of the Former Han capital of Ch'ang-an, was built soon afterwards. However, following the death of the first Sui Emperor in AD 605, Lo-yang became the new imperial residence. Although of brief duration (lasting only until AD 617), the Sui gave way to the long-lived T'ang dynasty.

This turbulent period of almost four centuries is covered by several dynastic histories, which (as at all periods down to the end of the Yuan) provide the most direct source of eclipse reports. In most of these histories, the major source of solar eclipse observations is the astrological treatise; the only exception is the *Sung-shu* ('History of the Liu-sung Dynasty'), which incorporates these data in its Five-Element section instead (see table 8.4). Records from the Chin, Liu-sung, Southern Ch'i, Northern Wei and Sui dynasties are contained in the treatises of the official histories devoted to these same dynasties. However, several official histories do not contain any treatises. The corresponding observations from the San-kuo (Three Kingdoms Period) are found in both the *T'ien-wen Chih* of the *Chin-shu* and the *Wu-hsing Chih* of the *Sung-shu* (this latter work also cites many Chin observations). Additionally, the *Sui-shu* astrological monograph covers the brief Liang, Ch'en and Northern Chou dynasties. This official history was compiled between AD 629 and 636 – soon after the end of the Sui dynasty – under Wei Cheng.

Four total eclipses (in AD 306, 454, 562 and 616) and two large partial obscurations of the Sun (in AD 360 and 429) are reported in the various treatises of the period. Of these, the eclipses of AD 360, 429, 454 and 616 are all recorded as total in the Imperial Annals of the period, but without any further details. In particular, the reports of totality in the Chin annals for AD 360 and in the Sung annals for AD 429 both conflict with the rather careful descriptions of partial eclipses found in the treatises. Two additional eclipses which occurred on AD 516 Apr 16 (this was generally annular on the Earth's surface) and 522 Jun 10 (generally total) were both described as complete – but without any further information – in the Basic Annals of the *Nan-shih* ('History of the Southern Dynasties'). The second record is also to be found in the Annals of the *Liang-shu*. These observations do not find support in any astronomical treatise. As for the Han dynasty, these observations – as well as the other eclipse reports in the various annals of the period – will not be considered further.

Of the six observations which either affirm or deny a central eclipse in the various treatises, three – AD 360, 429 and 454 – are especially detailed. All are to be found in a section of the Five-Element Treatise of the *Sung-shu* (chap. 34). I have included these in category 'A', and for each a full translation, together with commentary, is given below. A further eclipse, occurring in AD 494, was said to reach only a magnitude of one-third between 9 and 11 h. However, if the recorded local time of maximum is accurate, the computed magnitude would actually lie between 0.65 and 0.9. The fact that totality is denied may be of some consequence and this observation will be included in category 'B'. The remaining three rather brief records noted above (AD 306, 562 and 616) are assigned to class 'C' and will be discussed at the end of this section. Each simply notes the occurrence of a total obscuration of the Sun.

During the eclipse of AD 523 Nov 23 it is reported in the *Sui-shu* Astronomical Treatise (chap. 21) that Venus was seen by day, but the eclipse magnitude is not specified. This observation, although interesting, is of little significance in the present context since it does not necessarily imply a very large eclipse; numerous sightings of Venus in full daylight – sometimes for many days at a time – are recorded throughout Chinese history from the Han onwards.

(1) AD 360 Aug 28 (annular, mag. = 0.93): Chien-k'ang [class A]

> Sheng-p'ing reign period, 4th year, 8th month, day *hsin-ch'ou* [38], the first day of the month. The Sun was eclipsed. It was not complete and like a hook.
>
> [*Sung-shu*, chap. 34: section on Chin eclipse records.]

In part of the Astrological Treatise (chap. 12) of the *Chin-shu*, this eclipse is reported as follows:

> It was almost total and was in *Chueh*. Whenever an eclipse covers a small portion of the Sun the calamity it brings will be relatively small, but

when it covers a large portion of the Sun the consequences will be much more serious. *Chueh* forms the 'Celestial Entrance', and hence misfortune would fall upon the Head of State – the next year the Emperor died.

[Trans. Ho (1966, p. 159).]

A portion of the *Chin-shu* account has already been quoted (section 8.4) to illustrate the extent to which magnitude was regarded as affecting the omen-value of an eclipse. The young Emperor Mu died on 10 July in AD 361; he was then aged 19. Although the expression 'almost complete' in the *Chin-shu* text is obscure when applied to an eclipse which is generally annular, the *Sung-shu* terminology ('not complete and like a hook') is quite specific.

Results

For a partial eclipse at Chien-k'ang, either $\Delta T < 7120$ sec or > 9400 sec.
(NB computed local time of maximum = 11.3 h; solar altitude = 65 deg.)

(2) AD 429 Dec 12 (total, mag. = 1.04): Chien-k'ang [class A]

Yuan-chia reign period, 6th year, 11th month, day *chi-ch'ou* [26], the first day of the month. The Sun was eclipsed; it was not complete and like a hook. During the eclipse, stars were seen. At the hour of *fu* (= 15–17 h), then it disappeared (i.e. ended). In Ho-pei (province) the Earth was in darkness.

[*Sung-shu*, chap. 34.]

This eclipse was evidently total in Ho-pei (= Hebei) province but unfortunately a precise location is not specified. Situated far to the north of the Liu-Sung capital of Chien-k'ang, Ho-pei was then part of the Northern Wei Empire. Since the track of totality did not extend further south than 35 deg N latitude at any point on the Earth's surface, no alteration in ΔT will render this eclipse central at Ch'ien-k'ang. Hence no limits to ΔT can be derived from the observation that the phase was incomplete there.

(NB computed local time of maximum = 12.6 h; solar altitude = 34 deg.)

(3) AD 454 Aug 10 (total, mag. = 1.03): Chien-k'ang [class A].

Hsiao-chien reign period, first year, 7th month, day *ping-shen* [33], the first day of the month. The Sun was eclipsed; it was total; all the constellations (i.e. lunar lodges) were brightly lit.

[*Sung-shu*, chap. 34.]

The previous entry in chap. 34 of the *Sung-shu* also notes a total eclipse, this time on the day *hsin-ch'ou* (38), the first day of the seventh month in the 30th year of the Yuan-chia reign period. On this occasion it was said that 'the stars were all seen'. However, no eclipse was visible in China on this date – which corresponds to AD 453 Aug 20 – and apart from AD 454 Aug 10 itself, none could have been total in China at any time between AD 438 and 494. This latter record may well represent a further reference to the eclipse of AD 454, misplaced one calendar year by the compilers of the *Wu-hsing Chih*, and with the cyclical day adjusted retrospectively.

Whatever the explanation of the AD 453 record, in order for many stars to have appeared 'brightly lit' at the eclipse of AD 454 Aug 10, the eclipse must have been fully total.

RESULTS

For a total eclipse at Chien-k'ang in AD 454, either $+70 < \Delta T < 1800$ sec or $6130 < \Delta T < 7900$ sec.

As discussed in chapter 3, two possible ranges of ΔT are indicated by this observation. However, comparison with the ΔT values obtained from roughly contemporaneous Chinese observations in both this chapter and chapter 9 indicates that only the latter range (i.e. $6130 < \Delta T < 7900$ sec) is acceptable.

(N.B. computed local time of maximum = 8.9 h; solar altitude = 45 deg.)

(4) AD 494 Jun 19 (total: mag. = 1.04): Chien-k'ang [class B]

> Lung-ch'ang reign period, first year, 5th month, day *chia-hsu* [11], the first day of the month. At the hour of *szu* (9–11 h) the Sun was one-third eclipsed. At the hour of *wu* (11–13 h) it was restored to fullness.
>
> [*Nan-ch'i-shu*, chap. 12.]

No possible variation in ΔT will lead to a visible magnitude as small as 0.33 at the Southern Ch'i capital of Chien-k'ang. For any value of ΔT which places the Sun above the horizon at this site, the actual magnitude cannot have been less than 0.56. Furthermore, as noted above, if the recorded local time of maximum is accurate, the computed magnitude then would be at least 0.65. Hence, although the magnitude estimate must be erroneous, the fact that there is no mention of the total phase seems worth investigating. It seems clear from the text that the Sun was visible when the eclipse was at its height; there is nothing to suggest that cloud intervened. Computation shows that for values of ΔT in the range 5980 to 6600 sec the eclipse would be total at Chien-k'ang. Since the eclipse appears to have been only partial there, a value for ΔT outside these limits is required. In view of the textual error in recording the magnitude, the observation will be included in category 'B'.

Table 8.6 ΔT limits derived from 'C'-rated records of solar eclipses in the period from AD 220 to 616.

Julian date	Type	Mag.	Record	ΔT Range (sec) LL	UL
AD 306 Jul 27	A	0.94	total	6550	7890
AD 562 Oct 14	A	0.99	total	—	—
AD 616 May 21	A	0.97	complete	2300	3030

RESULTS

For a partial eclipse at Chien-k'ang, either $\Delta T < 5980$ or > 6600 sec.

(NB computed local time of maximum = 8.9 h; solar altitude = 48 deg.)

Each of the eclipses of AD 306, 562 and 616 was recorded as total (using the term *chin* in AD 616). The first of these records is to be found in chap. 34 of the *Sung-shu* while the two later observations are cited in different sections of chap. 21 of the *Sui-shu.* Each event occurred in a different dynasty (respectively Chin, Northern Chou and Sui). The appropriate capitals were Lo-yang (AD 306), Ch'ang-an (AD 562) and again Lo-yang (AD 616). Although the original Sui metropolis was Ta-hsing Ch'eng, by AD 616 it had been moved to Lo-yang. As it happens, all three eclipses were generally annular rather than total.

Assumption of a central eclipse in both AD 306 and 616 leads to specific limits for ΔT. However, no value of this parameter can render the eclipse of 562 central anywhere in China; the track of annularity never reached any further north than latitude +17 deg in the longitudes of China. It seems highly unlikely that an observation was communicated from some remote site outside China. Instead, there is a real possibility that the record represents a fairly successful prediction cited as though it were an observation. There is evidence of eclipse prediction in China, using numerical cycles based on past observation, from the San-kuo period onwards (Foley and Stephenson, 1997). For further details of Sui predictions, see chapter 9 below; examples of T'ang and Sung predictions are given in sections 8.6 and 8.7.

In table 8.6 are listed the following details for the three 'C' rated eclipses discussed above: Julian date; eclipse type; magnitude in central zone; observation; ΔT range to satisfy the record.

8.6 The T'ang dynasty and Wu-tai period (AD 617–960)

There are two official T'ang histories: the 'Old History of the T'ang Dynasty' (*Chiu-t'ang-shu*), written between AD 940 and 945 (soon after the dynasty had ended) under the direction of Liu Hsu, and the more systematically compiled 'New History of the T'ang Dynasty' (*Hsin-t'ang-shu*). The latter work was composed by a team of scholars supervised by Ou-yang Hsiu and Sung Ch'i between AD 1043 and 1060. The earlier history often contains details which are omitted in the revision – and this is true of several eclipse records.

The extensive astrological treatises of the *Chiu-t'ang-shu* and *Hsin-t'ang-shu* both contain numerous eclipse observations. These are to be found in chapters 36 and 32 of the respective works. Contrary to normal practice, the Annals of the *Chiu-t'ang-shu* sometimes give more extensive accounts of celestial phenomena than the Astrological Treatise of the same history. However, in recording solar eclipses, the treatises and annals of both histories tend to support one another; they do not contain mutually contradictory information as in earlier works. A roughly contemporaneous work, the *T'ang-hui-yao* ('Collected Essentials of T'ang History'), completed in AD 961, also cites many eclipse records but these do not yield any information which is not found in the dynastic histories.

Throughout most of the T'ang, the capital was Ch'ang-an, the former Sui metropolis of Ta-hsing Ch'eng. This city was re-named after the Former Han capital, although it lay a few kilometres toward the south-east of that site. The present-day city of Xi'an is centred on T'ang Ch'ang-an. Several large solar eclipses are noted in T'ang history (see below). However, the two official histories of the brief Wu-tai ('Five Dynasties') era, which commenced on the fall of the T'ang in AD 906, do not specify the magnitudes of the eclipses which they cite. In the remainder of this section, we shall concentrate exclusively on the T'ang observations.

Eight eclipses of the Sun are described as either total or very large in T'ang history. These were recorded in the years AD 756, 761, 879 and 888 (total) and 702, 729, 754 and 822 (partial). The observations of the eclipses of 702, 729, 754, 761 and 822 are fairly detailed and I have included all these in category 'A'. Since the account of the total obscuration in AD 888 is very brief and is only cited in the Annals of the *Hsin-t'ang-shu*, I have not considered it further.

Although the remaining two records (AD 756 and 879) are brief, both present unusual problems and will therefore be discussed first.

On a date equivalent to AD 756 Oct 28, both the *Hsin-t'ang-shu* Astrological Treatise (chap. 32) and the *Chiu-t'ang-shu* Annals (chap. 10) report that 'The Sun was eclipsed and it was total', the former work adding that 'the Sun was 10 deg in (the lunar lodge) *Ti*'. Although the date is accurately noted, the place of observation on this occasion cannot be established. Only a few months previously, the capital of Ch'ang-an had been captured by rebels led by An Lu-shan; the Emperor and his court had fled to the provinces. At the time of the eclipse there was turmoil. Ch'ang-an was not recovered until the following year.

Towards the end of the T'ang dynasty, on the date AD 879 Apr 25, another eclipse is recorded in the Astrological Treatise of the *Hsin-t'ang-shu*, in the following words: 'The Sun was eclipsed; it was total. It was 8 deg in *Wei*'. However, although the Moon and Sun prove to be in conjunction on that day, and the estimated RA of the Sun (32 deg) is in good accord with computation (36 deg), no eclipse in fact occurred. Further, none would be central in China for several years around this time. Rather than make a major adjustment to the date, it would seem most reasonable to assume that this entry represents an abortive prediction. The distance of the Moon from the node at conjunction was 18 deg, only just beyond the limit for a real eclipse to be possible. Direct evidence of solar eclipse prediction during the T'ang is to be found in the Astrological Treatise of the *Chiu-t'ang-shu*; thus eclipses in AD 778 and 790 were said to have been predicted but not seen. However, by making appropriate computations, it can be readily shown that the records of eclipses in both official T'ang histories contain many other false sightings which are cited as if they were observations. In some instances, the reported events did not occur; in several others, an eclipse was only visible in areas of the Earth's surface remote from China (for example in Antarctica). There is thus ample evidence of prediction on a large scale (Cohen and Newton, 1981–3; Foley and Stephenson, 1997).

The following translations of the five 'A' rated T'ang eclipse records are mainly based on the accounts in the Astrological Treatises of the *Chiu-t'ang-shu* and *Hsin-t'ang-shu*. However, when the Imperial Annals of the two histories provide additional information, this will also be considered.

(1) AD 702 Sep 26 (total, mag. = 1.04): Ch'ang-an [class A]

> Ch'ang-an reign period, 2nd year, 9th month, day *i-ch'ou* [2]. The Sun was eclipsed; it was almost complete. It was in *Chueh*.
>
> [*Hsin-t'ang-shu*, chap. 32.]

The Annals of the *Chiu-t'ang-shu* (chap. 6), although omitting the solar co-ordinate, state that the eclipse was 'not complete and like a hook; it was seen at the capital and in the four directions'. Here we have a rare instance where observation at the capital is expressly stated.

Results

For a partial eclipse at Ch'ang-an, either $\Delta T < 1460$ sec or > 2760 sec.
(NB computed local time of maximum = 15.6 h; solar altitude = 28 deg.)

(2) AD 729 Oct 27 (total, mag. = 1.03): Ch'ang-an [class A]

K'ai-yuan reign period, 17th year, 10th month, day *wu-wu* [55], the first day of the month. The Sun was eclipsed; it was not complete and like a hook. It was 9 deg in *Ti.*

[*Hsin-t'ang-shu*, chap. 32.]

This same account is found in the Annals of the *Chiu-t'ang-shu* (chap. 8), apart from the reference to the solar co-ordinate. The recorded RA of the Sun is equivalent to 206 deg, considerably less than the calculated value of 215 deg.

Results

For a partial eclipse at Ch'ang-an, either $\Delta T < 420$ sec or > 1190 sec.
(NB computed local time of maximum = 7.3 h; solar altitude = 8 deg.)

(3) AD 754 Jun 25 (total, mag. = 1.04): Ch'ang-an [class A]

T'ien-pao reign period, 13th year, 6th month, day *i-ch'ou* [2], the first day of the month. The Sun was eclipsed; it was almost total. It was 19 deg in *Tung-ching.*

[*Hsin-t'ang-shu*, chap. 32.]

The Annals of the *Chiu-t'ang-shu* (chap. 9), while omitting the position of the Sun, assert that the eclipse 'was not complete and like a hook'. The reported RA of the Sun, which corresponds to 97 deg, is identical to the calculated figure.

As the track of totality did not reach further north than latitude 30 deg, for any value of ΔT the eclipse would be no more than partial at Ch'ang-an. Hence calculation is in accord with observation but no limits on ΔT can be set.

(NB computed local time of maximum = 11.8 h; solar altitude = 79 deg.)

(4) AD 761 Aug 5 (total, mag. = 1.05): Ch'ang-an [class A]

Shang-yuan reign period, 2nd year, 7th month, day *kuei-wei* [20], the first day of the month. The Sun was eclipsed; the large stars were all seen. The Astronomer-Royal Ch'u T'an reported (to the Emperor): 'On day *kuei-wei* the Sun diminished. The loss began at 6 marks (*k'o*) after the hour of *ch'en.* At 1 mark after the hour of *szu* it was total. At 1 mark before the hour of *wu* it was restored to fullness. The eclipse was 4 deg in *Chang*'.

[*Chiu-t'ang-shu*, chap. 36.]

A block-print of the Chinese text of this account is shown in figure 8.5.

Both the Annals of the *Chiu-t'ang-shu* (chap. 10) and the Astrological Treatise of the *Hsin-t'ang-shu* affirm that the eclipse was total and that 'all the large stars were seen', as does the Astrological Treatise of the *T'ang-hui-yao* (chap. 42). The local times of the various phases recorded in the *Chiu-t'ang-shu* Treatise are ambiguously expressed – see chapter 9 below. It should be noted that the unit of time known as the *k'o* (here translated 'mark') represented 1/100 of a (24-hour) day, or 0.24 h; see also chapter 9. The RA of the Sun, reported in the Astrological Treatises of the *Hsin-t'ang-shu* and *Chiu-t'ang-shu* as 4 deg in *Chang* (i.e. 137 deg) is in close agreement with the calculated value of 138 deg.

For totality at Ch'ang-an, 1720 sec $< \Delta T <$ 3290 sec.

(NB computed local time of maximum = 9.8 h; solar altitude = 55 deg.)

(5) AD 822 Apr 25 (total, mag. = 1.07): Ch'ang-an [class A]

> Ch'ang-ch'ing reign period, 2nd year, 4th month, day *hsin-yu* [58], the first day of the month. The Sun was eclipsed. It was 12 deg in *Wei*; $\frac{1}{4}$ of it was not complete. In (the regions) Yen and Chao it was seen to be total.
>
> [*Chiu-t'ang-shu*, chap. 36.]

Yen and Chao represent the regions covered by the former Warring States bearing these names. These correspond roughly to the provinces of Ho-pei (= Hebei) and Shan-hsi (= Shanxi) in northern China. It is regrettable that more precise locations where totality was witnessed are not given. Evidently the compilers of the *Chiu-t'ang-shu* thought it unnecessary to state where the partial phase was seen and also considered this observation to be more important than the provincial reports of totality. There is thus a very strong presumption that the eclipse was partial (magnitude $\frac{3}{4}$) at the capital itself. The reported solar RA, equivalent to 37 deg, is close to the calculated figure of 36 deg.

Since totality was witnessed in north-east China, the central zone must have passed to the north of Ch'ang-an. Hence the observation from the capital sets a single (upper) limit to the value of ΔT.

Results

For a partial eclipse at Ch'ang-an, $\Delta T <$ 4020 sec.

The magnitude estimate of $\frac{3}{4}$ will be considered in chapter 9.

(NB computed local time of maximum = 12.2 h; solar altitude = 70 deg.)

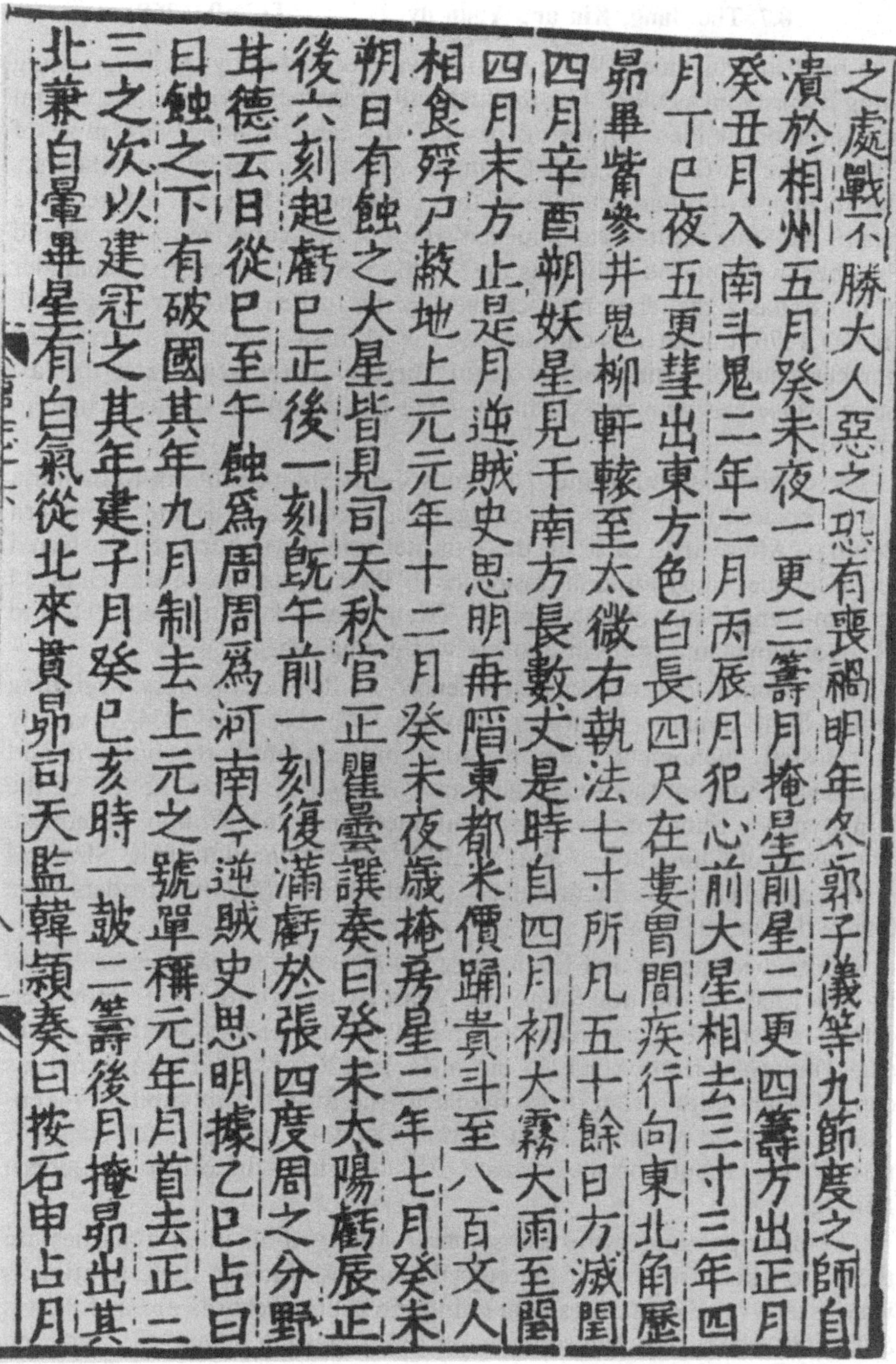

之處戰不勝大人惡之恐有喪禍明年冬郭子儀等九節度之師自
潰於相州五月癸未夜一更三籌月掩星前星二更四籌方出正月
癸丑月入南斗魁二年二月丙辰月犯心前大星相去三寸三年四
月丁巳夜五更彗出東方色白長四尺在婁胃間疾行向東北角歷
昴畢觜參井鬼柳軒轅至太微右執法七寸所凡五十餘日方滅閏
四月辛酉朔妖星見于南方長數丈是時自四月初大霧大雨至閏
四月末方止是月逆賊史思明再陷東都米價踊貴斗至八百文人
相食殍尸蔽地上元元年十二月癸未夜歲星掩房星二年七月癸未
朔日有蝕之大星皆見司天秋官正瞿曇譔奏曰癸未太陽虧辰正
後六刻起虧巳正後一刻既午前一刻復滿虧於張四度周之分野
甘德云日從巳至午蝕爲周周爲河南今逆賊史思明據乙巳占曰
日蝕之下有破國其年九月制去上元之號單稱元年月首去正二
三之次以建冠之其年建子月癸巳亥時一鼓二籌後月掩昴出其
北兼白暈畢星有白氣從北來貫昴司天監韓穎奏曰按石申占月

Fig. 8.5 Nineteenth century block print containing AD 761 total solar eclipse text.

8.7 The Sung, Kin and Yuan dynasties (AD 960–1368)

The brief and turbulent Wu-tai period was succeeded by the long-lasting Sung dynasty in AD 960. For the first half of the new dynasty, the capital was situated at Pien. However, in 1127 the capital, along with most of China lying north of the river Ch'ang-chiang (Yang-tze) fell to invaders – the Juchen – who had established the Kin dynasty 12 years before. As a result, the Sung capital was removed south of the Yang-tze to Lin-an and thus began the period known as the Southern Sung. During the remainder of this dynasty, the territories occupied by the Juchen were never regained. In AD 1276, Lin-an was captured by the Mongols, who had overrun the northern half of China half a century before. Three years later the last Sung prince was drowned during a naval battle and the dynasty came to an end.

The original capital of the Kin dynasty – at Shang-ching in Manchuria – was replaced in 1153 by Yen-ching (a city occupying the site of modern Beijing). Afterwards, in 1214, the Kin metropolis was further transferred to the former Sung imperial residence of Pien, whose name was changed to Pien-ching (*ching* means capital). Twenty years later, this city fell into Mongol hands and the Kin dynasty was extinguished.

Astronomical observations commenced at Pien at the very beginning of the Sung dynasty. However, between AD 1115 and 1234, a variety of celestial phenomena were recorded independently in southern and northern China by the Sung and Kin astronomers. After the fall of the Kin dynasty, Sung observations continued until the Mongol Conquest. The whole of China now remained under the domination of the Mongols – who established the Yuan dynasty – until 1368 when the invaders were finally driven out.

Kubilai became the first Yuan emperor. Although he had become Khan in AD 1260, he did not found the Yuan dynasty until 1271. However, eclipse observations are recorded in the official history of the dynasty (the *Yuan-shih*) from AD 1261 onwards. In 1267, Kubilai transferred his capital from Karakorum in Mongolia to the former Kin capital of Yen-ching, which he renamed Ta-tu ('Great Capital'). As its name suggests, this city was rebuilt on a vast scale. Its site is near the centre of modern Beijing.

The principal sources of eclipse observations in all three dynasties are the astrological treatises of the official histories. Each of these treatises is very extensive; the chapters containing the eclipse records are as follows: *Sung-shih* (chap. 52), *Kin-shih* (chap. 20), *Yuan-shih* (chap. 48). Between 1343 and 1345 the *Sung-shih* and *Kin-shih* were compiled by a team of Yuan scholars led by T'o T'o. The *Yuan-shih* itself was composed soon after the fall of the dynasty under the direction of Sung Lien. It is on record that this task took only 370 days. These various works describe total eclipses in AD 977, 1221 and 1275, and annularity in 1292. Large partial obscurations of the Sun were noted in 1022, 1054 and 1135. Two further eclipses of unspecified magnitude during which either 'the large stars were all visible' (AD 1214) or darkness occurred (1367) were also reported. No reference to the eclipse of AD 1022 occurs in the Sung Annals, but the other eclipses listed above are all mentioned in the annals of the appropriate history. An unusually vivid account of totality in AD 1275 is also to be found in the Five-Element Treatise of the *Sung-shih* (chap. 67).

During this same period, a total solar eclipse was also recorded without any further description in the Annals of the *Sung-shih* (chap. 18), but there is no corresponding entry in the Astrological Treatise. The recorded date, when converted to the Julian calendar (i.e. AD 1099 Nov 30), is in exact agreement with that of a total *lunar* eclipse which proves to have been visible in China. This entry must have been wrongly filed by the compilers of the history.

Complete obscurations of the Sun, on dates corresponding to AD 1005 Jan 13 and 1122 Mar 10 are briefly noted in the Annals (chaps. 14 and 29) of the *Liao-shih*, the official history of the Liao dynasty – also compiled under the direction of T'o T'o. A semi-nomadic kingdom, the Liao was established in the early tenth century in the extreme north of China. It was brought to an end by the Juchen in AD 1125. The *Liao-shih* does not contain an astronomical treatise. Even if the reports of totality were reliable, the place of observation cannot be established with any confidence; the Liao rulers had five separate residences scattered throughout Manchuria.

The careful accounts of totality in AD 1221 and 1275, of annularity in 1292, and of the partial phase in AD 1054 and 1135 will each be assigned to category 'A'. However, the observations in 1214 and 1367, although noting either the visibility of stars or darkness, do not mention the complete disappearance of the Sun directly. These and two other reports (AD 977 and 1022), which although merely noting that the phase was either 'total' or 'almost complete' require special comment, will be considered first in chronological order.

On a date corresponding to AD 977 Dec 13, both the Annals (chap. 4) and the Astrological Treatise (chap. 52) of the *Sung-shih* note the occurrence of a total eclipse of the Sun. Since the computed track of totality on this occasion did not reach further north than latitude 19 deg N, no variation in ΔT will render the eclipse central in any part of China. An 'almost complete' obscuration of the Sun on a date which is equivalent to AD 1022 Aug 1 is reported in the Astrological Treatise of the same history. Calculation shows that this eclipse, which is not mentioned in the Annals, would be visible mainly in the southern hemisphere and could not have been seen in China. In the *Sung-shih*, as in the official histories of earlier dynasties, a number of spurious sightings of eclipses are recorded along with genuine observations. The Sung astronomers are known to have attained a fair degree of success in predicting eclipses. Thus according to the *Sung-hui-yao chi-kao* (History of Sung Adminstration), the Sung Bureau of Astronomy predicted a total eclipse in AD 1021, but this was seen to be only partial (magnitude $\frac{4}{10}$) at the capital. Hence in both AD 977 and 1022, rather than assume a chance sighting at some remote location which somehow was communicated to the Sung capital, it would seem more reasonable to assume predictions recorded as if they were observations.

The astrological treatise of the *Kin-shih* (chap. 20) contains the following record, whose date corresponds to AD 1214 Oct 5: 'The Sun was eclipsed; the large stars were all visible'. The Annals of the *Kin-shih* (chap. 14) mention no more than the occurrence of this eclipse. If the brighter stars were indeed 'all visible', as the astrological treatise of the *Kin-shih* asserts, no phase less than totality would suffice. However, the fact that the record fails to mention the complete disappearance of the Sun considerably weakens its value. There is the additional complication that in the very year that the eclipse occurred the Kin capital was transferred from Yen-ching to Pien-ching so that the place of observation is uncertain.

Both the Annals of the *Yuan-shih* (chap. 47) and the *Ming-shih-lu* (chap. 24) note the occurrence of a solar eclipse which caused 'darkness by day' on a date corresponding to AD 1367 Jun 28. However, this eclipse was only partial on the Earth's surface and would be invisible in China. Presumably the record represents a faulty prediction which to the Ming historians who compiled the *Yuan-shih* became associated with the downfall of the Yuan dynasty very soon afterwards.

The four eclipses of AD 977, 1022, 1214 and 1367 will not be considered further. However, the more reliable records from AD 1054, 1135, 1221, 1275 and 1292 are translated in full below, along with comments. Also included in this section is a discussion of the solar obscuration of AD 1361. A vivid description of an almost complete eclipse in that year is quoted in a history of the town of Sung-chiang (near Shanghai). This is the earliest such account to appear in a surviving *fang-chih* or local history.

(1) 1054 May 10 (total, mag. = 1.03): Pien [class A]

> Huang-yu reign period, 6th year, 4th month, day *chia-wu* [31], the first day of the month. The Sun was eclipsed. The loss began on the south-west side. The Sun was more than $\frac{9}{10}$ eclipsed and it reached its maximum at 1 mark in the central half of the hour of *shen*.
>
> [*Yuan-shih*, chap. 53.]

This careful account of a large partial eclipse which fell short of totality at the Sung capital of Pien is recorded in the calendar treatise of the *Yuan-shih.* In this treatise, many past observations of eclipses were compared with the results derived from contemporary tables (see chapter 9). A very similar account to the above is to be found in chapter 283 of the *Wen-hsien T'oung-k'ao* ('Comprehensive History of Civilisation'), compiled *c.* AD 1300.

Since the track of totality did not reach further north than latitude +26 deg at any point on the Earth's surface, totality at Pien (lat. 34.8 deg) is out of the question; the magnitude cannot have exceeded 0.76. A range of ΔT cannot be derived. The timing of maximum phase, corresponding to 16.36 h, will be considered in chapter 9.

(NB computed local time of maximum = 16.5 h; solar altitude = 29 deg.)

(2) 1135 Jan 16 (annular, mag. = 0.93): Lin-an [class A]

> Shao-hsing reign period, 5th year, 1st month, the first day of the month. A man named Ch'en Te-i predicted that the Sun should be $8\frac{1}{2}$ tenths eclipsed with the beginning of loss in the initial half of the hour of *szu.* (These predictions) were verified by observation.
>
> [*Sung-shih*, chap. 81.]

This account is to be found in the calendar treatise of the *Sung-shih.* Although the observational record is brief, it would seem that a large partial eclipse was witnessed at Lin-an; there is no mention of annularity. The local time of beginning corresponds to some time between 9 and 10 h.

RESULTS

For a large partial eclipse at Lin-an, $\Delta T < 1840$ sec or > 3810 sec.

(NB computed local time of maximum = 11.4 h; solar altitude = 39 deg.)

(3) AD 1221 May 23 (total; mag. = 1.07): Kerulen River [class A]

An extensive record is to be found in one of the books which make up the *Tao-ts'ang* (Taoist Patrology). The book is entitled *Ch'ang-ch'un Chen-jen Hsi-yu-chi* (The Journey of the Adept Ch'ang-ch'un to the West). It is a detailed narrative of the travels of the elderly Taoist leader Ch'ang-ch'un from north-east China to Samarkand between AD 1220 and 1224 to meet Genghis Khan. The *Ch'ang-ch'un Chen-jen Hsi-yu-chi* (which is listed as No. 1410 in L. Wieger's catalogue of the *Tao-ts'ang*) was composed by Li Chih-ch'ang, one of the disciples of Ch'ang-ch'un who accompanied him on his long journey. A full translation of Li Chih-ch'ang's account has been made by Waley (1934). The observation of totality by the Kerulen

River in eastern Mongolia has been analysed in some detail by Stephenson and Yau (1992b) and the following discussion (including translations) is largely based on their investigation.

The eclipse is described in two separate sections of Li Chih-ch'ang's book. The year is given as *jen-wu*, the 19th year of the cycle, equivalent to AD 1221–2.

> (i) On the first day of the fifth month (May 23), at noon, the Sun was eclipsed and it was total. All the stars were therefore seen. A short while later the brightness returned. At that time we were on the southern bank of the river. The eclipse (began) at the south-west and (the Sun) reappeared from the north-east. At that place it is cool in the morning and warm in the evening; there are many yellow flowers among the grass. The river flows to the north-east. On both banks there are many tall willows. The Mongols use them to make their tents.
>
> [Trans. Stephenson and Yau (1992b).]

While Ch'ang-ch'un was passing the following winter at Samarkand, he discussed the eclipse with an astronomer:

> (ii) (Ch'ang-ch'un) asked (him) about the solar eclipse on the first day of the fifth month (May 23). The man replied: 'Here the Sun was eclipsed up to 6 *fen* ($\frac{6}{10}$) at the hour of *ch'en* (7–9 h)'. The Master continued, 'When we were by the Lu-chu Ho (Kerulen River), during the hour *wu* (11–13 h) the Sun was seen totally eclipsed and also south-west of the Chin-shan the people there said that the eclipse occurred at the hour *szu* (9–11 h) and reached 7 *fen*. At each of these three places it was seen differently. According to the commentary on the *Ch'un-ch'iu* by K'ung Ying-ta, when the body (of the Moon) covers the Sun, then there will be a solar eclipse. Now I presume that we must have been directly beneath it; hence we observed the eclipse to be total. On the other hand, those people on the sides (of the shadow) were further away and hence (their view) gradually became different. This is similar to screening a lamp with a fan. In the shadow of the fan there is no light or brightness. Further away from the sides (of the fan) then the light of the lamp gradually becomes greater.'
>
> [Trans. Stephenson and Yau (1992b).]

In both of the above descriptions, the standard term *chi* was used to indicate totality. Both the detail 'all the stars were then seen' and Ch'ang-ch'un's subsequent explanation of the cause of the loss of daylight provide emphatic evidence in favour of totality having been witnessed.

When the recorded date of the observation is converted to the Julian calendar (i.e. AD 1221 May 23), it agrees exactly with that of a calculated

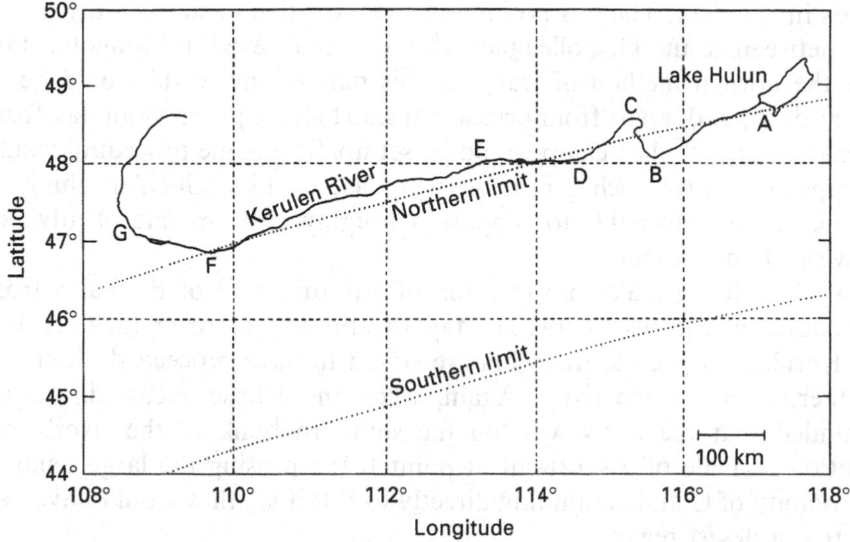

Fig. 8.6 Route of Ch'ang-ch'un up the Kerulen river valley in AD 1221. The computed track of the total solar eclipse of AD 1221 May 23 is also shown (based on a preliminary value for ΔT of 850 sec).

eclipse. The main difficulty lies in the fixing of the geographic location of Ch'ang-ch'un at the time. We know from the detailed narrative that on the 17th day of the fourth month (May 10) his party left the camp of Prince Tamuga, younger brother of Genghis Khan, having been given a farewell gift of several hundred oxen and horses, as well as ten wagons. Five days later – on the 22nd day of the month (May 15) – it is recorded that they reached Lake Hulun Nur at the place where the Kerulen River outflows. From here they proceeded in a westerly direction along the southern bank of the river. After the eclipse, they continued along the Kerulen valley for a further 16 days until they 'came to a point where the Kerulen makes a loop to the north-west, skirting some mountains'. Here, on a date corresponding to Jun 7, the party turned south-west.

In order to derive a fairly precise location for the travellers at the time of the eclipse, Stephenson and Yau (1992b) utilised the following maps of the area: (i) charts on a scale of 1:1 000 000 published by the US Army Map Service in 1944 (these were derived from Russian maps of the area on the same scale produced in 1936); (ii) a map on a scale of 1:2 500 000 published in the USSR in 1967; (iii) a 1:5 000 000 map in the *Times Atlas*. Using this source material, the course of the Kerulen River can be accurately delineated (see figure 8.6). It is apparent from the topography shown on the maps that the river has not changed its course appreciably on a time-scale of millennia.

In figure 8.6, the points A through G (working upstream) denote major bends in the river. There is nothing in the narrative to suggest any major halt between A and G. Colleagues who have journeyed in Mongolia state that the normal method of travel in this part of the world would be to rise at dawn and, apart from occasional short stops, progress all day from sunrise to sunset when camp would be set up. This same procedure would be repeated daily. Such a routine is implied by Li Chih-ch'ang himself. It thus seems reasonable to suppose a roughly uniform rate of advance between points A and G.

Li Chih-ch'ang makes no mention of any diversion of the party from the course of the river between May 15 and Jun 7. After they reached the Kerulen (at A), the travellers are stated to have proceeded along the southern bank of the river. Again, when the eclipse occurred, we are reminded that the party was 'on the southern bank of the river'. No mention is made of a short-cut at point B (by-passing the large bend in the vicinity of C and continuing directly to D); if so, they would have had to cross a desert region.

It may be estimated that if Ch'ang-ch'un indeed followed the entire course of the river from A to G, the total distance travelled would be about 830 km. This corresponds to an average daily rate of progress (over 23 days) of 36 km. Since at this time of year and latitude there would be nearly 16 hours of daylight per day, the rate of travel would seem quite reasonable. Although the members of the delegation travelled on horseback, their baggage was conveyed in ox-carts. Assuming that this average speed was maintained, when the eclipse occurred the party would be some $\frac{8}{23}$ of this distance from A or about 290 km from Lake Hulun. This site is roughly halfway between C and D. Over the whole of the section CD, the river flows north-east, as asserted in the narrative.

In addition, it is extremely unlikely that the eclipse could have taken place while Ch'ang-ch'un was still between A and B in figure 8.6. Point B is only about 145 km from A and thus only 0.17 of the full distance between A and G. If the party took 8 days to reach B, the average speed would only be 18 km per day, after which a mean daily rate of progress of as much as 46 km would be needed in order to arrive at G by Jun 7. Further, an observation between B and C is ruled out by the reference to the north-east flow of the Kerulen where the eclipse occurred. A location somewhere between C and D will thus be assumed.

In figure 8.7, which shows the eastern half of the Kerulen River on a larger scale, the computed northern boundary of the belt of totality using ΔT = 980 sec is marked. This boundary just touches point D. For any value of ΔT greater than 980 sec, the track would lie further south and the eclipse would not be total anywhere along the Kerulen westward of B. (As discussed above, a location to the east of B is highly unlikely.) Since the location of Ch'ang-ch'un at the time of the eclipse cannot be deduced

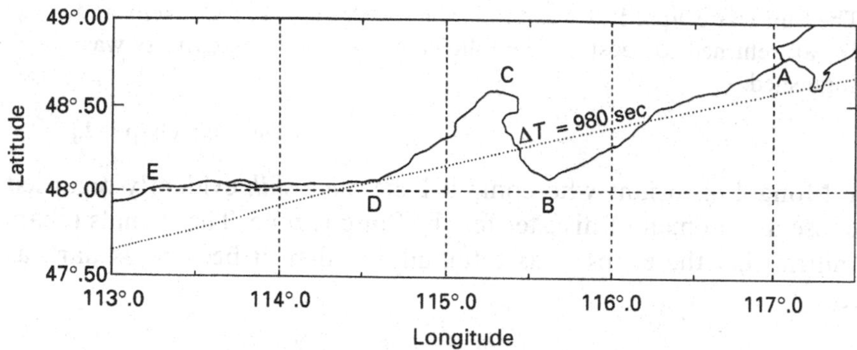

Fig. 8.7 Eastern half of Kerulen river valley showing the northern edge of the track of totality in AD 1221 computed for $\Delta T = 980$ sec.

more accurately than some point between C and D, it is not possible to derive a more refined upper limit for ΔT than this. The lower limit to ΔT for totality in this region is extremely negative (about -8100 sec). Hence only the upper limit of $+980$ sec is viable.

RESULTS

For totality to be observed by Ch'ang-ch'un on the Kerulen River, $\Delta T <$ 980 sec.

(NB computed local time of maximum at 48 deg N, 115 deg E = 12.9 h; solar altitude = 63 deg.)

Using the value of ΔT derived from equation (8.1) (i.e. 850 sec), both the computed magnitude (0.63) and local time (8.1 h) at Samarkand (lat. = 39.67 deg, long. = -66.95 deg) are in good accord with observation. Similar remarks apply to the report from near the Chin-shan (T'ien-shan) mountains in Central Asia (approx. lat. = 49 deg, long. = -88 deg). Here the computed magnitude was 0.73 and local time 10.1 h.

(4) 1275 Jun 25 (mag. = 1.07): Lin-an [class A]

> Te-yu reign period, 1st year, month VI, day *keng-tzu* [37], the first day of the month. The Sun was eclipsed; it was total. The sky and Earth were in darkness. People could not be distinguished within a foot. The chickens and ducks returned to roost. (It lasted) from the hour *szu* (9–11 h) to the hour *wu* (11–13 h); then it regained its brightness.
>
> [*Sung-shih*, chap. 67.]

The above account is from the Five-Element Treatise of the *Sung-shih*. A less extensive description of this event is to be found in the Astrological Treatise of the same history and this is also worth citing in full:

> The Sun was eclipsed; it was total. Stars were seen. The chickens and ducks all returned to roost. In the following year the Sung dynasty was extinguished.
>
> [*Sung-shih*, chap. 52.]

The Mongol historians who compiled the *Sung-shih* evidently regarded the eclipse as an omen of disaster for the Sung regime. The Annals (chap. 47) confirm that the eclipse was total, adding that 'it became as dark as night'.

Results

For totality at Lin-an $-700 < \Delta T < 1300$ sec.

By this period, ΔT had become so small that the set of values needed to render an eclipse total might well encompass negative figures, as is the case here.

(NB computed local time of maximum = 10.1 h; solar altitude = 63 deg.)

(5) AD 1292 Jan 21 (annular, mag. = 0.94): Ta-tu [class A]

> Chih-yuan reign-period, 29th year, first month, day *chia-wu* [31]. The Sun was eclipsed. A darkness invaded the Sun, which was not totally covered. It was like a golden ring. There were vapours like golden earrings on the left and right and a vapour like a halo completely surrounding it.
>
> [*Yuan-shih*, chap. 48.]

The description of the Sun as 'like a golden ring' is a very clear allusion to the ring phase – the only such example in East Asian history. The aptness of this description is clear from figure 8.8, which shows a photograph of a recent annular eclipse. Evidently a cloud of ice-crystals in the atmosphere was responsible for the halo display in AD 1292; the weather in Beijing (formerly Ta-tu) can be extremely cold at this time of year. The Annals (chap. 14) of the same history note that because of the eclipse the normal daily greetings to the Emperor were not offered by the court. However, in this latter account, the ring phase is not described, only the accompanying atmospheric phenomenon.

Results

For annularity at Ta-tu $-50 < \Delta T < 1830$ sec.

(NB computed local time of maximum = 12.9 h; solar altitude = 31 deg.)

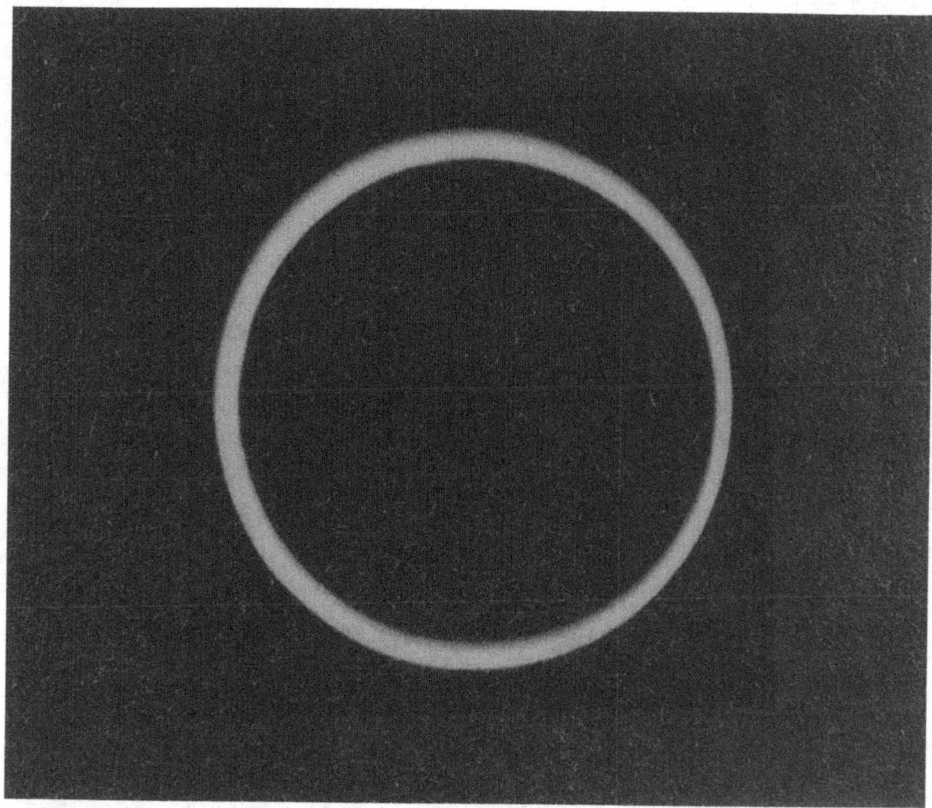

Fig. 8.8 Photograph by Loredano Ceccaroni of annular solar eclipse of 1994 May 10 showing the ring phase. (Courtesy: *l'astronomia.*)

(6) AD 1361 May 5 (total, mag. = 1.05): Sung-chiang [class A]

> Chih-cheng reign period, 21st year, 4th month, day *hsin-szu* [18], the first day of the month. As the Sun was about to sink (i.e. set), suddenly it lost its light. It took the shape of a plantain (*chiao*) leaf. The sky was as dark as night and the stars were shining brightly. For a short time (*literally*: for the duration of a meal), the sky became bright again. Then in a short time (the light) disappeared once more.
>
> [*Sung-chiang Fu-chih*, chap. 47.]

The *Sung-chiang Fu-chih* is a history of the town of *Sung-chiang* (pinyin: Songjiang) lying a little to the south-west of Shanghai. This history work was compiled by local government officials in the late Ming dynasty (some time between 1628 and 1644) using original documents.

The plantain leaf is oval in shape, with many fronds on its lower edge; it resembles a banana leaf and has often been used as a fan. In comparing the eclipsed Sun with a plantain leaf – rather than some other shape – the observer may have had in mind the very obvious fronds of this leaf.

Perhaps the most likely inference would seem to be that at maximum phase the Sun was reduced to a fine crescent edged with Baily's beads. However, it would probably be unwise to overstate this case, and hence I shall assume only a very large partial eclipse at Sung-chiang.

RESULTS

For a partial eclipse at Sung-chiang, $\Delta T < 500$ sec or > 1760 sec.

A result in the higher range would be discordant with many roughly contemporaneous observations; equation (8.1) leads to a value for ΔT of approximately 650 sec. Hence it seems reasonable to select only the limit of $\Delta T < 500$ sec. If it is only assumed that the Sun appeared crescent-shaped at maximum phase, it is not possible to obtain a more precise value for ΔT than this. However, since the description indicates an extremely large eclipse, a result for ΔT close to 500 sec is probably indicated.

(NB computed local time of maximum = 18.6 h; solar altitude = 2 deg.)

8.8 The Ming dynasty (AD 1368–1644)

The Ming dynasty, which was established in 1368 after the collapse of Mongol rule in China, had its first capital at Ying-t'ien (modern Nanjing). In 1420 the capital was removed to Pei-ching ('Northern Capital', i.e. Beijing). The astronomical treatise of the official history, the *Ming-shih*, does not contain any solar eclipse records; these are almost entirely confined to the Imperial Annals instead but there are no allusions to large eclipses in the annals. Although systematic observations are to be found in the *Ming-shih-lu* ('Veritable Records of the Ming Dynasty') – an extensive chronicle of events covering each reign of the dynasty – this compilation is also devoid of any references to either total or near-total solar obscurations. Hence we must turn to other Ming sources.

From about AD 1500 onwards, Ming provincial histories (*fang-chih*) often report large eclipses accompanied by darkness and the appearance of stars. However, by no means all of these records are reliable; frequently no eclipse took place on or near the stated date so that the event is unidentifiable. Possibly the compilers were careless in filing documents or individual reign periods became confused. Nevertheless there are many reliable accounts of total eclipses in the local histories, sometimes providing considerable detail. At least ten separate descriptions for each of the total solar eclipses of 1514 Aug 20, 1542 Aug 11, 1575 May 10 and 1641 Nov 3 are preserved from different parts of China. Although minor dating errors occur in certain texts, in most cases the date (when reduced to the Julian or Gregorian calendar) is in exact agreement with the results of modern computation.

The following descriptions of the total eclipse of AD 1514 Aug 20 are fairly typical. Both are translated from the texts quoted in the compilation by Beijing Observatory (1988). In each case the dates is correctly given as the ninth year of the Cheng-te reign period, eighth month, first day *hsin-mao* [28].

(i) At the hour of *wu* (i.e. between 11 and 13 h) the Sun was eclipsed. The sky and Earth became dark in the daytime. All the birds flew about in alarm. The domestic animals went into the forest. At the hour of *yu* (17–19 h) the light came back.

[*Fu-ning Chou-chih*: local history of Fu-ning county, Fu-chien (= Fujian) province.]

(ii) At the hour of *wu* suddenly the Sun was eclipsed; it was total. Stars were seen and it was dark. Objects could not be discerned at arm's length. The domestic animals were alarmed and people were terrified. After one (double-) hour it became light.

[*Tung-hsiang Hsien-chih*: local history of Tung-hsiang county, Chiang-hsi (= Jiangsi) province.]

The precise place of observation in each of the above accounts cannot be determined. It could have been anywhere in an area covering several thousand square kilometres. This is true for nearly all records in the Ming provincial histories. By the sixteenth century, ΔT would be so small (less than about 300 sec) that only a reliable observation at a precisely known place would be of any value in refining this parameter. Hence most observations are of little more than historical interest. The succeeding Ch'ing dynasty (AD 1644–1911) – during which ΔT was smaller still – lay entirely within the telescopic period and is thus beyond the scope of this book. Although the telescope was introduced into China around AD 1630 (D'Elia, 1960, pp. 41–2), eclipse and other astronomical observations continued to be made in China with the unaided eye until relatively recent times. These, of course, could not compete in precision with telescopic observations made in the West.

One of the very few Ming records of a total eclipse seen at a definite site comes from the town of Sung-chiang, near Shanghai. The observation is found in the *Sung-chiang Fu-chih*, the same source as the detailed Yuan account of the eclipse of AD 1361 (see section 8.7) and dates from 1575. The text may be translated as follows.

AD 1575 May 10 (total, mag. = 1.07): Sung-chiang [class A]

Fang-li reign period, 3rd year, 4th month, day *chi-szu* [6], the first day of the month. The Sun was eclipsed. At about the hour of *wu* (roughly noon) the eclipse was total. The bright day became dark.

[*Sung-chiang Fu-chih*, chap. 47.]

RESULTS

For totality at Sung-chiang, $-2150 < \Delta T < 1390$ sec.

The rather wide zone of totality ran nearly parallel to the equator in the longitudes of eastern China.

(NB computed local time of maximum = 14.8 h; solar altitude = 51 deg.)

8.9 Korea

The earliest history of Korea covers the interval from legendary times to AD 935. This work, entitled the *Samguk Sagi* (History of the Three Kingdoms), was compiled by Kim Pu-sik in AD 1145. It is written in Classical Chinese – a feature which it shares with virtually all other major Korean literary works composed before the present century. Very early in the first millennium AD the country was divided into the kingdoms of Koguryo, Paekche and Silla. Around AD 670 Silla subjugated its rivals and the era of Silla ascendancy lasted until AD 935. During both the Three Kingdoms (Samguk) period and the subsequent Silla domination, the *Samguk Sagi* chronicles many solar eclipses, using precisely the same terminology as found in Chinese records. However, the records are very brief and there are no references to either total or near-total obscurations.

In AD 935, the kingdom of Koryo, which had been founded some years before in northern Korea by Wang Kun, overthrew Silla. Koryo established a unified kingdom throughout the Korean peninsula which was to endure until AD 1392. During most of the Koryo dynasty, the capital was Songdo (lat. = 37.97 deg, long. = −126.57 deg), on the site of modern Kaesong. However, when the Mongols invaded the country in AD 1229, King Kojong and his court fled to the island of Kangwha (lat. = 37.73 deg, long. = −126.48 deg) where they remained in exile for many years. In 1270, Kojong's successor returned to the former capital of Songdo as a vassal of the Mongols. Not until 1364 were the Mongols finally driven out of Korea.

Early in the Koryo dynasty, a royal observatory was founded at Songdo and from this site numerous observations of celestial phenomena were made. These are reported in the extensive Astrological Treatise (chaps. 47–49) of the *Koryo-sa* (History of Koryo), the official history of the period. This work, compiled under the direction of Chong Inji in 1451, is modelled on a typical Chinese dynastic history. As in China, the motive for skywatching was largely divination, but no surviving eclipse record is accompanied by an astrological prognostication. Occasional lunar eclipse reports in the *Koryo-sa* describe ceremonies in which the king accompanied by his ministers attempted to rescue the Moon from its eclipse. Possibly these represented little more than the maintenance of a long-established tradition. However, no such *solar* rites are on record.

The *Koryo-sa* treatise contains a special section (in chap. 47) devoted to solar eclipse observations. Virtually identical records to these are to be found in the Basic Annals at the beginning of the same history (chaps. 1–46). In all, five solar eclipses (occurring in the years AD 1245, 1321, 1361, 1366 and 1390) were said to be complete. Each is briefly announced in both the Astrological Treatise and the Annals in the same form: namely that on a certain date there was a total obscuration of the Sun. There is nothing to suggest observation outside the capital. In each instance, totality is simply identified by the same term *chi* as in Chinese history (but read as *ki*); no further description – such as time of occurrence – is given. The many references to other solar eclipses in the *Koryo-sa* are devoid of any allusion to magnitude or other details

Since the pattern of the five Korean accounts of totality is very repetitive, a single example will suffice. The following observation was recorded on a date corresponding to AD 1245 Jul 25:

> King Kojong, 32nd year, 7th month, day *kuei-szu* [30], the first day of the month. The Sun was eclipsed and it was total.
>
> [*Koryo-sa*, chap. 47.]

As was customary from earliest times, the date follows the Chinese style, except that the name of the Korean king is used. (NB I have given the Chinese name of the cyclical day, rather than its Korean equivalent.) It was the practice each year for the Chinese emperor to present copies of the new calendar to the Korean envoy for use in his homeland. At this particular eclipse, the place of observation was presumably the island refuge of Kangwha. Although the phase was described as total, the above eclipse was in fact only annular. Here we have another instance of the use of *chi* to denote annularity.

Of the five eclipses said to be total in the *Koryo-sa*, only that of AD 1361 May 5 was also observed to be large in China. As noted above (section 8.7), this was described as virtually total just before sunset in Sung-chiang, with stars shining on account of the darkness. If this eclipse was indeed total in Songdo, the magnitude at Sung-chiang (nearly 1000 km to the south-west) would have been relatively small – less than about 0.8 – and maximum phase would occur more than two hours before sunset. Neither of these deductions are in accord with the circumstances so vividly described at Sung-chiang. Hence considerable doubt must be cast on the brief Korean allegation of totality. The reverse situation – an eclipse on the verge of totality at Sung-chiang and only partial at Songdo – is illustrated in figure 8.9.

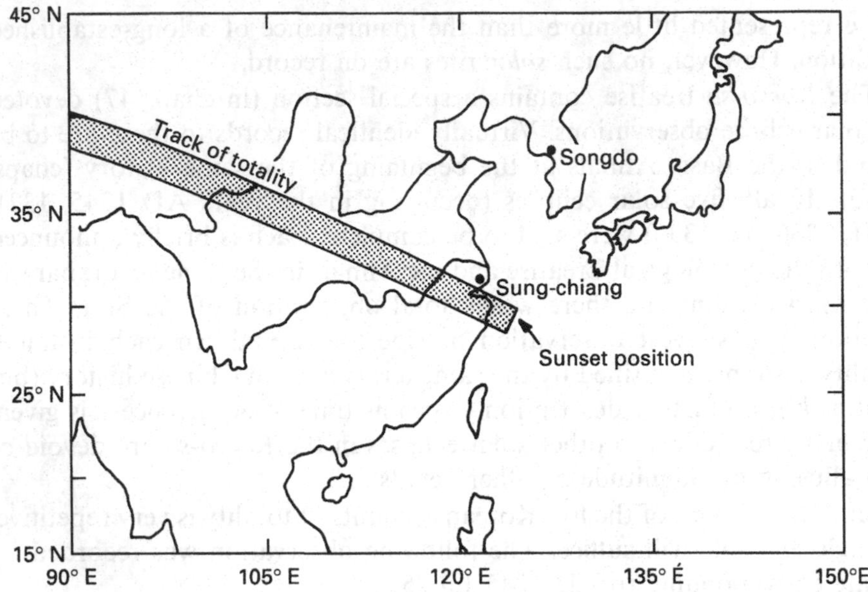

Fig. 8.9 Map showing the computed track of totality for the eclipse of AD 1361 May 5 in relation to Sung-chiang and Songdo (based on $\Delta T = 0$ sec).

On the remaining three occasions when the Sun was said to be completely obscured (AD 1321 Jun 26, 1366 Aug 7 and 1390 Oct 9) the belt of totality was extremely narrow – between about 10 and 20 km wide. The probability of these three marginal eclipses all being central at the capital of Songdo is very small. Further, it seems highly unlikely that chance observations of totality from random sites in the provinces would reach the capital and eventually be included in the official history. Hence, in view of the lack of reference to any phase but totality in the *Koryo-sa*, it seems plausible that at least some of the five records may merely represent abbreviated descriptions of eclipses which were originally described as almost complete. As in the case of the less detailed Chinese accounts discussed above, each of these observations will be included in category 'C'. Summary information is given in table 8.7.

The Koryo dynasty endured until AD 1392 when King Kongyangwang was deposed and General Yi Sung-gae established a new dynasty known as Yi. Soon afterwards, the capital was transferred from Songdo to Hanyang (modern Seoul, lat. = 37.55 deg, long. = −126.97 deg), where it has since remained. The Yi dynasty itself lasted until the Japanese annexation in 1910.

There is no official Yi history similar to the *Koryo-sa*. The most direct source of astronomical records, including solar eclipses, for the earlier part of this period is the *Choson Wangjo Sillok* (Veritable Records

Table 8.7 ΔT limits deduced from Korean observations of solar eclipses described as total (all 'C' rated).

					ΔT Range (sec)	
Julian date	Type	Mag.	Record	Capital	LL	UL
1245 Jul 25	A	0.95	total	Kangwha	20	1200
1321 Jun 26	T	1.002	total	Songdo	110	150
1361 May 05	T	1.05	total	Songdo	7000	8790
1366 Aug 07	T	1.004	total	Songdo	−910	−890
1390 Oct 09	T	1.005	total	Songdo	740	830
1397 May 27	T	1.06	total	Hanyang	280	1620
1460 Jul 18	T	1.06	total	Hanyang	−3070	−1750

of the Choson Dynasty), a detailed Korean chronicle resembling the *Ming-shih-lu* and extending down to the mid-nineteenth century. An even more extensive chronicle known as the *Sungjongwon Ilgi* (Diaries of the Court Secretariat) formerly covered the dynasty but owing to a serious of unfortunate events (notably the Japanese invasions of 1592 and 1596) very little of this work now survives from before AD 1625. Park (1979) gives an interesting comparison between the detail in seventeenth century records of the same celestial phenomena as found in the *Sungjongwon Ilgi* and *Choson Wangjo Sillok*; it is evident that the latter work omits many observations and condenses other reports. The loss of so many earlier volumes of the *Sungjongwon Ilgi* is much to be regretted.

In the *Choson Wangjo Sillok*, references to large partial obscurations of the Sun are lacking, while only two total eclipses are reported prior to AD 1600 – in the years AD 1397 and 1460. Both of these accounts are very brief, once again noting no more than the occurrence of totality using the term *ki* (i.e *chi*). Each observation will therefore be assigned a 'C' grading, and details summarised along with the Koryo material in table 8.7.

Since the paths of the umbral shadow in AD 1321, 1366 and 1390 were so narrow, only a very restricted range of ΔT can satisfy the record of totality in each case. Most of the results in table 8.7 are mutually discordant. In particular, the observation in AD 1361 indicates a very large positive set of values in which ΔT should lie whereas in 1460 the corresponding range is markedly negative. In summary, the contribution of the Korean records of total solar eclipses to the study of Earth's past rotation is virtually negligible.

8.10 Japan

Japanese observations of solar eclipses are also numerous, but unlike those from China and Korea they are scattered – along with references to other celestial phenomena – in a large number of diverse writings. These works include privately compiled histories, diaries and temple records as well as official histories such as the *Dainihonshi* (History of Great Japan). Fortunately, early in the present century Kanda (1934) made an exhaustive collection of astronomical observations from this varied material; this extends down to AD 1600. Kanda's extensive publication contains a major section devoted to solar eclipses. I have used this as my principal source of Japanese data (see also Kanda, 1935).

Japanese history emerges from legend around AD 600. The ancient chronicle known as the *Nihongi* (Annals of Japan, also known as the *Nihon Shoki*), which covers the period from earliest times to AD 693, contains several records of celestial phenomena, including a few solar obscurations from the seventh century. Among these, the eclipse of AD 628 Apr 10 is reported as total (using the character *chi*), but further details are absent. At this early period, the *precise* place of observation cannot be established; it was the custom to move the imperial residence to another site on the death of each ruler. However, it is known that the various capitals were all situated in the Yamato Plain, east of present-day Osaka (approximate lat. = 34.6 deg, long. = −135.8 deg).

By AD 710, the first fixed capital was built – at Nara (lat. = 34.68 deg, long. = −135.82 deg). This was modelled on the T'ang metropolis of Ch'ang-an, but its active life was short. In AD 784 the imperial residence was transferred to Heian-kyo (lat. = 35.03 deg, long. = −135.75 deg), later renamed Kyoto (the 'Capital'). Here it remained for more than a thousand years - until 1868, when it was replaced by Edo (Tokyo). In 1192, a Shogunate or military government was founded at Kamakura (lat. = 35.22 deg, long. = −139.55 deg). This was to last for 140 years, although Kyoto appears to have remained the more important cultural centre throughout this period. Edo (lat. = 35.67 deg, long. = −139.75 deg) did not attain prominence until AD 1590.

As was the practice in contemporary China during the T'ang dynasty, for several centuries after eclipse observations began in Japan, offices were closed on the day of a solar eclipse. The *Gehpei seiseiki* (which narrates the rise and fall of the Minamoto and Taira clans) relates that on a day corresponding to 1183 Nov 17 the Minamoto army fled, frightened by a solar eclipse (Nakayama, 1969, p. 51). This eclipse was only annular, 95 per cent of the solar diameter being covered in the central zone.

The following solar eclipses are reported as either total or very large in Japanese history before AD 1600: 628, 975, 1069, 1366, 1413, 1415 and

1460. Of these, both the eclipses of 1366 and 1460 were also said to be total in Korea (see section 8.9 above). Some of the Japanese descriptions are fairly detailed, and frequently several independent reports of the same event are available. The brief report in AD 628 most likely stems from the Yamato region while the various accounts of totality in AD 975 almost certainly originate from Heian-kyo, the capital of the time. However, because of the varied nature of the sources on the remaining dates there is considerable doubt regarding the place(s) of observation. I shall discuss the observations made in AD 628 and 975 in detail, but shall only abstract from the later texts.

(1) AD 628 Apr 10 (total, mag. = 1.05): Yamato Plain [class C]

> 36th year of Empress Suiko, spring, 2nd month, 27th day. The Empress took to her sick bed. 3rd month, 2nd day. There was a total eclipse of the Sun. 6th day. The Empress' illness became very grave and (death) was unmistakably near ... 7th day. The Empress died at the age of seventy-five.
>
> [Trans. Aston, 1972, p. 155.]

The recorded date proves to be in accurate accord with that of a computed solar eclipse. Although the obscuration of the Sun was said to be total, further description is lacking, so that I have included the observation in the lowest category. It is noteworthy that the event does not seem to be linked in any way with the Empress' death by the chronicler (unlike several similar occurrences in China – see above).

RESULTS

For totality in the Yamato Plain, $2270 < \Delta T < 2960$ sec.

(NB computed local time of maximum = 9.6 h; solar altitude = 48 deg.)

(2) AD 975 Aug 9 (total, mag. = 1.06): Heian-kyo [class A]

Kanda (1934) uncovered as many as 14 separate descriptions of this event. Three of these, which give detailed descriptions of totality, are translated below. In each case the date is correctly given as the 3rd year of the Ten-en reign period, 7th month, day *hsin-wei* [8], the first day of the month. Like the Koreans, the Japanese adopted the Chinese calendar from an early period in their history.

> (i) The Sun was eclipsed.... Some people say that it was entirely total. During the hours *mao* and *ch'en* (some time between 5 and 9 h) it was all gone. It was the colour of ink and without light. All the birds flew about in confusion and the various stars were all visible. There was a general amnesty (on account of the eclipse).
>
> [*Nihon Kiryaku*.]

(ii) At the hour *ch'en* (7–9 h), the Sun was eclipsed; it was completely total. All under heaven became entirely dark and the stars were all visible.

[*Fuso Ryakki.*]

(iii) The Sun was eclipsed; it was all gone. It was like ink and without light. The stars were all visible (*or*: stars were visible in the daytime – *text corrupt*).

[*Hyaku Rensho.*]

NB I have given the cyclical day and hour names in Chinese rather than their Japanese equivalents.

Brief comments on the above historical sources are as follows. The *Nihon Kiryaku* is a privately compiled history written about AD 1028. Of unknown authorship, it is believed to be historically important and trustworthy from about AD 887, and to be based on contemporary diaries of court officials at Heian-kyo. The *Fuso Ryakki* is a history of Buddhist affairs written by the monk Koen of Heizan (the mountains near Heian-kyo) in AD 1094. Finally, the *Hyaku Rensho* is a history of the imperial court from AD 960 to 1269. It was written at the end of this period.

RESULTS

For totality at Heian-kyo, $1230 < \Delta T < 4480$ sec.

This range of ΔT is very wide; the broad zone of totality ran nearly parallel to the equator in the longitudes of Japan.

(NB computed local time of maximum = 7.9 h; solar altitude = 31 deg.)

Darkness like that of night is briefly mentioned at the marginal (annular-total) eclipses of AD 1069 Jul 21 and 1366 Aug 7, but in neither case is the degree of obscuration of the Sun recorded. Both central zones would be extremely narrow. Darkness is also described at the eclipse of AD 1413 Feb 1, which was said to be total. There is a single report of totality for the eclipse of 1415 Jun 7, but no further details are given and – as in 1413 – the place of observation cannot be established from the text. The situation on Jul 18 in AD 1460 is more interesting. On this occasion, as many as ten reports note either the occurrence of totality and/or darkness. In the sources compiled by Kanda, individual descriptions include remarks such as: 'It was like a moonlight night and the stars appeared'; 'It was like a cloudless night for about half an hour'; 'There was a total eclipse and the stars were all visible'. It is unfortunate that the place where any of these records originated cannot be established. For example, there is nothing to relate them to the capital of Kyoto.

Table 8.8 ΔT limits derived from 'A' and 'B' rated East Asian records of solar eclipses.

Year[a]	Country[b]	Phase[c]	ΔT Range (sec)	
			LL[d]	UL[e]
−708	China	central	20 230	21 170
−187	China	partial	14 140	13 830
−180	China	central	11 800	12 720
−79	China	partial	8 520	8 200
−27	China	partial	9 530	8 090
−1	China	partial	2 150	2 050
+120	China	partial	8 970	8 150
+360	China	partial	9 400	7 120
+454	China	central	6 130	7 900
+494	China	partial	6 600	5 980
+702	China	partial	2 760	1 460
+729	China	partial	1 190	420
+761	China	central	1 720	3 290
+822	China	partial	—	4 020
+975	Japan	central	1 230	4 480
1135	China	partial	3 810	1 840
1221	Mongolia	central	—	980
1275	China	central	700	1 300
1292	China	central	−50	1 830
1361	China	partial	1 760	500
1575	China	central	−2 150	1 390

[a] Year.
[b] Country where the observation was made.
[c] Whether the eclipse was said to be central or partial.
[d] Lower limit (LL) to the value of ΔT deduced from the observation.
[e] Corresponding upper limit (UL) to ΔT.

8.11 Conclusion

The results of the investigation of 'A' and 'B' rated solar eclipse observations from China and Japan (there are no reports from Korea in this category) are listed in table 8.8. In table 8.9 are assembled the corresponding data obtained from the analysis of observations of lower reliability (class 'C'). Both of these tables use negative and positive integers for years

Table 8.9 ΔT limits derived from 'C' rated East Asian records of solar eclipses.

Year	Country	Phase	ΔT Range (sec)	
			LL	UL
−600	China	central	21 140	21 960
−548	China	central	16 150	21 710
−197	China	central	6 350	13 110
+2	China	central	8 570	8 590
+65	China	central	8 450	8 900
+306	China	central	6 550	7 890
+616	China	central	2 300	3 030
+628	Japan	central	2 270	2 960
1245	Korea	central	20	1 200
1321	Korea	central	110	150
1361	Korea	central	7 000	8 790
1366	Korea	central	−910	−890
1390	Korea	central	740	830
1397	Korea	central	280	1 620
1460	Korea	central	−3 070	−1 750

(rather than BC and AD) since their main purpose is to summarise the results obtained in earlier sections.

Following the adopted convention for a partial eclipse (see chapter 3), the lower ΔT limit in column 4 of table 8.8 is actually greater than the upper limit in column 5. This is because a range of ΔT values is excluded by the observation.

Most of the ΔT ranges in table 8.8 are plotted in figure 8.10. In this diagram, full vertical lines (usually short) represent the range of ΔT values derived from central solar eclipses. Broken vertical lines denote the range of ΔT values obtained from partial solar eclipses. NB in both cases, blank regions of solution space are void. For the partial eclipses in the years −187, −79 and −1, the excluded zones are so narrow that they cannot be satisfactorily displayed on the scale of the diagram. For all practical purposes, each observation is effectively redundant; virtually any value of ΔT would render each eclipse partial at the Chinese capital (Ch'ang-an) in accord with the record. It is evident that all of the data displayed in figure 8.10 form a fairly self-consistent set.

All of the ΔT limits in table 8.9 (category 'C') are depicted in figure 8.11. It is clear from this diagram that the Korean data (between the

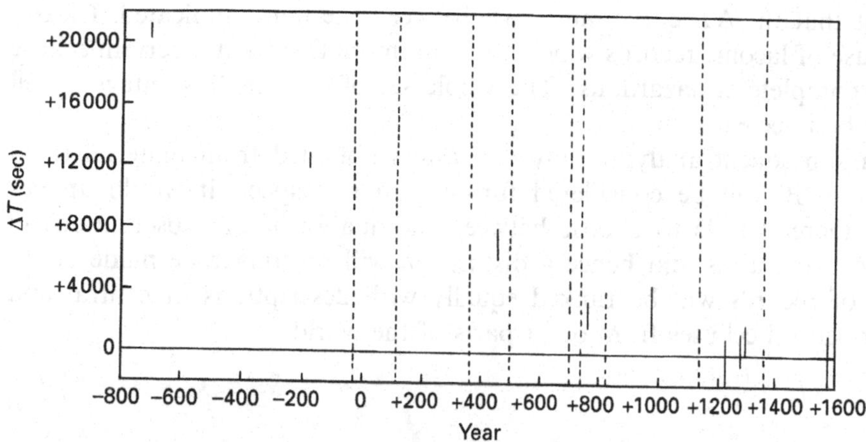

Fig. 8.10 ΔT ranges obtained from East Asian observations of central and large partial solar eclipses (categories 'A' and 'B').

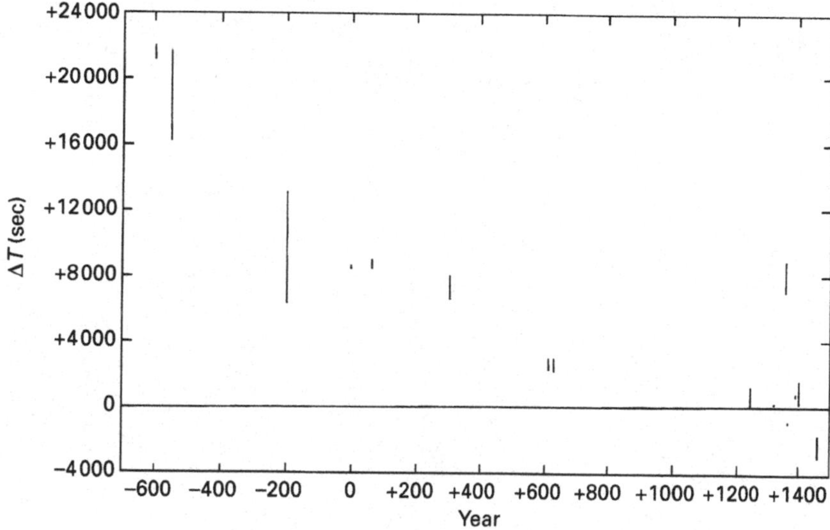

Fig. 8.11 ΔT ranges obtained from East Asian observations of central and large partial solar eclipses (category 'C' only).

years +1245 and +1460) lack mutual consistency. The results from the Chinese observations (between the years −600 and +607) and the solitary Japanese observation (+628) seem fairly self-consistent. However, it should be emphasised (as discussed in sections 8.5 to 8.7), that further Chinese records in category 'C' (in the years +562, +879, +977 and +1022) have been rejected because the eclipses proved to be invisible in China. In addition, several of the ΔT ranges shown in figure 8.11 (especially for the years +2 and +65) are extremely narrow and it would seem unrealistic to insist that the ΔT curve must pass between the limits indicated. Clearly the use of laconic records which assert no more than that a certain eclipse was complete is hazardous. The whole set of data in this category will thus be rejected.

In subsequent analysis, only the results obtained from material rated 'A' and 'B' will be considered further. In retrospect, it would appear that there is little to choose between the quality of the observations in these two classes and hence a distinction will no longer be made. Both sets of records will be ranked equally with descriptions of central and near-central eclipses from other parts of the world.

9

Other East Asian observations of solar and lunar eclipses

9.1 Introduction

In the previous chapter, East Asian records of total and near-total solar eclipses were analysed. Emphasis was placed on observations which either affirmed or specifically denied that the Sun was completely obscured. Other reported details – such as measurements of the times of the various contacts or rising or setting of the luminary whilst eclipsed – were not considered. The theme of the present chapter is the investigation of further observations of both solar and lunar eclipses from the same part of the world which in principle are of value in the determination of ΔT. These may be grouped in three main categories: (i) timed contacts; (ii) estimates of the proportion of the Sun covered at maximum phase; and (iii) instances where the Sun or Moon was said to rise or set whilst visibly obscured.

In the pre-telescopic period, there is a significant number of Chinese observations in each of the above categories. However, accounts of eclipses from Korea tend to be extremely brief and no useful records are preserved in the history of this country – whether in the *Koryo-sa* ('History of Koryo') or the *Choson Wangjo Sillok* ('Veritable Records of the Choson Dynasty'). Although many early Japanese observations of both solar and lunar eclipses give estimates of magnitude or local time, these are contained in a miscellany of sources. Not only is the place of origin often obscure (see also chapter 8), it is frequently difficult to distinguish observation from prediction. Despite the extensive compilation of Japanese eclipse data by Kanda (1934), a thorough investigation will be needed before useful ΔT results can be obtained from this material. The author eventually hopes to undertake such a project, but at present only the Chinese observations seem worthy of attention. The remainder of this chapter will thus be devoted exclusively to Chinese data.

9.2 General remarks on recorded eclipse timings

More than one hundred separate timings of eclipse contacts (both for the Sun and Moon) are recorded in Chinese history – the earliest dating from 134 BC. Yet compared with the total number of eclipses noted in the same sources, such measurements are still relatively rare. Sufficient measurements are preserved from certain dynasties (especially the Sung) to suggest that the timing of eclipses was practised fairly frequently. However, it would appear that most of the original observations have not survived. The majority of extant timings recorded in the official histories are not included in the astrological sections (in contrast to more general eclipse records) but tend to be found in calendar treatises (*li-shih*) instead. In these works, observations by the court astronomers were used to check the reliability of existing almanacs. Many further measurements of eclipse times (especially for the Moon) are noted in the *Wen-hsien T'ung-k'ao* ('Comprehensive History of Civilisation'). This work, compiled by Ma Tuan-lin (AD 1254–1323), was completed in AD 1307. Ma presumably obtained many of his detailed astronomical records from the now lost *Sung Shih-lu* (Annals of the Sung Dynasty).

Before considering individual measurements, some discussion of Chinese units of time seems necessary. Solar eclipse times recorded in Chinese history are almost invariably expressed in terms of the twelve *shih* or 'double hours'. Although lunar eclipse times are often cited in this same form, they are also frequently quoted relative to the five *keng* or night watches instead – especially before AD 600.

9.2.1 Double hours and their subdivisions

During the Former Han dynasty (*c.* 100 BC), the interval from one midnight to the next was divided into twelve equal parts known as *shih*. There is no evidence that the origin of the twelve *shih* is related to the occidental system of 24 hours in a day. Individual double hours were named according to the twelve *ti-chih* (terrestrial branches) of the sexagenary cycle – see chapter 8. The first double hour (*tzu*) was centred on midnight and thus began at 23 h local time. A full list of the *shih* and their equivalent local times is given in table 9.1. This table also includes the twelve symbolic animals with which the double hours were popularly associated; of course, these names were not used in astronomical practice.

From at least AD 1000, each double hour was bisected into an 'initial' and 'central' half. These intervals of time were named for the 'beginnings' (*ch'u*) and 'mid-points' (*cheng*) of the *shih*. The initial half of the first double hour (*tzu-ch'u*) corresponded to between 23 and 24 h, while the

Table 9.1 The twelve double hours

Name	Animal	Time (h)	Name	Animal	Time (h)
tzu	rat	23–01	wu	horse	11–13
ch'ou	ox	01–03	wei	sheep	13–15
yin	tiger	03–05	shen	monkey	15–17
mao	hare	05–07	yu	cock	17–19
ch'en	dragon	07–09	hsu	dog	19–21
szu	snake	09–11	hai	pig	21–23

central half (*tzu-cheng*) lasted from 24 h to 1 h – and so on throughout the other eleven *shih*.

An independent division of the day, known as a *k'o*, was also in regular usage. The literal meaning of this term is a 'notch' or 'mark' on a time-keeping scale (Needham, Lu, Combridge and Major, 1986, p. 9). Although *k'o* will be translated as 'mark' throughout this chapter, it should be emphasised that it relates specifically to an *interval* of time. During the Former Han, the natural day (midnight to midnight) was first divided into 100 *k'o*. Each of these units was thus equivalent to a 0.24 h so that 1 *shih* or double hour contained $8\frac{1}{3}$ marks. Despite the inconvenience of not having an integral number of marks to the double hour, in all but a few brief periods down to the late Ming dynasty there remained exactly 100 *k'o* to a 24-hour day.

In 6 BC Emperor Ai decreed that there should be 120 marks in a day, one double hour thus containing exactly ten *k'o*. However, this revision proved unpopular and only two months later the *k'o* was restored to its original length. Wang Mang, who usurped the Han throne in AD 9, repeated this experiment, but his revision was abolished on his assassination in AD 23. No further changes were made for nearly five centuries. In AD 507, the number of marks in a day was first reduced to 96, but afterwards (AD 544) it was increased to 108. When the traditional convention was reverted to by popular demand in AD 563, it was retained without any alteration until as late as AD 1628. In the latter year, Emperor T'ai-tsung ordained that there should be 96 marks in a day – each exactly one-quarter of an hour in length – and this definition continued to be adopted until the present century.

No accurate eclipse timings are reported in the early periods during which anomalous definitions of the *k'o* were in vogue, while the most recent period is after the introduction of the telescope and thus outside the scope of this book. Hence all subsequent discussion will be based on the basic figure of 100 of these units to a day. Measurements of eclipse

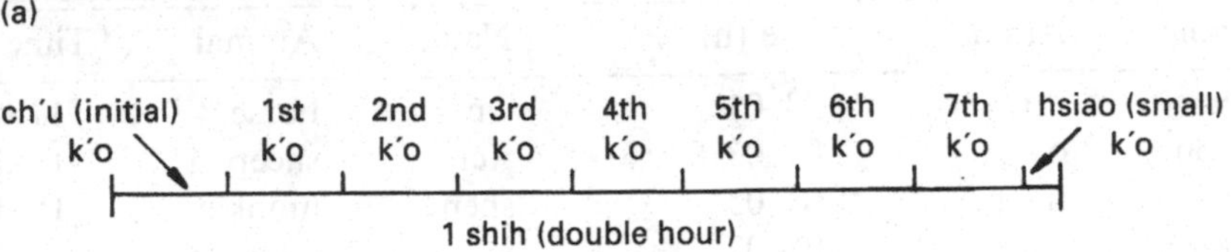

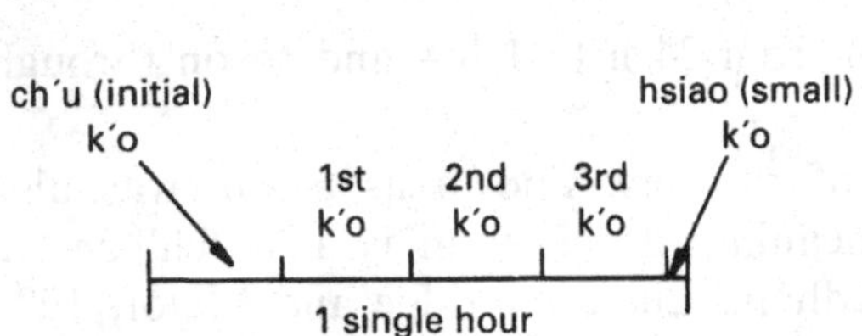

Fig. 9.1 Chinese double (a) and single (b) hours and their divisions.

times were made with the aid of water clocks – such devices were standard equipment at the imperial observatory from the Han dynasty onwards. Early descriptions of water clocks often mention graduations in both *shih* and *k'o* (Needham, Wang and Price, 1986, chaps. 4 and 6).

In most instances, *k'o* were counted from the beginning of each double hour. However, alternatively *k'o* could be measured relative to the start of each half-*shih*. Which of these systems was used in timing a particular eclipse is normally clear from the context. Individual 'hours' (whether double or single) commenced with an 'initial' (*ch'u*) mark, covering the first 0.24 h. This was followed by the first mark, which embraced the next 0.24 h, then the second mark and so on up to the eighth mark (for a double hour) or the fourth mark (for a single hour).

Since 1 *shih* was equal to $8\frac{1}{3}$ *k'o*, the eighth and last mark in a double hour was only one-third of the normal length, covering the final 0.08 h before the start of the next *shih*. This was often known as the *hsiao-k'o* ('small mark') – see figure 9.1a. When single hours were used, the fourth and last mark was extremely short, nominally equal to only 0.04 h; this was again known as the *hsiao-k'o* ('small mark') – see figure 9.1b. It is doubtful whether early water clocks could define either of the *hsiao-k'o* with any precision.

In order to illustrate the division of double hours into *k'o*, I have selected an example from chapter 53 of the *Yuan-shih*. Here it is stated that the lunar eclipse of AD 1106 Jan 21/22 reached maximum phase 'at 3

Table 9.2*a* Division of double hours and their associated mean local times.

Shih	0 k'o	1 k'o	2 k'o	3 k'o	4 k'o	5 k'o	6 k'o	7 k'o	8 k'o
	h	h	h	h	h	h	h	h	h
tzu	23.12	23.36	23.60	23.84	0.08	0.32	0.56	0.80	0.96
ch'ou	1.12	1.36	1.60	1.84	2.08	2.32	2.56	2.80	2.96
yin	3.12	3.36	3.60	3.84	4.08	4.32	4.56	4.80	4.96
mao	5.12	5.36	5.60	5.84	6.08	6.32	6.56	6.80	6.96
ch'en	7.12	7.36	7.60	7.84	8.08	8.32	8.56	8.80	8.96
szu	9.12	9.36	9.60	9.84	10.08	10.32	10.56	10.80	10.96
wu	11.12	11.36	11.60	11.84	12.08	12.32	12.56	12.80	12.96
wei	13.12	13.36	13.60	13.84	14.08	14.32	14.56	14.80	14.96
shen	15.12	15.36	15.60	15.84	16.08	16.32	16.56	16.80	16.96
yu	17.12	17.36	17.60	17.84	18.08	18.32	18.56	18.80	18.96
hsu	19.12	19.36	19.60	19.84	20.08	20.32	20.56	20.80	20.96
hai	21.12	21.36	21.60	21.84	22.08	22.32	22.56	22.80	22.96

marks in the hour of *yu*' and ended 'in the initial mark of the hour of *hsu*'. The first of these intervals corresponds to between 17.72 and 17.96 h local time (LT), while the second is from 19.00 h to 19.24 h. The alternative system, using single hours, is exemplified by an entry in chapter 283 of the *Wen-hsien T'ung-k'ao*. In this record, it is reported that the solar eclipse of AD 1052 Nov 24 ended 'at 1 mark in the central half of the hour of *wei*'. The equivalent LT is between 14.24 and 14.48 h.

Since the unit known as a *k'o* represented a time-interval, when a measurement was expressed to the nearest *k'o* I have assumed that the observation was made at any time during that interval. I have thus taken the average moment as the mid-point of the designated *k'o* (see Chen, 1983a, 1983b for a detailed discussion). Hence in the above examples, I shall make the approximate assumptions that the lunar eclipse of AD 1106 reached its maximum at a LT of 17.84 h and ended at 19.12 h while the solar eclipse of AD 1052 ended at 14.36 h.

On this basis, the equivalent LTs for each double hour and mark are summarised in table 9.2a, and for each single hour and mark in table 9.2b.

Very occasionally a measurement is quoted to the nearest half-*k'o*: for example '5$\frac{1}{2}$ marks in the hour of *wu*' (AD 1173). In such cases I shall still adopt the mid-point of the appropriate mark. Measurements expressed in half-marks are so few in number that it scarcely seems worthwhile assigning them a high weight. No surviving Chinese eclipse records from before the seventeenth century are expressed to better than the nearest half-*k'o*.

Table 9.2*b* Division of single hours and their associated mean local times.

	ch'u (initial) half					cheng (central) half				
Shih	0 k'o	1 k'o	2 k'o	3 k'o	4 k'o	0 k'o	1 k'o	2 k'o	3 k'o	4 k'o
	h	h	h	h	h	h	h	h	h	h
tzu	23.12	23.36	23.60	23.84	23.98	0.12	0.36	0.60	0.84	0.98
ch'ou	1.12	1.36	1.60	1.84	1.98	2.12	2.36	2.60	2.84	2.98
yin	3.12	3.36	3.60	3.84	3.98	4.12	4.36	4.60	4.84	4.98
mao	5.12	5.36	5.60	5.84	5.98	6.12	6.36	6.60	6.84	6.98
ch'en	7.12	7.36	7.60	7.84	7.98	8.12	8.36	8.60	8.84	8.98
szu	9.12	9.36	9.60	9.84	9.98	10.12	10.36	10.60	10.84	10.98
wu	11.12	11.36	11.60	11.84	11.98	12.12	12.36	12.60	12.84	12.98
wei	13.12	13.36	13.60	13.84	13.98	14.12	14.36	14.60	14.84	14.98
shen	15.12	15.36	15.60	15.84	15.98	16.12	16.36	16.60	16.84	16.98
yu	17.12	17.36	17.60	17.84	17.98	18.12	18.36	18.60	18.84	18.98
hsu	19.12	19.36	19.60	19.84	19.98	20.12	20.36	20.60	20.84	20.98
hai	21.12	21.36	21.60	21.84	21.98	22.12	22.36	22.60	22.84	22.98

It is of interest to compare a series of measured local times of sunrise and sunset expressed in these same units with their computed equivalents. Since an accurate knowledge of ΔT is not required, a direct comparison between observation and computation can be readily made. Maspero (1939) investigated a series of sunrise and sunset times measured around AD 600. These are recorded in chapter 17 of the *Sui-shu* and were for either Ch'ang-an or Lo-yang (both close to lat +34.5 deg N). They are given at 15-day intervals between the winter solstice and summer solstice and are expressed in double hours, marks and *fen*, the latter unit corresponding to $\frac{1}{60}$ of a *k'o*. All times are quoted to the nearest 5 *fen*. To give an example, the time of sunset at the winter solstice was said to be 7 *k'o*, 50 *fen* in the hour of *shen*, corresponding to 16.88 h; for comparison, the computed LT is 16.91 h. I have verified the accuracy of Maspero's reductions of the measurements to local time and his computations. His results reveal maximum observational errors of 0.09 h and standard errors of 0.06 h, the latter equivalent to about 15 *fen* or 1/4 *k'o*. It is apparent that the accuracy of the times of sunrise/sunset in the *Sui-shu* fall considerably below the expectations of the Sui astronomers. They possibly represent the best that could be achieved using the water clocks of the period.

Between AD 585 and 768, several solar eclipse timings are expressed relative to double hours using an unusual terminology. Most of these

Table 9.3 Anomalous solar eclipse records of uncertain interpretation.

Date AD	Ct	Measurement	Source
585 Jul 31	1	wu-hou 6 k'o	Sui-shu, 17
	M	wei-hou 1 k'o	
	4	wei-hou 5 k'o	
586 Dec 16	M	ch'en-hou 2 k'o	Sui-shu, 17
	4	ju-szu-shih 3 k'o	
594 Jul 23	1	wei-hou 3 k'o	Sui-shu, 17
721 Sep 26	M	wu-cheng-hou 3 k'o	Yuan-shih, 53
761 Aug 5	1	ch'en-cheng-hou 6 k'o	Chiu-t'ang-shu, 36
	T	szu-cheng-hou 1 k'o	
	4	wu-ch'ien 1 k'o	
768 Mar 23	M	wu-hou 1 k'o	Chiu-t'ang-shu, 36

observations incorporate the term *hou* (after, latter). Examples include *wu-hou* 3 *k'o* (AD 721) and *ch'en-cheng-hou* 6 *k'o* (AD 761). On two occasions, either *ju* (to enter) or *ch'ien* (former, before) replaces *hou* – for example *szu-ju* 3 *k'o* in AD 586. The meaning of all of these various compounds is somewhat ambiguous and different authorities disagree on their interpretation. Hence I shall not reduce any measurements employing these terms. I am grateful to Professors Chen Meidong and Chen Jiujin of the Institute for the History of Natural Sciences, Academia Sinica, Beijing for helpful discussion on this issue.

The dates and other details for the anomalous observations between AD 585 and 768 are summarised in table 9.3. This table lists the appropriate contact, etc. (1 = first contact, M = maximal phase, T = total, 4 = last contact), the measurement of time and the historical source. It should be noted that the time of first contact in AD 586 is given simply as *ch'en 2 k'o*, i.e. 2 marks in the hour of *ch'en*. This is the sole measurement in the whole set which uses a standard terminology and therefore the only one to be retained in subsequent investigation (see section 9.3.1).

9.2.2 Night-watches and their divisions

From the Sung dynasty onwards, *lunar* eclipses were usually timed in relation to the twelve double hours, in common with their solar counterparts.

However, in earlier periods (going as far back as the Han dynasty) it was customary to adopt the night-watches (*keng*) instead. Day was regarded as beginning at sunrise (*jih-ch'u*) and ending at sunset (*jih-ju*). However, night was defined as extending only from dusk (*hun*) to dawn (*tan* or *ming*). The interval from the end of evening twilight to the start of morning twilight (not sunset to sunrise) was divided into five equal watches. Unlike the double hours, the watches varied in length with the seasons.

Evening twilight was originally regarded as ending 3 marks (0.72 h) after sunset, while dawn began a similar time before sunrise. However, from the Eastern Han (AD 25 onwards) this figure was reduced to $2\frac{1}{2}$ marks (0.60 h). This revised convention remained unchanged for astronomical purposes at all later periods, despite movement of the capital to different latitudes (ranging from 30 to 40 deg N). At 0.6 h after sunset or before sunrise, the Sun attains a zenith distance of approximately 97.5 deg in central China (latitude 35 deg N). Hence the adopted convention agrees quite well with the modern definition for the limit of civil twilight – i.e. a solar zenith distance of 96 deg.

In a section of the calendar treatise of the *Sui-shu* (chap. 16) we find the following elementary rules:

> Double the number of marks reached at midnight (after twilight); this gives the number of marks in the (dark) night. Subtract this from 100 marks; what is left is the number of marks in the daytime (including twilight). Subtract 5 marks from the number of marks in the daytime or add it to the number of marks in the night. This gives the time in which the Sun is visible during the day or invisible during the night.

A useful series of explanatory diagrams is given in the *Hsin-i-hsiang Fa-yao* ('New Design for an Astronomical Clock') a work composed by Su Sung in the late eleventh century. These diagrams – one of which is illustrated by Needham, Wang and Price (1986, p. 203) – show the length of daylight, twilight, darkness and dawn in marks at the equinoxes, summer solstice and the winter solstice for the capital city of Pien (Kaifeng, latitude 34.8 deg N). In each case, the durations of evening and morning twilight are taken as $2\frac{1}{2}$ *k'o*. At the equinoxes, the duration of daylight is given as 50 marks and that of night 45 – a total of 100 marks including dusk and dawn. At the summer solstice, the lengths of daylight and night are 60 and 35 marks respectively while at the winter solstice the corresponding figures are 40 and 55 *k'o*.

Reference to the calendar treatises of the various official histories shows that the above figures were fairly standard at most periods. These values lead to minimum and maximum lengths for the five night watches of 7 and 11 marks respectively. For central China (latitude 35 deg N), the extreme lengths of the *keng* may be calculated as 6.9 marks (1.65 h) in summer and 10.8 marks (2.60 h) in winter – both very close to the traditionally adopted limits.

Each night watch was further subdivided into five equal graduations, whose length ranged from about 20 to 30 min, depending on the season. These divisions carried different names at different periods. Before the Sui dynasty, they were known as *ch'ang* ('calls'), after the periodic calls of the night watchmen. During the Sui, they were renamed *ch'ou* ('rods') implying the position of the pointing rod on a water clock. Later, the term *tien* ('points'), having a similar significance to rods, was used instead. However, at all periods there were five equal divisions to a *keng*.

The subdivisions of a night watch were simply numbered first, second, third, etc; there was no initial graduation, unlike in the case of marks within an hour or double hour. By definition, the third division of the third watch would thus be centred on midnight. Water clocks were designed to measure time in terms of the watches and their subdivisions, making allowance for the seasonal variation in the lengths of these units (Needham, Wang and Price, 1986, chaps. 4 and 6).

Figure 9.2a shows the five night-watches and their divisions at the winter solstice and figure 9.2b at the summer solstice for a latitude close to +35 deg.

Conversion of a measurement expressed in *keng* to LT requires a fair degree of routine calculation since it involves determination of the moments of sunset and sunrise for the particular latitude. For example, on a date equivalent to Dec 28/29 in AD 437, the *Sung-shu* (chap. 12) records a lunar eclipse observed at Chien-k'ang (Nanjing) as beginning 'at thc fourth call of the second watch'. The equivalent LT may be derived as follows. Firstly, the times of sunset and sunrise for this latitude (32.0 deg) are calculated as 17.03 h and 6.97 h respectively, taking into account mean refraction and the solar semi-diameter. (NB horizon profile effects are ignored – see section 9.7.) Allowing for the durations of both evening and morning twilight as 0.60 h gives the length of a watch as 2.55 h and a call as 0.51 h. The LT for the beginning of the second watch may thus be determined as 20.18 h, while the fourth call would last from 21.71 to 22.22 h. Obviously these are idealised results; in practice, uncertainties would be introduced by imperfections in timekeeping.

As in the case of *k'o*, if the time of a lunar eclipse is expressed to the nearest fifth of a night-watch, I have assumed the mid-point of the appropriate interval. Thus the beginning of the eclipse of AD 437 is quoted as occurring at the 'fourth call of the second watch'. Since this interval would extend from 21.71 to 22.22 h, I have estimated the time of first contact as approximately 21.97 h.

If the LT of a solar or lunar eclipse is only expressed to the nearest double or single hour or night watch, the result is probably too imprecise

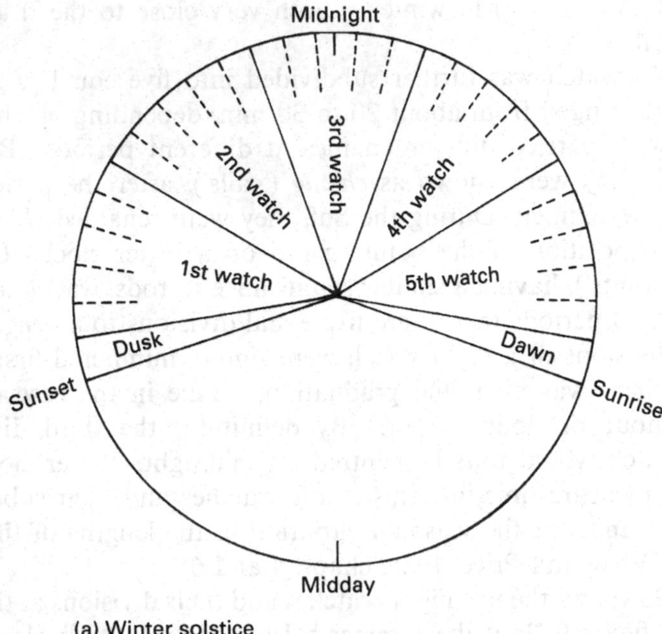

(a) Winter solstice

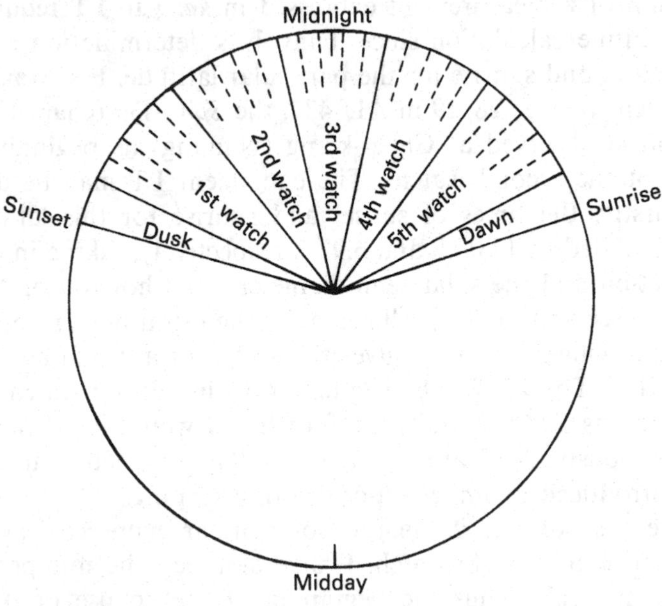

(b) Summer solstice

Fig. 9.2 Night watches and their divisions: (a) winter solstice; (b) summer solstice.

for a useful determination of ΔT to be made. Such crude observations may in some cases only represent casual estimates and thus will not be considered in this chapter. However, when a measurement is expressed to either the nearest *k'o* or fifth of a *keng*, in principle a fairly accurate value for the difference between TT and UT may be obtained. Although one-fifth of a night-watch is typically almost twice as long as a mark, I have considered it appropriate to assign equal weight to measurements using either type of unit. Observations in both categories which are preserved in the dynastic histories and other major sources are likely to be almost entirely the work of the imperial astronomers at the capital of the time. As noted in chapter 8, a regular watch for eclipses and other celestial phenomena was maintained at the capital; an alternative site is never mentioned in the records. Provincial observers would not normally possess timing facilities of comparable precision and in any case it is doubtful whether their measurements would be considered adequate for inclusion in an official compilation.

9.2.3 Estimates of lunar azimuth

In addition to times, a number of estimates of the azimuth of the Moon are preserved for eclipses which occurred during the Sui dynasty (late sixth century AD). Thus at the eclipse of AD 590 Apr 25/26 we find the following report:

> ...It was observed that when the Moon first rose from the south-east, it was already more than half eclipsed. 3 divisions (*fen*) after it reached the *ch'en* direction, 2 fifteenths remained eclipsed. It gradually recovered. By the end (*wei*) of the *ch'en* direction it was restored to fullness.
>
> [*Sui-shu*, chap. 17.]

Chinese astronomers divided the horizon into 24 zones of azimuth, each of width 15 deg. The first of these ranges was centred on due north and thus extended from azimuth 352.5 to 7.5 deg. Working in a clockwise direction, the odd-numbered zones – centred on 0, 30, 60 deg, etc. – were named after the *ti-chih* or terrestrial branches *tzu*, *ch'ou*, *yin* and so on. The twelve even-numbered directions – centred on 15, 45, 75 deg, etc. – were made up of eight of the names of the *t'ien-kan* or celestial stems (less *wu* and *chi*), together with four of the eight trigrams *ch'ien*, *k'un*, *sun* and *ken* – as in the *I-ching* (Book of Changes). In the above example from AD 590, the *ch'en* direction covered a range of azimuth from 115 to 127.5 deg.

In most eclipse records of this kind, it is simply noted that the Moon was 'above' (*shang*) a particular direction. Since the azimuth of the Moon typically increases by about 15 deg in an hour, this represents a fairly rough estimate. In other examples it is stated either that the Moon was at the 'end' (*mo*) of a zone or that an observation was made several *k'o* after the Moon reached a certain range of azimuth. These various records will be discussed in section 9.5.

9.2.4 Accuracy of early eclipse predictions

Early Chinese astronomers were well aware that their eclipse predictions were crude. For instance, in a section of the calendrical treatise of the *Sui-shu* (chap. 18) – compiled around AD 630 – it is remarked that:

> Eclipses of the Sun or Moon may begin or end early or late; they can deviate from normal in either direction. Therefore it is necessary to observe $12\frac{1}{2}$ marks (i.e. 3 hours) before and after the predicted time.

In the *Sui-shu* (chap. 17) and also in the somewhat earlier *Sung-shu* (chaps. 12 and 13), the recorded times of a number of eclipses are compared with their expected times. Discrepancies amounting to several hours are frequently noted. For example, the report of the lunar eclipse of AD 437 Jan 8/9 in chap. 12 of the *Sung-shu* runs as follows:

> Yuan-chia reign period, 13th year, 12th month, 16th day, full Moon. The Moon was eclipsed. The calculated time was the hour of *yu* (i.e. 17–19 h). The eclipse did not (actually) begin until the start of the hour of *hai* (i.e. 21–22 h)...

The above account ends with the measured time of totality expressed relative to the night watches.

Down the centuries, the quality of eclipse prediction slowly improved. Thus in the calendar treatise of the *Sung-shih* (chap. 75) it is recorded that the astronomer Chou Tsung (*c.* AD 1064) regarded eclipse predictions which came within 2 *k'o* (i.e. less than about half an hour) of observation as *ch'in* ('close'), those between 2 and 4 *k'o* as *chin* ('near') and those of 5 *k'o* or more as *yuan* ('far'). A similar system of assessment was developed by the astronomers of the Yuan dynasty – see below. Since Chinese attempts at predicting eclipses were largely restricted to the use of numerical cycles, little further progress was achieved until the arrival of the Jesuit astronomers in the seventeenth century. The Jesuits introduced western methods of eclipse calculation which were superior to indigenous techniques. Occasionally competitive eclipse prediction was arranged between them and the native Chinese astronomers, the Jesuits coming off best! (D'Elia, 1960, p. 41).

9.2.5 Sources of timed observations and terminology

Available sources of eclipse timings in the pre-Jesuit period are few in number. In the official histories, only the *Hou-han-shu* Five-Element Treatise (chap. 28), *Sung-shu* Calendar Treatise (chaps. 12 and 13), *Sui-shu* Calendar Treatise (chap. 17), *Chiu-t'ang-shu* Astrological Treatise (chap. 36), *Sung-shih* Calendar Treatise (chaps. 81 and 82) and *Yuan-shih* Calendar Treatise (chapter 53) contain accurate measurements. Many further data are preserved in the *Wen-hsien T'ung-k'ao* (chaps. 283 and 285), while a few others are contained in the *Sung-hui-yao Chi-kao* ('Collected Manuscripts of the Essentials of Sung History'). Of these latter works, the *Wen-hsien T'ung-k'ao* has already been mentioned above (see also chapter 8); the *Sung-hui-yao Chi-kao* was composed during the late Sung (*c.* AD 1250).

Only an isolated solar eclipse timing expressed to the nearest *k'o* is reported in the *Hou-han-shu*. This dates from AD 193. No further measurements are quoted so precisely until AD 585 (in the *Sui-shu*), although a few lunar eclipse times expressed to the nearest fifth of a night-watch are cited during this interval, beginning with AD 434 (in the *Sung-shu*).

The most prolific source of surviving measurements of eclipse times from pre-Jesuit China is chapter 53 of the *Yuan-shih*. This treatise, entitled *Shou-shih-li I* ('Discussion of the Season-Granting Calendar') was written by Li Ch'ien (AD 1233 – 1311) but was based largely on the work of the great Yuan astronomer Kuo Shou-ching (AD 1231–1316). It contains the earliest extensive investigation of historical eclipses (both solar and lunar) by Chinese astronomers, and includes a detailed discussion of the observations reported in the *Ch'un-ch'iu*. In particular, measured timings of solar eclipses from AD 221 to 1277 and of lunar obscurations from AD 434 to 1280 are compared with the results calculated from two separate almanacs. These are the contemporary *Shou-shih-li* ('Season-Granting Calendar') – devised by Kuo Shou-ching in AD 1281 – and its predecessor from the Kin dynasty, the *Ta-ming-li* ('Great Enlightenment Calendar') – produced by Chao Chi-wei in AD 1182. The relative success of the two systems in representing observation is evaluated in the *Yuan-shih* on a scale of five, depending on the magnitude of the discrepancy: *mi* ('very close': within 1 mark); *ch'in* ('close': 1 mark); *ch'i-ch'in* ('fairly close': 2 marks); *shu* ('off': 3 marks) and *shu yuan* ('far off': more than 3 marks). Kuo Shou-ching was able to show that in general the newer calendar better represented observation. However, it is not clear whether in making these various calculations adequate allowance was made for the different locations (especially longitudes) of the various observing stations.

The *Shou-shih-li I* first discusses solar eclipse observations and then continues with an investigation of reports of lunar obscurations. Although the original source of data is never specified, the text is careful to state the dynasty within which the observation was made. Some of the older timings are only expressed to the nearest *shih* or half-*shih*. The earliest solar eclipse mentioned in this treatise for which a measurement is expressed to the nearest *k'o* dates from AD 680. However, lunar eclipse observations which are timed to the nearest fifth of a *keng* are cited from AD 434 onwards. Many of the records are not preserved in other extant works; evidently Kuo Shou-ching and his colleagues had access to much material which did not find its way into the official histories of former dynasties.

Figure 9.3 shows a page from a nineteenth century printing of the *Shou-shih-li I*. This contains records of solar eclipses from AD 1053 and 1054 together with the results of calculation for each eclipse based on the *Ta-ming-li* and *Shou-shih-li*.

The *Wen-hsien T'ung-k'ao* also records many solar and lunar eclipses which are not found in other sources – especially in the period from about AD 1050 to 1100. This work contains an extensive astronomical treatise, separate sections of which are devoted to solar eclipses (chap. 283) and lunar obscurations (chap. 285).

Of the individual eclipse phases noted in the dynastic histories and other compilations mentioned above, the most commonly reported is the time of maximum – usually identified by the term *shen* ('extreme'). In only one case (AD 761) is the onset of totality (*chi*) reported for a solar obscuration but several lunar eclipse accounts note this phase. The start of an eclipse is often indicated by *k'uei-ch'u* ('beginning of loss') and the end by *fu-man* ('restoration of fullness'). Other expressions include *shih-sheng* ('beginning of emergence' – after maximum has passed), *sheng-kuang* ('emergence of light' – at the end of totality for a lunar eclipse), and *fu-yuan* ('restoration of roundness').

9.3 Careful solar eclipse timings recorded in Chinese history

In all, the LTs of about 60 solar eclipse contacts are quoted to the nearest *k'o* in the history of China prior to the telescopic era. Because of its very early date (AD 193 Feb 19), the oldest observation of this kind may well be of questionable reliability. In particular, it is not clear from the text whether first contact or maximum phase is referred to. The remaining observations fall into three distinct chronological groups: between AD 585 and 768; from 1052 to 1277; and after 1572. Throughout this section, emphasis will be placed on the second group of data. As noted in section 9.2.1, many of the measurements made in the earliest of the three periods are difficult to interpret and will not be considered here. The few surviving measurements which are preserved between 1572 and 1644 – when Jesuit astronomers commenced regular eclipse observations in China – are still only expressed to the nearest mark. By this period, ΔT was so small (of the order of 100 sec) that such rough determinations are valueless. For a discussion of the accuracy of Jesuit solar eclipse

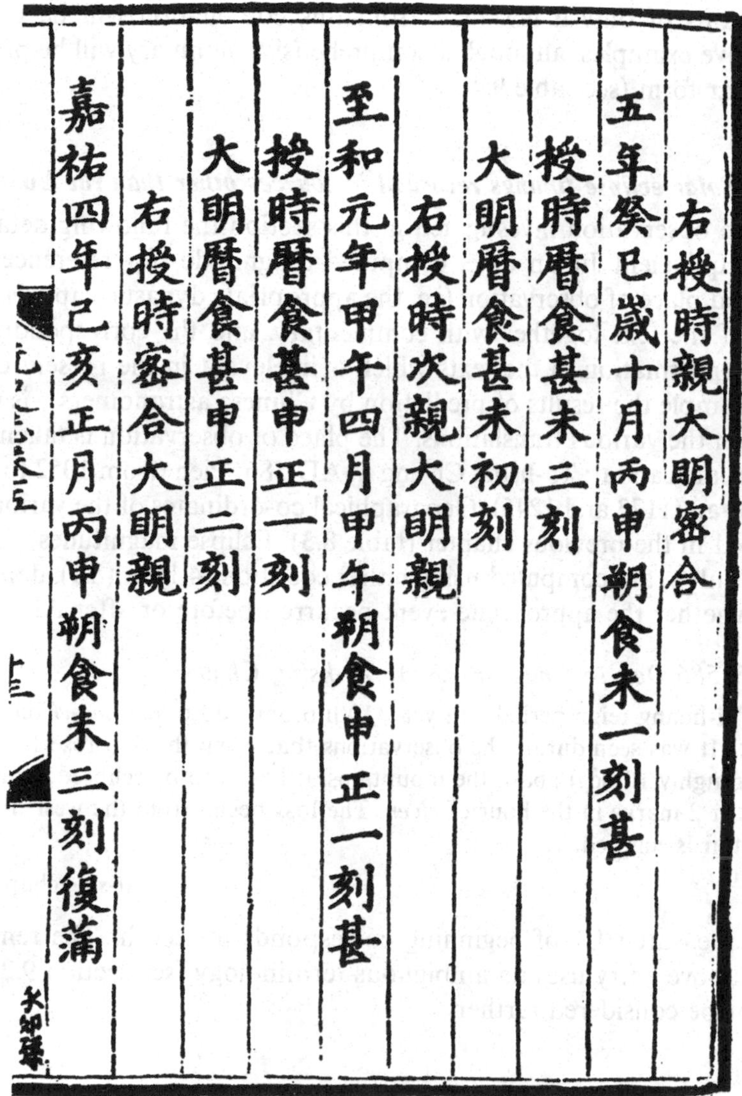

右授時親大明密合
五年癸巳歲十月丙申朔食未一刻甚
授時曆食甚未三刻
大明曆食甚未初刻
右授時次親大明親
至和元年甲午四月甲午朔食申正一刻甚
授時曆食甚申正一刻
大明曆食甚申正二刻
右授時密合大明親
嘉祐四年己亥正月丙申朔食未三刻復滿

Fig. 9.3 Page from a nineteenth century printing of the *Shou-shih-li I*, containing records of solar eclipses in AD 1053 and 1054.

observations made at Beijing from 1644 to 1785, see Stephenson and Fatoohi (1995).

Translations of the few texts in works other than the *Yuan-shih* which cite eclipse timings to the nearest *k'o* are given below. These show an interesting variety. Many more solar eclipse timings are reported in the *Yuan-shih* Calendar Treatise than in the other sources listed above (see section 9.2.5), but the form of these records is rather stereotyped. Hence translations of only a few texts from this compilation will be given as illustrative examples, although a comprehensive summary will be provided in tabular form (see table 9.4).

9.3.1 Solar eclipse timings recorded in sources other than the Yuan-shih

For each observation investigated in this section the following details are given: equivalent Julian date, computed magnitude (for reference only), presumed place of observation (i.e. the appropriate dynastic capital), translation of the text together with commentary, and the corresponding ΔT results. Information in the texts which is irrelevant in the present context – for example the results of prediction by Chinese astronomers – is not included in the various translations. The place of observation is taken as the appropriate capital: Ta-hsing Ch'eng in AD 586; Pien (from 1052 to 1100); and Lin-an (1173 and 1245). Geographical co-ordinates of the various sites are listed in the previous chapter (table 8.3). Eclipse magnitudes – for reference only – are computed using either equation (4.1) or (8.1), depending upon whether the appropriate event occurred before or after AD 950.

(1) AD 586 Dec 16 (mag. = 0.84): Ta-hsing Ch'eng

> K'ai-huang reign period, 6th year, 10th month, 30th day *ting-ch'ou* [14]... It was seen during the observations that when the Sun rose 1 *chang* (i.e. roughly 10 deg) above the mountains, it began to be eclipsed; (this was) at 2 marks in the hour of *ch'en*. The loss began from the west; it was two-thirds eclipsed....
>
> [*Sui-shu*, chap. 17.]

The measured LT of beginning corresponds to 7.60 h. The remainder of the above entry uses an ambiguous terminology (see section 9.2.1) and will not be considered further.

RESULTS

LT of first contact = 7.60 h, UT = 0.35 h. Computed TT = 1.37 h, thus ΔT = 3650 sec.

(2) AD 1040 Feb 15 (mag. = 0.69): Pien

> Pao-yuan reign period, first year, first month, day *ping-ch'en* [53], the first day of the month. The Sun was eclipsed by 6 divisions (*fen*) and it was restored at 1 mark in the hour of *shen*.
>
> [*Wen-hsien T'ung-k'ao*, chap. 283.]

There was no eclipse on or near the stated day in the first year of the Pao-yuan reign period (which corresponds to AD 1038). Within two years on either side of the recorded date, only the eclipse of AD 1040 Feb 15 would be visible at Pien. This event occurred in the *third* year of the Pao-yuan reign period, but otherwise both the month and day are correct. Hence the date 1040 Feb 15 can be confidently assumed.

As will be discussed in section 9.6, the unit of magnitude translated as 'division' (*fen*) was equal to one-tenth of the solar (or lunar) apparent diameter.

The LT of end corresponds to 15.36 h.

RESULTS

LT of last contact = 15.36 h, UT = 7.99 h. Computed TT = 8.79 h, thus ΔT = 2850 sec.

(3) AD 1046 Apr 9 (mag. = 0.46): Pien

> Ch'ing-li reign period, 6th year, 3rd month, day *hsin-szu* [18], the first day of the month. The Sun was eclipsed by $4\frac{1}{2}$ divisions (*fen*). At 3 marks in the hour of *shen* it was restored.
>
> [*Wen-hsien T'ung-k'ao*, chap. 283.]

The LT of end corresponds to 15.84 h. In chapter 53 of the *Yuan-shih*, the time is given as '3 marks in the central half of the hour of *shen*' – i.e. 16.84 h. Because of this disagreement, the observation will not be considered further.

(4) AD 1052 Nov 24 (mag. = 0.11): Pien

> Huang-yu reign period, 4th year, 11th month, day *jen-yin* [39], the first day of the month. The Sun was eclipsed by more than 2 divisions (*fen*). At 1 mark in the central half of the hour of *wei* it was restored.
>
> [*Sung-shih*, chap. 81.]

The LT of end corresponds to 14.36 h.

RESULTS

LT of last contact = 14.36 h, UT = 6.58 h. Computed TT = 7.39 h, thus ΔT = 2900 sec.

(5) AD 1053 Nov 13 (mag. = 0.66): Pien

> Huang-yu reign period, 5th year, 10th month, day *ping-shen* [33], the first day of the month. At 1 mark in the central half of the hour of *wu*, the Sun was eclipsed by $4\frac{1}{2}$ divisions (*fen*).
>
> [*Wen-hsien T'ung-k'ao*, chap. 283.]

The LT of maximum corresponds to 12.36 h. In chapter 53 of the *Yuan-shih*, the time is given as '1 mark in the hour of *wei*' – i.e. 13.36 h. Because of this discrepancy between the two sources, the observation will be rejected.

(6) AD 1054 May 10 (mag. = 0.71): Pien

> Chih-ho reign period, 1st year, 4th month, day *chia-wu* [31], the first day of the month. The Sun was eclipsed by more than 9 divisions (*fen*) and the loss began on the south-west side. It reached its maximum at 1 mark in the central half of the hour of *shen*. On this day it was cloudy and rain fell...
>
> [*Wen-hsien T'ung-k'ao*, chap. 283.]

The LT of maximum, which corresponds to 16.36 h, is confirmed by the record in chapter 53 of the *Yuan-shih*.

RESULTS

LT of last contact = 16.36 h, UT = 8.65 h. Computed TT = 9.17 h, thus ΔT = 1900 sec.

(7) AD 1059 February 15 (mag. = 0.36): Pien

> Chia-yu reign period, 4th year, first month, day *ping-shen* [33], the first day of the month. The Sun was eclipsed by more than 3 divisions (*fen*). At 3 marks in the initial half of the hour of *wei* it was restored.
>
> [*Wen-hsien T'ung-k'ao*, chap. 283.]

The LT of last contact, which corresponds to 13.84 h, is endorsed by the record in chapter 53 of the *Yuan-shih*.

RESULTS

LT of last contact = 13.84 h, UT = 6.47 h. Computed TT = 7.13 h, thus ΔT = 2400 sec.

(8) AD 1068 Feb 6 (mag. = 0.55): Pien

> Hsi-ning reign period, 1st year, 1st month, day *chia-hsu* [11], the first day of the month. The Sun was eclipsed. According to the astronomers, on this day at 8 marks in the hour of *szu* the Sun was seen to diminish; the loss began on the south-west side. After (*hou*) 5 marks in the hour of *wu*, the eclipse reached six divisions (*fen*). Not until 3 marks in the hour of *wei* was it restored to roundness.
>
> [*Sung-hui-yao Chi-kao.*]

In recording the time of maximum, the use of (*hou*) after the figure 5 will be assumed to be an alternative for *shang*, implying a moment at the end of the 5th mark.

The LTs of beginning, maximum and end correspond respectively to 10.96 h, 12.44 h and 13.84 h.

RESULTS

(i) LT of first contact = 10.96 h, UT = 3.60 h. Computed TT = 4.05 h, thus ΔT = 1600 sec.

(ii) LT of maximal phase = 12.44 h, UT = 5.07 h. Computed TT = 5.40 h, thus ΔT = 1200 sec.

(iii) LT of last contact = 13.84 h, UT = 6.48 h. Computed TT = 6.71 h, thus ΔT = 850 sec.

(9) AD 1094 Mar 19 (mag. = 0.81): Pien

> Yuan-yu reign period, 9th year, 3rd month, day *jen-shen* [9], the first day of the month. The Astronomer-Royal reported that the Sun should have been eclipsed, but on account of cloud cover it was not (fully) seen. It began (to be visible) at 3 marks in the hour of *wei*. It was seen through clouds that the Sun was eclipsed by more than one division (*fen*) on the south-west side. At 6 marks in the hour of (*wei*) it reached its maximum of 7 divisions. Then it became invisible on account of the covering of clouds.
>
> [*Wen-hsien T'ung-k'ao*, chap. 283.]

Since the eclipse was said to increase from $\frac{1}{10}$ to $\frac{7}{10}$ in 3 marks, it is a reasonable approximation to assume that first contact occurred $\frac{1}{2}$ mark before the phase of $\frac{1}{10}$ was observed, corresponding to a LT of 13.72 h. Maximum phase was at 14.56 h, after which the eclipse was lost to view in the clouds. Evidently the degree of obscuration of the Sun had ceased to noticeably increase by the time the Sun entered the clouds. The LT of maximum is confirmed by the report in chapter 53 of the *Yuan-shih*.

RESULTS

(i) LT of first contact = 13.72 h, UT = 6.20 h. Computed TT = 6.62 h, thus ΔT = 1500 sec.

(ii) LT of maximal phase = 14.56 h, UT = 7.04 h. Computed TT = 7.78 h, thus ΔT = 2650 sec.

(10) AD 1100 May 11 (mag. = 0.34): Pien

> Yuan-fu reign period, 3rd year, 4th month, day *ting-mao* [4], the first day of the month. The Astronomer-Royal said, 'In the initial half of the hour of *ch'en*, the Sun was eclipsed at the north-west by 4 divisions (*fen*). At 3 marks in the central half of the hour of *szu* it was restored to fullness...
>
> [*Sung-hui-yao Chi-kao.*]

The LT of beginning (7–8 h) is only roughly expressed. The LT of end corresponds to 10.84 h.

RESULTS

LT of last contact = 10.84 h, UT = 3.12 h. Computed TT = 2.84 h, thus ΔT = −1000 sec.

(11) AD 1173 Jun 12 (mag. = 0.58): Lin-an

> Ch'ien-tao reign period, 9th year, 5th month, the first day of the month. Both officials and students from this Bureau observed that the Sun was eclipsed $4\frac{1}{2}$ divisions (*fen*). It began from the north-west at $5\frac{1}{2}$ marks in the hour of *wu*. The eclipse reached its maximum at the north side at 2 marks in the initial half of the hour of *wei*. It was restored to roundness at the north-east at 1 mark in the initial half of the hour of *shen*.
>
> [*Sung-shih*, chap. 81.]

Assuming that '$5\frac{1}{2}$ marks in the hour of *wu*' means the middle of the appropriate mark, the time of first contact is equivalent to 12.32 h. The LTs of maximum and end correspond respectively to 13.60 h and 15.36 h.

RESULTS

(i) LT of first contact = 12.32 h, UT = 4.30 h. Computed TT = 4.55 h, thus ΔT = 900 sec.
(ii) LT of maximal phase = 13.60 h, UT = 5.57 h. Computed TT = 6.12 h, thus ΔT = 2000 sec.
(iii) LT of last contact = 15.36 h, UT = 7.33 h. Computed TT = 7.72 h, thus ΔT = 1400 sec.

(12) AD 1202 May 23 (mag. = 0.58): Lin-an

> Chia-t'ai reign period, 2nd year, 5th month, day *chia-ch'en* [41], the first day of the month. The Sun was eclipsed. The loss began at 1 mark in the initial half of the hour of *wu*. It was restored to roundness in the initial half of the hour of *wei*... the eclipse reached 3 divisions (*fen*) at $3\frac{1}{2}$ marks in the initial half of the hour of *wu*.
>
> [*Sung-shih*, chap. 81.]

The LT of first contact corresponds to 11.36 h; this is confirmed by the record in chapter 53 of the *Yuan-shih*. In the case of last contact, the LT (13–14 h) is only approximately expressed. Assuming that '$3\frac{1}{2}$ marks in the initial half of the hour of *wu*' means the middle of the appropriate mark, the LT of maximum phase is equivalent to 11.84 h.

RESULTS

(i) LT of first contact = 11.36 h, UT = 3.28 h. Computed TT = 3.47 h, thus ΔT = 700 sec.
(ii) LT of maximum = 11.84 h, UT = 3.76 h. Computed TT = 4.23 h, thus ΔT = 1700 sec.

(13) AD 1245 Jul 25 (mag. = 0.74): Lin-an

Shun-yu reign period, 5th year. According to the calculations of an astronomical official named Cheng, the Sun should have been eclipsed at 3 marks in the initial half of the hour of *wei*. It was observed at 4 marks in the central half of the hour of *wei*. The magnitude should have been 8 divisions (*fen*) but it was actually 6 divisions.

[*Sung-shih*, chap. 82.]

No more than the year is given, but the eclipse of Jul 25 is the only obscuration of the Sun visible during that time.

The observed LT of beginning corresponds to 14.98 h.

RESULTS

LT of first contact = 14.98 h, UT = 7.05 h. Computed TT = 7.42 h, thus ΔT = 1350 sec.

The above investigations are summarised in table 9.4 together with the analysis of measurements which are recorded only in the *Yuan-shih* (see section 9.3.2).

9.3.2 Solar eclipse timings recorded in the Calendar Treatise of the Yuan-shih

In citing the solar eclipse timings in its Calendar Treatise, the *Yuan-shih* draws upon material from a number of earlier dynasties as well as contemporary data. As already noted, the records in the *Shou-shih-li I* follow a very repetitive style. Evidently the original observations were re-formulated to conform to a standardised pattern, in the process rejecting all unnecessary detail. Observations of eclipses of the Sun are contained in a section entitled 'Solar Eclipses from the Three Kingdoms Period Onwards'. In each individual entry, the word Sun is never mentioned since the overall title renders it superfluous.

Sexagenary years, rare in Chinese texts, are consistently employed in addition to regnal years in this treatise, evidently to facilitate counting intervals. Incidentally, the use of sexagenary years began around 200 BC. The same names were used as for cyclical days (see chapter 8). The enumeration of the system is such that AD 1 was the 58th year of a cycle. Hence the cyclical year for any AD date may be found by adding 57 to the AD year, dividing by 60, then multiplying the remainder by 60. For example, 1997 is the 14th year (*ting-ch'ou*) of the present cycle – i.e. the 'Year of the Ox'. (Obviously in the BC period, a correction of one year would have to be made on account of the absence of a year zero.)

In many cases, a report of an eclipse quoted in the *Shou-shih-li I* may also be found in the Astrological Treatise of the appropriate dynastic history. However, in no instance is a precise time quoted in one of these alternative sources. Rather surprisingly, the detailed measurements between AD 585 and 768 which are found in the Calendar Treatise of the *Sui-shu* or the Astrological Treatise of the *Chiu-t'ang-shu* are not mentioned in the *Shou-shi-li I*. Just possibly, the Yuan astronomers were uncertain as to the interpretation of the times quoted in these texts (in common with modern investigators). Several examples of duplication between the data in the *Shou-shih-li I* and *Wen-hsien T'ung-k'ao* or *Sung-shih* have already been noted – as well as two cases of discord (AD 1046 and 1053).

The date of the earliest observation for which the time is quoted to the nearest *k'o* in chapter 53 of the *Yuan-shih* is equivalent to AD 680 Nov 27. The text may be translated as follows:

> T'ang dynasty, Yung-lung reign period, epochal year *keng-ch'en* [17], 11th month, day *jen-shen* [9], the first day of the month. (The Sun) was eclipsed. It reached its maximum at 4 marks in the hour of *szu*. (According to) the Season-Granting System, the eclipse maximum was at 7 marks in the hour of *szu*, (whilst according to) the Great Enlightment System the eclipse maximum was at 5 marks in the hour of *szu*. The Season-Granting (System) is off, the Great Enlightment (System) is close.

As noted above (section 9.2.5), a tabular error of 3 marks was regarded by the Yuan astronomers as 'off' whilst a discrepancy of 1 mark was described as 'close'. Both of the official T'ang histories simply report the occurrence of a solar eclipse on the stated day, without mentioning the time of the contacts.

In common with other T'ang records of eclipses, the place of observation on this occasion was evidently the capital of Ch'ang-an. The observed LT of greatest phase corresponds to approximately 10.08 h, a UT of 2.67 h, leading to a value for ΔT of 2500 sec.

Several other observations quote the measured times of all three phases, an example being given by the record of the last eclipse in the series: that of AD 1277 Oct 28:

> Present dynasty, Chih-yuan reign period, 14th year, 10th month, day *ping-ch'en* [53], the first day of the month. (The Sun) was eclipsed. Beginning of loss in the initial mark of the central half of the hour of *wu*; maximum at 1 mark in the initial half of the hour of *wei*; restoration of fullness at 2 marks in the central half of the hour of *wei* ...
>
> [*Yuan-shih*, chap. 53.]

There follows the usual comparison with predictions according to the Season-Granting and Great-Enlightment systems. By this time Ta-tu had been the Yuan capital for more than a decade and in the previous year Kuo Shou-ching had commenced building the Imperial Observatory there. The approximate observed LTs of beginning, greatest phase and end correspond respectively to 12.12 h, 13.36 h and 14.60 h., i.e. UTs of 4.10, 5.34 and 6.58 h. Since the computed TTs are 4.18 h, 5.48 h and 6.75 h, ΔT values of 300, 500 and 600 sec are required to satisfy these observations.

Since the eclipse records in the *Shou-shih-li I* follow a highly stereotyped pattern, no further translations will be given in this section. Calculations have been made for the following locations: Ch'ang-an (AD 680–702); Pien (AD 1066–1243); and Ta-tu (AD 1277). A further eclipse timing is reported on a date corresponding to AD 1260 Apr 12. However, the observation was made several years before the Yuan capital of Ta-tu was established. Hence the place of origin of the record (which states that the eclipse reached its maximum at 2 marks in the central half of the hour of *shen*) is very doubtful. The record will thus be rejected.

The various results are summarised in table 9.4, along with the results from solar eclipse timings analysed in section 9.3.1. In this table are given the following details, column by column: (1) Julian date (year, month and day); (2) phase observed (1 = first contact, M = maximum; 4 = last contact); (3) double hour; (4) (where appropriate) initial (I) or central (II) half of this double hour; (5) the number of marks within the designated double or single hour; (6) the equivalent LT in hours and decimals; (7) the corresponding UT; (8) the computed TT; and (9) the resultant value of ΔT (in sec) necessary to satisfy the observation. Full Julian dates are given in column 1 since most of these are not cited elsewhere in this section.

NB in AD 1068, maximum phase was recorded 'after' (*hou*) 5 marks in the hour of *wu*. Hence the observation will be presumed to have been made at the end of this division; this is indicated by a + sign in column 5.

The ΔT values listed in table 9.4 are plotted in figure 9.4. Following the usual practice for such diagrams in this book, positive numbers are adopted for years rather than AD. Nearly all of the observations represented in this diagram date from between AD 1040 and 1277. Bearing in mind that each measurement is quoted to the nearest *k'o* (i.e. some 860 sec), the scatter of points within this date range is considerably larger than might be expected.

9.4 Lunar eclipse timings

Like the preserved solar eclipse timings discussed above, these observations fall into three distinct chronological groupings: the date ranges are: from AD 434 to 596; between AD 948 and 1280; and after 1577. Measurements in the latest of these periods (which like their solar counterparts are all

Table 9.4 ΔT values from more reliable solar eclipse timings recorded in Chinese history.

Date AD	Ct	Hour	Half	Mark	LT (h)	UT (h)	TT (h)	ΔT (sec)
586 Dec 16	1	ch'en	–	2	7.60	0.35	1.37	3650
680 Nov 27	M	szu	–	4	10.08	2.67	3.36	2500
691 May 4	M	mao	–	2	5.60	22.25	22.53	1000
702 Sep 26	M	shen	–	3	15.84	8.42	9.07	2350
1040 Feb 15	4	shen	–	1	15.36	7.99	8.79	2850
1052 Nov 24	1	wei	II	1	14.36	6.58	7.39	2900
1054 May 10	M	shen	II	1	16.36	8.65	9.17	1900
1059 Feb 15	4	wei	I	3	13.84	6.47	7.13	2400
1066 Sep 22	M	wei	–	2	13.60	5.82	6.29	1700
1068 Feb 6	1	szu	–	8	10.96	3.60	4.05	1600
	M	wu	–	5+	12.44	5.07	5.40	1200
	4	wei	–	3	13.84	6.48	6.71	850
1069 Jul 21	M	ch'en	–	3	7.84	0.30	0.44	500
1080 Dec 14	M	szu	–	6	10.56	2.93	2.93	0
1094 Mar 19	1	wei	–	2.5	13.72	6.20	6.62	1500
	M	wei	–	6	14.56	7.04	7.78	2650
1100 May 11	4	szu	II	3	10.84	3.12	2.84	–1000
1107 Dec 16	1	wei	–	2	13.60	5.99	6.28	1050
	M	wei	–	8	14.96	7.35	7.74	1400
	4	shen	–	6	16.56	8.95	9.27	1150
1173 Jun 12	1	wu		5.5	12.32	4.30	4.55	900
	M	wei	I	2	13.60	5.57	6.12	2000
	4	shen	I	1	15.53	7.33	7.72	1400
1183 Nov 17	M	szu	II	2	10.60	2.39	2.61	800
1195 Apr 12	1	wu	I	2	11.60	3.57	3.81	850
1202 May 23	1	wu	I	1	11.36	3.28	3.47	700
	M	wu	I	3.5	11.84	3.76	4.23	1700
1216 Feb 19	M	shen	II	4	16.98	9.20	9.31	400
1243 Mar 22	M	szu	I	2	9.60	1.67	1.77	350
1245 Jul 25	1	wei	II	4	14.98	7.05	7.42	1350
1277 Oct 28	1	wu	II	0	12.12	4.10	4.27	300
	M	wei	I	1	13.36	5.34	5.48	500
	4	wei	II	2	14.60	6.58	6.75	600

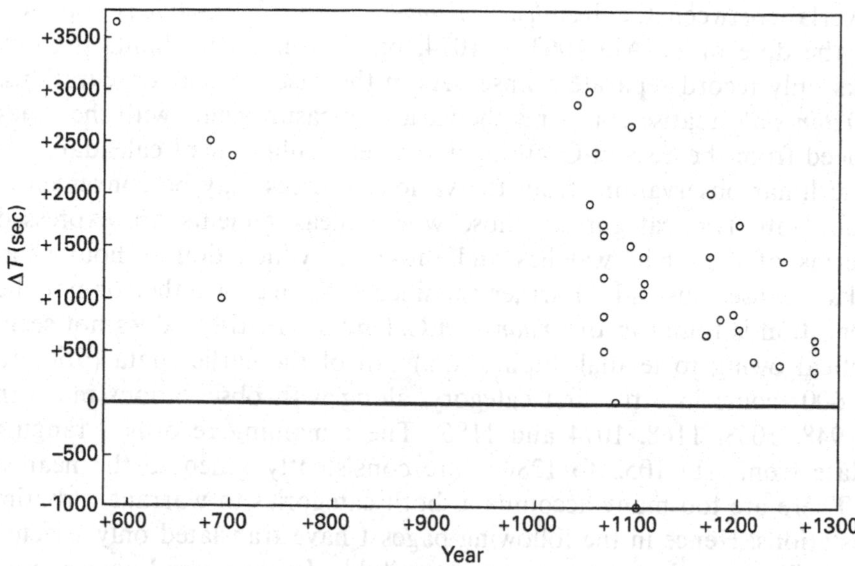

Fig. 9.4 ΔT results derived from Chinese solar eclipse timings.

quoted to the nearest mark until 1644) will not be analysed since by then ΔT was very small (see also section 9.3).

During the earliest interval (AD 434–596), there are several useful observations in the calendar treatises of both the *Sung-shu* (chaps. 12 and 13) and *Sui-shu* (chap. 17); each date range is fairly narrow – respectively from AD 434 to 440 and from AD 585 to 596. In contrast with the solar eclipse records from the same period, most of the accounts of lunar obscurations which are to be found in the *Sung-shu* and *Sui-shu* are summarised in the *Shou-shih-li I* of the *Yuan-shih*. There is exact agreement between the two sets of LTs in every case. It should be pointed out that there are minor differences in terminology between the *Shou-shih-li I* and the *Sui-shu* when describing the same observation. Thus, while the time-units for the eclipses reported in the *Sui-shu* are consistently quoted as 'rods' (*ch'ou*), the *Shou-shih-li I* uses the term 'calls' (*ch'ang*) in AD 592, 'points' (*tien*) in 595 and 'rods' only in 596. However, these terms all relate to the same units – i.e. one-fifth of a night-watch (see section 9.2.2).

Between AD 948 and 1280, numerous lunar eclipses are reported. Most are contained in the *Wen-hsien T'ung-k'ao* (chap. 285) and the *Yuan-shih* (chap. 53). A very few additional observations are cited in the *Sung-shih* Calendar Treatise (chaps. 81 and 82). In our recent paper on the Earth's past rotation (Stephenson and Morrison, 1995), I had overlooked the *lunar* observations recorded in the *Wen-hsien T'ung-k'ao*. I am grateful to Professor Xu Zhentao of Purple Mountain Observatory, Nanjing – who lately spent a year at Durham University – for drawing my attention to them.

Overlap between the *Wen-hsien T'ung-k'ao* and *Yuan-shih* is common over the date range AD 1063 to 1074, but beyond these limits the two works only record separate eclipses. As in the case of solar obscurations, the *Yuan-shih* treatise compares the various measurements with the times deduced from the Season-Granting and Great Enlightment calendars.

The lunar observations from the various sources may be conveniently divided into two categories: those where measurements are expressed in terms of the night watches and those for which double hours and marks are used instead. (Further classification – e.g. whether or not the observation is found in the *Yuan-shih* Calendar Treatise – does not seem practical owing to textual duplication.) All of the earlier data (prior to AD 600) come into the first category, along with observations made in AD 948, 1068, 1168, 1074 and 1185. The remaining records – ranging in date from AD 1052 to 1280 – are consistently timed to the nearest *k'o*. There are too many accounts in both categories to warrant verbatim translations. Hence in the following pages I have translated only selected texts to illustrate the type of material available. In each case I have chosen duplicate entries relating to the same event from different sources. As in the case of solar eclipses, all reference to prediction in the texts is omitted.

Following the practice adopted throughout this book, all Julian (or Gregorian) dates for lunar eclipses will be double – e.g. Jun 23/24.

9.4.1 Lunar eclipses timed in fifths of a night-watch

Most records merely assert that a particular phase occurred at a certain division of a watch. Under these circumstances, the mid-point of the specified division will normally be assumed (a similar convention was adopted for solar eclipse timings – see above). However, the eclipse of AD 595 Dec 22/23 was stated to begin 'after' (*hou*) the fourth rod of the first watch. In this case the start of the fifth rod will be assumed.

The very earliest lunar eclipse for which an accurate time is specified is recorded in chapter 12 of the *Sung-shu* on the night of AD 434 Sep 4/5:

> Yuan-chia reign period, 11th year, 7th month, 16th day, full Moon. The Moon was eclipsed. The calculated time was the hour of *mao*. The Moon began to be eclipsed at the second call of the fourth watch, in the initial half of the hour of *ch'ou*. The eclipse was total at the fourth call. The Moon was at the end of the 15th degree of *Ying-shih* (lunar lodge).

The text ends with an estimate of the RA of the Moon. The *Shou-shih-li I* gives a standardised version of the above account, leaving out all superfluous detail.

> Yuan-chia reign period, 11th year, *chia-hsu* [11], 7th month, day *ping-tzu* [13], full Moon. The Moon was eclipsed. Beginning of loss at the second

call of the fourth watch. The eclipse was total at the fourth call of the fourth watch.

There follows a comparison with the times deduced from the Season-Granting and Great Enlightment calendars. Interestingly, in reporting the results of their calculations for this and all subsequent lunar eclipses which were timed relative to the night-watches, the Yuan astronomers used the contemporary unit 'point' (*tien*), while retaining more archaic names in summarising the observational record itself.

The equivalent LTs of first contact and beginning of totality (taken as the mid-points of the second and fourth calls of the fourth watch) were respectively 1.65 h and 2.46 h. These correspond to UTs of 17.65 h and 18.42 h. Comparing these with the computed TTs of 18.14 h and 19.20 h leads to ΔT values of respectively 1750 and 2800 sec.

The remaining records in the *Sung-shu* relate to the eclipses of AD 437 Jan 8/9, 437 Dec 28/29, and 440 Oct 26/27. These have much in common with the account of the eclipse of AD 434 in the same source. However, reports from the Sui dynasty tend to be more detailed, noting which part of the Moon was obscured, as well as specifying the lunar azimuth at the various phases. A useful example is provided by the account of the total eclipse of AD 596 Dec 10/11 in the Calendar Treatise of the *Sui-shu* (chap. 17):

> K'ai-huang reign period, 16th year, 11th month, 16th day, *i-ch'ou* [2]...Not until the first rod of the third watch was the Moon seen in the clouds above the direction *ping* (roughly SSE, azimuth approximately 165 deg). It was already about 3 fifteenths eclipsed and the loss began from the east side. Above the direction *ting* (roughly SSW, azimuth approximately 195 deg), the eclipse was total. Afterwards it reappeared from the SE side. It was not until the third rod of the fourth watch that it was restored to fullness; the Moon was then at the end of the direction *wei* (azimuth approximately 217.5 deg).

A very brief summary of the above account, giving only the date and time of last contact is given in chap. 53 of the *Yuan-shih*:

> K'ai-huang reign period, 16th year, *ping-ch'en* [53], 11th month, day *ping-tzu* [13], full Moon. (The Moon) was eclipsed. At the third rod of the fourth watch it was fully restored.

The measurement of last contact (assumed to relate to the middle of the third rod) is equivalent to an LT of 2.60 h. Although ignored by the Yuan astronomers, the observation made near the start of the eclipse is of comparable precision. Adjusting the eye estimate of $\frac{3}{15}$ (0.20) for the likely systematic error of observation (see chapter 3) leads to only about 0.10 of the lunar diameter being obscured when the Moon first became visible 'through clouds'. This eclipse, which eventually became total, would reach a phase of 0.10 only about 0.15 h after first contact. On this basis, the eclipse may be considered to have started 0.15 h before the middle of the 'first rod of the fourth watch', which occurred at 22.99 h. The equivalent LT of first contact may thus be estimated as 22.84 h.

If the above LTs are reduced to UT, and subtracted from the computed TT values, the corresponding ΔT values may be derived (see table 9.5).

In making the necessary computations for the data in this section, the places of observation have been taken as Chien-k'ang (AD 434–543), Ta-hsing Ch'eng (AD 585–596), Pien-chou (AD 948), Pien (AD 1074) and Lin-an (AD 1185). Pien-chou was the former name for Pien.

In table 9.5 are summarised the various observations for which measurements are specified relative to the night-watches, together with equivalent LTs (deduced from the computed LTs of sunrise and sunset), equivalent universal times, and resulting ΔT values.

In AD 595, first contact was recorded 'after' (*hou*) the fourth division of the first watch. Hence the observation will be presumed to have been made at the end of this division; this is indicated by a + sign in column 4.

The various results listed in table 9.5 (with the exception of the highly discordant ΔT value of −2300 sec in AD 1168) are plotted in figure 9.5. In this diagram, the distribution of dates is quite dissimilar from that in figure 9.4, with relatively few observations after AD 800. Although the two results derived from the eclipse of AD 434 are obviously discordant, most of the ΔT values between AD 434 and 596 form a fairly self-consistent set. This is a period when few useful data from other parts of the world are preserved.

9.4.2 *Lunar eclipses timed in double hours and marks*

Most of the extant observations in this category are recorded in either chapter 53 of the *Yuan-shih* or chapter 285 of the *Wen-hsien T'ung-k'ao*. In addition, attention has been drawn in chapter 3 to an isolated report, dating from AD 1168 Mar 25/26, which is preserved in chapter 82 of the *Sung-shih*. The account of this eclipse, which is particularly detailed may be translated as follows:

> Ch'ien-tao reign period, 4th year, 2nd month, 14th day *ting-wei* [44], full Moon. The Moon was eclipsed...(predictions)...On this evening when the Moon rose there was light cloud. Until the fall of darkness it could not be seen that the Moon had been totally eclipsed. When the 3rd mark of the initial half of the hour *hsu* was reached, as expected it was shining and so

Table 9.5 ΔT values from lunar eclipse timings expressed in terms of the night watches.

Date AD	Ct	Watch	Div.	LT (h)	UT (h)	TT (h)	ΔT (sec)
434 Sep 4/5	1	IV	2	1.65	17.65	18.14	1750
	2	IV	4	2.42	18.42	19.20	2800
437 Jan 8/9	2	I	3	18.95	11.23	13.30	7450
437 Dec 28/29	1	II	4	21.97	14.16	15.83	6000
	2	III	1	22.99	15.18	17.00	6550
440 Oct 26/27	1	II	1	20.70	12.54	14.26	6200
	M	II	3	21.64	13.49	15.74	8100
543 May 4/5	1	III	3	0.00	15.99	17.27	4600
585 Jan 21/22	1	I	1	17.99	10.98	12.65	6000
	M	I	4	19.49	12.49	14.07	5700
	4	II	1	20.49	13.49	15.49	7200
592 Aug 28/29	1	I	3	20.05	12.78	14.42	5900
593 Aug 17/18	1	V	1	3.05	19.82	21.11	4650
595 Dec 22/23	1	I	4+	19.59	12.38	13.89	5450
	M	II	3	21.40	17.80	15.32	4050
	4	III	1	22.96	15.76	16.73	3500
596 Dec 10/11	1	III	1–	22.86	15.56	16.73	4200
	4	IV	3	2.60	19.30	20.63	4800
948 Jan 28/29	1	IV	4	2.94	19.59	20.81	4450
1074 Oct 7/8	1	IV	5	3.19	19.34	19.78	1600
	2	V	3	4.56	20.71	20.85	500
1168 Mar 25/26	4	II	2	21.52	13.58	12.93	–2300
1185 Apr 16/17	M	III	3	0.00	15.94	15.97	100

it could be known that this eclipse had been total on rising. It was restored to fullness at the 3rd mark in the central half of the hour of *hsu*. It was the 2nd point of the 2nd watch.

[*Sung-shih*, chap. 82.]

This observation is unusual in that it gives the time of last contact in both systems: hours and marks, and also night-watches and divisions. The LT of third contact corresponds to 19.84 h and that of last contact (expressed in double hours) to 20.84 h. These two results are very self-consistent; the computed interval between third and last contact (independent of ΔT) was 1.03 h. Converting these to UT (respectively 11.90 h and 12.90 h) and comparing with the computed TTs of 11.90 h

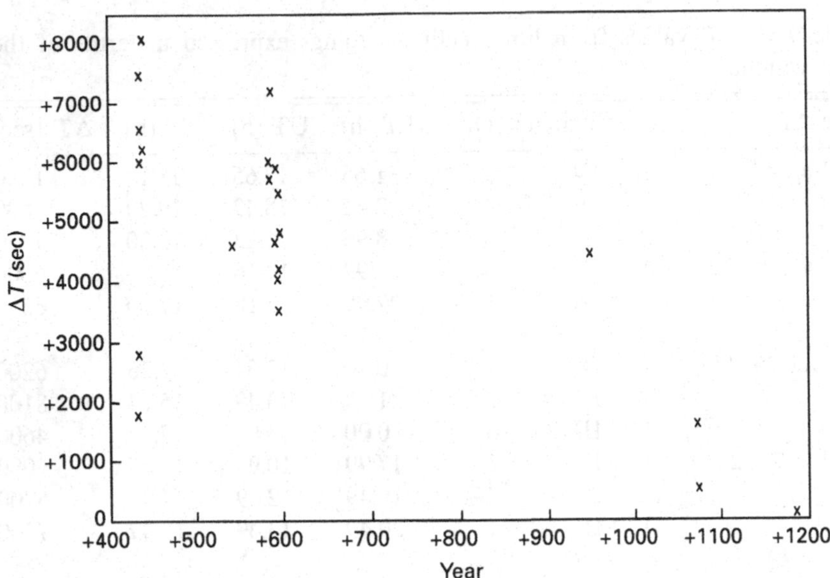

Fig. 9.5 ΔT results obtained from lunar eclipse times which were expressed in terms of the night watches.

and 12.93 h, enable values for ΔT of 0 and +100 sec to be deduced. The LT for the end of the eclipse expressed in terms of the night-watches, equivalent to 21.52 h (rather than 20.84 h), is clearly discordant. Presumably a scribal error has occurred here; the equivalent value of ΔT is −2300 sec.

The numerous *Yuan-shih* timings expressed to the nearest *k'o* extend from AD 1052 to 1280, while the corresponding *Wen-hsien T'ung-k'ao* data cover the relatively brief period from AD 1063 to 1099. About 40 individual timings are preserved in each source. Two examples of the type of observation available are as follows, one from the Sung dynasty (AD 1069 Dec 30/31) and the other from the Yuan itself (1277 May 18/19):

> (AD 1069.) Sung dynasty, Hsi-ning reign period, 2nd year, *chi-yu* [46], intercalary 11th month, day *ting-wei* [44], full Moon. (The Moon) was eclipsed. Beginning of loss at 6 marks in the hour of *hai*; eclipse maximum at 5 marks in the hour of *tzu*; restoration of fullness at 4 marks in the hour of *ch'ou*.
>
> [*Yuan-shih*, chap. 53.]

The above LTs are equivalent to 22.56 h (start), 0.32 h (maximum) and 2.08 h (end). Expressing them in UT and subtracting from the computed TT values leads to ΔT results of 2650, 1750 and 800 sec (see table 9.7 below).

> (AD 1277.) Present Dynasty, Chih-yuan reign period, 14th year, *ting-ch'ou* [14], 4th month, *kuei-yu* [10], full Moon. (The Moon) was eclipsed. Beginning of loss at 6 marks in the hour of *tzu*; the eclipse was total at 3 marks in the hour of *ch'ou*; maximum at 5 marks in the hour of *ch'ou*; reappearance of light at 7 marks in the hour of *ch'ou*; restoration of fullness at 4 marks in the hour of *yin*.
>
> [*Yuan-shih*, chap. 53.]

The various measurements may be reduced to the following LTs: 0.56 h (start); 1.84 h (beginning of totality); 2.20 h (maximum eclipse); 2.68 h (end of totality); 4.08 h (end). All of these times apart from the LT of maximum phase may be used in the determination of ΔT. However, the moment of greatest phase for a total lunar eclipse cannot be fixed by direct observation; the recorded time is merely the mean of the determinations for the beginning and end of totality.

Several of the records in the *Wen-hsien T'ung-k'ao* are rather detailed and most give an estimate of magnitude as well as citing the LTs of the various phases. For example, the record of the partial lunar eclipse of AD 1071 Dec 9/10 may be translated as follows:

> Hsi-ning reign period, 4th year, 11th month, day *ping-shen* [33]. The Moon was eclipsed. At 2 marks in the hour of *mao* the eclipse began to be seen in the west. It commenced at the south-east side. Not until 6 marks in the hour of *mao* was the eclipse at its maximum, reaching less than $4\frac{1}{2}$ divisions (*fen*). It was rather less than 1 deg in *Tung-ching* (lunar lodge). Before it became bright it set eclipsed and the restoration was not seen.

The *Yuan-shih* also reports this eclipse, quoting the same times as above but ignoring the other observational details. First contact occurred at a LT of 5.60 h and maximal phase at 6.56 h.

The eclipse of AD 1063 Nov 8/9 is somewhat ambiguously recorded. This is described in the *Wen-hsien T'ung-k'ao* as follows:

> Chia-yu reign period, 8th year, 10th month, day *kuei-wei* [20]. The Moon was totally eclipsed. It reached its maximum at 7 marks in the hour of *mao*. Then it set eclipsed and was not seen.

There is nothing in the text to indicate that the Moon reappeared after totality so that the time of maximum may have been incorrectly interpolated by the observers. Hence the measurement will be rejected. The same LT of greatest phase is also reported in the *Yuan-shih* but without any reference to moonset.

In deriving ΔT values, the places of observation have been taken as the Sung capitals of Pien (AD 1052 to 1106) and Lin-an (AD 1168), and the Mongol capital of Ta-tu (AD 1270 to 1279).

Table 9.6*a* Analysis of anomalous lunar eclipse records in the Wen-hsien T'ung-k'ao.

Date AD	Ct	Hour	Mark	LT (h)	UT (h)	TT (h)	ΔT(sec)	Remarks
1067 Mar 3/4	1	ch'ou	4	2.08	18.65	21.84	11 500	set ecl.
1068 Aug 15/16	1	ch'ou	5	2.32	18.73	21.67	10 600	set ecl.
1069 Dec 30/31	1	hai	1	21.36	13.86	15.80	7 000	—
	M	tzu	0	23.12	15.62	17.30	6 050	—
	4	ch'ou	3	1.84	18.34	18.80	1 650	—
1073 Oct 18/19	1	ch'ou	4	2.08	18.20	20.74	9 150	set ecl.
	M	yin	1	3.36	19.48	22.16	9 650	set ecl.
1074 Oct 7/8	1	ch'ou	1	1.36	17.52	19.78	8 150	set ecl.
	2	ch'ou	6	2.56	18.72	20.85	7 650	set ecl.
1077 Feb 10/11	1	tzu	3	23.84	16.48	18.44	7 050	—
	M	tzu	7	0.80	17.44	20.05	9 400	—
	4	ch'ou	2	1.84	18.48	21.66	11 450	—

Most of the observations in the *Wen-hsien T'ung-k'ao* lead to results for ΔT between about 0 and 1000 sec – see table 9.7 below. Values in this region are in satisfactory agreement with the results obtained from roughly contemporaneous solar eclipse observations – see table 9.4 and figure 9.4. However, the data from two discrete periods (AD 1067 to 1069 and also AD 1073 to 1077) are highly discordant, ΔT values ranging from around 6000 to 12 000 sec. Analysis of these observations is summarised in table 9.6a. As it happens, four of the records in these two brief intervals (AD 1067, 1068, 1073 and 1074) state that the Moon set eclipsed, but this assertion cannot be reconciled with the timed measurements. The values of ΔT indicated by the timed contacts, as listed in table 9.6a, lead to altitudes of the Moon (in the west) at last contact of respectively +5, +11, +21 and +13 deg. Hence in no case would the Moon set eclipsed. It appears that serious errors have occurred in the transmission of the original timings for all four eclipses and this presumably applies to the two other dates in the table (AD 1069 and 1077). Hence all six observations will be rejected.

Two of the eclipses cited above (AD 1069 and 1074) are also reported in chapter 53 of the *Yuan-shih*, but with considerably differing times from those in the *Wen-hsien T'ung-k'ao*. For these dates, the compilers of the

Table 9.6*b* Comparison between anomalous lunar eclipse records in the Wen-hsien T'ung-k'ao and counterparts in the Yuan-shih.

Date AD	Ct	Hour	Mark	LT (h)	ΔT(sec)	Hour	Mark	LT (h)	ΔT(sec)
1069 Dec 30/31	1	hai	1	21.36	7000	hai	6	22.56	2650
	M	tzu	0	23.12	6050	tzu	5	0.32	1750
	4	ch'ou	3	1.84	1650	ch'ou	4	2.08	800
1074 Oct 7/8	1	ch'ou	1	1.36	8150	IV w	5p	3.19	1600
	2	ch'ou	6	2.56	7650	V w	3p	4.56	500

Yuan-shih treatise evidently relied on a different source from that consulted by Ma Tuan-lin. The two sets of data are compared in table 9.6b. It should be noted that although most times are in hours and marks, the *Yuan-shih* specifies night-watches (w) and points (p) in AD 1074.

In each case the *Yuan-shih* timings lead to values of ΔT which are in good accord with the roughly contemporaneous solar data in figure 9.4. In particular, the ΔT figures obtained in AD 1074 are commensurate with the Moon setting while still eclipsed – as implied in the corresponding *Wen-hsien T'ung-k'ao* account. Hence for both AD 1069 and 1074, the *Yuan-shih* measurements will be retained in preference to the obviously faulty information in the *Wen-hsien T'ung-k'ao.*

Observations timed to the nearest mark which are reported in the *Yuan-shih* (YS), *Sung-shih* (SS) and *Wen-hsien T'ung-k'ao* (WH) – as selected above – are summarised in table 9.7. (NB in the last column of this table, SC stands for source.) With very few exceptions, all times were expressed relative to the start of double hours rather than single hours. Of the several eclipses for which the time of maximum phase (i.e. mid-eclipse) is recorded, most were partial. For these, it may be confidently assumed that the quoted time represents a direct measurement. However, three further eclipses (in the years AD 1088, 1099 and 1277) were total and here the reported time of mid-eclipse apparently represents an interpolation between the measurements for second and third contact. On the first two dates, the times of immersion and emersion are no longer preserved. Hence it seems permissible to use the interpolated moment of mid-eclipse. However, in AD 1277, the reported time of this phase is superfluous since the second and third contact times are available (see also above). NB all times of greatest phase in table 9.5 (expressed with respect to the night-watches) relate to partial obscurations of the Moon.

Figure 9.6 is a plot of the ΔT values listed in table 9.7. All results are included apart from the two highly discordant data in AD 1099 (ΔT =

Table 9.7 ΔT values from lunar eclipse timings expressed in terms of double hours and marks.

Date AD	Ct	Hour	Half	Mark	LT (h)	UT (h)	TT (h)	ΔT(sec)	SC
1052 Dec 8/9	1	yin	—	4	4.08	20.41	20.99	2100	YS
1069 Dec 30/31	1	hai	—	6	22.56	15.06	15.80	2650	YS
	M	tzu	—	5	0.32	16.82	17.30	1750	YS
	4	ch'ou	—	4	2.08	18.58	18.80	800	YS
1071 Jun 15/16	4	hsu	—	5	20.32	12.69	12.82	500	WH
1071 Dec 9/10	1	mao	—	2	5.60	21.93	22.02	300	YS,WH
	M	mao	—	6	6.56	22.89	23.15	950	YS,WH
1073 Apr 24/25	1	hai	—	1	21.36	13.67	13.92	900	YS,WH
	M	hai	—	6	22.56	14.87	15.26	1400	YS,WH
	4	tzu	—	4	0.08	16.39	16.60	800	YS,WH
1078 Jul 27/28	2	hsu	—	2	19.60	12.05	12.13	300	WH
	4	hai	—	3.5	21.84	14.29	14.62	1200	WH
1081 May 25/26	4	hsu	—	6	20.56	12.87	13.25	1350	WH
1082 Nov 8/9	1	yu	—	2	17.60	9.75	10.21	1650	WH
	M	yu	—	7	18.80	10.95	11.27	1150	WH
	4	hsu	—	3	19.84	11.99	12.32	1200	WH
1085 Sep 6/7	1	hsu	—	3	19.84	12.15	12.19	150	WH
	2	hsu	—	7	20.80	13.11	13.13	50	WH
	4	tzu	—	1	23.36	15.67	15.73	200	WH
1088 Jul 6/7	1	hai	—	5	22.32	14.76	14.94	650	WH
	M	tzu	—	6	0.56	17.00	16.75	-900	WH
	4	ch'ou	—	4	2.08	18.52	18.56	150	WH
1089 Jun 25/26	4	ch'ou	—	6	2.56	18.97	19.27	1100	WH
1092 Apr 24/25	1	hai	I	1	21.36	13.67	13.87	700	WH
	2	hai	—	7	22.80	15.11	14.80	−1100	WH
	4	tzu	—	7	0.80	17.11	17.40	1050	WH
1097 Jan 30/31	M	hsu	—	1	19.36	12.00	12.29	1050	WH
1099 Jun 5/6	2	tzu	—	3	23.84	16.17	17.58	5100	WH
	3	ch'ou	—	2	1.60	17.93	19.00	3850	WH
	4	yin	—	2	3.60	19.93	20.11	650	WH

Table 9.7 – *continued.*

Date AD	Ct	Hour	Half	Mark	LT (h)	UT (h)	TT (h)	ΔT(sec)	SC
1099 Nov30/Dec1	1	hai	—	4	22.08	14.34	14.66	1150	WH
	M	tzu	—	4	0.08	16.34	16.54	700	WH
	4	ch'ou	—	4	2.08	18.34	18.41	250	WH
1106 Jan 21/22	M	yu	—	3	17.84	10.47	10.53	200	WH
	4	hsu	—	0	19.12	11.75	12.01	950	WH
1168 Mar 25/26	3	hsu	I	3	19.84	11.90	11.90	0	SS
	4	hsu	II	3	20.84	12.90	12.93	100	SS
1270 Apr 7/8	1	ch'ou	—	3	1.84	18.08	17.72	−1300	YS
	M	yin	—	0	3.12	19.36	19.23	−450	YS
	4	yin	—	6	4.56	20.79	20.73	−200	YS
1272 Aug 10/11	1	ch'ou	—	0	1.12	17.41	17.35	−200	YS
	M	ch'ou	—	6	2.56	18.84	18.63	−750	YS
	4	yin	—	3	3.84	20.13	19.92	−750	YS
1277 May 18/19	1	tzu	—	6	0.56	16.72	16.90	650	YS
	2	ch'ou	—	3	1.84	18.00	17.99	−50	YS
	3	ch'ou	—	7	2.80	18.97	19.22	900	YS
	4	yin	—	4	4.08	20.25	20.31	200	YS
1279 Mar 29/30	1	tzu	—	5	0.32	16.60	16.72	450	YS
	M	ch'ou	—	2	1.60	17.84	17.85	50	YS
	4	ch'ou	—	7	2.80	19.08	18.97	−400	YS
1279 Sep 21/22	1	ch'ou	—	5	2.32	18.40	18.59	700	YS
	M	yin	—	0	3.12	19.20	19.45	900	YS
	4	yin	—	4	4.08	20.16	20.31	550	YS
1280 Sep 10/11	4	hsu	—	1	19.36	11.50	11.79	1050	YS

5100 and 3850 sec). It seems very likely that scribal errors are responsible for these anomalous values.

9.5 Azimuth measurements for lunar eclipses

Examples of the use of lunar azimuths in marking eclipse contacts have already been given earlier in this chapter (e.g. section 9.2.3). Extant records of eclipses described in this way originate exclusively from the

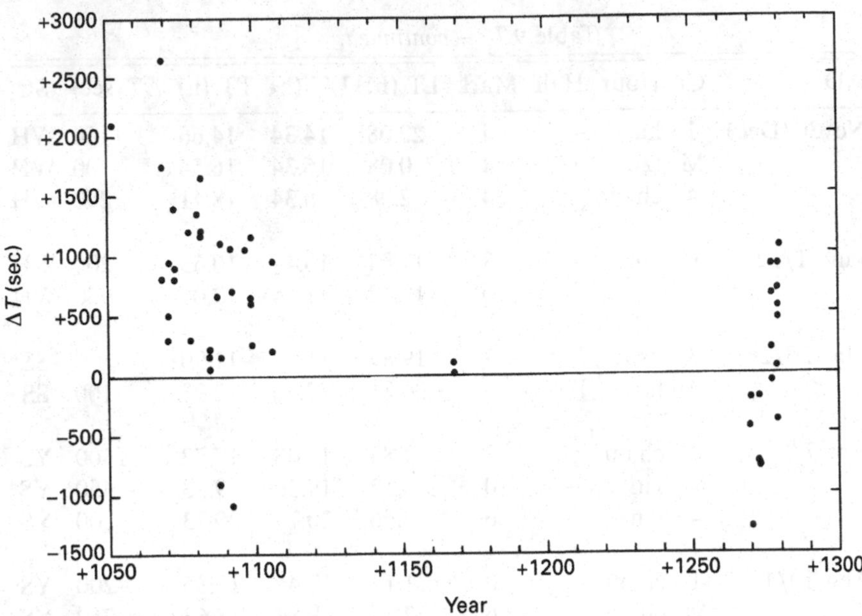

Fig. 9.6 ΔT results derived from lunar eclipse times expressed in double hours and marks.

Sui dynasty. Several predictions follow this form, but there are only four observations: AD 590 Apr 25/26, 590 Oct 18/19, 595 Dec 22/23 and 596 Dec 10/11. The azimuth measurements will be discussed in turn below; timings or horizon phenomena have already been considered above. All of the following records are taken from chapter 17 of the *Sui-shu.* Although some have been cited earlier in this chapter, they are assembled here for direct reference.

(1) AD 590 Apr 25/26 (mag. = 0.69): Ta-hsing Ch'eng

> K'ai-huang reign period, 10th year, 3rd month, 16th day *kuei-mao* [40]. According to the almanac...(various predictions)...It was observed that when the Moon first rose from the south-east, it was already more than half eclipsed. 3 divisions (*fen*) after it reached the *ch'en* direction (azimuth roughly 115 deg), 2 fifteenths remained eclipsed. It gradually recovered. By the end (*wei*) of the *ch'en* direction (azimuth approximately 127.5 deg) it was restored to fullness.

The interpretation of the expression '3 divisions (*fen*) after it reached the *ch'en* direction' is obscure. Only the azimuth measurement at last contact will be considered. Since it is not clear how precise this measurement is, I have assumed the very end of the zone.

RESULTS

Last contact when Moon at azimuth 127.5 deg. Equivalent LT = 20.72 h (Apr 25), UT = 13.41 h. Computed TT = 15.09 h, thus ΔT = 6050 sec.

(2) AD 590 Oct 18/19 (mag. = 0.78): Ta-hsing Ch'eng

K'ai-huang reign period, 10th year, 9th month, 16th day *keng-tzu* [37]. According to the almanac... (predictions)... The observations now showed that 2 *k'o* (i.e. 0.48 h) after the Moon reached the direction *wu* (172.5 to 187.5 deg), the eclipse began from the east side. Afterwards it moved towards the south. When it was above the direction *wei* (202.5 to 217.5 deg), the southern part was four-fifths eclipsed. It gradually recovered. $1\frac{1}{2}$ *k'o* (0.36 h) after reaching the direction *shen* (232.5 to 247.5 deg) it was restored to fullness.

RESULTS

(i) First contact 0.48 h after Moon reached azimuth 172.5 deg. Equivalent LT = 0.21 h (Oct 19), UT = 16.72 h (Oct 18). Computed TT = 18.10 h, thus ΔT = 4950 sec.

(ii) Mid-eclipse when Moon at approximate azimuth 210 deg. Equivalent LT $\sim$ 0.88 h (Oct 19), UT $\sim$ 17.39 h (Oct 18). Computed TT = 19.55 h, thus ΔT $\sim$ 7800 sec.

(iii) Last contact 0.36 h after Moon reached azimuth 232.5 deg. Equivalent LT = 2.12 h (Oct 19), UT = 18.63 h (Oct 18). Computed TT = 21.00 h, thus ΔT = 8550 sec.

(3) AD 595 Dec 22/23 (mag. = 0.63): Ta-hsing Ch'eng

K'ai-huang reign period, 15th year, 11th month, 16th day *keng-wu* [7]. According to the almanac... On that night after the 4th rod (*ch'ou*) of the first watch, the Moon began to be eclipsed above the direction *ch'en* (112.5 to 127.5 deg). It began from the south-east side. At the third rod of the second watch, the Moon was above the direction *szu* (142.5 to 157.5 deg); it was about two-thirds eclipsed. It gradually recovered. At the first rod of the third watch, the Moon was above the direction *ping* (157.5 to 172.5 deg); it was restored to fullness.

RESULTS

(i) First contact when Moon at approximate azimuth 120 deg. Equivalent LT $\sim$ 22.83 h (Dec 22), UT $\sim$ 15.62 h. Computed TT = 13.90 h, thus ΔT $\sim$ −6200 sec.

(ii) Mid-eclipse when Moon at approximate azimuth 150 deg. Equivalent LT $\sim$ 23.63 h, UT $\sim$ 16.43 h. Computed TT = 15.32 h, thus ΔT $\sim$ −4000 sec.

(iii) Last contact when Moon at approximate azimuth 165 deg. Equivalent LT $\sim$ 23.86 h, UT $\sim$ 16.65 h. Computed TT = 16.73 h, thus ΔT $\sim$ +300 sec.

(4) AD 596 Dec 10/11 (mag. = 1.76): Ta-hsing Ch'eng

> K'ai-huang reign period, 16th year, 11th month, 16th day, *i-ch'ou* [2]. According to the almanac... Not until the first rod of the third watch was the Moon seen in the clouds above the direction *p'ing* (157.5 to 172.5 deg). It was already about 3 fifteenths eclipsed and the loss began from the east side. Above the direction *ting* (187.5 to 202.5 deg), the eclipse was total. Afterwards it reappeared from the south-east side. It was not until the third rod of the fourth watch that it was restored to fullness, the Moon being then at the end (*wei*) of the direction *wei* (azimuth approximately 217.5 deg).

As discussed above (section 9.4.1), I have assumed that the first observation was made close to first contact. Last contact will be taken to have occurred when the Moon was at the very end of the azimuth zone *wei*.

RESULTS

(i) First contact when Moon at approximate azimuth 165 deg. Equivalent LT ~ 23.72 h (Dec 10), UT ~ 16.42 h. Computed TT = 16.73 h, thus ΔT ~ 1100 sec.

(ii) Mid-eclipse when Moon at approximate azimuth 195 deg. Equivalent LT ~ 0.28 h (Dec 11), UT ~ 16.99 h (Dec 10). Computed TT = 18.68 h, thus ΔT ~ 6100 sec.

(iii) Last contact when Moon at azimuth 217.5 deg. Equivalent LT = 0.65 h (Dec 11), UT = 17.36 h (Dec 10). Computed TT = 20.63 h, thus ΔT = 11 750 sec.

It is clear from the large scatter in the derived ΔT values (ranging from −6200 sec to +11 750 sec in only 6 years) that this type of observation is of little value in the study of the Earth's past rotation. Accordingly, the results obtained in this section will thus not be considered further.

9.6 Estimates of solar eclipse magnitude

The oldest determination of an eclipse magnitude (as a fraction of the solar diameter) which is preserved from China dates from 134 BC. On this occasion it was stated that 'more than half' of the Sun was obscured. Over the next few centuries several eclipses were said to be 'like a hook' (see chapter 8) – suggesting a magnitude of at least $\frac{2}{3}$. However, not until AD 489 do we find the next recorded estimate of the fraction of the Sun which was covered by the Moon. Quite possibly, several further observations were made in the intervening time but if so all have been lost.

Between AD 489 and 586, eight estimates of solar eclipse magnitude are preserved, while a further observation (in AD 523) notes the degree of obscuration of the Sun at sunrise. After AD 586 there is a gap of more than 150 years without any further magnitude estimates. For six of the observations made between AD 489 and 586, magnitudes are expressed to the nearest $\frac{1}{15}$ of the solar diameter. Over a similar time-interval (actually from AD 438 to 596) we find much the same situation for lunar eclipses. Hence $\frac{1}{15}$ of the diameter of the appropriate luminary seems to have been a fairly standard unit for specifying eclipse magnitudes at this period.

No further observations of this kind – whether of the Sun or Moon – are preserved in Chinese history between AD 597 and 767. From this latter date until AD 1020, there are only isolated magnitude estimates (in AD 768, 792, 822, 927, 937 and 994), after which much more frequent data become available. For those eclipses between AD 768 and 994, no magnitudes are expressed in fifteenths of the solar or lunar diameter; tenths or other fractions are used instead. From AD 1020 onwards, most recorded eclipse magnitudes in Chinese history (whether for the Sun or Moon) are expressed in *fen* ('divisions') – each equal to one-tenth of the apparent diameter of the luminary. This convention continued in the Jesuit era. As pointed out in chapter 3, the practice of quoting magnitude in fifteenths or tenths reflects an independent tradition from that adopted in Babylon and elsewhere in the occidental world; there the standard was the digit or twelfth part of the disk.

Observations of eclipse magnitude represent crude eye estimates rather than measurements, while the computed magnitude is not very responsive to changes in ΔT. Hence isolated data (as between AD 768 and 994) or groups of observations from relatively recent epochs are of little value in the determination of ΔT. Attention will thus be mainly confined to the data from the fifth and sixth centuries since they form a useful chronological set. The very early observation in 134 BC will also be included – mainly on account of its historical interest.

It is rarely stated whether the northern or southern limb of the Sun was obscured. Hence in principle each individual observation should lead to two widely spaced results for ΔT. Fortunately – as discussed in chapter 3 – of the two possible solutions only one is usually valid since the other places the Sun below the horizon at the required phase.

On one occasion (AD 490 Mar 7), it was reported in the *Wei-shu* (chap. 105) that the Sun was $\frac{1}{15}$ eclipsed. However, the text also stated that the Sun was covered by clouds from time to time. It is thus possible that the eclipse reached a much larger phase during the period of cloud cover without the observers being aware of it. Since the magnitude estimate could well lead to a misleading value for ΔT, the observation will be rejected. For the remaining eclipses discussed below, there is nothing in the text to indicate that visibility of maximal phase was prevented by cloud – or the Sun reaching the horizon.

(1) BC 134 Aug 19: Ch'ang-an

> Yuan-kuang reign period, first year, 7th month, day *kuei-wei* [20], the day before the last day of the month. The Sun was eclipsed and it was 8 deg in Yi... At noon the eclipse began from the north-east and reached more than half of the disk. At the hour of *pu* (= 15 h) it was restored.
>
> [*Han-shu*, chap. 27c.]

It will be assumed that 'more than half' meant somewhere between $\frac{1}{2}$ and $\frac{2}{3}$ – a mean of 0.59.

RESULTS

For a magnitude of 0.59 near noon at Ch'ang-an, $\Delta T = 13\,500$ sec.

NB the above value of ΔT would lead to a LT of greatest phase of 12.5 h, in close accord with the text. A result for ΔT of -350 sec would also lead to a magnitude of 0.59. However, here the LT of maximum would be much later: 17.3 h. Hence only the first value for ΔT need be considered.

(2) AD 489 Mar 18: P'ing-ch'eng

> T'ai-ho reign period, 13th year, 2nd month, day *i-hai* [12], the first day of the month. The Sun was 8 fifteenths eclipsed.
>
> [*Wei-shu*, chap. 105.]

RESULTS

No value of ΔT will produce a visible magnitude smaller than about 0.55 at P'ing-ch'eng. Hence although the observed figure of $\frac{8}{15}$ (i.e. 0.53) may well be a fair estimate, a result for ΔT cannot be deduced.

NB P'ing-ch'eng, near modern Ta-t'ung (= Datong) was the capital of the Northern Wei Dynasty until AD 493, after which Lo-yang became the imperial residence for this dynasty.

(3) AD 493 Jan 4: Chien-k'ang

> Yung-ming reign period, 10th year, 12th month, day *kuei-wei* [20], the first day of the month. The calculated time (for the eclipse) was half way through the hour of *wu* (11–13 h). It was not until the start of the *wei* hour (13–15 h) that the Sun began to be eclipsed. The loss began from the north-west side. It was eclipsed 4 tenths. At the hour of *shen* (15–17 h) its brightness and colour were fully restored.
>
> [*Nan-ch'i-shu*, chap. 12.]

The generally low precision of eclipse prediction at this early period has already been discussed (section 9.2.4).

RESULTS

No value of ΔT can render the visible magnitude of this eclipse less than 0.44 at Chien-k'ang. Hence the observation cannot be used to yield a value for ΔT.

(4) AD 494 Jun 19: Chien-k'ang

Lung-ts'ang reign period, first year, 5th month, day *chia-hsu* [11], the first day of the month. The Sun was one-third eclipsed at the hour of *szu* (9–11 h). It was restored at the hour of *wu* (11–13 h).

[*Nan-ch'i-shu*, chap. 12.]

RESULTS

No value of ΔT can render the visible magnitude of this eclipse less than 0.57 at Chien-k'ang. For values of ΔT which lead to maximal phase between 9 and 11 h (as reported) the computed magnitude lies between 0.64 and 0.90. Clearly there is a scribal error in the text; the true magnitude must have been considerably greater than the recorded figure (see also chapter 8).

(5) AD 512 Jun 29: Lo-yang

Yen-chang reign period, first year, 5th month, day *chi-wei* [56], the last day of the month. The Sun was 9 fifteenths eclipsed.

[*Wei-shu*, chap. 105.]

RESULTS

For a computed magnitude of 0.60 at Lo-yang, either ΔT = 6050 or 35 000 sec.

NB for both of these values the Sun would be above the horizon at maximum eclipse; however, the latter result is clearly discordant and will be discarded.

(6) AD 585 Jul 31: Ta-hsing Ch'eng

K'ai-huang reign period, 5th year, 6th month, 30th day... The observations show that the Sun did not begin to be eclipsed until more than (*shang*) 6 marks after (*hou*) the hour of *wu*. The loss began from the north-west side; it (reached a magnitude of) 6 fifteenths. Not until 1 mark after (*hou*) the hour of *wei* did it begin to reappear. At 5 marks [after the hour of *wei*] it was restored to fullness.

[*Sui-shu*, chap. 17.]

Most renderings of the recorded times in this and the following entry are open to conjecture (see section 9.2.1).

Table 9.8 ΔT values from solar eclipse magnitude estimates recorded in Chinese history before AD 750.

Year	Est. mag.	ΔT (sec)
−133	$>\frac{1}{2}$	13 500
+489	$\frac{8}{15}$	3 450
+512	$\frac{9}{15}$	6 050
+585	$\frac{6}{15}$	6 200
+586	$\frac{2}{3}$	7 400

RESULTS

For a computed magnitude of 0.40 at Ta-hsing Ch'eng, ΔT = 6200 sec.

NB no other value for ΔT would produce a visible eclipse of magnitude 0.40 at Ta-hsing Ch'eng.

(7) AD 586 Dec 16: Ta-hsing Ch'eng

> K'ai-huang reign period, 6th year, 10th month, 30th day *ting-ch'ou* [14]... It was seen during the observations that when the Sun rose 1 *chang* (i.e. roughly 10 deg) above the mountains, it began to be eclipsed; (this was) at 2 marks in the hour of *ch'en*. The loss began from the west; it was two-thirds eclipsed. At 2 marks after (*hou*) the hour of *ch'en* it began to reappear. At more than (*shang*) 3 marks from the start (*ju*) of the hour of *szu* it was restored to fullness.
>
> [*Sui-shu*, chap. 17.]

RESULTS

For a computed magnitude of 0.67 at Ta-hsing Ch'eng, ΔT = 7400 sec.

NB no other value for ΔT would produce a visible eclipse of magnitude 0.67 at Ta-hsing Ch'eng.

The above results are summarised in table 9.8, which lists the year (− or +), recorded magnitude and the derived ΔT values.

The ΔT results listed in column 3 of table 9.8 are in fair accord with one another and with the figures obtained from roughly contemporaneous eclipse timings (see above). However, since the results are derived from crude naked-eye estimates, and the ΔT values obtained from similar observations made elsewhere (chapters 5 and 13) show a much larger scatter, they will not be considered further.

9.7 Sun rising or setting eclipsed

Only a very few events in this category are recorded in Chinese history during the whole of the pre-telescopic period. Possibly many further observations were made down the centuries but if so the details have not been preserved. Instances in which the Sun was eclipsed at sunrise or sunset were reported in the years 35 BC, AD 529, 532, 937 and 1042, but after the last date there are no similar records until the Jesuit era. To the above observations may be added that of the eclipse of 89 BC, the account of which implies that the Sun was fully recovered before sunset.

Translations and discussions of the various records – most of which are to be found in the astrological treatises of official histories – are given below. Computed eclipse magnitudes at maximal phase are cited at the head of each entry; these are for reference only.

The eclipses of both 89 and 35 BC were described as almost complete ('like a hook') and this led to their inclusion in chapter 8 above. However, for convenience translations will be repeated here. In calculating the LTs of sunrise, sunset, etc., the effect of horizon profile will be ignored. Apart from Ch'ang-an (where the two earliest observations were made) all of the capitals mentioned in this and the following two sections (namely Chien-k'ang, Lo-yang and Pien) were located in the great plain of north-east China. The site of Ch'ang-an (also known as Ta-hsing Ch'eng during the Sui dynasty) is located in a more mountainous region, but the surrounding terrain in the Wei River valley is fairly level over several tens of kilometres – especially towards the west. At this early period, ΔT was so large ($\sim$ 10 000 sec) that minor horizon irregularities are relatively unimportant.

Estimates of the proportion of the Sun obscured at sunrise or sunset will be considered separately (section 9.9).

(1) BC 89 Sep 29 (mag. = 0.91): Ch'ang-an

> Cheng-ho reign period, 4th year, 8th month, day *hsin-yu* [58], the last day of the month. The Sun was eclipsed; it was not complete and like a hook. It was 2 deg in *K'ang*. At the hour of *pu* (= 15–17 h) the eclipse began from the north-west. Towards the hour of sunset it was restored.
>
> [*Han-shu*, chap. 27.]

It is clear from the text that last contact was observed before the Sun reached the horizon.

RESULTS

Last contact on Sep 29 before sunset. LT of sunset = 17.99 h. Hence LT of contact $<$ 17.99 h, UT $<$ 10.59 h. Computed TT = 12.61 h, thus ΔT $>$ 7250 sec.

(2) BC 35 Nov 1 (?) (mag. = 0.83): Ch'ang-an

Chien-chao reign period, 5th year, 6th month, day *jen-shen* [9], the last day of the month. The Sun was eclipsed. It was not complete and like a hook; then (*yin*) it set.

[*Han-shu*, chap. 27.]

The error in the date of this observation has already been considered in chapter 8. There it was shown that although the recorded date is equivalent to Aug 23 in 34 BC, the only viable date for the eclipse is Nov 1 in the previous year. If it is only assumed that the Sun set eclipsed, the following limits (i) and (ii) may be derived.

RESULTS

(i) First contact on Nov 1 before sunset. LT of sunset = 17.41 h. Hence LT of contact < 17.40 h, UT < 9.92 h. Computed TT = 12.00 h, thus ΔT > 7600 sec.

(ii) Last contact after sunset. LT of contact > 17.40 h, UT > 9.92 h. Computed TT = 13.94 h, thus ΔT < 13 500 sec.

Combining these limits yields 7600 < ΔT < 13 500 sec.

(iii) Alternatively, the text seems to imply that the Sun had the appearance of a hook when it reached the horizon. Hence probably at least two-thirds of the solar disk would have been obscured at sunset. On this assumption, a value of ΔT in the narrower range 10 000 < ΔT < 12 000 sec may be deduced, although these limits can only be approximate.

(3) AD 529 Nov 17 (?): Lo-yang

Hung-an reign period, 2nd year, 10th month, day *chi-yu* [46], the first day of the month. The Sun rose from beneath the Earth already eclipsed by 7 fifteenths. It began from the south-west side.

[*Wei-shu*, chap. 105.]

The recorded date cannot be correct; there was no eclipse on or near Nov 17 in AD 529. During the second year of the Hung-an reign period (AD 529 Jan 26 to AD 530 Jan 14) there were two solar eclipses but neither was visible in China. Computations using equation (4.1) show that within seven years before and after the recorded date only a single eclipse could have been visible near sunrise at Lo-yang: AD 532 Nov 13. This event, which is considered immediately below, also occurred in the 10th month of a year but on the day *hsin-yu* [58]. It is possible that this latter eclipse was mistakenly reported twice in the same history – both on the correct date and also three years beforehand. However, it may be calculated that for any value of ΔT which renders the eclipse of AD 532 visible at sunrise at the Wei capital of Lo-yang, the magnitude at maximum phase cannot have exceeded 0.3 there. This is significantly less than the $\frac{7}{15}$ (i.e. 0.47) said to have been obscured in the earlier year.

(4) AD 532 Nov 13 (mag. = 0.21): Lo-yang

> Tai-chang reign period, 1st year, 10th month, day *hsin-yu* [58], the first day of the month. The Sun rose eclipsed from beneath the Earth. It began from the south-west side.
>
> [*Wei-shu*, chap. 105.]

In contrast to the previous entry, the recorded date of this event is exactly correct. The reference to the eclipse beginning from the south-west side possibly suggests that the phase was still small at sunrise. However, no estimate of the degree of obscuration of the solar disk is given, which may imply that only the last stages of the eclipse were visible. If so, the position angle of first contact may have merely been interpolated. Under these circumstances, it seems best to assume only that the Sun rose at some time between first and last contact.

RESULTS

(i) First contact on Nov 13 before sunrise. LT of sunrise = 6.83 h. Hence LT of contact < 6.83 h, UT < 23.14 h (Nov 12). Computed TT = 23.81 h, thus $\Delta T > 2400$ sec.

(ii) Last contact after sunrise. LT of contact > 6.83 h, UT > 23.14 h (Nov 12). Computed TT = 1.05 h (Nov 13), thus $\Delta T < 6850$ sec.

Combining these limits yields $2400 < \Delta T < 6850$ sec.

(5) AD 937 Feb 14 (mag. = 0.52): Pien

> T'ien-fu reign period, 2nd year, first month, day *i-mao* [52]. The Sun was eclipsed... It was 3 tenths eclipsed... When the Sun rose in the east it was 3 tenths eclipsed. It gradually (re-)emerged. At the hour of *mao* it was restored to fullness.
>
> [*Chiu-wu-tai-shih*, chap. 139.]

The text seems to imply that the phase was already declining when the Sun rose (i.e. between greatest phase and last contact). On the day in question, sunrise would occur at Pien at 6.48 h – a time already towards the end of the hour of *mao*.

RESULTS

(i) Maximum before sunrise. LT of sunrise = 6.48 h. Hence LT of mid-eclipse < 6.48 h, UT < 23.12 h (Feb 13). Computed TT = 23.25 h, thus $\Delta T > 500$ sec.

(ii) Last contact after sunrise. LT of contact > 6.48 h, UT > 23.12 h (Feb 13). Computed TT = 0.10 h (Feb 14), thus $\Delta T < 3500$ sec.

Combining these limits yields $500 < \Delta T < 3500$ sec.

(6) AD 1042 Jun 20 (mag. = 0.41): Pien

> Ch'ing-li reign period, 2nd year, 6th month, day *jen-shen* [9], the first day of the month. The Sun was eclipsed by 5 divisions (*fen*). At 6 marks in the hour of *yu* (18.56 h) the eclipse was 2 divisions and then the Sun set beneath the Earth and was invisible.
>
> [*Wen-hsien T'ung-k'ao*, chap. 283.]

The above account clearly indicates that the Sun set after maximal phase. Sunset would actually occur at 19.22 h, so that the recorded LT is about 2/3 of an hour in error.

RESULTS

(i) Maximum before sunset. LT of sunset = 19.22 h. Hence LT of mid-eclipse < 19.22 h, UT < 11.60 h. Computed TT = 11.36 h, thus $\Delta T > -850$ sec.

(ii) Last contact after sunset. LT of contact > 19.22 h, UT > 11.60 h. Computed TT = 12.17 h, thus $\Delta T < 2050$ sec.

Combining these limits yields $-850 < \Delta T < 2050$ sec.

The ΔT limits obtained in this section are summarised in table 9.9, along with the results from lunar eclipses to be derived in section 9.8.

9.8 Moon rising or setting eclipsed

Between AD 438 and 590, six instances where the Moon rose or set eclipsed are noted in Chinese history. Subsequently, no similar accounts are preserved until the late eleventh century, when a further ten observations (between AD 1042 and 1081) are extant. After this latter date, only a single report in this category (from AD 1168) is accessible until the Jesuit period.

The earliest records mentioned above are cited in the dynastic histories, but those from the eleventh century are to be found in chapter 285 of the *Wen-hsien T'ung-k'ao*. Finally, the eclipse of AD 1168 is reported in the *Sung-shih*. For all of these events – as in the case of the other eclipses discussed in this chapter – there is no valid reason for assuming a place of observation other than the capital of the time.

The accounts of the eclipses of AD 571 Oct 18/19, 1063 Nov 8/9, 1074 Oct 7/8, 1081 May 25/26 and 1168 Mar 25/26 all seem to imply that the Moon rose eclipsed whilst in total shadow. As discussed in chapter 3, such observations are of dubious reliability since the totally eclipsed Moon is probably not visible when on the horizon. Hence they will not be considered further. An additional record from Ta-hsing Ch'eng (AD 590 Apr 25/26), which implies that the Moon rose more than half eclipsed, will also be rejected since the eastern horizon as seen from this site is somewhat mountainous – see, for example, entry (7) in section 9.6.

A full translation of each of the remaining records is given below, together with a commentary and the calculated ΔT limits. Although only a few records lead to critical limits for ΔT, the observations are of interest since they give a good indication of the variety of lunar eclipse reports which are preserved in Chinese history. NB as emphasised in chapter 3, the computed LT of moonrise or moonset for any given place and date is a function of ΔT.

Estimates of the fraction of the Moon eclipsed at moonrise or moonset will be discussed separately in section 9.9.

(1) AD 438 Jun 23/24 (mag. = 1.13): Chien-k'ang

> Yuan-chia reign period, 15th year, 5th month 15th day, full Moon. When the Moon first rose it was already eclipsed. Its brightness had already regained one-quarter (of itself). It was 16 deg in *(Nan-)tou* (lunar lodge).
>
> [*Sung-shu*, 12.]

Since the eclipse was actually total, it seems clear that by the time that the Moon rose, the phase was declining.

RESULTS

(i) Third contact before moonrise on Jun 23. LT of moonrise = 19.13 h. Hence LT of contact < 19.13 h, UT < 11.19 h. Computed TT = 12.53 h, thus $\Delta T > 4800$ sec.

(ii) Last contact after moonrise. LT of moonrise = 19.19 h. Hence LT of contact > 19.19, UT > 11.25 h. Computed TT = 13.71 h, thus $\Delta T <$ 8850 sec.

Combining these limits yields $4800 < \Delta T < 8850$ sec.

(2) AD 503 Jun 25/26 (mag. = 1.08): Lo-yang

> Ching-ming reign period, 4th year, 5th month, day *ting-mao* [4]. The Moon was in *(Nan-)tou* (lunar lodge). It rose eclipsed from beneath the Earth. It was eclipsed 12 fifteenths.
>
> [*Wei-shu*, 105.]

There is no mention of this eclipse being total, suggesting that the Moon rose after the end of totality.

RESULTS

(i) Third contact before moonrise on Jun 25. LT of moonrise = 19.20 h. Hence LT of contact < 19.20 h, UT < 11.69 h. Computed TT = 13.08 h, thus $\Delta T > 5000$ sec.

(ii) Last contact after moonrise. LT of moonrise = 19.23 h. Hence LT of contact > 19.23, UT > 11.73 h. Computed TT = 14.45 h, thus $\Delta T <$ 9800 sec.

Combining these limits yields $5000 < \Delta T < 9800$ sec.

(3) AD 513 Jun 4/5 (mag. = 1.35): Lo-yang

> Yen-ch'ang reign period, 2nd year, 4th month, day *chi-hai* [36]. The Sun was in *Tsui* (lunar lodge) and the Moon in *Wei.* It (i.e. the Moon) rose from beneath the Earth with 3 fifteenths remaining (eclipsed). It gradually became full.
>
> [*Wei-shu*, chap. 105.]

The recorded date actually corresponds to Jun 25, but taking the cyclical day as *i-hai* [16] rather *chi-hai* gives Jun 5 instead. The written characters *i* and *chi* are very similar and confusion between them is fairly common.

Tsui (the 20th lunar lodge) and *Wei* (the 6th lodge) were roughly 180 deg apart. Since the eclipse was actually total, the phase evidently was already waning at moonrise.

RESULTS

(i) Third contact before moonrise on Jun 4. LT of moonrise = 19.17 h. Hence LT of contact < 19.17 h, UT < 11.59 h. Computed TT = 12.22 h, thus $\Delta T > 2250$ sec.

(ii) Last contact after moonrise. LT of moonrise = 19.22 h. Hence LT of contact > 19.22, UT > 11.65 h. Computed TT = 13.40 h, thus $\Delta T <$ 6300 sec.

Combining these limits yields $2250 < \Delta T < 6300$ sec.

(4) AD 514 May 24/25 (mag. = 0.92): Lo-yang

> Yen-ch'ang reign period, 3rd year, 4th month, day *kuei-szu* [30]. The Moon was in *Wei* (lunar lodge). It rose eclipsed from beneath the Earth. It was eclipsed 14 fifteenths.
>
> [*Wei-shu*, 105.]

The observed degree of obscuration (0.93) is very close to the calculated magnitude. Comparing this entry with others in the same source (see above) suggests that on this occasion moonrise occurred around the time of maximal phase. However, it is also possible that the Moon rose only slightly eclipsed and eventually reached a magnitude of 14/15. Because of this uncertainty, it seems best to assume only that moonrise took place between first contact and approximately maximum eclipse.

RESULTS

(i) First contact before moonrise on May 24. LT of moonrise = 19.02 h. Hence LT of contact < 19.02 h, UT < 11.41 h. computed TT = 11.35 h, thus $\Delta T > -250$ sec.

(ii) Maximum phase after moonrise. LT of moonrise = 19.09 h. Hence LT of contact > 19.09 UT > 11.48 h. Computed TT = 13.02 h, thus ΔT < 5500 sec.

Combining these limits yields $-250 < \Delta T < 5500$ sec.
Obviously the upper limit can only be very rough.

(5) AD 1042 Jul 5/6 (mag. = 1.11): Pien

Ch'ing-li reign period, 2nd year, 6th month, day *ting-hai* [24]. The Moon was eclipsed by 6 divisions. Then it set eclipsed and the restoration was not seen.

[*Wen-hsien T'ung-k'ao*, chap. 285.]

This eclipse was total, and it seems clear from the text that the Moon set before the total phase had been reached.

RESULTS

(i) First contact before moonset on Jul 6. LT of moonset = 4.80 h. Hence LT of contact < 4.80 h, UT < 21.23 h (Jul 5). Computed TT = 20.76 h, thus $\Delta T > -1700$ sec.

(ii) Second contact after moonset. LT of moonset = 4.87 h. Hence LT of contact > 4.87 h, UT > 21.30 h (Jul 5). Computed TT = 22.02 h, thus $\Delta T < 2600$ sec.

Combining these limits yields $-1700 < \Delta T < 2600$ sec.

(6) AD 1067 Mar 3/4 (mag. = 1.43): Pien

Chih-ping reign period, 4th year, 2nd month, day *chia-wu* [31]. The Moon was eclipsed. It began to be seen at 4 marks in the hour of *ch'ou* in the west. It was was 15 deg in *I* (lunar lodge). At 6 marks in the hour of *ch'ou* the eclipse reached its maximum. It was more than 8 divisions. It set eclipsed in the direction *yu* and was not seen.

[*Wen-hsien T'ung-k'ao*, chap. 285.]

As in the previous example, it is apparent from the text that the Moon set before the onset of totality.

RESULTS

(i) First contact before moonset on Mar 4. LT of moonset = 6.15 h. Hence LT of contact < 6.15 h, UT < 22.71 h (Mar 3). Computed TT = 21.84 h, thus $\Delta T > -3150$ sec.

(ii) Second contact after moonset. LT of moonset = 6.17 h. Hence LT of contact > 6.17 h, UT > 22.73 h (Mar 3). Computed TT = 22.94 h, thus $\Delta T < 750$ sec.

Combining these limits yields $-3150 < \Delta T < 750$ sec.

(7) AD 1068 Aug 15/16 (mag. = 0.38): Pien

> Hsi-ning reign period, first year, 7th month, day *i-yu* [22]. The Moon was eclipsed at 5 marks in the hour of *ch'ou*. It was 10 deg in *Wei* (lunar lodge). It began to be seen at the north-eastern side and was eclipsed by $2\frac{1}{2}$ divisions. Then it set eclipsed and was not seen.
>
> [*Wen-hsien T'ung-k'ao*, chap. 285.]

The estimated fraction of the lunar diameter covered at moonset is significantly less than the computed magnitude at greatest phase. Hence it may be inferred that the Moon set before the eclipse reached its maximum.

RESULTS

(i) First contact before moonset on Aug 16. LT of moonset = 5.32 h. Hence LT of contact < 5.32 h, UT < 21.73 h (Aug 15). Computed TT = 21.67 h, thus $\Delta T > -200$ sec.

(ii) Maximal phase after moonset. LT of moonset = 5.38 h. Hence LT of contact > 5.38 h, UT > 21.79 h (Aug 15). Computed TT = 22.73 h, thus $\Delta T < 3400$ sec.

Combining these limits yields $-200 < \Delta T < 3400$ sec.

(8) AD 1071 Jun 15/16 (mag. = 0.95): Pien

> Hsi-ning reign period, 4th year, 5th month, day *chi-hai* [36]. The Moon was eclipsed. At 1 mark in the hour of *hsu* it began to be seen at the south-east. It rose eclipsed; it was not round but 6 divisions (were covered). It was within the degrees of *Tung-ching* (lunar lodge). After 5 marks it was restored.
>
> [*Wen-hsien T'ung-k'ao*, chap. 285.]

At the time, the Sun (not the Moon) would be in *Tung-ching*, which is situated in Gemini. It seems clear from the text that the Moon rose when the phase was declining. For example, there is no mention of any increase in phase despite the fact that the estimated obscuration at moonrise (0.60) is much less than the maximum (0.95). Furthermore, the reported period of visibility after moonrise (5 marks or 1.2 h) is only about one-third of the computed total duration.

RESULTS

(i) Maximum phase before moonrise on Jun 15. LT of moonrise = 19.22 h. Hence LT of contact < 19.22 h, UT < 11.59 h. Computed TT = 11.09 h, thus $\Delta T > -1800$ sec.

(ii) Last contact after moonrise. LT of moonrise = 19.29 h. Hence LT of contact > 19.29, UT > 11.66 h. Computed TT = 12.82 h, thus $\Delta T <$ 4200 sec.

Combining these limits yields $-1800 < \Delta T < 4200$ sec.

(9) AD 1071 Dec 9/10 (mag. = 0.43): Pien

Hsi-ning reign period, 4th year, 11th month, day *ping-shen* [33]. The Moon was eclipsed. At 2 marks in the hour of *mao* (the eclipse) began to be seen in the west. It commenced at the south-east side. Not until 6 marks in the hour of *mao* was the eclipse at its maximum, reaching less than $4\frac{1}{2}$ divisions. It was rather less than 1 deg in *Tung-ching* (lunar lodge). Before it became bright it set eclipsed and the restoration was not seen.

[*Wen-hsien T'ung-k'ao*, chap. 285.]

The estimated fraction of the Moon in shadow (0.45) is very close to the computed magnitude. Nevertheless, it is apparent from the text that the eclipse was declining when the Moon set.

RESULTS

(i) Maximum before moonset on Dec 10. LT of moonset = 7.17 h. Hence LT of contact < 7.17 h, UT < 23.51 h (Dec 9). Computed TT = 23.15 h, thus $\Delta T > -1150$ sec.

(ii) Last contact after moonset. LT of moonset = 7.22 h. Hence LT of contact > 7.22 UT > 23.56 h (Dec 9). Computed TT = 0.28 h (Dec 10), thus $\Delta T < 2600$ sec.

Combining these limits yields $-1150 < \Delta T < 2600$ sec.

(10) AD 1073 Oct 18/19 (mag. = 0.56): Pien

Hsi-ning reign period, 6th year, 9th month, day *i-mao* [52]. The Moon was eclipsed. At 4 marks in the hour of *ch'ou* the eclipse began to be seen at the north-east side. At 1 mark in the hour of *yin* it was at its maximum, reaching 6 divisions. Then it began to be restored. When it was 3 divisions short of roundness it set eclipsed and the restoration was not seen.

[*Wen-hsien T'ung-k'ao*, chap. 285.]

It is clear that the Moon set between greatest phase and last contact.

RESULTS

(i) Maximum before moonset on Oct 19. LT of moonset = 6.54 h. Hence LT of contact < 6.54 h, UT < 22.67 h (Oct 18). Computed TT = 22.16 h, thus $\Delta T > -1850$ sec.

(ii) Last contact after moonset. LT of moonset = 6.59 h. Hence LT of contact > 6.59 h, UT > 22.72 h (Oct 18). Computed TT = 23.58 h, thus $\Delta T < 3100$ sec.

Combining these limits yields $-1850 < \Delta T < 3100$ sec.

Table 9.9 ΔT limits from horizon eclipses.

Year[a]	Type[b]	Rise/set[c]	ΔT Range (sec) LL[d]	UL[e]
−88	Sun	rise	7 250	—
−34	Sun	set	~10 000	~12 000
+438	Moon	rise	4 800	8 850
+503	Moon	rise	5 000	9 800
+513	Moon	rise	2 250	6 300
+514	Moon	rise	−250	~5 500
+532	Sun	rise	2 400	6 850
+937	Sun	rise	500	3 500
1042	Sun	set	−850	2 050
1042	Moon	rise	−1 700	2 600
1067	Moon	set	−3 150	750
1068	Moon	set	−200	3 400
1071a	Moon	rise	−1 800	4 200
1071b	Moon	set	−1 150	2 600
1073	Moon	set	−1 850	3 100
1081	Moon	rise	−850	2 750

[a] The year (− or +).
[b] Whether the eclipse was of the Sun or Moon.
[c] Whether the eclipse occurred near the rising or setting of the appropriate luminary.
[d] The derived lower limit to ΔT in sec (LL).
[e] The derived upper limit to ΔT in sec (UL).

(11) AD 1081 Nov 19/20 (mag. = 1.63): Pien

Yuan-feng reign period, 4th year, 10th month, day *hsin-szu* [18]. The Moon was eclipsed. At dusk it rose eclipsed by 7 divisions. It was in the degrees of *Pi* (lunar lodge). After 2 marks it was restored.

[*Wen-hsien T'ung-k'ao*, chap. 285.]

The Moon evidently rose considerably after the end of totality.

Results

(i) Third contact before moonrise on Nov 19. LT of moonrise = 17.04 h. Hence LT of contact < 17.04 h, UT < 9.23 h. Computed TT = 8.99 h, thus $\Delta T > -850$ sec.

(ii) Last contact after moonrise. LT of moonrise = 17.07 h. Hence LT

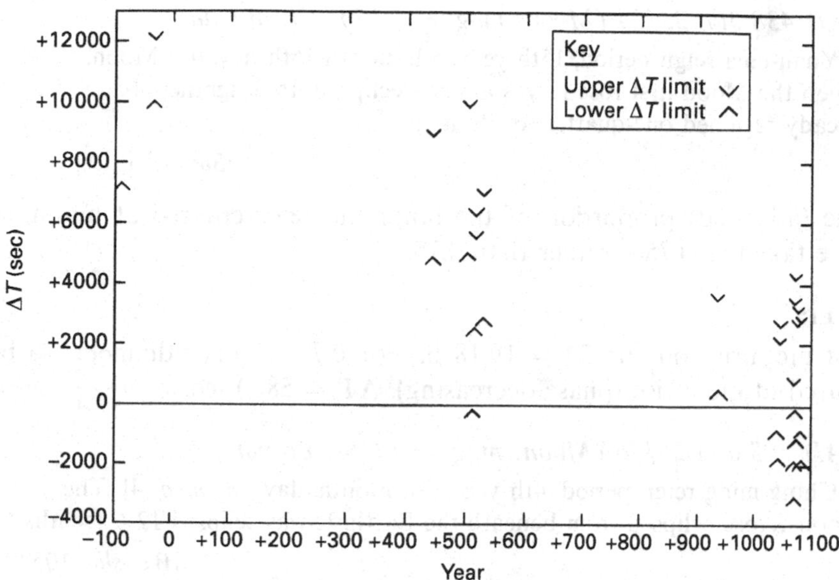

Fig. 9.7 ΔT limits obtained from sunrise/sunset and moonrise/moonset eclipse observations.

of contact > 17.07 h, UT > 9.26 h. Computed TT = 10.03 h, thus $\Delta T <$ 2750 sec.

Combining these limits yields $-850 < \Delta T < 2750$ sec.

The above results, along with those obtained from the solar obscurations discussed in section 9.7, are listed in table 9.9.

The above ΔT limits are shown in figure 9.7. Although most limits are clearly redundant, a few others are more critical. However, it should be noted that the upper limit to ΔT set by the observation in +514 (5500 sec) is only approximate.

9.9 Estimates of the proportion of the Sun or Moon obscured at its rising or setting

In all of the following cases, the fact that the Sun or Moon rose or set whilst eclipsed has already been considered in sections 9.7 or 9.8. Hence only abbreviated translations will be cited below. In each case the magnitude at greatest phase is given for reference. As discussed in chapter 3 (see especially figure 3.23), corrections to the proportion of the lunar diameter covered at moonrise or moonset will be made in order to allow for systematic errors in estimating magnitude; these are most significant for small eclipses. No such amendment will, however, be made for the two solar observations.

(1) AD 438 Jun 23/24 (Moon: mag. = 1.13): Chien-k'ang

> Yuan-chia reign period, 15th year, 5th month 15th day, full Moon. When the Moon first rose it was already eclipsed. Its brightness had already regained one-quarter (of itself)...
>
> [*Sung-shu*, 12.]

The (adjusted) proportion of the lunar diameter covered at moonrise will be taken as 0.78 – rather than 0.75.

RESULTS

LT of moonrise on Jun 23 = 19.18 h. For 0.78 of lunar diameter to be obscured at moonrise (phase decreasing), ΔT = 5800 sec.

(2) AD 503 Jun 25/26 (Moon: mag. = 1.08): Lo-yang

> Ching-ming reign period, 4th year, 5th month, day *ting-mao* [4]. The Moon...rose eclipsed from beneath the Earth. It was eclipsed 12 fifteenths.
>
> [*Wei-shu*, 105.]

Since there is no mention of this eclipse being total, the Moon apparently rose after the end of totality. The (adjusted) proportion of the lunar diameter covered at moonrise will be taken as 0.84.

RESULTS

LT of moonrise on Jun 25 = 19.23 h. For 0.84 of lunar diameter to be obscured at moonrise (phase decreasing), ΔT = 6000 sec.

(3) AD 513 Jun 4/5 (Moon: mag. = 1.35): Lo-yang

> Yen-ch'ang reign period, 2nd year, 4th month, day *chi-hai* [36]...It (i.e. the Moon) rose from beneath the Earth with 3 fifteenths remaining (eclipsed). It gradually became full.
>
> [*Wei-shu*, chap. 105.]

As the eclipse was total, when the Moon rose the phase evidently was already on the decline. The (adjusted) proportion of the lunar diameter covered at moonrise will be taken as 0.12.

RESULTS

LT of moonrise on Jun 4 = 19.24 h. For 0.12 of lunar diameter to be obscured at moonrise (phase decreasing), ΔT = 5750 sec.

(4) AD 514 May 24/25 (Moon: mag. = 0.92): Lo-yang

> Yen-ch'ang reign period, 3rd year, 4th month, day *kuei-szu* [30]. The Moon...rose eclipsed from beneath the Earth. It was eclipsed 14 fifteenths.
>
> [*Wei-shu*, 105.]

RESULTS

The observed phase (0.93: adjusted value approximately 0.98) is very close to – and actually exceeds – the computed magnitude. Hence it is not possible to derive a useful value for ΔT from this observation.

(5) AD 937 Feb 14 (Sun: mag. = 0.52): Pien

T'ien-fu reign period, 2nd year, 1st month, day *i-mao* [52]... When the Sun rose in the east it was 3 tenths eclipsed. It gradually (re-)emerged...

[*Chiu-wu-tai-shih*, chap. 139.]

RESULTS

LT of sunrise on Feb 14 = 6.50 h. For 0.30 of solar diameter to be obscured at sunrise (phase decreasing), ΔT = 2100 sec.

(6) AD 1042 Jun 20 (Sun: mag. = 0.41): Pien

Ch'ing-li reign period, 2nd year, 6th month, day *jen-shen* [9], the first day of the month. The Sun was eclipsed by 5 divisions. At 6 marks in the hour of *yu* (18.56 h) the eclipse was 2 divisions (i.e. 2 tenths) and then the Sun set...

[*Wen-hsien T'ung-k'ao*, chap. 283.]

RESULTS

LT of sunset on Jun 20 = 19.24 h. For 0.20 of solar diameter to be obscured at sunset (phase decreasing), ΔT = 1200 sec.

(7) AD 1042 Jul 5/6 (Moon: mag. = 1.11): Pien

Ch'ing-li reign period, 2nd year, 6th month, day *ting-hai* [24]. The Moon was eclipsed by 6 divisions (i.e. 6 tenths). Then it set eclipsed ...

[*Wen-hsien T'ung-k'ao*, chap. 285.]

The proportion of the lunar diameter covered at moonset will be taken as 0.60. From figure 3.23, negligible amendment is necessary at this particular magnitude.

RESULTS

LT of moonset on Jul 6 = 4.81 h. For 0.60 of lunar diameter to be obscured at moonset (phase increasing), ΔT = 700 sec.

(8) AD 1067 Mar 3/4 (Moon: mag. = 1.43): Pien

Chih-ping reign period, 4th year, 2nd month, day *chia-wu* [31]. The Moon was eclipsed... at 4 marks in the hour of *ch'ou*... It was more than 8 divisions (i.e. 8 tenths) at 6 marks in the hour of *ch'ou*... It set eclipsed...

[*Wen-hsien T'ung-k'ao*, chap. 285.]

The (adjusted) proportion of the lunar diameter covered at moonset will be taken as 0.83.

RESULTS

LT of moonset on Mar 4 = 6.14 h. For 0.83 of lunar diameter to be obscured at moonset (phase increasing), ΔT = 100 sec.

(9) AD 1068 Aug 15/16 (Moon: mag. = 0.38): Pien

> Hsi-ning reign period, first year, 7th month, day *i-yu* [22]. The Moon was eclipsed at 5 marks in the hour of *ch'ou*... It was eclipsed by $2\frac{1}{2}$ divisions (i.e. $2\frac{1}{2}$ tenths)... Then it set eclipsed...
>
> [*Wen-hsien T'ung-k'ao*, chap. 285.]

The (adjusted) proportion of the lunar diameter covered at moonset will be taken as 0.18.

RESULTS

LT of moonset on Aug 16 = 5.31 h. For 0.18 of lunar diameter to be obscured at moonset (phase increasing), ΔT = 950 sec.

(10) AD 1071 Jun 15/16 (Moon: mag. = 0.95): Pien

> Hsi-ning reign period, 4th year, 5th month, day *chi-hai* [36]. The Moon... rose eclipsed; it was not round but 6 divisions (i.e. tenths) (were covered)... After 5 marks it was restored.
>
> [*Wen-hsien T'ung-k'ao*, chap. 285.]

The (adjusted) proportion of the lunar diameter covered at moonrise will be taken as 0.60.

RESULTS

LT of moonrise on Jun 15 = 19.29 h. For 0.60 of lunar diameter to be obscured at moonrise (phase decreasing), ΔT = 1300 sec.

(11) AD 1071 Dec 9/10 (Moon: mag. = 0.43): Pien

> Hsi-ning reign period, 4th year, 11th month, day *ping-shen* [33]. The Moon was eclipsed... Not until 6 marks in the hour of *mao* was the eclipse at its maximum, reaching less than $4\frac{1}{2}$ divisions... Before it became bright it set eclipsed...
>
> [*Wen-hsien T'ung-k'ao*, chap. 285.]

RESULTS

The estimated phase (0.45: adjusted value 0.42) is so close to the computed magnitude that it is not possible to derive a useful value for ΔT from this observation. Even quite large changes in ΔT would produce little alteration in the computed phase at moonset.

(12) AD 1073 Oct 18/19 (Moon: mag. = 0.56): Pien

> Hsi-ning reign period, 6th year, 9th month, day *i-mao* [52]. The Moon was eclipsed... At 1 mark in the hour of *yin* it was at its maximum, reaching 6 divisions. Then it began to be restored. When it was 3 divisions (i.e. tenths) short of roundness, it set eclipsed...
>
> [*Wen-hsien T'ung-k'ao*, chap. 285.]

The (adjusted) proportion of the lunar diameter covered at moonset will be taken as 0.24.

RESULTS

LT of moonset on Oct 19 = 6.55 h. For 0.24 of lunar diameter to be obscured at moonset (phase decreasing), ΔT = 1850 sec.

(13) AD 1081 Nov 19/20 (mag. = 1.63): Pien

> Yuan-feng reign period, 4th year, 10th month, day *hsin-szu* [18]. The Moon... rose eclipsed by 7 divisions (i.e tenths)... After 2 marks it was restored.
>
> [*Wen-hsien T'ung-k'ao*, chap. 285.]

The (adjusted) proportion of the lunar diameter covered at moonrise will be taken as 0.72.

RESULTS

LT of moonrise on Nov 19 = 17.07 h. For 0.72 of lunar diameter to be obscured at moonrise (phase decreasing), ΔT = 100 sec.

Table 9.10 summarises the ΔT values obtained in this section.

The results in table 9.10 are plotted in figure 9.8. The range of individual values is best illustrated by the set of observations in the narrow interval between AD 1042 and 1081 (i.e. ΔT between 100 and 1850 sec). As comparison with figure 9.6 shows, the scatter is similar to that for contemporaneous timed data. This diagram also includes the few ΔT results obtained in section 9.6 from estimates of solar eclipse magnitude at greatest phase (see table 9.8).

9.10 Other records of lunar eclipses

At least six lunar eclipses are known to be alluded to on the oracle bones of the Shang dynasty. However, as yet there is no general agreement on the dates of these events, largely because the extant inscriptions are incomplete.

Xu *et al.* (1989) and Xu *et al.* (1995) give translations of all the known Shang records of lunar eclipses. From these I have selected the following

Table 9.10 ΔT values derived from eclipses in which the phase was estimated on the horizon.

Year[a]	Type[b]	Rise/set.[c]	Mag. est.[d]	ΔT (sec)[e]
+438	Moon	rise	$\frac{3}{4}$	5800
+503	Moon	rise	$\frac{12}{15}$	6000
+513	Moon	rise	$\frac{3}{15}$	5750
+937	Sun	rise	$\frac{3}{10}$	2100
1042	Sun	set	$\frac{2}{10}$	1200
1042	Moon	set	$\frac{6}{10}$	700
1067	Moon	set	$\frac{8}{10}$	100
1068	Moon	set	$\frac{2.5}{10}$	950
1071	Moon	rise	$\frac{6}{10}$	1300
1073	Moon	set	$\frac{3}{10}$	1850
1081	Moon	set	$\frac{7}{10}$	100

[a] Year.
[b] Type of eclipse (Sun or Moon).
[c] Whether the eclipsed object was rising or setting.
[d] Original estimate of phase.
[e] Computed value of ΔT.

two examples:

(i) The divination on the day *kuei-wei* [20] was performed by Cheng: 'Will there be no disaster in the next ten days?' On the third day *i-yu* [22], at night, an eclipse of the Moon was reported. In the 8th month.

(ii) Day *jen-shen* [9], at night, the Moon was eclipsed.

The first of these texts is the only surviving Shang record of an eclipse which gives both the cyclical day and the month. As many inscriptions of a non-astronomical nature reveal, it was the custom to place the lunar month at the end of a text. Unfortunately little is known about how the calendar was regulated at this early period or at what stage during each night the date was changed. Hence dating of this and other Shang eclipse observations is problematical. For instance, Tung Tso-pin (1945) derived a series of dates between 1362 and 1218 BC, while Dubs (1947) preferred an interval roughly 150 years later. Recently, on the assumption that all of the known records originate from around the same time, Pang *et al.* (1988b) deduced a set of dates between about 1100 and 1080 BC.

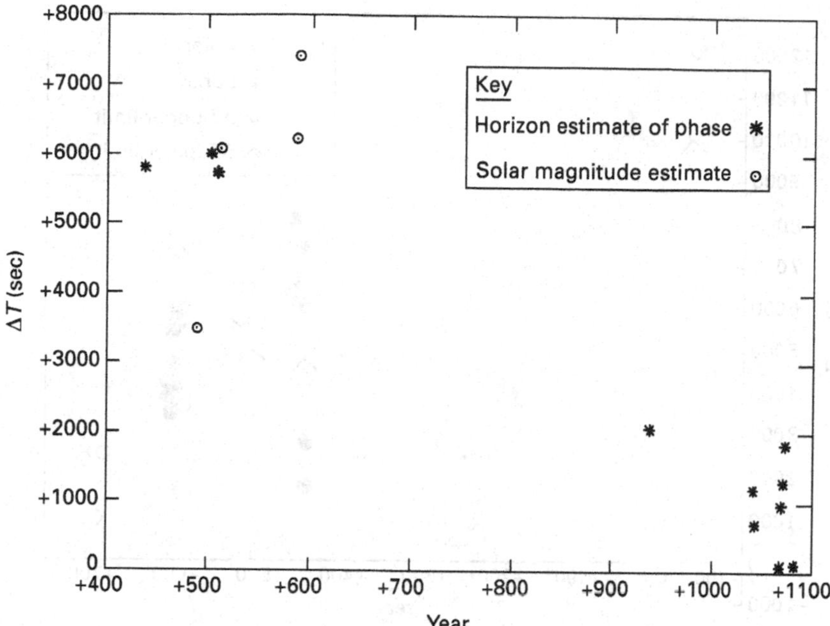

Fig. 9.8 ΔT values derived from estimates of eclipsed proportion of Sun and Moon when on horizon and solar eclipse magnitudes.

Since an *unambiguous* date has yet to be derived for any of the Shang lunar eclipse texts, and individual accounts give little or no information on the time of night when an observation was made, the various inscriptions are at present of negligible value in the study of the Earth's past rotation. Further research on this issue would seem very desirable.

Although there are numerous references to *total* lunar eclipses throughout Chinese history (mainly using the character *chi*), allusions to the colour of the Moon when completely obscured are very rare. The earliest identifiable observation dates from AD 499. At the total lunar eclipse of Sep 5/6 in that year, the Astrological Treatise (chap. 12) of the *Nan-ch'i-shu* noted that 'its colour was all red'. The same source, in describing an eclipse of the Moon alleged to occur on a date corresponding to AD 498 May 18 stated that 'its colour was red like blood'. There was no eclipse on or near the stated day and the true date cannot be isolated. No comparable descriptions are found in subsequent Chinese history until the sixteenth century. The allusion to blood in AD 498 is particularly interesting since numerous medieval European accounts of the totally eclipsed Moon make similar comparisons (see chapter 11).

Among the many occultations of stars and planets noted in Chinese history, only one (in AD 712) was said to have involved the eclipsed Moon. This is recorded in the Astrological Treatise (chap. 36) of the *Chiu-t'ang-*

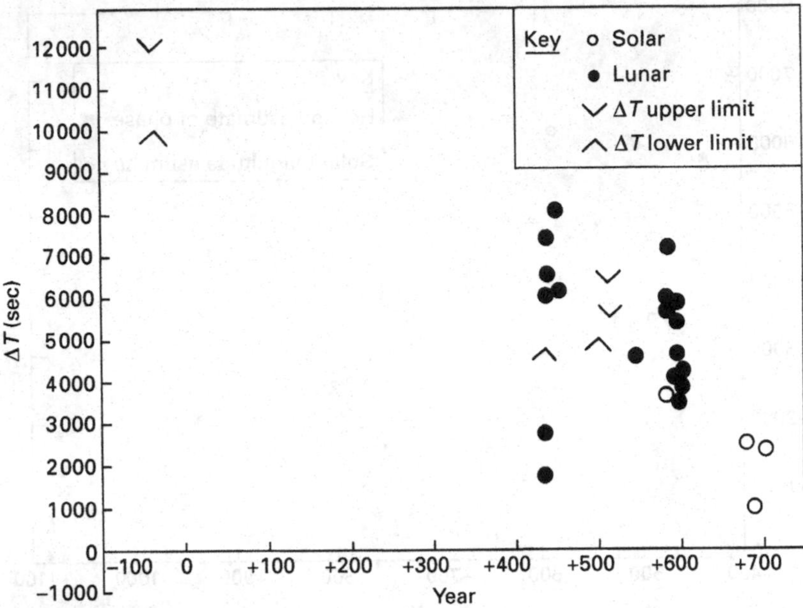

Fig. 9.9 ΔT results and more critical limits obtained in this chapter from Chinese observations of solar and lunar eclipses between the dates −100 and +750.

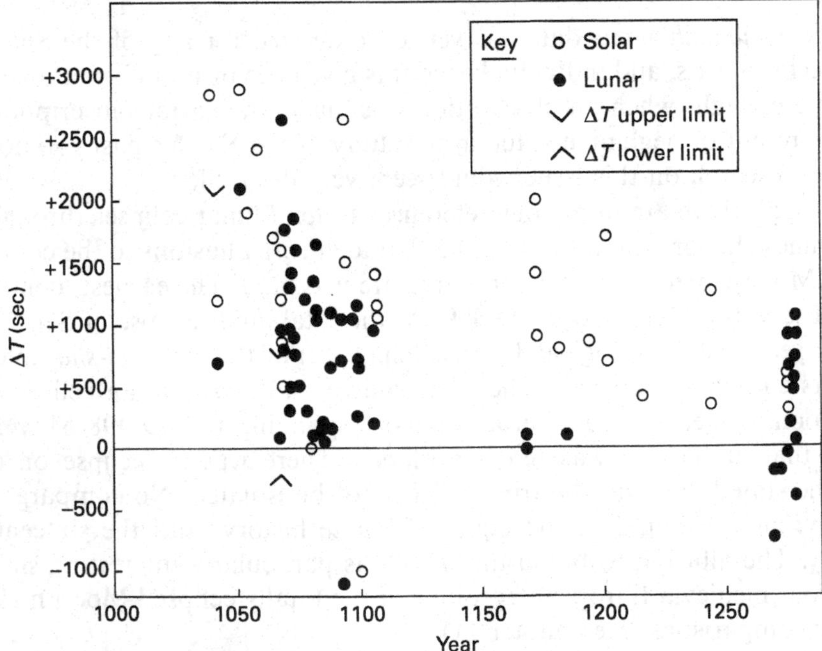

Fig. 9.10 ΔT results and more critical limits obtained in this chapter from Chinese observations of solar and lunar eclipses between +1000 and +1300.

shu. The following observation is described on a date corresponding to AD 712 Sep 19/20:

> Hsien-t'ien reign period, first year, 8th month, 14th day. At night the Moon was eclipsed and it was complete. A star entered (*ju*) within the spirit (*p'o*) of the Moon.

Evidently the 'spirit' of the Moon was its dimly lit disk during the total phase (mag. = 1.39). It would seem unlikely that the 'star' was a meteor; there was a special term for such a rapidly moving object – *liu-hsing* ('flowing star'). The term *ju* ('to enter') does not necessarily imply any movement of the occulted object, only apparent motion relative to the Moon. This was one of several standard terms to describe the occultation of a star or planet by the Moon.

The longitude and latitude of the Moon (corrected for parallax) at the middle of the total phase were respectively 356.4 deg and −0.2 deg. Correcting for precession and converting to RA and dec, the approximate celestial co-ordinates at the epoch J2000 are 13.24 h, +5.5 deg. There was no planet in the vicinity of the eclipsed Moon on the night in question, although the rather faint star σ Vir (mag. +4.8) was very close. This was presumably the star alluded to in the text.

9.11 Conclusion

Chinese records are obviously a prolific source of valuable observations of both solar and lunar eclipses. The various ΔT results and more critical limits obtained in this chapter (apart from the azimuth data and solar magnitudes which are discarded) are plotted in figures 9.9 and 9.10. Figure 9.9 covers the date range from −100 to +750 and figure 9.10 from +1000 to +1300. There are no useful observations in the intervening interval (actually from +703 to +936).

10

Records of eclipses in ancient European history

10.1 Introduction

Compared with the careful observations of similar age which are recorded on the Late Babylonian astronomical texts, many of the eclipse records in ancient Greek and Roman history come as something of an anticlimax. Although numerous descriptions of both solar and lunar obscurations are preserved in these sources, commencing as early as the seventh century BC, most accounts are too vague to be suitable for investigating the Earth's past rotation. The majority of writings which mention eclipses are literary rather than technical, and include historical works, biographies and even poems. Late nineteenth and early twentieth century authors such as Nevill (e.g. 1906a), Ginzel (1899), Cowell (e.g. 1906b), and Fotheringham (e.g. 1920b) paid much attention to these observations. However, this was largely because little other material was available at the time.

As noted in chapter 3, the mainstay of investigations made around the beginning of the present century was undoubtedly untimed observations of large solar eclipses. Attempts to date the various records and identify the places of observation proved an almost irresistible challenge to Fotheringham and his contemporaries, and much effort was expended in these pursuits. Considerable interest was also shown in using ancient eclipses to date historical events. (For a recent summary, see Stephenson, 1993.)

There seems little doubt that many records in Greek and Roman history relate to eclipses which were either total or fell not much short of this phase. Thus Fotheringham (1908) noted that although total obscurations of the Sun are slightly less frequent than their annular counterparts on the Earth's surface as a whole, the majority of the solar eclipses mentioned in the Greek and Roman classics were in fact generally total. The implications are that where these events were observed, most exceeded the magnitude of a typical central annular eclipse (i.e. greater than about 0.95). However, such a statistical conclusion, although interesting, is insufficient to allow useful information on ΔT to be obtained from individual records.

The most valuable observations for this purpose are those which either affirm that the central phase (whether annular or total) was seen or expressly deny this – by stating that the Sun was reduced to a crescent.

Careful timings of the contacts for only a single *solar* eclipse – dating from AD 364 – are extant from ancient Europe, and indeed there are no further examples from this part of the world until as late as the fourteenth century (see chapter 11). Although lunar eclipses are also recorded in abundance in ancient European literature, most are of no value in the determination of ΔT since they merely note the occurrence of an eclipse. Several Greek timings dating from between 201 BC and AD 125, and also one instance of the Moon rising partially obscured, are reported in a single source: Ptolemy's *Almagest*. In addition to this material, an observation of the Moon rising eclipsed in 331 BC quoted by Pliny is also worth considering.

In this chapter, ancient European records of solar and lunar eclipses which (at least in principle) are of value for studying Earth's past rotation will be considered under the following categories:

(i) Solar eclipses analysed by Fotheringham in his 1920b paper (section 10.2).

(ii) Other descriptions of large solar eclipses in which such details as darkness or the visibility of stars is noted (10.3).

(iii) Timed lunar eclipses reported in the *Almagest* (10.4).

(iv) The moonrise eclipse of 331 BC (10.5).

The geographical co-ordinates of the principal places mentioned in this chapter are listed in table 10.1.

Figure 10.1 shows the remarkably self-consistent ΔT results obtained from the Late Babylonian timings which were assigned double weight in chapters 5 and 6. These observations are mainly of lunar eclipses, along with a few solar events. The abcissa is graduated in negative and positive years, rather than their BC equivalents; on the scale of the diagram, any differences are negligible. Also included in this diagram is the range in ΔT set by the highly reliable observation of a total solar eclipse in 136 BC. The linear fit to the timed data (a useful first approximation) has the equation

$$\Delta T = 9520 - 1612\tau, \qquad (10.1)$$

where τ is measured in centuries from the epoch 0 (i.e. 1 BC). This equation is denoted by a dashed line in figure 10.1. Where appropriate, both this diagram and equation (10.1) will be utilised to help establish dates of the ancient European eclipses discussed in this chapter.

Table 10.1 Principal sites at which eclipses were recorded in ancient Europe.

Place	Latitude (deg)	Longitude (deg)
Alexandria	+31.22	−29.92
Athens	+37.98	−23.73
Chaeroneia	+38.40	−22.90
Palermo	+38.13	−13.30
Paros	+37.07	−25.10
Rhodes	+36.43	−28.23
Rome	+41.90	−12.50
Sparta	+37.08	−22.42
Syracuse	+37.07	−15.30
Thasos	+40.77	−24.70
Thebes	+38.32	−23.32

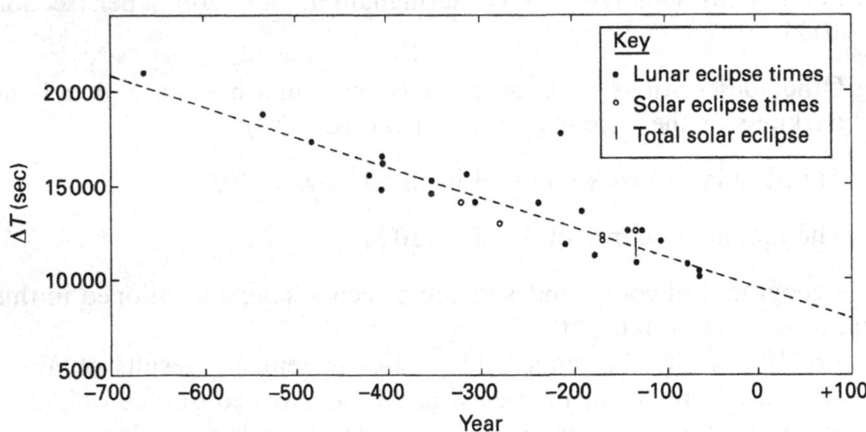

Fig. 10.1 Best fitting straight line to the ΔT results obtained from the Late Babylonian timings which were assigned double weight in chapters 5 and 6.

10.2 Solar eclipses analysed by Fotheringham

Although the series of observations which Fotheringham discussed in his 1920b paper is probably no more reliable than any of the other ancient data investigated in this chapter, it has attracted such attention among geophysicists – notably Munk and MacDonald (1960, pp. 188–191) and Dicke (1966) – that it seems appropriate to consider it first.

Fotheringham analysed ten observations of either real or supposed central eclipses of the Sun together with timings of the various phases of a

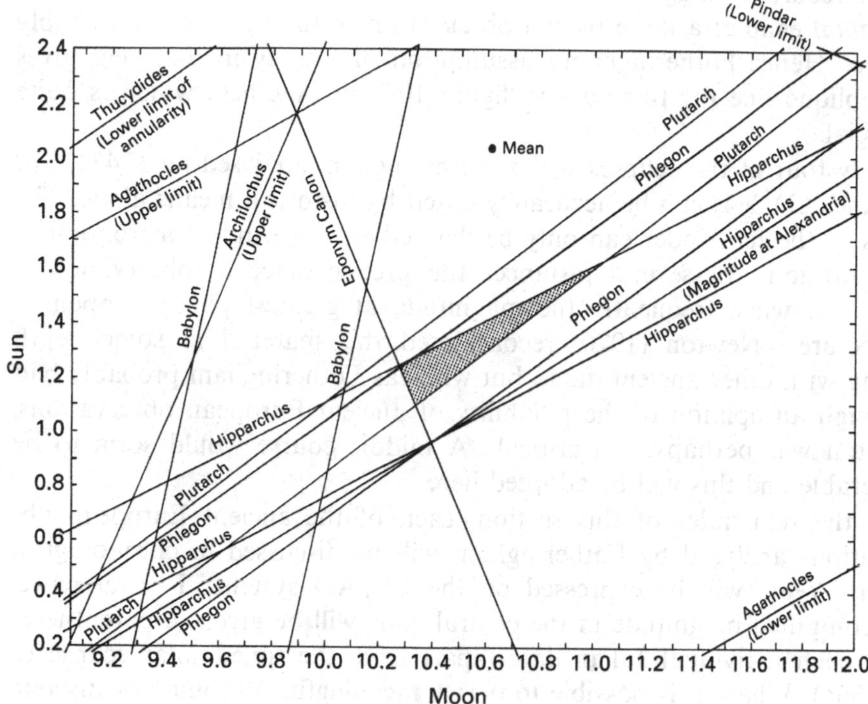

Fig. 10.2 Fotheringham's (1920b) diagram showing the solar and lunar accelerations on UT derived from ancient eclipses.

single partial obscuration. Most of these are recorded in Greek literature, and will be considered here. However, the two earliest observations – which have already been considered in chapters 4 and 5 – were from Babylonia (*c*. 1000 BC: uncertain identification) and Assyria (763 BC).

From each of the various eclipse records which he analysed, Fotheringham derived linear equations relating the lunar and solar accelerations (C^* and c') on UT and plotted them in a diagram which has since become well known to geophysicists. A copy of this diagram is shown in figure 10.2. (For a discussion of the relation between the accelerations of the Moon and Sun on a UT framework and their relation to changes in the length of the day on TT, see chapter 1.)

Fotheringham selected a small triangular area of solution space, bounded by the limits obtained from three eclipses (−762, −128 and +71: i.e. 763 BC, 129 BC and AD 71) as the most likely result. This area is shown shaded in figure 10.2. However, Fotheringham's technique was heavily dependent on the reliability of each individual observation which he used. Rejection of even a single observation would have considerably enlarged his preferred area. As discussed in chapter 4, there is nothing in the record to suggest that the Assyrian observation in 763 BC relates to a *total* eclipse; a large partial obscuration of the Sun would probably suffice. Hence Fotheringham's assumption of totality in that year gives the oblique line for this date in figure 10.2 an exactness which is quite artificial.

Only four of the eclipses which Fotheringham analysed (763, 431, 310 BC and AD 364) can be accurately dated by historical means alone. The dates of the remainder can only be derived with the aid of astronomical computation. In several instances the precise place of observation is in doubt, while frequently the magnitude at greatest phase is open to conjecture. Newton (1970) reconsidered this material in some depth (along with other ancient data), but whereas Fotheringham probably had too high an opinion of the reliability of ancient European observations, Newton was perhaps over-critical. A middle course would seem to be preferable and this will be adopted here.

In the remainder of this section, each of the ancient European observations analysed by Fotheringham will be discussed in chronological order. Years will be expressed on the BC/AD system. For reference, the computed magnitude in the central zone will be given in parentheses after each individual Julian date (apart from the timed solar eclipse of AD 364). Where it is possible to derive meaningful ΔT limits or discrete values, these will be quoted towards the end of each entry.

(1) BC 648 Apr 6 ? (total, 1.07): Paros or Thasos?

Archilochus, one of the earliest Greek poets after Homer and Hesiod, makes a clear reference to a very large solar eclipse in one of his poems:

> Nothing can be surprising any more or impossible or miraculous, now that Zeus, father of the Olympians has made night out of noonday, hiding the bright sunlight (*apokruphas phaos heliou lampontos*), and ... fear has come upon mankind. After this, men can believe anything, expect anything. Don't any of you be surprised in future if land beasts change places with dolphins and go to live in their salty pastures, and get to like the sounding waves of the sea more than the land, while the dolphins prefer the mountains.
>
> [Archilochus, fragment 122; trans. Barron and Easterling (1985, p. 127).]

A more accurate rendering of the phrase *apokruphas phaos heliou lampontos*, here translated as 'hiding the bright sunlight', would be 'hiding the light of the gleaming Sun' – much as in the quotation cited by Fotheringham. Combining this expression with the description of the loss of daylight 'night out of noonday' strongly suggests that the eclipse was either total or fell only slightly short of this phase. Evidently the phenomenon left a singularly profound impression on its beholders, as illustrated by Archilochus' subsequent use of hyperbole.

Fotheringham regarded the date BC 648 as the only possible one and he assumed that the eclipse was witnessed at either Paros or Thasos – islands respectively in the extreme south and north of the Aegean – where Archilochus is known to have spent most of his life. However, both the date and place of observation require further consideration.

Although the dates of both the birth and death of Archilochus are unknown, two of his poems provide useful chronological information (see Rankin, 1977, chap. II). In one of these, Archilochus mentions King Gyges of Lydia (*c.* 687–652 BC) and in the second he alludes to the destruction of the city of Magnesia by Treres (*c.* 655–650 BC). That Archilochus was a contemporary of King Gyges (as asserted by Herodotus, I, 12) is clear from the context of his poem which begins:

> I have no interest in the property of golden Gyges. Envy has never taken hold of me...

It is known that Archilochus was born on Paros. He later moved to Thasos where he served as a soldier (possibly a mercenary). After occasional travels elsewhere in the region (including Sparta) he eventually returned to Paros, where he was later killed helping to defend the island against invasion. Archilochus often mentions Paros or Thasos by name in his poetry. The locations of both islands (and also Sparta) are shown in figure 10.3.

Newton (1970, p. 92) regarded the solar phenomenon to which Archilochus refers as merely a 'literary eclipse'. However, that the poet himself saw the eclipse is suggested by the following remarks of Barron and Easterling (1985, pp. 117–119):

> (Archilochus) is the first Greek writer to take his material from what he claims to be his own experience and emotions, rather than from the stock of traditions...It would be absurd to claim that a poet composing songs for performance in a small community in which everyone knew everyone else would not exploit the audience of that society and its relationships.

Based on the above considerations, I shall assume that the eclipse was either total or nearly so at either Paros or Thasos. Reference to the global eclipse maps of von Oppolzer (1887) indicates that between 700 and 610 BC only the following solar obscurations could have been total in this region: BC 691 Jul 28, 657 Apr 15, 648 Apr 6, 646 Sep 8 and 637 Aug 29. Several annular eclipses would also have beeen visible in the Aegean during this same period (in 689, 662, 661, 651, 650, 641, 635 and 633 BC) but in each case the magnitude within the central zone was between about

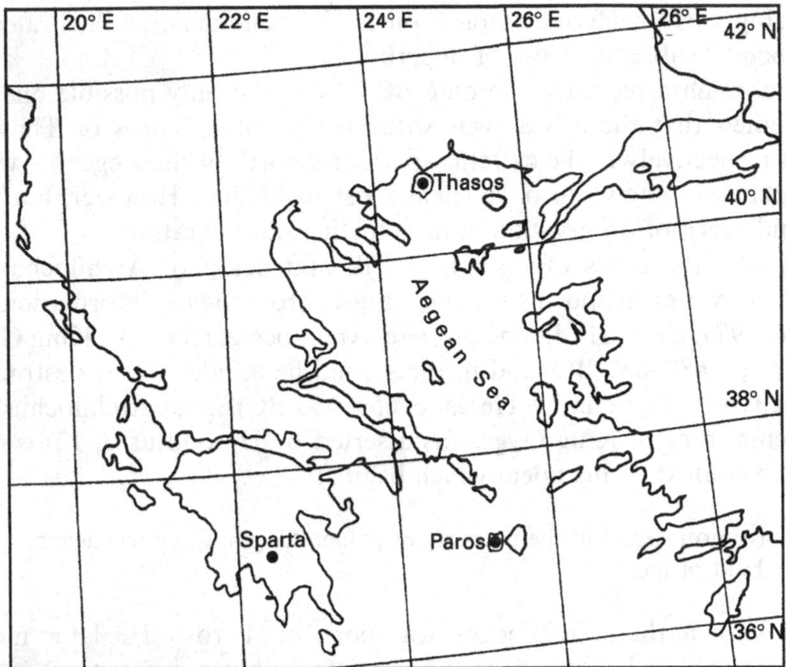

Fig. 10.3 The locations of the islands of Paros and Thasos (and also Sparta), where Archilochus is known to have lived.

0.93 and 0.96. None of these events could have caused the impressive loss of daylight described so vividly by Archilochus.

On historical grounds, it is probably impossible to distinguish between the five dates of total eclipses noted above so that astronomical computation must be resorted to – see table 10.2.

None of the eclipses listed in table 10.2 could have been total around *noon* on either island – contrary to what Archilochus implies – but in each case the computed time of totality was within 2 or 3 hours of midday, the Sun being high in the sky (altitude between about 40 and 50 deg). In principle, any of these circumstances would probably satisfy the poetic reference. However, not all of the indicated ΔT ranges are plausible. For the eclipses of 691, 646 and 637 the appropriate limits are in marked discord with the results derived from roughly contemporaneous Babylonian observations (see figure 10.1). Computations for these same three eclipses based on equation (10.1) effectively eliminate them from further consideration. In each case, small magnitudes are indicated at both Paros and Thasos (between 0.62 and 0.77), while all three local times of maximum phase would be late in the afternoon (after about 17.0 h) with low solar altitude (10 or 20 deg). The calculated circumstances elsewhere in the Aegean would be very similar. Almost certainly, a choice must rest between 657 and 648 BC.

Table 10.2 ΔT ranges necessary for seventh century BC eclipses to attain totality at either Paros or Thasos.

Year BC[a]	Mag.[b]	Place[c]	ΔT Range (sec) LL[d]	UL[e]	LT (h)[f]	Alt. (deg)[g]
691	1.06	Paros	25 900	26 800	15.2	45
		Thasos	26 900	27 850	14.9	49
657	1.07	Paros	17 050	18 000	10.3	52
		Thasos	16 050	17 150	10.7	52
648	1.07	Paros	19 450	21 050	9.3	41
		Thasos	17 800	19 350	10.1	45
646	1.04	Paros	27 150	28 050	15.2	38
		Thasos	28 800	29 700	14.5	44
637	1.05	Paros	25 100	25 950	15.0	42
		Thasos	25 850	26 700	14.7	45

[a] Year BC.
[b] Magnitude in the central zone.
[c] Assumed place of observation (either Paros or Thasos).
[d] Lower limit to ΔT in sec (LL) to render the eclipse total at the selected place.
[e] The corresponding upper limit to ΔT (UL).
[f] Computed local time of greatest phase to the nearest 0.1 h, as deduced from the mean ΔT of columns 4 and 5.
[g] Calculated solar altitude in deg at this moment.

In table 10.3 are listed the magnitude, local time of greatest phase and altitude at Paros and Thasos for the eclipses of 657 and 648 BC, calculated using equation (10.1) (ΔT values respectively 20 100 and 19 950 sec).

From table 10.3, it would appear that 648 BC, the date favoured by Fotheringham, is the better choice for the eclipse of Archilochus – whether the place of observation was Paros or Thasos (or elsewhere in the region). However, a date of 657 BC cannot be entirely eliminated; as is evident from figure 10.1, data points at this early period are too sparse to rule out the ΔT range necessary to achieve totality at either Paros or Thasos ($16\,050 < \Delta T < 18\,000$ sec).

The poem can thus be confidently dated to within a decade (i.e. between 657 and 648 BC), but possibly no better than this. Although this result is of some importance to the historian, because of the uncertain date the record is of little value in the investigation of the Earth's past rotation.

Table 10.3 Local circumstances for the eclipses of 657 and 648 BC at Paros and Thasos computed using equation (10.1).

Year BC	Place	Mag.	LT (h)	Alt. (deg)
657	Paros	0.94	9.8	47
	Thasos	0.88	9.8	46
648	Paros	0.97	9.9	46
	Thasos	1.07	9.9	44

(2) BC 585 May 28 ? (total, 1.07): Asia Minor

Herodotus, who has been aptly described as the 'Father of History', alludes to the interruption of a battle in Asia Minor by a marked loss of daylight:

> After this, seeing that Alyattes would not give up the Scythians to Cyraxes at his demand, there was war between the Lydians and the Medes five years... They were still warring with equal success, when it chanced, at an encounter which happened in the sixth year, that during the battle the day was turned to night. Thales of Miletus had foretold this loss of daylight to the Ionians, fixing it within the year in which the change did indeed happen. So when the Lydians and the Medes saw the day turned to night, they ceased from fighting, and both were the more zealous to make peace. Those who reconciled them were Syennesis the Cicilian and Labnetus the Babylonian....
>
> [Herodotus, I, 74; trans. Godley (1975, vol. I, pp. 91–93).]

The first century AD writer Pliny makes no mention of the battle, but in alluding to Thales he makes it clear that the celestial phenomenon mentioned by Herodotus was understood to be a solar eclipse. Pliny also specifies the year when it occurred:

> The original discovery (of the cause of eclipses) was made in Greece by Thales of Miletus, who in the fourth year of the 48th Olympiad (585/4 BC) foretold the eclipse of the Sun that occurred in the reign of Alyattes, in the 170th year after the foundation of Rome (584/3 BC).
>
> [Pliny, *Naturalis Historia*, II, 53; trans. Rackham (1938, vol. I, p. 203).]

Although Herodotus (*c.* 484–425 BC) wrote some 150 years after the events which he describes, his account of the various historical events occurring at this time is fairly detailed – for example he is able to give the names of the intercessors who made peace between the two sides. However, Herodotus also records a similar fall of darkness a century later – in his own time – which cannot be explained by an eclipse. A translation of Herodotus' description of this event is as follows:

> The army (of Xerxes) then wintered, and at the beginning of spring was ready and set forth from Sardis to march to Abydos. When they had set forth, the Sun left his place in the heavens and was unseen, albeit the sky was without clouds and very clear, and day was turned into night.
>
> [Herodotus, VII, 37; trans. Godley, (1971, vol. II, p. 351).]

On this latter occasion, no suitable eclipse can be identified. The date when Xerxes left Sardis is fixed as 480 BC by Herodotus' reference to the Olympic Games being held in the summer of the same year (VII, 206). However, the only large eclipse visible in Asia Minor around this time was that of BC 478 Feb 17 and this was only annular – as little as 0.90 of the Sun being covered in the central zone. Hence either the darkness was of atmospheric origin or Herodotus has mistakenly synchronised two quite separate events.

It thus seems that Herodotus' account of the battle between the Lydians and the Medes being interrupted by the onset of darkness by day may also be of doubtful reliability, but further discussion seems worthwhile.

As noted above, the year 170 AUC given by Pliny corresponds to 584/3 BC Allowing some uncertainty in the date, Fotheringham selected the eclipse of BC 585 May 28. Judging from the charts of von Oppolzer (1887), no other eclipse could have turned the day into night in Asia Minor – where the battle occurred – between about 602 and 558 BC, and this is confirmed by detailed investigation using equation (10.1). Two annular eclipses during this period (588 and 581 BC) were probably fairly large in Asia Minor but in each central zone the Moon would cover no more than 0.96 of the Sun. Hence these events could scarcely produce the effects described by Herodotus.

The calculated date of 585 BC is remarkably close to that indicated by Pliny. However, the precise place where the battle was fought is unknown; as Fotheringham stressed, it could have been almost anywhere in Asia Minor. Because of the uncertainty in the interpretation of the record and the place of observation, it would seem hardly feasible to use the observation in the investigation of ΔT.

Whether the eclipse was indeed predicted by the Greek astronomer Thales of Miletus (who died at an advanced age in 548 BC) may well be doubted. It appears that the first person to give the true explanation of eclipses was Anaxagoras (500 – 428 BC) – rather than Thales (Heath, 1932, pp. xxiii and 27). At this early period, Babylonian astronomers, although probably furnished with centuries of past observations, attained very poor success in anticipating eclipses for a given location (see chapter 3). To quote Neugebauer (1975, p. 604),

> In the early days of classical studies one did not assume that in the sixth century BC a Greek philosopher had at his disposal the astronomical and

mathematical tools necessary to predict a solar eclipse. But then one could invoke the astronomy of the 'Chaldeans' from whom Thales could have received whatever information was required. This hazy but convenient theory collapsed in view of the present knowledge about the chronology of Babylonian astronomy in general and the lunar theory in particular. It is now evident that even three centuries after Thales no solar eclipse could be predicted to be visible in Asia Minor – in fact not even for Babylon. There remains another vague hypothesis: the prediction by means of cycles (if need be again available upon request from Babylon). But unfortunately there exists no manageable cycle of solar eclipses visible at a given locality...

A recent revival of the idea that Thales indeed predicted an eclipse – inferring the most likely date as BC 582 Sep 21 – is due to Panchenko (1994). However, it should be emphasised that the track of totality on this occasion did not reach further north than latitude +34.0 deg at any point on the Earth's surface. Thus even on the southern coast of Asia Minor (lat. = +36.0), the magnitude cannot have been greater than 0.93, regardless of the value of ΔT adopted. In fact, using equation (10.1) (ΔT = 18 950 sec), the magnitude at any point in the peninsula cannot have exceeded 0.85.

In brief, the assertions by Herodotus and Pliny probably owe their origin to the various legends which accumulated around the personality of Thales.

(3) BC 463 Apr 30 ? (total, 1.06): Thebes

In the odes of Pindar, who is generally regarded as the greatest lyric poet of ancient Greece, there are two allusions to the effects of a very large solar eclipse:

(i) Beam of the Sun! O thou that seest from afar, what will thou be devising? O mother of mine eyes! O star supreme, reft from (*kleptomenon*) us in the daytime (*en amerai*)! Why hast thou perplexed the power of man and the way of wisdom, by rushing forth on a darksome track? Art thou bringing on us some new and strange disaster? Yet by Zeus, I implore thee, thou swift driver of steeds! Do thou, O queen! Change this world-wide portent into some painless blessing for Thebes ...

But art thou bringing a sign of some war, or wasting of produce, or an unspeakably violent snow-storm, or fatal faction, or again, some overflowing of the sea on the plain, or frost to bind the earth, or heat of the south wind streaming with raging rain? Or wilt thou, by deluging the land, cause the race of men to begin anew? I in no wise lament whate'er I shall suffer with the rest!

[Pindar, *Paean*, ix; trans. Sandys (1978, pp. 547–549).]

(ii) God can cause unsullied light to spring out of black night. He can also shroud in a dark cloud of gloom the pure light of day.

[Pindar, fragment 142; trans. Sandys (1978, p. 577).]

Both of these texts may refer to one and the same event, but the first, entitled 'To the Thebans', is particularly important since it makes specific reference to the city which was both the birthplace of Pindar and his main domicile. In an article specially devoted to this eclipse (1920c), Fotheringham gives a fuller discussion than in his better known 1920b paper. However, in both works he considered only the eclipse of BC 463 Apr 30. Whether there were other large eclipses visible in Greece during Pindar's lifetime needs to be carefully examined.

Pindar was born in a village adjacent to Thebes in either 522 or 518 BC, the latter date being more likely (Race, 1986, p. 1). Since he is believed to have attained the age of 80, the most probable year of his death is 438 BC. It is possible to date many of his poems with confidence, not least because some are written in honour of the victors at various Olympic Games. Unfortunately Pindar's ninth paean (excerpted above) does not come into this category. Scholars are agreed that he wrote his earliest ode in 498 BC (while still a young man) and his last in 446 (Race, *ibid.*).

Bowra (1964, pp. 83–4 and pp. 378–9) had no doubt that Pindar saw the eclipse, remarking that there is a sense of urgency in his paean. It evidently left a deep impression on him, and although he seems to have heard that the phenomenon was visible at other places too ('worldwide'), he regarded it as a major portent for Thebes. In describing the apparent theft of the Sun and the resulting darkness, Pindar clearly refers to either a complete solar eclipse or one in which the Sun all but disappeared. Although the former alternative would seem more in keeping with his description and subsequent anxiety, Fotheringham (1920b) pointed out that use of the present tense of the verb translated as 'reft from' (i.e. *kleptomenon*) is consistent with incomplete action.

From von Oppolzer (1887), only four eclipses could have been large in Greece between 506 BC and the death of Pindar – a conclusion supported by computation using equation (10.1). The dates are: BC 493 Nov 24 (total), 488 Sep 01 (annular), 478 Feb 17 (annular) and 463 Apr 30 (total). All were thus during his active lifetime. Since any attempt to judge his degree of maturity when he produced *Paean* ix is subjective, it is not possible to separate these four dates on literary grounds. (I am grateful to Professor C. Cary of Royal Holloway College, Egham for helpful discussion on this matter.) However, astronomical techniques prove more useful.

The eclipse of 478 BC can be fairly easily eliminated from further consideration. On this occasion, no more than 0.89 of the solar diameter would have been obscured (even in the zone of annularity) so that the event could not have caused darkness; it might well have passed unnoticed. For totality at Thebes in 493 BC, a value of ΔT in the range $22\,150 < \Delta T < 22\,850$ sec would be required. However, this narrow zone is in marked conflict with several of the roughly contemporary Babylonian limits in figure 10.1. Calculation using equation (10.1) ($\Delta T = 17\,450$ sec) indicates a magnitude of only 0.82 at Thebes exactly at noon.

Both the eclipses of 488 and 463 BC are more promising candidates. Although the former event was also annular, the Moon was sufficiently close to the Earth to cover almost 0.99 of the solar diameter. Values of ΔT which render the eclipse central at Thebes ($17\,900 < \Delta T < 18\,450$ sec) are quite plausible, as reference to figure 10.1 shows. However, they yield a time of greatest phase just at sunrise (local time 5.4 h). Although the darkness would probably be heightened under these circumstances, it is debatable whether a sunrise eclipse would be described as 'in the daytime' (*en amerai*). In 463 BC, the ΔT limits required to produce totality at Thebes ($18\,100 < \Delta T < 20\,600$ sec) are also fairly reasonable. Here, the corresponding local time would be 13.5 h with the solar altitude 58 deg – very much a daytime event.

From the above arguments, it seems highly likely that the eclipse of Pindar is that of 463 BC, but since a date of 488 BC cannot be entirely eliminated, the observation recorded by Pindar cannot be effectively utilised to set limits to the value of ΔT.

(4) BC 431 Aug 3 (annular, 0.98): Athens

In his *History of the Peloponnesian War*, the contemporary Greek historian Thucydides notes three eclipses: two of the Sun and one of the Moon. Only the first solar obscuration appears to have been unusually large, but this is especially interesting since the account is the earliest in European history to mention the visibility of stars during an eclipse. Among the records of other civilisations, only a single Chinese observation, dating from 444 BC, makes an earlier allusion to stars seen under these circumstances (see chapter 8). Thucydides' account may be translated as follows:

> The same summer, at the beginning of a new lunar month (the only time by the way at which it appears possible), the Sun was eclipsed after noon. After it had assumed the form of a crescent (*menoeides*), and some of the stars had come out, it returned to its natural shape.
>
> [Thucydides, II, 28; trans. Livingstone (1949, p. 109).]

This same eclipse is probably alluded to by both Cicero (*De Republica*, I, 16) and Plutarch (*Life of Pericles*, XXXV) in a story concerning Pericles, a pupil of Anaxagoras. It is related that a solar eclipse occurred during the Peloponnesian War, bringing on darkness and creating fear. However, using his cloak, Pericles demonstrated the cause of the eclipse, thus dispelling the alarm. The details of the tale as given by Cicero and Plutarch differ considerably, but both writers lived fully 500 years after the event. According to Plutarch the story was told in the schools of philosophy. Only the contemporary account by Thucydides merits consideration here.

It is well established that the Peloponnesian War between Athens and Sparta began in 431 BC and lasted for 27 years. Thucydides (II, 1) notes that the opening battle occurred in the spring, towards the end of the year when Pythodorus was archon (chief magistrate) at Athens. Complete lists of the Athenian archons are preserved from 481 to 292 BC (Bickerman, 1980, pp. 138–9) and these indicate that Pythodorus was archon in 432/1 BC. The eclipse itself occurred only a few months after the start of the war, and thus the date must be BC 431 Aug 3. This was the largest eclipse visible in Greece for several years around this time.

Thucydides was an Athenian. Born probably in the early 450s, he assumed command of a section of the Athenian forces in 424 BC, the eighth year of the war (IV, 104). The other two eclipses which he mentions are of little *astronomical* value, but the solar obscuration acts as a useful chronological aid. Occurring in the summer of the eighth year, this is described by Thucydides (IV, 52) as 'small'. It may be confidently dated as BC 424 Mar 21. Using equation (10.1) ($\Delta T = 16\,350$ sec), the computed magnitude at Athens was 0.72. Only one other solar eclipse was visible in Greece over a period of ten years beginning in 430 BC This took place on 426 Nov 4, during the *tenth* year of the war, and could scarcely be designated a 'summer' event. (NB it was Thucydides' practice to divide each year into two conventionalised seasons, 'summer' and 'winter', each lasting about six months.) Finally, the lunar eclipse was said to happen in the summer of the 19th year (VII, 50). Of the two lunar eclipses in BC 413 (Mar 4 and Aug 27), the latter better fits the events discussed at the time.

As in the record from the first year of the war, Thucydides notes that the solar eclipse in the eighth year occurred at the time of new Moon, while he remarks that the lunar obscuration happened at full Moon. This may illustrate his knowledge of the true cause of eclipses, which had only recently been explained by the Athenian Anaxagoras (500–428 BC).

It seems quite likely that Thucydides saw the eclipse of 431 BC himself, although this cannot be proved. He seems to have had a special interest in such phenomena, remarking (I, 23) that solar eclipses were unusually numerous during the Peloponnesian War. In 431 BC, Thucydides was presumably in his home city of Athens. He makes no mention of any travels from there until the eighth year of the war (IV, 104), while in the year after the eclipse he tells us (II, 48) that he caught the plague in Athens. We know (IV, 105) that Thucydides had the right of working gold mines in Thrace (lat. *c.* +41 deg, long. *c.* −26 deg). However, there is nothing in his text to indicate that he was there when the eclipse occurred.

For values of ΔT which render the eclipse of 431 BC large in Greece, greatest phase would occur in the late afternoon, and thus long after midday. However, Thucydides probably only gave a very general indication of the time of day. Applying equation (10.1) (ΔT = 16 450 sec), the eclipse reached its maximum of 0.86 at 17.4 h, the solar altitude then being 19 deg. In Thrace, the circumstances would be very similar, although the computed magnitude would be rather greater (0.91). Thucydides describes only a very large partial eclipse (the term *menoeides* means 'crescent-shaped'), suggesting that the ring phase was not witnessed.

Even if the eclipse had been central (magnitude 0.98), Thucydides' reference to the visibility of several stars is difficult to explain. Although Venus (mag. −3.4), 19 deg to the east of the Sun, would no doubt have been fairly prominent, no other planet or star should have been detectable. In the late afternoon, both Jupiter (41 deg to the west of the Sun) and Sirius (62 deg to the south-west) would be below the horizon. Mercury, fairly close to Venus, would be faint (mag. +1.3). As Fotheringham noted, after Venus the brightest star or planet above the horizon would be Vega (mag. +0.4). Perhaps some allowance for exaggeration should be made here.

RESULTS

For a partial eclipse at Athens, $\Delta T < 11\,800$ or $> 12\,650$ sec.

NB only a narrow zone of avoidance is thus indicated. In Thrace the corresponding ΔT limits are $\Delta T < 14\,100$ or $> 14\,750$ sec, but assumption of an observation from there is purely speculative.

(5) BC 310 Aug 15 (total, 1.05): Sicily

In his *Library of History*, the first century BC historian Diodorus Siculus (i.e. Diodorus of Sicily) records two eclipses. The earlier of these (364 BC) is reported without any details, but the description of the later event is remarkably vivid. This occurred after the tyrant Agathocles, with a fleet of sixty ships, had made his escape from a blockade at Syracuse harbour by the Carthaginians.

> Agathocles, who was already at the point of being overtaken and surrounded, gained unhoped for safety as night closed in. On the next day there occurred such an eclipse of the Sun that utter darkness set in and the stars were seen everywhere; wherefore Agathocles' men, believing that the prodigy portended misfortune for them, fell into even greater anxiety about the future. After they had sailed for six days and the same number of nights, just as day was breaking, the fleet of the Carthaginians was unexpectedly seen not far away... When Libya came into sight....
>
> [Diodorus, XX, 5–6; trans. Geer (1954, vol. X, pp. 155–157).]

Diodorus noted (XX, 3) that the above events occurred during the archonship of Hieromemnon in Athens, corresponding to 310/09 BC (Bickerman, 1980, p. 139). Hence there can be no question that the correct date is BC 310 Aug 15. Diodorus is careful to name the Athenian archon for each year at the appropriate point in his history. From von Oppolzer, no other eclipse could have been total in Sicily between 346 and 303 BC – a result confirmed by calculations based on equation (10.1). Nothing short of totality could have caused 'utter darkness', with the stars being seen 'everywhere'. It is unfortunate that the precise place of observation cannot be determined; otherwise this would have been the most valuable record of a total solar eclipse from ancient Europe.

Whether Agathocles sailed northwards or southwards after escaping from Syracuse is not known, although he was probably still off the coast of eastern Sicily when the eclipse occurred. There seems to be no reason for doubting Diodorus' assertion that the eclipse occurred on the day after the escape. Values of ΔT (around 15 000 sec) which produce totality in the neighbourhood of Syracuse indicate that the eclipse happened early in the morning (local time 7.4 h). Hence although the time of day when the escape took place cannot be ascertained, Agathocles would probably have been at sea for less than 24 hours (perhaps considerably less) when the Sun was obscured.

The voyage from Syracuse to the Libyan coast (possibly in the neighbourhood of Cape Bon) took six days and six nights. There is little likelihood of the fleet calling at some other port en route for most of Sicily was in enemy (i.e. Carthaginian) hands (Diodorus, XX, 3). If the ships travelled northwards (through the straits of Messina), the total distance covered to reach Libya would be about 600 km; if southwards only about 400 km. Hence the average speed – assuming continuous travel – would be respectively about 4 or 3 km per hour. Captain R. Hardy of South Tyneside College assures me that such a low mean rate is quite reasonable for primitive sailing vessels. Agathocles' ships were equipped with oars, which would almost certainly be used during the chase. If we assume even double the above rate between the escape from Syracuse and the eclipse, the fleet may only have covered some 100 km northward or 70 km toward the south – significantly less if the ships escaped from the harbour when the day was already well advanced.

The two extreme positions (roughly 100 km to the north or 70 km to the south of Syracuse) are thus perhaps (lat. =+37.8 deg, long. =−15.3 deg) and (lat. =+36.6 deg, long. =−15.0 deg). In figure 10.4, these are indicated respectively by NL and SL.

Fotheringham derived limits of: (lat. =+38.42 deg, long. =−15.50 deg) and (lat. =+36.58 deg, long. =−15.00 deg). The southern location

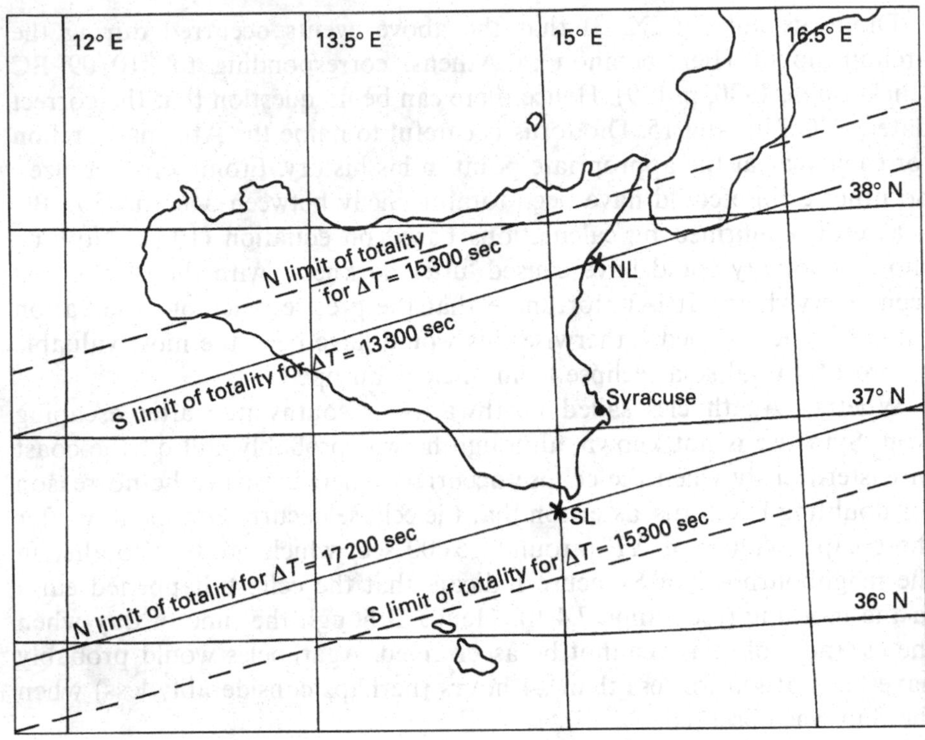

Fig. 10.4 The eclipse of Agathocles (BC 310 Aug 15). The edges of the computed belts of totality for specific ΔT limits are shown.

is similar to that derived here. However, the distance to his suggested northern site (150 km) would seem excessive.

During totality, Venus (mag. −3.4), Mercury (mag. −1.2), Jupiter (mag. −1.2), and Sirius (mag. −1.4) would all be well placed for visibility.

Results

For totality at Syracuse itself, $14\,350 < \Delta T < 16\,500$ sec, which is in excellent accord with the results obtained from Babylonian data (see figure 10.1). If the fleet of Agathocles had reached the estimated northern limit of 100 km from Syracuse, as long as the *southern* limit of totality extended as far south as this point, the required conditions would be fulfilled. A value of $\Delta T > 13\,350$ sec would thus be needed. Equally, if the fleet was 70 km from Syracuse in a southerly direction, it would be necessary for the *northern* limit of totality to extend at least as far north as this. A value of $\Delta T < 17\,200$ sec is thus indicated. Combining these results, we find that a reasonable range for ΔT would appear to be $13\,300 < \Delta T < 17\,200$ sec. However, it is most unfortunate that more definite results cannot be obtained.

In figure 10.4 the southern limit of totality for $\Delta T = 13\,300$ sec and the northern limit for $\Delta T = 17\,200$ sec are shown by full lines. Also depicted (by dashed lines) are the limits of totality computed with a mean value for ΔT of 15 250 sec (i.e. the average of 13 300 and 17 200 sec) to illustrate the true geographical width of the track.

(6) BC 129 Nov 20 ? (total, 1.01): Hellespont

Hipparchus, the eminent Greek astronomer of the second century BC, is renowned for his skill in both observation and theory. One of his numerous achievements was to calculate the distance of the Moon using two simultaneous observations of an unidentified solar eclipse. This was seen to be fully total in the Hellespont, while at Alexandria $\frac{4}{5}$ of the solar diameter was obscured. From these observations, he deduced that the distance of the Moon was approximately $67\frac{1}{3}$ Earth radii, a result which compares well with the true mean of 60.25 in the same units.

Most of Hipparchus' writings survive only in quotations and paraphrases by other ancient authors. A brief account of his use of the eclipse is given by Ptolemy in the *Almagest*. The fullest description is by Pappus of Alexandria (*c.* AD 320), who was probably the last great geometer of ancient Greece. Further information (without mentioning Hipparchus by name) is supplied by the first century BC astronomer Cleomedes. Translations of these three accounts are as follows:

(i) Hipparchus tries to demonstrate the Moon's distance by guessing at the Sun's. First he supposes that the Sun has the least perceptible parallax, in order to find its distance, and then he uses the solar eclipse which he adduces; at one time he assumed that the Sun has no perceptible parallax, at another that it has a parallax big enough [to be observed]. As a result, the ratio of the Moon's distance came out different for him for each of the hypotheses he put forward; for it is altogether uncertain in the case of the Sun, not only how great its parallax is, but even whether it has any parallax at all.

[Ptolemy, *Almagest*, V, 11; trans. Toomer (1984, pp. 243-244).]

(ii) So Hipparchus, being uncertain concerning the Sun, not only how great a parallax it has but whether it has any parallax at all, assumed in his first book of 'On Sizes and Distances' that the Earth has the ratio of a point and center to the Sun [i.e. the Sun's sphere]. And at one time using the eclipse he adduced, he assumed that it had the least parallax, and at another time a greater parallax. Hence the ratios of the Moon's distances came out different. For in Book 1 of 'On Sizes and Distances' he takes the following observation: an eclipse of the Sun, which in the

Hellespontine region was an exact eclipse of the whole Sun, such that no part of it was visible, but at Alexandria by Egypt approximately four-fifths of the diameter was eclipsed. By means of the above he shows in Book 1 that, in units of which the radius of the Earth is one, the least distance of the Moon is 71, and the greatest 83. Hence the mean is 77... Then again he himself in Book 2 of 'On Sizes and Distances' shows from many considerations that, in units of which the radius of the Earth is one, the least distance of the Moon is 62, the mean $67\frac{1}{3}$ and the Sun's distance 490. It is clear that the greatest distance of the Moon will be $72\frac{2}{3}$.

[Pappus, Commentary on the *Almagest*, V, 11; trans. Toomer (1974).]

(iii) Moreover, such an observation has been made in the case of an eclipse of the Sun. Once when the Sun was wholly eclipsed in the Hellespont, it was observed in Alexandria to be eclipsed except for the fifth part of its diameter, which is, according to the sight, except for two digits and a little more... Now since it is 5000 stades from Alexandria to Rhodes and just so many from thence to the Hellespont, necessarily a digit will be seen at Rhodes; besides, proceeding hence to the Hellespont, this will also decrease in proportion, since when the Hellespont is reached, it will entirely vanish....

[Cleomedes, *De Motu Circularis Corporum*, II, 3; trans. auct.]

Toomer (1974) explains that the reason for Hipparchus obtaining two 'mean' results for the lunar distance of 77 and $67\frac{1}{3}$ Earth radii is that in Book 1 he assumed a solar parallax of zero (i.e. placing the Sun at an infinite distance away) while in Book 2 he adopted a parallax of 7 arcmin.

There is nothing in any of the above records to suggest that Hipparchus witnessed the eclipse, or even that it occurred during his lifetime. The date of his treatise is unknown. On historical evidence alone, all that can be established is that he was able to gain access to two careful records of the same eclipse from the region of the Hellespont and from Alexandria. Since Alexandria was founded in 331 BC, no date earlier than this can be considered. In principle, the eclipse might have occurred at any time between 331 BC and the death of Hipparchus (*c.* 120 BC) – an interval of more than two centuries.

Fotheringham (1909 and 1920b) considered only the dates BC 310 Aug 15 (i.e. the eclipse of Agathocles) and 129 Nov 20, rejecting the former since he was of the opinion that a magnitude estimate from Alexandria at such an early period was unlikely. However, more recent investigations by Newton (1970, pp. 104–110) and Toomer (1974) have discussed several alternative dates, ranging from 310 to 125 BC These will all be considered below, along with other possible identifications.

It should be emphasised that four-fifths (the estimate from Alexandria) is an unusual way of expressing an eclipse magnitude. As discussed in chapter 3, the Greeks – following the Babylonians – normally used digits or twelfths of the solar disk. However, Newton (1970, p. 134) made an interesting suggestion. A magnitude of 9 digits of *area* corresponds closely to 9.5 digits of diameter, as is evident from Ptolemy's conversion table (*Almagest*, VI, 8) – see also chapter 3. When converted to a fraction, 9.5 digits is virtually equal to $\frac{4}{5}$ (actually 0.79). This may well be the true explanation of the value used by Hipparchus.

Although the eclipse might well have taken place before Hipparchus' own time, it is useful to summarise details about his life. The following information is derived mainly from Dicks (1960, pp. 3ff.).

Hipparchus was born in Bithynia in the north-west of Asia Minor, probably around 194 BC. He was a citizen of Nicaea. It is known from Ptolemy that Hipparchus made observations of the risings and settings of stars in Bithynia (Dicks, 1960, p. 3). Subsequently, when he was aged about thirty, he left Bithynia for Rhodes, where he seems to have spent the rest of his life. What is probably Hipparchus' earliest known observation at Rhodes (of the autumnal equinox) dates from 162 BC (*Almagest*, III, 1). His latest surviving measurements (of the positions of the Sun and Moon) were made in 127 BC (*Almagest*, VI, 5). He probably died about 120 BC, although the exact year of his death is unknown.

Totality of the eclipse investigated by Hipparchus is very carefully described by Pappus in the following words: 'an exact (*akribos*) eclipse of the whole Sun, such that no part of it was visible'. That the eclipse was indeed complete is confirmed by Cleomedes. It is noteworthy that both writers omit superfluous details such as darkness or the appearance of stars; only the knowledge that the eclipse was complete was important to Hipparchus. Pappus asserts that the observation was made 'in the Hellespontine region' (*en tois peri ton Hellesponton topois*). Rather than simply meaning the strait itself (now known as the Dardanelles), this denotes one of the *climata* or belts of latitude 400 stades (approximately 0.6 deg) wide, within which celestial phenomena did not appear to change appreciably (Dicks, 1960, pp. 154–162). For all practical purposes, inhabitants of the same *clima* were considered to be in the same latitude.

In his *Commentary on Aratus*, his only fully extant work, Hipparchus states that 'in the Hellespontine regions' (*en tois peri ton Hellesponton topois* – precisely the expression used by Pappus), the longest day is 15 hours. Ignoring refraction and the semidiameter of the Sun, this gives a latitude of +41.07 deg. For comparison, Ptolemy obtained 40;56 (i.e. 40.93) deg. There is obviously a considerable error in both results since the latitude of the Hellespont strait itself ranges from +40.0 to +40.4 deg. From Strabo – the first century BC geographer who derived much of his information from the work of Hipparchus – it is possible to fix the latitude limits of the Hellespontine *clima* with fair accuracy. Towards the south, Alexandria in the Troad (lat. = +39.8 deg) was included in the same *clima*, but the zone did not quite reach as far north as Lysimachia (lat. = +40.6 deg), a little beyond the eastern end of the strait (Strabo, *Geography*, II, 5.40). A latitude range between about +39.8 and +40.4 deg thus seems most likely.

Hipparchus considered Alexandria in Egypt (at which the magnitude of $\frac{4}{5}$ was estimated) to be 10 deg south of the Hellespont and thus close to latitude 31 deg N – a result closely confirmed by Ptolemy, who gave 30;58 (or 30.97) deg. These values are quite near the true figure of +31.22 deg.

In his calculation of the the lunar distance, Hipparchus needed to assume that both places where the eclipse was observed lay under the same meridian. Following Eratosthenes, he believed that Alexandria in Egypt (long. = −29.9 deg), Rhodes (−28.2 deg), the Troad (−26.2 deg), Lysimachia (−26.6 deg) and Byzantium (−29.0 deg) were on the same meridian of longitude (Strabo, II, 5.7). Hence it is reasonable to infer that the observation in the Hellespont was made in the vicinity of the strait itself, rather than much to the east or west (but still in the same latitude zone). As illustrated in figure 10.5, Hipparchus' assumption entails errors of several degrees in longitude but in ancient times determination of this parameter presented serious difficulties. In alluding to the magnitude of the eclipse at Rhodes, as well as at the Hellespont and Alexandria, Cleomedes is making an indirect reference to Hipparchus. However, this detail is purely incidental.

From von Oppolzer, several eclipses could possibly have been total in the Hellespont region between 331 BC and 120 BC These are BC 310 Aug 15, 303 Apr 2, 282 Aug 6, 228 Sep 7, 221 May 5, 190 Mar 14, 188 Jul 17, 183 Oct 19, 176 Jun 6, 174 Oct 10, 163 Mar 15, 136 Apr 15 and 129 Nov 20. After the last date there is no further candidate until BC 116 Aug 29 – fully ten years after Hipparchus' last known observation. Newton (1970, p. 109) also considered the eclipse of BC 125 Sep 7, but this was only annular. In the central zone, 99.7 per cent of the Sun would have been covered by the Moon, leaving a narrow ring of light still visible. This is not consistent with the careful description of totality by Pappus. Toomer (1974) made calculations for the eclipse of BC 217 Feb 11. However, the track of totality on this occasion could not have reached further north than about 34 deg N latitude anywhere on the Earth's surface. Hence, regardless of the precise value of ΔT adopted, the eclipse could not have been central anywhere near the Hellespont.

In table 10.4, I have listed for each of the eclipses cited above – apart from 217 BC – the computed magnitudes (using equation (10.1)) at both the Hellespont and Alexandria and also the approximate ranges of ΔT

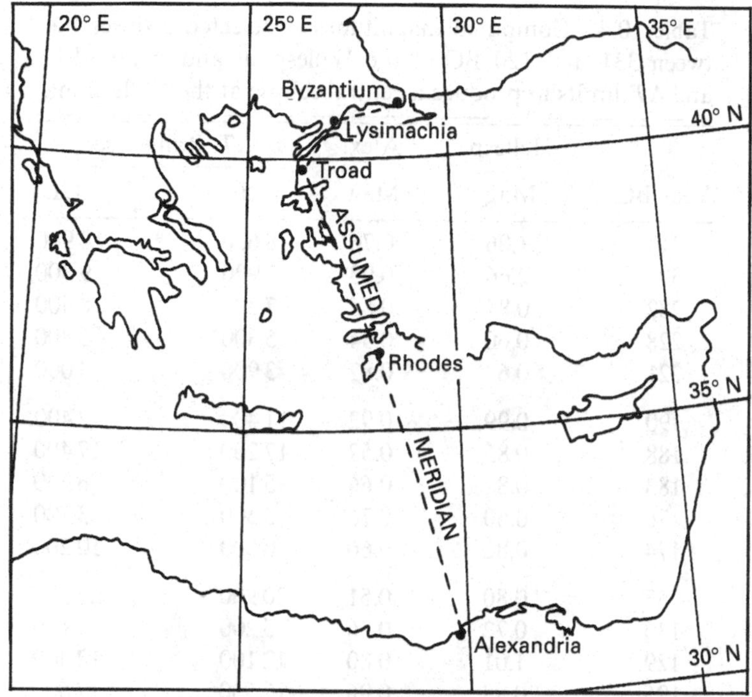

Fig. 10.5 Hipparchus' assumed meridian through Alexandria and Byzantium.

necessary to produce totality at the Hellespont (annularity in the case of 125 BC). For this purpose I have taken the mean co-ordinates of the Hellespont as: lat. =+40.1 deg, long. =−26.5 deg. The indicated lower limit (LL) and upper limit (UL) to ΔT are estimated to the nearest 100 sec in each case. It will be noted that in 125 BC, the ΔT limits to achieve annularity are effectively identical. This is because the apparent lunar and solar diameters were almost equal.

Reference to figure 10.1 shows that only three of the dates in table 10.4 can be considered viable: BC 310 Aug 15, 190 Mar 14 and 129 Nov 20. For each of the three selected eclipses, the ΔT ranges derived for the Hellespont are very close to the mean straight line through the Babylonian timed data. However, on all other occasions, the results are extremely discordant.

In order to attempt to distinguish between the dates 310, 190 and 129 BC, it is necessary to consider both the historical and astronomical circumstances in more detail.

A date as early as 310 BC seems most unlikely for a variety of reasons. What Hipparchus required for his calculation was a reliable – i.e. 'exact' – observation of totality in the Hellespont region combined with a careful magnitude estimate ($\frac{4}{5}$) for the same eclipse at Alexandria. He felt able to

Table 10.4 Computed magnitudes of selected eclipses between 331 and 120 BC at the Hellespont and Alexandria and ΔT limits to produce a central eclipse at the Hellespont.

	Hellesp.	Alex.	ΔT Range (sec)	
Year BC	Mag.	Mag.	LL	UL
310	1.06	0.76	8 600	14 900
303	0.66	0.93	7 900	9 300
282	0.84	0.55	7 300	8 400
228	0.60	0.34	5 300	5 800
221	0.63	0.82	3 900	4 000
190	0.99	0.92	11 400	12 400
188	0.85	0.57	17 200	17 400
183	0.81	0.66	5 100	6 700
176	0.60	0.76	3 500	3 700
174	0.82	0.66	10 000	10 200
163	0.80	0.51	20 900	22 200
136	0.72	0.86	5 200	6 100
129	1.01	0.80	12 100	12 400
125	0.88	0.96	17 500	17 500

put sufficient trust in both observations to use them in a determination of the lunar distance, and – as we know in retrospect – they were sufficiently accurate to enable him to obtain a reasonably sound result. However, there is nothing in the available historical records to suggest that he could have obtained effective data long before his own time.

Hipparchus was in possession of numerous Babylonian observations (see chapter 4). By comparison, he seems to have had access to only scattered groups of Greek data, other than his own measurements. Apart from an isolated solstice determination by Meton in 432 BC (which Hipparchus rejected as unreliable), the earliest Greek observations which Hipparchus is known to have used date from between 295 and 272 BC These were nearly all by the Alexandrian astronomer Timocharis and were exclusively of conjunctions of either the Moon or Venus with certain stars (*Almagest*, VII, 3; X, 4). The next set of early Greek data which Hipparchus analysed were of measurements of planetary position, ranging in date from 272 to 241 BC (*Almagest*, IX, 7; IX, 10). Here, the observer cannot be established with certainty. Finally, Hipparchus investigated a series of four observations of lunar eclipses made at Alexandria between 201 and 174 BC (*Almagest*, IV, 11 and VI, 5; see also section 10.4 below). Once again, the observer is unknown.

Judging from his analysis of these various observations, as outlined in the *Almagest*, if Hipparchus had possessed records of many further Greek observations, he would have had no particular motive for dividing them into discrete chronological groupings. This suggests that he had little else but the above to draw on. In particular, although the first and third sets of observations were from Alexandria, Toomer (1974) doubts whether there was ever a 'school of astronomy' there at this period. He remarks that we can name very few people in Alexandria who made astronomical observations or calculations during this interval.

The solar eclipse of 310 BC occurred more than a century before the first eclipse record which definitely originates from Alexandria and fully 15 years before the first known astronomical observation made at this city. Further, unless a special watch for solar eclipses was maintained there, the eclipse of 310 BC might well have escaped notice. Computations using the mean value for ΔT to produce totality at the Hellespont, as listed in table 10.4 ($\Delta T = 11\,800$ sec), indicate that at maximum phase the solar altitude would be as much as 40 deg so that if the sky was clear the radiance of the Sun would be dazzling. As it happens, similar calculations imply that the Sun would be much lower in the sky both in 190 BC (altitude 22 deg) and 129 BC (13 deg).

If we turn now to the observation at the Hellespont, as noted above, the descriptions of totality at this location by both Pappus and Cleomedes are remarkably precise and devoid of any irrelevant details such as the visibility of stars. Whatever its date, the 'eclipse of Hipparchus' is the only known example from ancient Europe which clearly describes the complete disappearance of the Sun. Most early writers were much more concerned with relating the effects of the darkness which accompanied a large eclipse. Hipparchus required – and clearly obtained – more than this. The eclipse of 310 BC occurred several generations before his own era. Hence for any report of this event from the Hellespont he would probably have to rely on an account in some historical or literary work. If this resembled the material discussed elsewhere in this chapter, he would have no justification for assuming that the whole of the Sun was obscured – only that the phase must have been very large. Many further early records, compiled by Ginzel (1899), are of even poorer quality than this.

Only two other ancient writers are known to have obtained access to separate accounts of the same eclipse from long before their own time. In each case the reports relate to the lunar obscuration of 331 BC. Pliny (*Natural History*, II, 180) noted that at Arbela (lat. 35.9 deg N, long. 44.3 deg E), the eclipse of Sep 20 in that year occurred at the second hour of night whilst in Sicily the rising Moon was obscured (see section 10.5 below). On the other hand, Ptolemy (*Geography*, I, 4) stated that the same event was seen at Arbela in the *fifth* hour but took place at Carthage in the second hour. Neither author gives any other information. Just how Pliny and Ptolemy synchronised the dates of the separate observations which they describe – i.e. at Arbela and another location – is not known. Nevertheless, it is clear that at least one of their sources was unreliable. Hipparchus was himself well aware of the value of independent observations of the same lunar eclipse in determining the difference between the longitudes of two sites (Strabo: *Geography*, I, 1.12 – see also chapter 13). However, there is no evidence that he was ever able to put this technique to effective use because 'the total absence in antiquity of any scientific organisation deprived the whole method of its practical importance' (Neugebauer, 1975, p. 667).

In summary, the likelihood of Hipparchus obtaining two unusually reliable descriptions of the same eclipse from at least 150 years before he made his calculation of the lunar distance must be minimal. The eclipse of 310 BC can thus almost certainly be discounted, and in the subsequent discussion only the dates 190 and 129 BC will be considered. In both cases the historical circumstances are much more favourable.

Toomer (1974) made a strong case for a date of 190 BC for the eclipse of Hipparchus. He remarked that this fell in the middle of the period from which Hipparchus obtained four reports of lunar eclipses from Alexandria (201 to 174 BC) and thought that it was possible that the same person made all four observations – as well as estimating the magnitude of the solar eclipse. With regard to the record of totality at the Hellespont, Toomer noted that the historian Livy (XXXVII, 9) gives details of a Roman naval attack on the towns of the Hellespont strait which occurred in the spring of 190 BC – about the time that the eclipse took place. If the fleet had indeed witnessed totality, the commander, Livius Salinator, could well have sent a detailed report back to Rome.

Toomer was able to further deduce that if Hipparchus had access to a record of totality from the Hellespont in 190 BC and a magnitude estimate of $\frac{4}{5}$ at Alexandria, he could easily have obtained a result for the lunar distance very close to that cited by Pappus. Toomer considered that this would not be possible in either 310 or 129 BC. In making his calculations, Toomer assumed that on account of the limited capabilities of ancient mathematics Hipparchus would have had to suppose that the Sun was in the meridian at both places at the time of the eclipse. However, this would be a very poor approximation. For values of ΔT (around 12 000 sec) which produce totality at the Hellespont, greatest phase at both sites would occur fully 3.5 hours before noon. Since all four lunar eclipses observed from Alexandria around this period were carefully timed (see section 10.4), it seems reasonable to suppose that if the solar eclipse of 190 BC was indeed seen there, the time of day would also be recorded.

Under these circumstances, Hipparchus could conceivably have made at least some allowance for the local time of observation.

As it happens, calculation shows that if the eclipse of 190 BC was fully total at the Hellespont, the magnitude at Alexandria must necessarily have been between 0.86 and 0.90. Hence the width of the unobscured portion of the solar limb would have been almost twice the amount assumed by Hipparchus. As a result, if he had made even a rough correction for the time of day he would have obtained a bad value for the lunar distance.

Similar considerations for the eclipse of 129 BC (ΔT around 12 250 sec) yield a local time of around 15.9 h at both locations. However, if the eclipse was total at the Hellespont, the magnitude at Alexandria would be 0.80 – exactly the observed figure. This would lead to a sound result for the lunar distance if Hipparchus had made even an approximate allowance for the time of day. The eclipse of 129 BC has the attraction that it occurred during Hipparchus' active lifetime. Hence if he observed it to be fairly large at Rhodes he would have ample opportunity to enquire (i) just where – towards the north – it was total and (ii) if the magnitude had been estimated further south – i.e. as far from the central zone as possible. However, the situation is hypothetical; we cannot establish on historical grounds whether the eclipse of 129 BC took place before or after Hipparchus wrote his treatise.

Since we can only speculate on the method of analysis used by Hipparchus and whether or not he might have seen the eclipse in question, it is probably impossible to decide between the dates 190 and 129 BC Unfortunately, these dates are too far apart to enable deduction of a useful ΔT range by combining the separate ΔT intervals on the assumption that totality was definitely observed at the Hellespont region on one or other of these two dates.

(7) AD 29 Nov 24? (total, 1.02): Nicaea?

Eusebius, the great theologian and historian of the fourth century AD, preserves a description of a major solar eclipse by the second century writer Phlegon of Tralles. Very little of Phlegon's works now survive.

> And Phlegon also who compiled the *Olympiads* writes about the same things in his 13th book in the following words: 'In the fourth year of the 202nd Olympiad (AD 32–33), an eclipse of the Sun took place greater than any previously known, and night came on at the sixth hour of the day, so that stars actually appeared in the sky; and a great earthquake took place in Bithynia and overthrew the greater part of Nicaea'.
>
> [Phlegon, *Olympiades*, fragment 17 – quoted by Eusebius, *Chronicon*; trans. Fotheringham, (1920b).]

Both the date and place of observation present difficulties. Fotheringham, following Kepler, identified the eclipse as that of AD 29 Nov 24 (the *first* year of the 202nd Olympiad). This was the only large eclipse visible in the eastern Mediterranean for many years around this time. The computed time of maximal phase in this region (approximately 11.0 h using equation (10.1)), is in fair accord with that recorded by Phlegon (i.e. the sixth hour). Although the date thus seems firmly established, the place of observation is unknown. Phlegon, a citizen of Tralles (lat. = 37.85 deg, long. = −27.85 deg) wrote more than a century after the event. In selecting Nicaea (lat. = 40.42 deg, long. = −29.60 deg), where the earthquake occurred, as the possible place where the eclipse was seen, Fotheringham considered this inference 'weaker than any of the the other presumptions' which he had used.

Computation using equation (10.1) (ΔT = 9050 sec) indicates that the eclipse of AD 29 was indeed fully total at Nicaea (at 10.8 h), while at Tralles the phase would be virtually total (0.997) at 11.0 h. Quite possibly the eclipse recorded by Phlegon may have been witnessed in this region but the precise place of observation is purely a matter of speculation. A reliable solution for ΔT thus cannot be made.

(8) AD 71 Mar 20? (total: 1.01): Chaeroneia?

The Greek philosopher and biographer Plutarch gives a vivid account of a total eclipse in one of his dialogues entitled *The Face on the Moon.* In this same work he also makes a brief reference to the corona.

(i) (Lucius) smiled thereat and said...'Now grant me that nothing that happens to the Sun is so like its setting as a solar eclipse. You will if you call to mind this conjunction recently which, beginning just after noonday, made many stars shine out from many parts of the sky and tempered the air in the manner of twilight. If you do not recall it, Theon here will cite us Mimnermus and Cydias, Archilochus and Stesichorus besides, and Pindar, who during eclipses bewail "the brightest star bereft" and "at midday night falling" and say that the beam of the Sun <is sped> the path of shade'.

[Plutarch, *The Face on the Moon*, 931D–E; trans. Cherniss and Helmbold (1957, vol. XII, pp. 117–119).]

(ii) Even if the Moon, however, does sometimes cover the Sun entirely, the eclipse does not have duration or extension; but a kind of light is visible about the rim which keeps the shadow from being profound and absolute.

[Plutarch, *The Face on the Moon*, 932E; trans. Cherniss and Helmbold (1957, vol. XII, p. 121).]

The descriptions of eclipses by Archilochus and Pindar mentioned in (i) have already been discussed above. Regrettably, the relevant works by the poets Mimnermus, Cydias and Stesichorus – all of whom lived *c.* 600 BC – no longer survive.

Cherniss and Helmbold (1957, vol. XII, p. 11) argued that if Plutarch was indeed referring to the corona in (ii), his description is 'remarkably tame'. They suggested instead that the account is more likely to refer to an annular eclipse. However, surprising as it often seems to modern astronomers who have witnessed total obscurations of the Sun, the corona does not appear to have left much of an impression on observers in ancient and medieval times. Before AD 1600, only one other account of a total eclipse (AD 968) definitely mentions the corona, even though many detailed descriptions of total eclipses are preserved in medieval European and Arabic chronicles (see chapters 11 and 12). The record from AD 968, which originates from Constantinople, likens the corona to 'a certain dull and feeble glow, like a narrow headband, shining around the extreme portion of the edge of the disk' (see chapter 11). Plutarch's description is, in fact, not too dissimilar from this much later record. By comparison, during a central annular eclipse the unobscured portion of the Sun is dazzling in brightness. The ring phase could scarcely be described as keeping 'the shadow from being profound and absolute' since often there is hardly any noticeable reduction in daylight.

Fotheringham, and later Cherniss and Helmbold, regarded the first statement – (i) above – as a reference to a real event (rather than some product of Plutarch's imagination as suggested by Newton (1970, pp. 115–117). Muller (1975), who had personally witnessed several total eclipses from various sites, thought that account had 'the definite flavour of personal experience and eye-witness description... the probability that this is a real record is very high'. There are sound reasons for believing that the eclipse was authentic. Plutarch is the only known ancient writer to allude to the corona, and he is unique among classical authors in noting a fall in temperature during a total eclipse. His whole account, which also describes the appearance of stars in different parts of the sky, is so original that it seems highly likely that either Plutarch himself or one of his close associates was an eye-witness to the events which he so vividly describes. In Plutarch's dialogue, the eclipse was the basis of an intellectual – if somewhat entertaining – discussion. Most of the characters identified in the text are known to have been associates of Plutarch himself (see Cherniss and Helmbold (1957, pp. 3ff.)).

On the basis of a careful consideration of the content of Plutarch's numerous dialogues, Fotheringham considered it 'a reasonable presumption that the eclipse was total at either Delphi or Chaeroneia'. He further regarded AD 71 Mar 20 as the only viable date. However, although there can be little doubt that a total obscuration of the Sun is described above – a lesser phase could scarcely cause many stars to 'shine out from many parts of the sky' – both the date and place of observation require careful discussion.

Plutarch was born at Chaeroneia in Boeotia around AD 46 and died after AD 119. Although he was normally resident in Chaeroneia throughout his life, he is known to have travelled throughout central Greece. He also paid several official visits to Rome and lectured on philosophy there. Plutarch had close links with the Athenian Academy, while from about AD 95 he held a priesthood for life at Delphi – not far from Chaeroneia.

The beginning of Plutarch's dialogue entitled *The Face on the Moon* is lost and, with it, any indication of date or place. Fotheringham pointed out that on this occasion, Plutarch's brother Lamprias – who also resided at Chaeroneia – presided over the assembled company. However, Lucius, to whom the reference to the eclipse is attributed, appears in two other dialogues based in Rome. Both of these details should probably be regarded as circumstantial. Plutarch's many dialogues are usually set in various places in central Greece, but sometimes in Rome – the places with which he himself was familiar. In view of the fact that the 'recent conjunction' was so clearly remembered, it seems highly likely that totality was witnessed at one of these locations. The place of observation of the eclipse will thus be assumed to be either central Greece or Rome. More remote places, such as Alexandria – which Plutarch is only known to have visited once – would seem much less likely. In particular, Plutarch seldom appeals to his Alexandrian experiences in his numerous dialogues.

During the lifetime of Plutarch, only four eclipses could have been total in the central or eastern Mediterranean: AD 59 Apr 30, 71 Mar 20, 75 Jan 5 and 83 Dec 27 – as may be judged from the charts of von Oppolzer (1887). In all probability, a choice must be made between one of these alternatives. The eclipse of AD 59 occurred when Plutarch was only about 13 years old and thus would no longer be 'recent' when he wrote his dialogue. However, although less likely than the other selected dates, it cannot be discounted on these grounds alone.

The magnitude, local time and solar altitude for each eclipse at Chaeroneia (taken as representative of central Greece) and Rome – as computed from equation (10.1) – and the ΔT limits (LL and UL) in seconds necessary to produce totality at either place are listed in table 10.5.

With reference to table 10.5, no value of ΔT will produce a central eclipse at either Chaeroneia or Rome in AD 59 or at Rome in AD 83. As clearly depicted by Ginzel (1899), the track of totality in AD 59 ran almost parallel to the equator throughout much of the Mediterranean. Regardless of the precise value of ΔT at the time, the eclipse can never

Table 10.5 Investigation of solar eclipses during the life of Plutarch.

Year AD	Place	Mag.	LT	Alt.	ΔT Range (sec) LL	UL
59	Chaeroneia	0.92	15.3	+40	—	—
	Rome	0.81	14.3	+49	—	—
71	Chaeroneia	1.01	11.0	+48	9 500	9 600
	Rome	0.80	10.1	+40	6 250	6 350
75	Chaeroneia	0.86	16.1	+ 6	12 000	12 200
	Rome	0.94	15.3	+11	8 000	8 200
83	Chaeroneia	0.80	14.2	+21	3 250	3 650
	Rome	0.71	13.1	+23	—	—

have been total north of latitude 36.3 deg N. The central zone would thus pass far to the south of Rome and the Italian peninsula and also a little to the south of the Grecian peninsula. Even at Sparta (the scene of other dialogues by Plutarch), which was much better placed than Chaeroneia in AD 59, the magnitude cannot have exceeded 0.96, which is far from sufficient to produce the effects described.

For values of ΔT which indicate totality at Rome in AD 83, the Sun would set before greatest phase was reached. At Chaeroneia, a highly discordant range of ΔT (around 3500 sec) would be needed to produce totality and here maximum eclipse would occur only a few minutes before sunset – with the solar altitude no more than 3 deg. Thus the description 'beginning just after noonday' would be far from appropriate.

The circumstances in AD 75 are also unfavourable. Computations using the ΔT results listed in table 10.5 indicate that totality at both Rome and Chaeroneia would again occur in the late afternoon. At Rome, the local time would be 15.7 h with a solar altitude of only 7 deg (compared with the meridian altitude of 25 deg). In Chaeroneia, the local time would be 15.2 h and the altitude of the Sun 14 deg (meridian altitude 29 deg). Since it is likely that the eclipse would not be noticed until it was already well advanced, neither of these situations could be described as close to midday. Although the ΔT range required to achieve totality at Rome (around 8100 sec) is in fair agreement with figure 10.1, that for Chaeroneia (some 12 100 sec) is discordant; ΔT had already decreased to between 11 210 and 12 120 sec by 136 BC, as indicated by the Babylonian observation of a total solar eclipse at that date.

The eclipse of AD 71 is a much more promising candidate. Reference to figure 10.1 shows that the ΔT limits necessary to secure totality at

Chaeroneia (close to 9550 sec) are quite acceptable. However, for Rome, a value for of only around 6300 sec would be needed. Anticipating the results from the analysis of the careful timed observations at Alexandria in AD 364 – see the immediately following entry – a figure for ΔT of as much as 8300 sec is indicated at this late date.

The only difficulty with the eclipse of AD 71 as seen from Chaeroneia is that for totality at this site greatest phase would occur around 11.0 h, rather than 'just after noonday'. However, the Sun would then be almost at its height – altitude 48 deg, or only 3 deg less than the meridian altitude. There is no suggestion in the record that time was carefully measured; to the casual bystander this eclipse would be regarded as occurring close to midday. Although the duration of totality would not exceed 40 sec, this would be sufficient to render several stars visible, with Venus (mag. −3.5) prominent, 37 deg to the east of the Sun.

In summary, a date for the 'eclipse of Plutarch' of AD 71 Mar 20 seems very likely. However, since the historical details are somewhat tenuous and the above date has been derived only as the result of extensive astronomical computations, it would seem scarcely justifiable to use the eclipse to make any deductions on the value of ΔT.

(9) AD 364 Jun 16: Alexandria

In his commentary on the *Almagest*, Theon of Alexandria (*c.* AD 370), records a solar eclipse which he observed during the afternoon.

> ...the exact ecliptic conjunction which we have discussed, and which took place according to the Egyptian calendar in the 1112th year from the reign of Nabonassar, $2\frac{5}{6}$ equal or equinoctial hours after midday on the 24th of Thoth, and according to the Alexandrian calendar reckoned by simple civil days in the 1112th year of the same reign, $2\frac{5}{6}$ equal or equinoctial hours after midday on the 22nd of Payni... And moreover we observed with the greatest certainty the time of the beginning of contact, reckoned by civil and apparent time, as $2\frac{5}{6}$ equinoctial hours after midday, and the time of the middle of the eclipse as $3\frac{4}{5}$ hours, and the time of complete restoration as $4\frac{1}{2}$ hours approximately after the said midday on the 22nd of Payni.
>
> [Theon of Alexandria, 332; trans. Fotheringham, (1920b).]

This account follows very much the style of the eclipse records in the *Almagest* itself (see chapter 4 and section 10.4 below). Fotheringham converted the recorded date of the conjunction to the Julian calendar (AD 364 Jun 16) and this corresponds exactly with that of a tabular solar eclipse (e.g. as listed by von Oppolzer, 1887). For a more recent discussion of the eclipse, see Tihon (1976–77). There seems no reason for doubting that Alexandria, where Theon lived, was the place of observation – especially since the first person plural is used in referring to the measurements. Fotheringham regarded the event as 'the only ancient eclipse of the Sun for which an astronomically observed time is recorded'. This remark is no longer valid in view of the many similar Babylonian reports which have since become available – see chapter 5. However, it is still the only solar eclipse for which careful measurements of time are available from ancient Europe.

The local times of the various phases are all conveniently expressed in equal hours after midday.

RESULTS

(i) Local time of first contact on Jun 16 = 14.83 h, UT = 12.79 h. Computed TT = 15.09 h, thus ΔT = 8100 sec.

(ii) LT of mid-eclipse (i.e. greatest phase) = 15.64 h, UT = 13.76 h. Computed TT = 16.01 h, thus ΔT = 8300 sec.

(iii) LT of last contact = 17.39 h, UT = 14.46 h. Computed TT = 16.79 h, thus ΔT = 8400 sec.

These three results are remarkably self-consistent – evidence of Theon's care. (NB at Alexandria the computed magnitude = 0.38.)

In summary, of the material investigated by Fotheringham in his 1920b paper, the last observation – the timed eclipse of AD 364 – is the most reliable. In almost every other case there are severe difficulties of interpretation.

10.3 Other ancient reports of large solar eclipses

There are many further allusions to solar eclipses in ancient Greek and Latin writings which were not considered by Fotheringham in his 1920b paper. Most of these were discussed by Ginzel (1899). However, only in a few instances is there any reason for believing that an unusually large obscuration of the Sun occurred at a known place and on an unambiguous date. Sources which relate events occurring centuries beforehand are often unreliable. In particular, Plutarch recorded several solar eclipses which were seen long before his own time. He had a habit of including allusions to darkness which were not mentioned in accounts of the same event by earlier writers (notably in 431 and 364 BC). Perhaps he was influenced by his experiences of the total eclipse of AD 71 (see section 10.2).

From the large array of material, I have selected only the following three records since they seem to indicate a particularly large eclipse of the Sun and the place and date are at least reasonably well established.

(1) BC 394 Aug 14 (annular, 0.94): near Chaeroneia

This eclipse is recorded as partial both by the contemporary writer Xenophon in his *Hellenica* and by Plutarch in his *Life of Agesilaus.* When the eclipse occurred, the army of King Agesilaus II of Sparta was on its way to engage in battle with the Thebans near Chaeroneia, having marched southwards through Thrace.

(i) Next day he (Agesilaus) crossed the mountains of Achaea Phthiotis and for the future continued his march through friendly territories until he reached the confines of Boeotia. Here at the entrance of that territory, the Sun seemed to appear in a crescent shape....

[Xenophon, *Hellenica* IV, 3, 10; trans. Dakyns (1892, vol. II, p. 54).]

(ii) Agesilaus now marched through the pass of Thermopylae, traversed Phocis, which was friendly to Sparta, entered Boeotia, and encamped near Chaeroneia. Here a partial eclipse of the Sun occurred...After advancing as far as Coroneia and coming in sight of the enemy...

[Plutarch, *Life of Agesilaus*, XVII; trans. Perrin (1961, vol. V, p. 47).]

The date of the arrival of Agesilaus in Boeotia is fixed by Diodorus Siculus (XIV, 82) as the archonship of Diophantos, the second year of the 96th Olympiad, and thus 395/4 BC (Bickerman, 1980, p. 138).

Xenophon accompanied Agesilaus on his campaign and thus probably witnessed the eclipse himself. He does not give the place of observation more precisely than the border of Phocis and Boeotia. However, the subsequent battle took place in the plain of Coroneia, just within the border of Boeotia. Plutarch, who clearly relies mainly on Xenophon's narrative, names the place as Chaeroneia, his own home town. He evidently deduced this from a consideration of the route followed by Agesilaus. From the map published by Stier and Kirsten (1956, p. 19), Chaeroneia was almost on the border of Boeotia and Phocis and about 10 km to the NW of the site of the battle. The eclipse must thus have been seen within a few kilometres of Chaeroneia.

Calculations using equation (10.1) (ΔT = 15 200 sec) indicate that at Chaeroneia the eclipse would reach a magnitude of 0.92 at 9.6 h with a solar altitude of 52 deg. Although the Sun would be high in the sky at the time, Agesilaus would be marching in a south-easterly direction, facing into the Sun, so that the eclipse would be more easily noticeable.

Results

For a partial eclipse at Chaeroneia, $\Delta T < 12\,710$ or $> 14\,250$ sec.

(2) BC 188 Jul 17 (total, 1.01): Rome

The great Roman historian Livy (Titus Livius) in his work entitled '*From the Founding of the City*' records an unusual darkness in the city of Rome soon after the Ides of March.

> Then, when Marcus Valerius Messala and Gaius Livius Salinator had been inaugurated as consuls on the Ides of March, they consulted the senate... Before the new magistrates (i.e. the consuls) departed for their provinces, a three-day period of prayer was proclaimed in the name of the College of Decemvirs at all the street corner shrines because in the daytime, between about the third and fourth hours, darkness had covered everything (*tenebrae obortae fuerant*).
>
> [Livy, XXXVIII 36, 4; trans. Sage (1936, vol. XI, pp. 117–119).]

Marcus Valerius Messala and Gaius Livius Salinator were consuls in 566 AUC or 188/7 BC (Bickerman, 1980, p. 147). Although Livy (*c.* 60 BC – AD 17) wrote long after the eclipse, he researched his material with considerable care. Livy offers no explanation for the darkness, but the eclipse of Jul 17 in that year was certainly extremely large in Rome. As is well known, the Roman calendar was in chaos around this period so that it is quite plausible that the Ides of March could have taken place in the summer.

Although a more accurate rendering of the phrase *tenebrae obortae fuerant* is 'darkness had arisen', it seems clear from the declaration of the three-day period of prayer at shrines throughout the city of Rome that the darkness was awe-inspiring. If the eclipse interpretation is correct, the lack of reference to such an event could possibly be explained by its occurrence on an overcast day. Computations using equation (10.1) (ΔT = 12 500 sec) indicate that at Rome the eclipse was virtually total (mag. = 0.995) at 6.3h, the solar altitude then being 17 deg. Since sunrise was at 4.48 h, greatest phase would actually be reached during the second (seasonal) hour of the day, but it is clear that the recorded time is only intended to be very approximate.

Although the eclipse of 188 BC was almost certainly the cause of the darkness by day in Rome, the circumstances are too vague to warrant deduction of any ΔT limits from the record.

(3) AD 484 Jan 14 (total, 1.03): Athens

A very large eclipse of the Sun occurring in AD 484 was regarded as an omen of the death of the eminent Greek philosopher, Proclus. This event is described by Marinus Neapolitanus in his *Life of Proclus.* Marinus became head of the Athenian School of Philosophy in AD 485 on the death of Proclus.

> A year before his death there were various omens. There was an eclipse of the Sun which was so pronounced as to turn day into night and the darkness was deep enough for the stars to become visible; it occurred in the eastern horn of the sign of Capricorn. And the almanacs predicted another eclipse that would occur after the first year. They say that such events that are observed to happen in the heavens are indicative of things that happen on the earth; so that these eclipses clearly foretold us of the privation and departure as it were of the light of philosophy.
>
> [Marinus, *Life of Proclus*, chap. 37; trans. Rosan (1949, p. 34).]

According to Marinus (chap. 36), Proclus died in Athens 124 years after the accession of Emperor Julian (AD 361) and in the year when Nicagoras the Younger was the archon at Athens (AD 484/5). The exact date of Proclus' death is specified as the 17th day of the Athenian month Munychion and the 17th day of the Roman month April – i.e. AD 485 Apr 17.

The only large eclipse visible in Greece around the stated date occurred on AD 484 Jan 14. The Sun (long. = 295 deg) would indeed be in Capricorn at the time. If we calculate on the basis of equation (10.1) (ΔT = 5500 sec) for Athens, the eclipse would reach greatest phase (mag. 0.997) at 7.2 h, with the Sun barely above the eastern horizon (alt. = 0.8 deg). It is unfortunate that the record does not specify whether or not the eclipse was total. However, such ambiguity is not unexpected. In January, the Sun remains invisible for more than half an hour after sunrise at Athens, since it is obscured by Mt Hymettus (1 km high and about 8 km ESE of the city). Thus only the last stages of the eclipse would actually be seen.

As it happens, no star or planet brighter than about mag. 0 would be visible during this eclipse. The altitude of Venus would be very similar to that of the Sun so that the planet would also be hidden by Mt Hymettus when the eclipse was at its height. Both Jupiter and Sirius would be far below the western horizon. Mercury (mag. +0.3), 26 deg to the west of the Sun would be well placed for visibility, but although both Mars and Saturn would both be above the horizon, they would be faint (magnitudes respectively +1.5 and +0.8). Hence, although the solar altitude was so low, a very large magnitude would be required to render several stars visible. Unfortunately, because of the lack of direct reference to the degree of obscuration of the Sun, it is not possible to use the observation to set any firm limits to the value of ΔT at this epoch.

10.4 Greek lunar eclipse observations recorded in Ptolemy's *Almagest*

The timed Babylonian observations of eclipses recorded in Ptolemy's *Mathematike Syntaxis*, or *Almagest*, have already been discussed in detail in chapter 4. Similar measurements by Greek astronomers between 201

BC and AD 145 (all relating to eclipses of the Moon) are also reported in various sections of the *Almagest* (IV, 9; IV, 11; and VI, 5). These data were considered by Fotheringham (1920a) and later by Newton (1970, pp. 100 ff.), but it seems desirable to make a full re-investigation here.

Five of the Greek observations were made at Alexandria, while one (141 BC) is from Rhodes. Very probably, the observer at Rhodes was Hipparchus himself. Ptolemy (*Almagest*, III, 1) alludes to two further lunar eclipses in 146 and 135 BC which Hipparchus presumably observed, but no times are preserved.

As in the case of the Babylonian eclipses recorded by Ptolemy, the dates of the Greek observations are expressed in terms of the 365-day Egyptian year. This enabled the exact number of days between any two selected epochs to be readily calculated. Some years are numbered from the era of Nabonassar of Babylon (747 BC), but on several occasions the Kallippic cycle is used. This cycle was named after the fourth century BC Athenian astronomer Kallippos, and was closely equal to four Metonic cycles – each of 19 years. The first Kallippic cycle began in 330 BC. Ptolemy makes reference to the two subsequent cycles, starting in 254 and 178 BC. Each commenced around the time of the summer solstice.

In the various translations given below, which are all quoted from Toomer (1984), the Egyptian month numbers are given in Roman numerals after the month names. Toomer converted each date to the Julian calendar, using negative years in the BC period. In every case, the reduced date is in exact accord with that of a tabular eclipse.

In 201 BC, the Moon was seen to rise already eclipsed. Analysis of this observation follows the pattern outlined in chapter 7; note that the computed local time of moonrise is a function of ΔT, as discussed in earlier chapters.

(1) BC 201 Sep 22/23 (mag. = 0.73): Alexandria

> We will pass to the second set of three eclipses he (Hipparchus) set out, which he says were observed in Alexandria. He says that the first of these occurred in the 54th year of the Second Kallippic Cycle, Mesore [XII] 16 in the Egyptian calendar [–200 Sep 22]. In this eclipse the Moon began to be obscured half an hour before it rose, and its full light was restored in the middle of the third hour [of night]...
>
> [*Almagest*, IV, 11; trans. Toomer (1984, p. 214).]

The time of onset may have been estimated from the degree of obscuration of the Moon when it rose. Both the statement that the Moon rose eclipsed and the measurement that it ended $2\frac{1}{2}$ hours after sunset (i.e. the middle of the third hour) are of value in the determination of ΔT. It is evident from the magnitude of the time-interval between sunset and last contact that the Moon rose before maximal phase was reached.

Use of ordinal numerals in expressing the time of last contact implies the use of seasonal hours and this will be assumed here. However, as the eclipse took place so close to the equinox (which would occur on Sep 26), whether seasonal or equinoctial hours is used is unimportant. Since the seaport of Alexandria lies at the head of the flat plain of the Nile Delta, no allowance for horizon profile is necessary.

RESULTS

(i) First contact on Sep 22 before moonrise. LT of moonrise = 18.02 h, hence LT of contact < 18.02 h, UT < 15.93 h. Computed TT = 18.85 h, thus $\Delta T > 10\,500$ sec.

(ii) Maximal phase after moonrise. LT of moonrise = 18.06 h, hence LT of contact > 18.06 h, UT > 15.97 h. Computed TT = 20.39 h, thus $\Delta T < 15\,900$ sec.

Combining these limits, $10\,500 < \Delta T < 15\,900$ sec.

(iii) Last contact at 2.5 seasonal hours after sunset. LT of sunset = 18.12 h, hence length of night = 11.76 h, 1 seasonal hour = 0.98 h. LT of last contact = 20.57 h, UT = 18.48 h. Computed TT = 21.92 h, thus $\Delta T = 12\,400$ sec.

(2) BC 200 Mar 19/20 (mag. = 1.39): Alexandria

> ...He says that the next eclipse occurred in the 55th year of the same cycle, Mechir [VI] 9 in the Egyptian calendar [–199 Mar 19], that it began when $5\frac{1}{3}$ hours of night had passed, and was total...
>
> [*Almagest*, IV, 11; trans. Toomer, p. 214.]

The terminology again suggests the use of seasonal hours, but so close to the equinox the precise units are unimportant.

RESULTS

First contact on Mar 19 at 5.33 seasonal hours after sunset. LT of sunset = 17.99 h, hence length of night = 12.02 h, 1 seasonal hour = 1.00 h. LT of first contact = 23.32 h, UT = 21.48 h. Computed TT = 0.76 h (Mar 20), thus $\Delta T = 11\,800$ sec.

(3) BC 200 Sep 11/12 (mag. = 1.59): Alexandria

> ...He says that the third eclipse occurred in the same (55th) year of the Second Cycle, on Mesore [XII] 5 in the Egyptian calendar [–199 Sep 11] and that it began when $6\frac{2}{3}$ hours of the night had passed, and was total. He also says that mid-eclipse occurred at about $8\frac{1}{3}$ hours of night, that is $2\frac{1}{3}$ seasonal hours after midnight....
>
> [*Almagest*, IV, 11; trans. Toomer, p. 215.]

The time of mid-eclipse was presumably derived from measurements made at second and third contact.

RESULTS

(i) First contact on Sep 12 at 6.67 seasonal hours after sunset. LT of sunset = 18.30 h, hence length of night = 11.40 h, 1 seasonal hour = 0.95 h. LT of first contact = 0.64 h, UT = 22.60 h (Sep 11). Computed TT = 2.12 h (Sep 12), thus ΔT = 12650 sec.

(ii) Mid-eclipse at 8.33 seasonal hours after sunset. LT of mid-eclipse = 2.21 h, UT = 0.17 h. Computed TT = 3.94 h, thus ΔT = 13 550 sec.

(4) BC 174 Apr 30/May 1 (mag. = 0.62): Alexandria

> In the seventh year of Philometor, which is the 574th from Nabonassar, on Phamenoth [VII] 27/28 in the Egyptian Calendar [–173 May 0/1], from the beginning of the eighth hour till the end of the tenth in Alexandria, there was an eclipse of the Moon which reached a maximum obscuration of 7 digits from the north, so mid-eclipse occurred $2\frac{1}{2}$ seasonal hours after midnight, which corresponds to $2\frac{1}{3}$ equinoctial hours....
>
> [*Almagest*, VI, 5; trans. Toomer, p. 283.]

RESULTS

(i) First contact on May 1 at 7 seasonal hours after sunset. LT of sunset = 18.62 h, hence length of night = 10.76 h, 1 seasonal hour = 0.90 h. LT of first contact = 0.90 h, UT = 22.82 h (Apr 30). Computed TT = 1.88 h (May 1), thus ΔT = 11 000 sec.

(ii) Last contact at 10 seasonal hours after sunset. LT of contact = 3.59 h, UT = 1.51 h. Computed TT = 4.50 h, thus ΔT = 10 750 sec.

NB Ptolemy's deduction that $2\frac{1}{2}$ seasonal hours was equal to $2\frac{1}{3}$ equinoctial hours is only approximate. A better equivalence would be $2\frac{1}{4}$ seasonal hours.

(5) BC 141 Jan 27/28 (mag. = 0.26): Rhodes

> ...Again, in the thirty-seventh year of the Third Kallippic Cycle, which is the 607th from Nabonassar, Tybi [V] 2/3 in the Egyptian Calendar [–140 Jan 27/28], at the beginning of the fifth hour [of night] in Rhodes, the Moon began to be eclipsed; the maximum obscuration was 3 digits from the south. Here then, the beginning of the eclipse was 2 seasonal hours before midnight, which corresponds to $2\frac{1}{2}$ equinoctial hours....
>
> [*Almagest*, VI, 5; trans. Toomer, p. 284.]

RESULTS

First contact on Jan 27 at 4 seasonal hours after sunset. LT of sunset = 17.07 h, hence length of night = 13.86 h, 1 seasonal hour = 1.16 h. LT of first contact = 21.69 h, UT = 20.09 h. Computed TT = 22.47 h, thus ΔT = 8550 sec.

NB 2 seasonal hours corresponded better to $2\frac{1}{3}$ equinoctial hours than $2\frac{1}{2}$.

(6) AD 125 Apr 5/6 (mag. = 0.15): Alexandria

> The first eclipse we used is the one observed in Babylon... The second eclipse we used is the one observed in Alexandria in the ninth year of Hadrian, Pachon [IX] 17/18 in the Egyptian Calendar [125 Apr 5/6], $3\frac{3}{5}$ equinoctial hours before midnight. At this eclipse too the Moon was obscured $\frac{1}{6}$ of its diameter from the south.
>
> [*Almagest*, IV, 9; trans. Toomer, p. 206.]

In the subsequent account it is implied that the time refers to mid-eclipse. The date of the 'first eclipse' corresponds to 491 BC (see chapter 3).

RESULTS

Mid-eclipse on Apr 5 at 3.60 equinoctial hours before midnight, hence LT = 20.40 h, UT = 18.45 h. Computed TT = 21.32 h, thus ΔT = 10 350 sec.

10.5 The moonrise eclipse of BC 331 Sep 20/21

Several Greek and Roman writers record that an eclipse of the Moon happened about the time of a battle between the army of Alexander the Great and the Persian forces at Arbela (near the site of Nineveh, lat. 35.9 deg N, long. 44.3 deg E). Only Pliny the Elder (Gaius Plinius Secundus) notes that on the same evening the Moon was seen to rise eclipsed in Sicily. The observation at Arbela is of considerable historical importance since it enables the exact date of the battle to be fixed. However, only the Sicilian report is of any value in determining ΔT. In a section of his *Natural History* discussing the effect of the curvature of the Earth (and, in particular, of terrestrial longitude) on the visibility of eclipses, Pliny gives the following account:

> Consequently inhabitants of the East do not perceive evening eclipses of the Sun and Moon, nor do those dwelling in the West see morning eclipses, while the latter see eclipses at midday later than we do. The victory of Alexander the Great is said to have caused an eclipse of the Moon at Arbela at 8 p.m. (*noctis secunda hora*) while the same eclipse in Sicily was when the Moon was just rising (*exoriens*)... this was because the curve of the globe discloses and hides different phenomena for different localities.
>
> [Pliny, *Natural History*, II, 72; trans. Rackham (1938, vol. I, p. 313).]

In the same passage, Pliny also describes a relatively recent solar eclipse (AD 59 April 30) that was visible in Campania (southern Italy, long. 15 deg E) between the 7th and 8th hours, while in Armenia (long. 45 deg E) it occurred between the 10th and 11th hours.

Writing *c.* AD 70, and thus fully four centuries after the lunar obscuration, Pliny does not specify his source. As noted above (section 10.2),

Table 10.6 ΔT results obtained from ancient Greek timings of lunar and solar eclipses.

Year	Type	Ct	ΔT (sec)
−200	lunar	1	12 400
−199a		1	11 800
−199b		1	12 650
−199b		M	13 550
−173		1	11 000
−173		4	10 750
−140		1	8 550
+125		M	10 350
+364	solar	1	8 100
+364		M	8 300
+364		4	8 400

Ptolemy (*Geography*, I, 4) stated that the eclipse was seen at Arbela in the *fifth* hour but took place at Carthage in the second hour.

The date of the battle is given by Arrian (*Anabasis*, II, 7.6) as during the month Pyanopsion when Aristophanes was archon at Athens. Aristophanes held the position of archon in 331/0 BC (Bickerman, 1980, p. 139). Pyanopsion was the fourth month in the Athenian calendar so that, since the first month of the year began in midsummer, the date of the lunar eclipse can be firmly established as BC 331 Sep 20/21. According to Plutarch (*Life of Alexander*, XXXI) the eclipse preceded the battle by 11 days. Hence the battle itself would take place on Oct 1.

It is a pity that the precise place of observation in Sicily is not recorded. I have taken as centres for calculation the major cities of Syracuse in the east of Sicily and Panormus (i.e. Palermo, lat. = 38.13 deg, long. = −13.30 deg) in the west of the island. How far advanced the eclipse was at moonrise is not stated. Hence it can only be assumed that the Moon rose at some time between first and last contact. Of course, no allowance for horizon profile can be made.

Results *Syracuse*

(i) First contact on Sep 20 before moonrise. LT of moonrise = 18.12 h, hence LT of contact < 18.12 h, UT < 17.02 h. Computed TT = 20.77 h, thus $\Delta T > 13\,500$ sec.

(ii) Last contact after moonrise. LT of moonrise = 18.20 h, hence LT

Table 10.7 ΔT limits obtained from ancient European untimed observations of lunar and solar eclipses.

		ΔT Range (sec)	
Year	Type	LL	UL
–430	solar	12 650	11 800
–393	solar	14 250	12 710
–330	lunar	13 000	25 050
–309	solar	13 300	17 200
–200	lunar	10 500	15 900

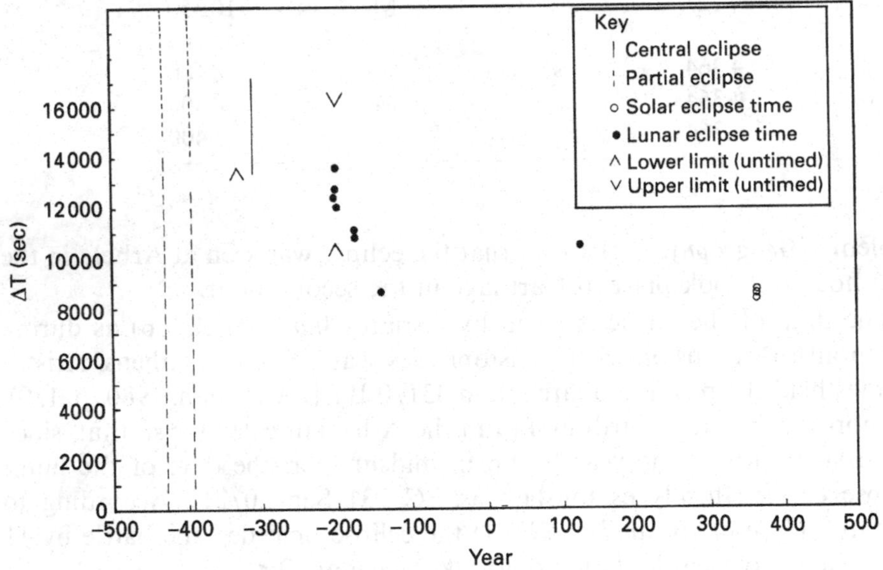

Fig. 10.6 ΔT values and limits derived from the more reliable ancient European eclipse observations.

of contact > 18.20 h, UT > 17.11 h. Computed TT = 0.07 h (Sep 21), thus ΔT < 25 050 sec.

Combining these limits yields 13 500 < ΔT < 25 050 sec.

RESULTS *Panormus*

(i) First contact before moonrise. LT of moonrise = 18.12 h, hence LT of contact < 18.12 h, UT < 17.16 h. Computed TT = 20.77 h, thus ΔT > 13 000 sec.

(ii) Last contact after moonrise. LT of moonrise = 18.21 h, hence LT of contact > 18.21 h, UT > 17.25 h. Computed TT = 0.07 h (Sep 21), thus ΔT < 24 550 sec.

Combining these limits yields $13\,000 < \Delta T < 24\,550$ sec. Taking the widest possible limits for both places yields $13\,000 < \Delta T < 25\,050$ sec.

An interesting account of the eclipse by Alexander's troops at Arbela is recorded by Curtius in his *History of Alexander*. This gives one of the few known allusions in ancient history to the red colour of the totally eclipsed Moon.

> But about the first watch the Moon in eclipse, hid at first the brilliance of her heavenly body, then all her light was sullied and suffused with the hue of blood.
>
> [Curtius, IV, 10, 1; trans. Rolfe (1946, vol. I, p. 253).]

Curtius relates that Alexander's soldiers regarded the phenomenon as a bad omen and accordingly Alexander ordered his Egyptian soothsayers to give their opinion. They foretold the defeat of the Persians, and Alexander's troops, heartened by these disclosures, proved victorious in the subsequent combat.

10.6 Conclusion

In table 10.6 are summarised the ΔT results obtained from the various timed measurements for both lunar and solar eclipses made at Alexandria and Rhodes between −200 (i.e. 201 BC) and +364 (AD 364). As in earlier chapters, negative dates are preferred to their BC equivalents.

The few usable results obtained from the untimed observations discussed in this chapter are listed in table 10.7. In this table, since the solar eclipses of −430 and −393 were described as partial, a range of ΔT is excluded. This is indicated by the lower limit being numerically greater than the upper limit. For example, in −393 either $\Delta T < 12\,700$ or $> 14\,250$ sec.

The results listed in both tables are plotted in figure 10.6. Clearly the timed data form a useful set, only the observation from Rhodes in −140 (probably by Hipparchus!) being discordant. As will be seen later, the three measurements by Theon of Alexandria in +364 lie in a gap in which few other useful observations are preserved from elsewhere.

11
Eclipse records from medieval Europe

11.1 Introduction

For the purpose of this chapter, the term 'medieval' will be loosely interpreted to mean the period between the close of the classical age in Europe (*c.* AD 500) and the beginning of the telescopic era. It thus includes the Renaissance. Numerous observations of both solar and lunar eclipses were recorded in Europe during this period, especially after AD 1000. In most cases, the date is accurately reported and the precise place of observation is known. However, with only a few exceptions, times were only crudely estimated to the nearest hour or so. Thus most observations may effectively be regarded as untimed.

As discussed in chapter 3, untimed *lunar* eclipses are of negligible value for the determination of ΔT (unless it is clearly implied that the Moon rose or set eclipsed). Most of this chapter will be devoted to the investigation of *solar* obscurations in which the Sun was either totally or very largely covered. In section 11.9, a few careful timings of solar eclipses from the early fourteenth century will also be analysed. Finally, in section 11.10, an eighth century report of an occultation of the planet Jupiter by the eclipsed Moon – a very rare event – will be discussed.

11.2 Historical sources

European reports of solar eclipses (as well as lunar obscurations) from the Middle Ages and Renaissance are mainly found in historical works, notably chronicles; only a few observations are reported in astronomical treatises. During this period the chronicle was a major literary form and numerous works of this kind were compiled at a wide variety of monasteries and towns scattered throughout much of Europe. Most texts were composed in Latin, although the vernacular was frequently used – especially after about AD 1400. Often chronicles extended over several centuries, sometimes being continued by scribes of widely differing interests. Many annalists were in the habit of including references to eclipses and other striking celestial phenomena (e.g. bright comets, meteors, and displays of the aurora borealis) along with matters of local and national concern. Despite the rarity of total solar eclipses at any one place, numerous reports of these events are preserved since there was such a large number of potential observing sites in Europe. In fact, more than two-thirds of all known medieval accounts of complete obscurations of the Sun originate from Europe.

As will be evident from some of the records translated below, few chroniclers (in common with many educated people today) had much understanding of the cause of eclipses. Further, most observers would have no advance warning when such an event was about to take place and might not notice an eclipse until it was already well under way. Nevertheless, medieval accounts are often remarkably detailed, sometimes giving a careful description of the complete disappearance of the Sun or noting the effects of the loss of daylight on people and animals.

A large number of European chronicles have been published in their original language by editors such as Muratori (1723–) and Pertz (1826–). Although I have consulted several manuscript sources (in the Vatican Library), the search for records of specific phenomena occurring at widely spaced intervals is a very tedious operation. Hence I have relied almost exclusively on the printed texts. The compilations of Muratori (*Scriptores Rerum Italicarum*) and Pertz (*Monumenta Germaniae Historica, Scriptores*) will be subsequently abbreviated to *SRI* and *MGH* respectively. Other sources consulted will be cited in full – including secondary sources where a primary source is relatively inaccessible.

Celoria (1877a, 1877b) and Ginzel (1884a, 1884b, 1918) made extensive searches of published chronicles for accounts of solar eclipses, whether of large or small magnitude. They were able to uncover numerous records which they quoted *in extenso*, leaving them in their original language. Other literature searches were made about half a century ago by Thorkelsson (1933), who concentrated on Icelandic chronicles, and Vyssotsky (1949), who restricted his attention to Russian annals. Recent careful work by Dr G. R. Levi-Donati of Perugia, Italy and by Dr Marek Zawilski of the Planetarium and Astronomical Observatory, Lodz, Poland have brought to light a number of important records of eclipses from the fourteenth and fifteenth centuries. Levi-Donati's principal papers on this theme were published in 1987 and 1989, while Zawilski's compilation – revised in 1994 – is as yet unpublished. I have no doubt that further literature searches would reveal additional important observations, but any such effort would be a major undertaking and progress seems likely to be slow.

Newton (1972b), who provided useful historical notes about many chronicles, translated most of the records cited by Celoria and Ginzel. However, several of his quotations are incomplete and his translations occasionally lack technical accuracy. In this chapter, I have either re-translated or cited reliable translations of all of the medieval European texts known to me which either (a) assert that a central eclipse was witnessed or (b) specifically deny totality or the ring phase.

Chronicles of medieval Europe can be divided into two main types: regional and local. Regional annals, typified by the *Anglo-Saxon Chronicle* and many other British chronicles, drew on a wide area. These often noted eclipses, but usually the place of observation is very uncertain. In contrast to the regional annals, monastic and town chronicles – especially those of continental Europe – were mainly concerned with their own locality. It is thus usually quite reasonable to assume that eclipses noted in these works were observed at the centre where the chronicle was compiled. Sometimes this same place is specifically mentioned in the report of an eclipse, while occasionally the annalist tells us that he was an eyewitness. In general, if an obscuration of the Sun was particularly large, descriptions from even quite neighbouring locations differ in so many details as to be obviously independent. Annals of monasteries and towns provide the main source of the observations discussed in this chapter.

A few untimed reports of total and annular eclipses from the sixteenth and early seventeenth century are contained in astronomical treatises by Clavius, Kepler, and others. These observations will also be considered, along with a few unusually careful timings of solar eclipse contacts by the fourteenth century French astronomer Johannes de Muris.

11.3 Chronological discussion

The Venerable Bede (early eighth century) was the first historian to systematically number years relative to the Christian Era, and this practice was adopted by most medieval European chroniclers. However, Spanish historians often used the inception of the Julian calendar (38 BC) instead, while both Byzantine and Russian annalists counted years from the assumed era of the creation, i.e. Sep 1 in BC 5509. Michael the Syrian, who was Patriarch of Antioch in the twelfth century, recorded many celestial phenomena in his chronicle – including total eclipses of the Sun in AD 812 and 1176. He followed the practice of the Eastern Orthodox Church by counting years from the 'era of Alexander', actually a standardised Seleucid era; on this scheme, year 1 was taken to commence on Oct 1 in 312 BC. (NB This date is very close to the Seleucid era on the ancient Macedonian calendar - i.e. Oct 7 in 312 BC.)

The Julian calendar, with its associated Kalends (*Kalendae*), Nones (*Nonae*) and Ides (*Idus*), was widely adopted in medieval Europe. In the ancient Roman calendar, the Ides occurred around the middle of the lunar month, while the Nones was so-called because it was the ninth day before the Ides. The Kalends was the first of each month. In every month of the year except March, May, July and October the Nones occurred on the 5th day and the Ides on the 13th. However, in the four months just mentioned, both the Nones and Ides took place two days later than usual (i.e. on the 7th and 15th).

Dates in a month were counted retrospectively from each of these three reference days. The day before the Kalends (i.e. the last day of the previous month) was referred to as *Pridie Kalendas.* Since inclusive counting was used, the day before the *Pridie* was known as the 'third day before the Kalends' (usually abbreviated to *III Kalendas*) and so on. For example, Christmas Day was the eighth day before the Kalends of January (*VIII Kalendas Januarias*). Similar rules applied for the Nones and Ides, the day before the Nones being *Pridie Nonas* and before the Ides *Pridie Idus.* However, in these cases the date specified occurred in the same month as the Nones or Ides. Thus the date of the eclipse of AD 840 May 5 was recorded as *III Nonas Maias* – see section 11.6.1.

The operational rules of the Julian calendar are summarised in table 11.1. The general layout of this table is based on table 12.82.1 in the *Explanatory Supplement to the Astronomical Almanac* (Seidelmann, 1992). It should be noted that the details listed for the month of February are for a common year (not divisible by 4), when February would have 28 days. Under these circumstances, Feb 28 would be *Pridie Kalendas Martias,* while in a bissextile or leap year Feb 28 would be *III Kalendas Martias* and Feb 29 *Pridie Kalendas Martias.*

In medieval Italy, days were often counted forwards from the beginning of a month (*intrante*) or backwards from the end (*stante*), depending on which was nearer. However, some chroniclers anticipated our present-day system of counting straight through the month. The weekday is often recorded in European chronicles. Sunday was the first day of the week; this was normally termed *Dies Dominica.* However, the other weekdays were usually simply rendered as *feria* 2 (Monday), *feria* 3 (Tuesday), and so on – see table 11.2.

Most recorded dates when converted to the Julian calendar (or the Gregorian calendar from AD 1582) are in *exact* accord with the tabular dates of eclipses which according to calculation were large in Europe. In particular, for all of the records translated in this chapter, the weekday is invariably given correctly. Saint's days and other Christian festivals were often used to mark days; the equivalent Julian dates are specified in individual entries below.

Although the months of the Syrian calendar – as used by Michael the Syrian – were corrupted versions of Babylonian names (see chapter 4),

Table 11.1 The Julian calendar.

	January August December	April June September November	February	March May July October
1	Kalendae	Kalendae	Kalendae	Kalendae
2	IV Nonas	IV Nonas	IV Nonas	VI Nonas
3	III Nonas	III Nonas	III Nonas	V Nonas
4	Pridie Nonas	Pridie Nonas	Pridie Nonas	IV Nonas
5	Nonae	Nonae	Nonae	III Nonas
6	VIII Idus	VIII Idus	VIII Idus	Pridie Nonas
7	VII Idus	VII Idus	VII Idus	Nonae
8	VI Idus	VI Idus	VII Idus	VIII Idus
9	V Idus	V Idus	V Idus	VII Idus
10	IV Idus	IV Idus	IV Idus	VI Idus
11	III Idus	III Idus	III Idus	V Idus
12	Pridie Idus	Pridie Idus	Pridie Idus	IV Idus
13	Idus	Idus	Idus	III Idus
14	XIX Kalendas	XVIII Kalendas	XVI Kalendas	Pridie Idus
15	XVIII Kalendas	XVII Kalendas	XV Kalendas	Idus
16	XVII Kalendas	XVI Kalendas	XIV Kalendas	XVII Kalendas
17	XVI Kalendas	XV Kalendas	XIII Kalendas	XVI Kalendas
18	XV Kalendas	XIV Kalendas	XII Kalendas	XV Kalendas
19	XIV Kalendas	XIII Kalendas	XI Kalendas	XIV Kalendas
20	XIII Kalendas	XII Kalendas	X Kalendas	XIII Kalendas
21	XII Kalendas	XI Kalendas	IX Kalendas	XII Kalendas
22	XI Kalendas	X Kalendas	VIII Kalendas	XI Kalendas
23	X Kalendas	IX Kalendas	VII Kalendas	X Kalendas
24	IX Kalendas	VIII Kalendas	VI Kalendas	IX Kalendas
25	VIII Kalendas	VII Kalendas	V Kalendas	VIII Kalendas
26	VII Kalendas	VI Kalendas	IV Kalendas	VII Kalendas
27	VI Kalendas	V Kalendas	III Kalendas	VI Kalendas
28	V Kalendas	IV Kalendas	Pridie Kalendas	V Kalendas
29	IV Kalendas	III Kalendas		IV Kalendas
30	III Kalendas	Pridie Kalendas		III Kalendas
31	Pridie Kalendas			Pridie Kalendas

Table 11.2 Days of the week in medieval Latin chronicles.

Name	Equivalent
Dies Dominica	Sunday
Feria 2	Monday
Feria 3	Tuesday
Feria 4	Wednesday
Feria 5	Thursday
Feria 6	Friday
Feria 7	Saturday

they had exact parallels with the months of the Julian calendar. However, the first month of the year corresponded with October. Only two instances of the use of this calendar will be found in the present chapter (in sections 11.6 and 11.8). The Syrian calendar was often used by medieval Arab astronomers – see chapter 13.

Seasonal hours (12 to the day and 12 to the night), following the ancient Greek and Roman practice, were commonly used throughout medieval times. On this system, the first hour commenced with sunrise, while the seventh started at noon. Very occasionally, hours were counted from sunset or midnight instead. As already emphasised above, estimates of time tended to be informal, no matter which system was used.

11.4 Interpretation of observations of large solar eclipses

In medieval Latin texts, the phrase *eclipsis solis* is frequently used to describe a solar eclipse. The term *eclipsis* is a derivation from the Greek *ekleipsis.* Expressions such as *sol obscuratus est* ('the Sun was obscured') or *sol defectus est* ('the Sun failed') are also common. Medieval accounts of large solar eclipses often mention the occurrence of darkness or the visibility of stars. It is important to stress here that neither feature is sufficient by itself to establish totality. Sensations of loss of daylight are particularly subjective, while the appearance of stars is often noted at eclipses which were only annular on the Earth's surface. An account of the eclipse of AD 1191 Jun 23 in a contemporary English chronicle would seem to suggest a particularly large magnitude:

> 1191. In the month of June on Sunday, the Vigil of the Nativity of St John the Baptist (Jun 23), the 9th day before the Kalends of July, on the 27th day of the Moon, at the 9th hour of the day, the Sun was eclipsed

> and it lasted for three hours; the Sun was so obscured that darkness arose over the Earth and stars appeared in the sky. And when the eclipse withdrew, the Sun returned to its original beauty.
>
> [*Gesta Regis Henrici II et Ricardi I*; Stubbs (1867, I, p. 339).]

The date of the eclipse, including the weekday, is correctly given. The precise place of observation in England cannot be established, but it can be calculated that even where the ring phase was visible no more than 0.95 of the solar diameter was obscured by the Moon. Although the allusion to darkness and stars would seem to imply a major eclipse, the diminution in daylight would actually be fairly small.

It is of interest to investigate the astronomical circumstances relating to those annular eclipses in which stars were reported by medieval European chroniclers. In addition to 1191, the dates of these events are as follows: AD 891 Aug 8 (Constantinople), 990 Oct 21 (Constantinople), 1033 Jun 29 (Beze, France), 1147 Oct 26 (Gembloux, Belgium), and 1310 Jan 31 (Avignon, France). Only a single star was noticed in 1147, but on each of the other occasions several stars were said to be seen. In 1033, it was stated that 'stars shone behind and in front of (i.e. to the east and west of) the Sun' (*Annales Besuenses*: *MGH*, V, 19). The following eyewitness description of the eclipse of 1310 Jan 31 from Avignon is unusually careful:

> 1310. On the last day of January at the 8th hour of the day at Avignon there was an eclipse of the Sun, and it was eclipsed in an extraordinary manner, and was notably sparkling. There appeared as if at nightfall a single star, a second was the opinion of the crowd (*vulgi*). Then a remarkable semicircle was seen and it lasted until past the ninth hour.
>
> [*Ptolomaei Lucensis Hist. eccles.* lib. XXIV; *SRI*, XI, 1232.]

Since annularity is not alleged, the proportion of the Sun actually covered may have been appreciably less than in the central zone. This remark could also apply to all the other observations whose dates are given above.

In table 11.3a are listed the following details for each of the above six eclipses: year, place of observation, magnitude in the central zone, computed magnitude at the place of observation, local time and solar altitude at greatest phase. Calculated details are obtained using a preliminary expression for ΔT – as derived from equation (8.1). It will be noted that only in one case (AD 1033) can more than about 95 per cent of the solar diameter have been obscured – even where the ring phase was visible. On this occasion, the zone of annularity ran almost parallel to the equator in the longitudes of France, and did not reach further north than latitude +46.4 deg – see figure 11.1. It may be deduced that the magnitude at Beze

Table 11.3*a* Medieval annular eclipses at which stars were seen.

Year AD	Place	Cent.	Mag.	LT (h)	Alt. (deg)
891	Constantinople	0.95	0.93	11.6	64
990	Constantinople	0.92	0.84	12.9	35
1033	Beze	0.99	0.96	11.6	65
1147	Gembloux	0.95	0.92	10.2	21
1191	England	0.95	0.95	10.9	57
1310	Avignon	0.94	0.84	13.7	27

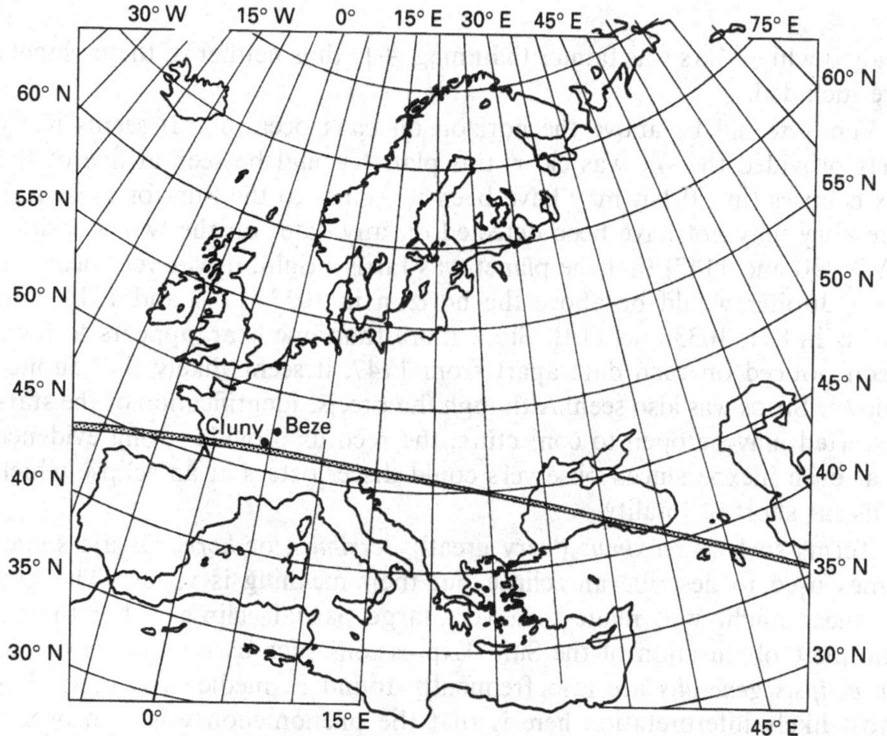

Fig. 11.1 Map showing belt of annularity at the solar eclipse of AD 1033 Jun 29 in relation to Beze and Cluny (after Schroeter, 1923).

(lat = +47.47 deg) cannot have exceeded 0.96, regardless of the precise value of ΔT. At this same eclipse, the Sun was said to become like the crescent Moon as seen from Cluny – see section 11.7.

Table 11.3b lists the magnitude and elongation from the Sun of Mercury, Venus and Jupiter and the elongation of the bright star Sirius (mag. −1.5) on each of the above dates. On every occasion, Saturn was fainter than

Table 11.3*b* Bright planets and stars at medieval annular eclipses.

	Mercury		Venus		Jupiter		Sirius
Year AD	Mag.	Elong.	Mag.	Elong.	Mag.	Elong.	Elong.
891	+0.5	27°E	−4.2	40°E	−1.5	68°E	53°W
990	−0.6	4°E	−3.6	14°E	−2.3	131°E	121°W
1033	+3.2	3°W	−3.3	28°E	−1.7	94°E	13°W
1147	−0.8	4°W	−3.6	42°W	−2.1	123°W	125°W
1191	+2.8	5°E	−3.5	1°E	−2.2	126°W	6°W
1310	+1.5	10°E	−3.4	28°E	−1.5	3°E	132°E

mag. 0 while Mars was fainter than mag. +1; thus neither of these planets are included.

Venus would be above the horizon on each occasion. It seems likely that provided the sky was clear, this planet would be seen at five of the six eclipses (in 1191 it may have been too close to the Sun for visibility). Mercury may not have been detected on any date: on the two occasions (AD 990 and 1147) that the planet was fairly bright, it was very near the Sun. Jupiter would be above the horizon in 1033, 1147 and 1310, and Sirius in 891, 1033 and 1191. Since more than one 'star' appears to have been noticed on each date apart from 1147, it seems likely that Jupiter and/or Sirius was also seen. Although the precise identification of the stars reported may be open to conjecture, the records provide strong evidence that even inexperienced observers could discern stars at an eclipse which falls far short of totality.

Terms such as *maximus* ('very great'), *terribilis*, or *horribilis* are sometimes used to describe an eclipse but their meaning is vague. The first of these might well relate to a very large partial eclipse rather than a complete obscuration of the Sun. Expressions such as *eclipsis universalis* or *eclipsis generalis* are also frequently found in medieval annals. The most likely interpretation here is that the phenomenon was seen over a wide area – as indicated by reports from travellers – rather than that the whole Sun was covered; expressions of this kind are seldom accompanied by allusions to darkness or the other characteristic features of totality. The phrase *tenebrae in toto orbe* is also ambiguous; in principle, it can either be rendered 'darkness over the whole world' or 'darkness in the whole disk (of the Sun)'. The former alternative – a hyperbole similar to the equally common term *tenebrae in universa terra* ('darkness over the whole Earth') – would seem to be more likely. This could simply mean that the whole sky was darkened, giving the impression that the phenomenon was worldwide.

In view of the above remarks, I have not assumed totality unless the complete disappearance of the Sun is clearly described. This means rejecting a number of unusually vivid records, such as the following from Toledo which relates to the eclipse of AD 1239 Jun 3:

> The Sun was obscured on Friday at the 6th hour of the day, and it lasted for a while between the 6th and 9th hours and it lost all its strength (*toda su fuerza*) and there was as though night. There appeared many stars, and then the Sun grew bright again of its own accord, but for a long time it did not regain the strength that it usually has. Era 1277 (Julian = AD 1239).
>
> [*Anales Toledanos Segundos*; in Spanish; Florez (1747), XXIII, 409.]

Although the above account relates that the Sun lost *all* its strength, this could simply mean that what remained of it could be viewed without dazzling the eyes – rather than necessarily implying totality. Other vivid descriptions in which either totality is not clearly asserted or the place of observation is in doubt are cited in section 11.8.

Some accounts of large eclipses may well have been inspired by the Passion narrative in the Synoptic Gospels. Indeed several texts closely resemble the wording in the Vulgate (Latin) versions of these gospels. In particular, it is well known that St Luke (XXIII, 45) attributes the darkness at the Crucifixion to a solar eclipse (Sawyer, 1972). The fact that this darkness lasted three hours may account for some of the excessive durations of darkness found in a number of medieval records of large solar eclipses – extreme examples being three hours (Coimbra, AD 1239) and four hours (Reichersberg, AD 1241) – see section 11.6.1 below. A possible alternative explanation is provided by the following report of the eclipse of 1860 Jul 18, as observed in Sudan by Bey (1860):

> But at the moment of totality, all became silent and dumb. Neither a cry nor a rustling, nor even a whisper (was heard), but everywhere there was anxiety and consternation. To everyone the two minutes of the eclipse were like two hours. I do not exaggerate or imagine any of these details. Several people whom I questioned after the eclipse regarding the duration of totality replied that it had lasted for two hours...

Evidently the shock caused by the onset of sudden and intense darkness can cause the unsuspecting observer to temporarily lose all sense of the passage of time.

In marked contrast to accounts of this kind, some medieval reports of total solar eclipses merely state that the Sun was completely obscured – without including any descriptive remarks. For instance in AD 1485, the *Annales Mellicenses* from Melk (Austria) only report that

On the 16th day of March at the 4th hour there was a complete (*plena*) eclipse of the Sun.

[*MGH*, IX, 524.]

Because of the lack of supporting evidence in such cases, I have felt that it is unwise to make use of the ΔT limits derived from such brief records. It seems quite possible here that the annalist summarised a more detailed description and inferred totality even when this was not originally alleged.

When describing the appearance of the Sun at eclipses which were generally annular, medieval European accounts often use ambiguous terminology. As noted in chapter 3, even at a considerable distance from a zone of annularity the angle between the cusps of the crescent is fairly small. In consequence, it is not always possible to decide whether a complete or broken ring is referred to. For instance, records from Cologne (*Annales Colonienses Maximi*: *MGH*, XVII, 822) and Hirschau (*Trithemii Annales Hirsaugienses*: *MGH* I, 512) both state that at the eclipse of AD 1207 Feb 27 'many testified that they saw a human head in the Sun'. These two reports probably share a common origin. Nevertheless, a rather similar metaphor occurs in the description of the eclipse of AD 1147 Oct 25 from Gembloux (*Continuatio Gemblacensis Sigeberti Chronicon*: *MGH* VI, 389). It is tempting to assume that there is an oblique reference to the ring phase in such cases, but a very large partial eclipse might well suffice instead. When assessing medieval European records of annular eclipses, I have preferred to adopt a very cautious view, bearing in mind that the observers were not astronomers.

11.5 Place co-ordinates

The eclipse observations discussed in this chapter were made at a wide variety of locations. Table 11.4 lists the geographical co-ordinates (in degrees and decimals) of the various places (towns, monasteries, etc.) which are referred to. I have followed the convention adopted throughout this book that northern latitudes and westerly longitudes are positive. In most cases I have given the modern names of towns. A useful reference work which gives the modern equivalents of early European place names is the *Orbis Latinus*, edited by Plechl (1972). If the name of a city has an established English equivalent – such as Florence, Prague or Vienna – I have adopted this. For historical reasons, I have preferred to use Antioch (rather than the modern name of Antakya) and also Constantinople (rather than Istanbul). The former city was also called Antakya in the medieval Islamic period (see also chapter 13). Obviously, no naming scheme can be fully satisfactory. I have determined nearly all geographic co-ordinates from careful measurements on 1:50 000 maps. I am grateful to Mr B. Allaker – formerly of the Map Library, University of Newcastle upon Tyne – for much assistance in the use of the excellent map collection which he supervised.

11.6 Eclipses which were central at a known location

In this section, both total and annular eclipses are considered. In accordance with my comments in section 11.4, I have divided the records into two categories: those which give a detailed account of a central eclipse (section 11.6.1) and those less reliable reports which simply mention the occurrence of the total phase (11.6.2). As it happens, all entries in the latter category date from either AD 1433 or 1485. Although ΔT limits will be calculated in each case, only the results obtained in section 11.6.1 will be considered further.

Each entry discussed below begins with the calculated Julian date (including the weekday). This is followed by a statement of the type of eclipse (together with the computed magnitude in the central zone for reference) and the place of observation. A full translation of the relevant part of each text is given, together with its source. Most records are in Latin; otherwise the language is specified. Unless stated to the contrary, all translations both here and elsewhere in this chapter are by the author – with occasional help from colleagues. Following the translation are brief comments, together with the derived range of ΔT. Finally, I have given the computed local time and solar altitude at maximum phase based on the provisional ΔT formula adopted earlier – equation (8.1). These data are for reference only, and do not influence the selection of the material in any way. It should be noted that if two or more accounts of the same eclipse are preserved from different locations, I have listed the places alphabetically.

11.6.1 Detailed descriptions

(1) AD 840 May 5 [Wednesday] (total, mag. = 1.08): Bergamo (Italy)

> In the third year of the Indiction, the Sun was hidden from this world and stars appeared in the sky as if it were midnight, on the third day before the Nones of May (May 5) during the Litanies of Our Lord. There was great distress, and while the people beheld it, many thought that this age would last no longer. But while they were contemplating these simple things, the Sun shone again and trembling as it were began to escape from its former shade.
>
> [*Andreas Bergomatis Chronicon*; *MGH*, III, 235.]

The Indiction was a 15-year fiscal period, the first of which commenced in AD 312. The third year of the appropriate Indiction began on AD 839

Table 11.4 Geographic co-ordinates for medieval European sites.

Place	Latitude (deg)	Longitude (deg)
Antioch	36.20	−36.17
Arezzo	43.47	−11.90
Augsburg	48.37	−10.90
Avignon	43.93	−4.80
Belvoir	52.90	−0.73
Bergamo	45.40	−9.67
Bergen	60.38	−5.33
Beze	47.47	−5.27
Braunschweig	52.25	−10.50
Brauweiler	49.87	−7.75
Celle	52.63	−10.06
Cesena	44.13	−12.25
Cluny	46.42	−4.65
Coimbra	40.22	+8.42
Constantinople	41.02	−28.98
Erfurt	50.97	−11.03
Evreux	49.06	−1.18
Farfa	42.22	−12.70
Florence	43.77	−11.25
Foligno	42.95	−12.72
Forcalquier	43.93	−5.77
Gembloux	50.57	−4.70
Hamburg	53.55	−10.00
Heilsbronn	49.35	−10.83
Karlstejn	49.95	−14.20
Kerkrade	50.87	−6.07
Kobrin	52.27	−24.37
Krakow	50.05	−19.92
Lanercost	54.95	+2.72
Liege	50.63	−5.58
Magdeburg	52.17	−11.63
Malmesbury	51.60	+2.10
Marola	44.50	−10.50
Marseille	43.30	−5.37
Melk	48.23	−15.35

Table 11.4 – *continued.*

Place	Latitude (deg)	Longitude (deg)
Mytho	53.70	−25.10
Montpellier	43.62	−3.88
Naples	40.83	−14.25
Neider Alteich	48.87	−12.70
Novgorod	58.50	−31.33
Nurnberg	49.45	−11.08
Orkney	58.98	+2.97
Perugia	43.12	−12.38
Prague	50.10	−14.42
Pskov	57.80	−28.43
Reichersberg	48.07	−13.37
Rome	41.90	−12.48
St Blasien	47.77	−8.13
St Germain	48.87	−2.33
Salzburg	47.80	−13.05
Siena	43.32	−11.33
Split	43.50	−16.45
Stade	53.62	−9.48
Toledo	39.87	+4.03
Vienna	48.20	−16.37
Vigeois	45.38	−1.52
Voormezele	50.82	−2.87
Vysehrad	50.07	−14.42
Wroclaw	51.08	−17.00
York	53.97	+1.08
Zbraslav	49.97	−14.40
Zwiefalten	48.23	−9.45

Sep 1. This vivid eye-witness record is unusual in that it not only alludes to the complete disappearance of the Sun but also notes its emergence after the total phase. The 'trembling' may be a reference to the formation of Baily's beads after totality (Muller, 1975, p. 8.47). Although the chronicle of Andreas of Bergamo covers much of Lombardy (N Italy), he was a contemporary and may well be reporting his own observation.

RESULTS

For totality at Bergamo, $1610 < \Delta T < 6800$ sec.

In northern Italy, the zone of totality ran almost parallel to the equator so that the observation at Bergamo is satisfied by a very wide range of ΔT.

(NB local time of maximum phase = 13.4 h; solar altitude = 57 deg.)

(2) AD 968 Dec 22 [Tuesday] (total, mag. = 1.03): Constantinople

> When the Emperor was waging war in Syria, at the winter solstice there was an eclipse of the Sun such as has never happened apart from that which was brought on the Earth at the Passion of our Lord on account of the folly of the Jews... The eclipse was such a spectacle. It occurred on the 22nd day of December, at the 4th hour of the day, the air being calm. Darkness fell upon the Earth and all the brighter stars revealed themselves. Everyone could see the disc of the Sun without brightness, deprived of light, and a certain dull and feeble glow, like a narrow headband, shining round the extreme parts of the edge of the disc. However, the Sun gradually going past the Moon (for this appeared covering it directly) sent out its original rays and light filled the Earth again... At the same time I myself was also staying in Constantinople, undertaking a liberal course of study...
>
> [*Leonis Deaconis Historiae*, lib. IV, cap. 11; Niebuhr (1828, p. 72); in Greek, with a rendering by the editor into Latin.]

Niebuhr, in his commentary, gives the year as AD 968. Apart from the eclipse of AD 968 Dec 22 (the month and day of which are correctly given in the text), no other winter obscuration of the Sun could have been total at Constantinople for several decades. Leo, who was aged 38 at the time, includes the above eye-witness description in his history of Constantinople. He stresses that he was there when the eclipse occurred; the reference to Syria is merely an aside. Leo's splendid account contains the earliest known reference to the corona which is definitely datable. It is also one of the remarkably few surviving allusions to this phenomenon before the eighteenth century. From Leo's description of the corona as a narrow band, it would appear that the Sun was near minimum activity at the time. Later historians of Constantinople (such as Cedrenus) mention this same eclipse but in much less detail – see Ginzel (1884a). Presumably they are merely drawing on Leo's account.

In the central and eastern Mediterranean, the track of totality ran almost parallel to the equator (see figure 3.4) and hence this eclipse would be total at Constantinople for a wide range of ΔT. Since the zone of totality came no further south than 40.1 deg N at any point on the Earth's surface, an observation from Syria is out of the question.

RESULTS

For totality at Constantinople, $1580 < \Delta T < 5400$ sec.
(NB local time of maximum phase = 11.4 h; solar altitude = 25 deg.)

(3) AD 968 Dec 22 [Tuesday] (total, mag. = 1.03): Farfa (Italy)

> 969. The Sun was in darkness (*fuit in tenebris*) on the 22nd day of the the month of December.
>
> [*Annales Farfenses, MGH*, XI, 589.]

Although brief, this record from the chronicle of the monastery of Farfa still seems to assert that the Sun was completely obscured. There is clearly an error of one year in the date; both the month and day are correct. No other eclipse would be large in Italy for many years before or after AD 968.

As for Constantinople, a wide range of ΔT would render this eclipse total at Farfa.

RESULTS

For totality at Farfa, $-3200 < \Delta T < 2600$ sec.
(NB local time of maximum phase = 9.8 h; solar altitude = 18 deg.)
Combining the results from the Constantinople and Farfa observations yields $1580 < \Delta T < 2600$ sec.

(4) AD 1124 Aug 11 [Monday] (total, 1.05): Novgorod (Russia)

> (6632.) In the month of August on the 11th day, before the evening service, the Sun began to diminish and perished completely. Great fright and darkness were everywhere. And the stars appeared and the Moon (*sic*). And the Sun began to augment and became full again and everyone in the town was very glad.
>
> [*Novorodskaya I Letopic*; in Russian; trans. Vyssotsky (1949, p. 8).]

A copy of the manuscript version of the above text is illustrated in figure 11.2.

The year 6632 – numbered from the Byzantine era – corresponds to AD 1123/4. Other chronicles of Novgorod also give similar descriptions of the eclipse. The 'town' mentioned in the text can be no other than Novgorod itself, the Russian capital of the time.

RESULTS

For totality at Novgorod, $960 < \Delta T < 2700$ sec.
(NB local time of maximum phase = 14.4 h; solar altitude = 37 deg.)

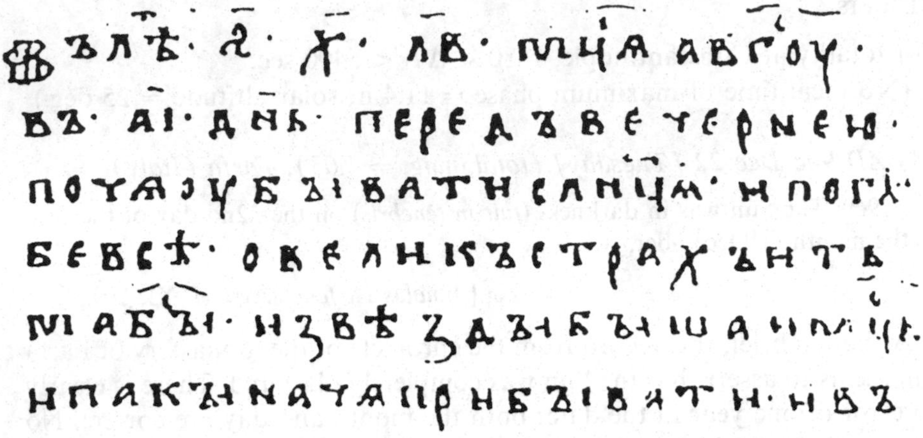

Fig. 11.2 Account of the total solar eclipse of AD 1124 Aug 11 from Novgorod as reported in the *Novorodskaya I Letopic*. (Courtesy: Dr Marek Zawilski.)

(5) AD 1133 Aug 2 [Wednesday] (total, 1.07): Augsburg (Germany)

(1133.) Duke Frederick... set fire to the town of Augsburg and killed many of its citizens... An eclipse of the Sun occurred on the 4th day before the Nones of August at midday for about one hour, such as is not seen in a thousand years. Eventually the whole sky was dark like night, and stars were seen over almost the whole sky. At length the Sun, emerging from the darkness, appeared like a star, afterwards in the form of a new Moon; finally it assumed its original form.

[*Honorii Augustodensis: Summa Totius et Imagine Mundi*; *MGH* X, 131.]

This is the last entry in the chronicle of Honorius, who at the time was Presbyter of Augsburg. He may well have seen the eclipse himself; certainly his account is based on an eye-witness description.

RESULTS

For totality at Augsburg, $-60 < \Delta T < 1150$ sec.

(NB local time of maximum phase = 12.5 h; solar altitude = 57 deg.)

(6) AD 1133 Aug 2 [Wednesday] (total, 1.07): Heilsbronn (Germany)

In the year of the Incarnation of our Lord 1133 ... on the 4th day before the Nones of August (Aug 2), the 4th day of the week (Wednesday) when the day was declining towards the ninth hour, the Sun in a single moment became as black as pitch, day was turned into night, very many (*plurimae*) stars were seen, objects on the ground appeared as they usually do at night...

[*Notae Halesbrunnenses*; *MGH*, XVI, 13.]

The *Notae Halesbrunnenses* consists of no more than three entries by an unknown hand on the last page of another chronicle. All relate to major natural events: an earthquake (AD 1117), eclipse (1133) and a locust plague (1338). In view of the title of this brief series of records, there is no particular reason for doubting that it originates from the monastery of Heilsbronn.

The description of the eclipse, whose date and weekday (Wednesday) are correctly assigned, is clearly that of an eye-witness. Several accounts from various parts of Europe mention that an unusual number of stars were visible on this occasion, while various English annals – although not alleging totality – remark that 'stars were seen around the Sun'. In particular, Venus (mag. −3.5) and Mercury (mag. −0.5) would be prominent on opposite sides of the Sun (respectively 29 deg W and 13 deg E), while several of the brighter stars would be well placed for visibility.

RESULTS

For totality at Heilsbronn, 320 sec $< \Delta T <$ 1570 sec.
(NB local time of maximum phase = 12.5 h; solar altitude = 56 deg.)

(7) AD 1133 Aug 2 [Wednesday] (total, 1.07): Reichersberg (Austria)

1133. That great eclipse of the Sun occurred on the 4th day before the Nones of August, the 27th day of the Moon, the 13th year of the Indiction. After midday, between the 7th and 8th hours, an eclipse of the Sun was seen in Leo ... Very many (*plurimae*) stars were seen near the Sun; the hearts of many were transfixed, despairing of the light. The Sun, as if it did not exist, was entirely concealed; for about half an hour it was like night. The face of the world was sad, terrible, black, wonderful...

[*Chronicon Magni Presbyterii*; *MGH*, XVII, 454.]

The annalist Magnus, who was attached to the monastery of Reichersberg, died in 1195. In the above account, he correctly states that the Sun (whose longitude was 136 deg) was in Leo.

RESULTS

For totality at Reichersberg, 540 sec $< \Delta T <$ 1740 sec.
(NB local time of maximum phase = 12.8 h; solar altitude = 56 deg.)

(8) AD 1133 Aug 2 [Wednesday] (total, 1.07): Salzburg (Austria)

1133. In this year on the 4th day before the Nones of August in the heat of midday the Sun suddenly disappeared (*subito sol disparuit*) and a little afterwards it seemed terribly darkened over (*obtenebratus*) like sackcloth of hair (*saccus cylicinus*); and stars also appeared in the sky.

[*S. Rudperti Salisburgensis Annales Breves*; *MGH*, IX, 758.]

The *Annales Breves* is just one of the chronicles of the monastery of St Peter in Salzburg. It is considered to be original after AD 1060. Newton (1972b, p. 266), reading *saccus cyclinus* for the *saccus cylicinus* of the text, inferred that the eclipsed Sun looked like 'a round bag' and both he and Muller (1975, pp. 8.52 and 8.53) regarded this as an allusion to the corona. This is no more than a supposition, but even if the original expression *saccus cylicinus* is retained, both the sudden disappearance of the Sun and its dark appearance are indicative of totality.

RESULTS

For totality at Salzburg, 280 < ΔT < 1460 sec.

(NB local time of maximum phase = 12.8 h; solar altitude = 57 deg.)

Combining the ΔT ranges deduced from the observations in AD 1133 yields 540 < ΔT < 1150 sec.

(9) AD 1147 Oct 26 [Sunday] (annular, 0.96): Brauweiler (Germany)

> 1147. On Sunday, the 7th day before the Kalends of November (Oct 26), a solar eclipse occurred at the 3rd hour and persisted until after the 6th. This eclipse stood fixed and motionless for a whole hour, as noted on the 'clock' (*horologio*)... During this hour a circle (*circulus*) of different colours and spinning rapidly (*maximo rotatu*) was said to be in the way.
>
> [*Annales Brunwilarenses*; *MGH*, XVI, 727.]

Once again, the recorded date (including the weekday) is accurate. Immediately after the word *horologio*, the text is damaged. Although a little obscure, this report from the monastery of Brauweiler seems to make a direct reference to the ring phase; the obstacle in the way of the Sun (i.e. the Moon) was said to be circular. The duration of annularity is grossly exaggerated, possibly because the attention of the observers was intently fixed on the Sun; the 'clock' may have been a sundial.

RESULTS

For annularity at Brauweiler, 300 sec < ΔT < 1190 sec.

(NB local time of maximum phase = 10.9 h, solar altitude = 23 deg.)

(10) AD 1176 Apr 11 [Sunday] (total, mag. = 1.06): Antioch

> In this year 1487 (Seleucid), on New Sunday, the 11th of the month of Nisan (= April), at daybreak, at the end of Office, that is, after the reading of the Gospel, the Sun was totally obscured; night fell and the stars appeared; the Moon itself was seen in the vicinity of the Sun. This was a sad and terrifying sight, which caused many people to lament with weeping; the sheep, oxen and horses crowded together in terror. The darkness lasted for two hours; afterwards the light returned. Fifteen days after, in this month of Nisan at the decline of Monday, at dusk, there was

an eclipse of the Moon in the part of the sky where the eclipse of the Sun had taken place...

[*Chronicle* of Michael the Syrian, Book XX, chap. 3; in Syriac; trans. from the rendering into French by Chabot (1905, vol. III, p. 367).]

The Syriac text, as printed in Chabot's edition, is illustrated in figure 11.3.

The year 1487 Seleucid is equivalent to AD 1175/6. For both the solar and lunar eclipses, the recorded dates are accurate – equivalent to AD 1176 Apr 11 and 26 respectively. The principality of Antioch had been founded by European Christians in AD 1098 in territory captured from the Muslims during the First Crusade. It endured until 1268. Michael, who had been Patriarch of Antioch (now known as Antakya) since AD 1166, was writing as a contemporary. In view of the ecclesiastical references in the eyewitness account of the solar eclipse, he probably experienced the phenomenon himself. There is no particular reason for assuming a place of observation other than Antioch.

RESULTS

For totality at Antioch, $-190 < \Delta T < 1600$ sec.
(NB local time of maximum phase = 7.1 h; solar altitude = 20 deg.)

(11) AD 1185 May 1 [Wednesday] (total, 1.07): Novgorod

Several chronicles of Novgorod give similar descriptions of this eclipse, but the following accounts are especially detailed:

(i) (6693.) On the first day of the month of May, on the day of the Saint Prophet Jeremiah, on Wednesday, during the evening service, there was a sign in the Sun. It became very dark, even the stars could be seen; it seemed to men as if everything were green, and the Sun became like a crescent of the Moon, from the horns of which a glow similar to that of red-hot charcoals was emanating. It was terrible to see this sign of the Lord.

[*Lavrentievskaya Letopis*; in Russian; trans. Vyssotsky (1949, p. 9).]

(ii) (6693.) On the first day of the month of May, during the ringing of the bells for the evening service, there was a sign in the Sun. It became very dark for an hour or longer and the stars were visible and to men everything seemed as if it were green. The Sun became like a crescent of the new Moon and from its horns a glow like a roasting fire was coming forth and it was terrible to see the sign of the Lord. Then the Sun cleared and we were happy again.

[*Novgorodskaya II Letopis*; in Russian; trans. Vyssotsky (1949, p. 10).]

Fig. 11.3 Copy of the Syriac text showing Michael the Syrian's account of the total solar eclipse of AD 1176 Apr 11 and a total lunar eclipse 14 days later as seen at Antioch. (Chabot, 1905.)

The year 6693 (Byzantine) corresponds to AD 1185/6. The above records, although both originating from Novgorod, differ to some extent. The 'roasting fire' or 'glow similar to that of red-hot charcoals' emanating from the horns of the crescent are almost certainly allusions to the chromosphere. This only becomes visible when an eclipse is on the verge of totality. In this context, it is intriguing that Captain Bullock – see Ranyard (1879, p. 114) – described the total solar eclipse of AD 1868 Aug 18 in closely similar terms:

> About two seconds before (emersion), and at the position where the Sun reappeared, there suddenly broke out a thin rim of beaded fire like intensely glowing coals, which I first took to be the Sun's limb, and was hesitating whether or not I should call 'time' when the Sun itself burst out in its splendour.

Although the eclipse of AD 1185 is not described as total, the most likely interpretation of the records in the *Novgorodskaya* chronicle is that the emergence of an unusually bright chromosphere signalled the end of the darkness of totality.

In the longitudes of western Russia, the track of totality ran almost parallel to the equator and hence this eclipse would be fully complete at Novgorod for an extremely wide range of ΔT (see also Fig. 3.13).

RESULTS

For totality at Novgorod, $-2200 < \Delta T < 10\,500$ sec.

(NB local time of maximum phase = 16.7 h; solar altitude = 25 deg.)

(12) AD 1239 Jun 3 [Friday] (total, 1.08): Arezzo (Italy)

> While I was in the city of Arezzo, where I was born, and in which I am writing this book, in our monastery, a building which is situated towards the end of the fifth latitude zone (*clima*), whose latitude from the equator is 42 and a quarter degrees and whose westerly longitude is 32 and a third, one Friday, at the 6th hour of the day, when the Sun was 20 deg in Gemini and the weather was calm and clear, the sky began to turn yellow and I saw the whole body of the Sun covered step by step and it became night. I saw Mercury close to the Sun, and all the animals and birds were terrified; and the wild beasts could easily be caught. There were some people who caught birds and animals, because they were bewildered. I saw the Sun entirely covered for the space of time in which a man could walk fully 250 paces. The air and the ground began to become cold; and it (the Sun) began to be covered and uncovered from the west.
>
> [Ristoro d'Arezzo: *Delle composizione del mondo*, lib. I, cap. XVI; in Italian; quoted by Celoria (1877a).]

This is just one of several detailed accounts of the total eclipse of AD 1239 from southern Europe: Portugal, Spain, France, Italy and Croatia – see below.

Although the dates of birth and death of the philosopher Ristoro d'Arezzo are not known, he completed his book entitled *Delle composizione del mondo* in AD 1282. This is the oldest surviving scientific treatise in Italian. Although he gives no date for the eclipse, it may be readily identified as 1239 Jun 3 (a Friday). The solar longitude was then 79 deg (i.e. 19 deg in Gemini) – almost exactly the value given in the text. The only other large eclipse in Italy for many years occurred on Oct 6 (a Sunday) in AD 1241.

Ristoro's estimate of his latitude is more than a degree in error (correct figure 43.47). Possibly the longitude was measured from Baghdad.

This vivid account is the earliest known which gives a meaningful estimate of the duration of totality; near the central line the computed duration of totality is around 5 min 45 sec, during which time a man might perhaps walk 250 Roman (double) paces – Muller (1975, p. 8.62). On this occasion, Mercury (mag. +1.0) would be 21 deg west of the Sun, but the much more brilliant planet Venus (mag. −3.5) would be only 0.8 deg from the solar limb. Presumably Ristoro confused the two planets; he would be aware that Mercury is normally closer to the Sun than Venus.

In the longitudes of Italy, the track of totality ran almost parallel to the equator, and hence this eclipse would be total at Arezzo and other Italian sites (see below) for a wide range of ΔT. This is unfortunate, in view of both the quantity and quality of the available observations.

RESULTS

For totality at Arezzo $-4350 < \Delta T < 3600$ sec.
(NB local time of maximum phase = 13.2 h; solar altitude = 64 deg.)

(13) AD 1239 Jun 3 [Friday] (total, 1.08): Cesena (Italy)

1239. On Friday at the beginning of June, after the 9th hour, the Sun was covered with darkness and it became completely black. It remained like this for the space of an hour, and the Moon was in front of it. Almost all of the stars were manifestly seen in the sky and this appeared plainly to everyone. There was also a certain fiery aperture (*foramen ignitum*) in the Sun's disc on the lower part. The Moon itself was on the 29th day. Night arose over the whole Earth. In verse:

'In the year one thousand, two hundred and thirty-nine
When June was beginning; on the third day:
The Sun was obscured, with its disc covered with darkness,
In full daylight the Sun became without light.

For a whole hour the Sun was dead and remote from us,
This marvel happened on the sixth day of the week'.

[*Annales Caesenates*; *SRI*, XIV, 197.]

The 'fiery aperture' almost certainly refers to a major prominence on the Sun, one of the very few such observations in the entire pre-telescopic period.

RESULTS

For totality at Cesena $-3300 < \Delta T < 2700$ sec.
(NB local time of maximum phase = 13.3 h; solar altitude = 64 deg.)

(14) AD 1239 Jun 3 [Friday] (total, 1.08): Coimbra (Portugal)

On the 3rd day before the Nones of June (Jun 3), on the same day that Christ suffered, namely the 6th day of the week (Friday), and at the same time that darkness occurred over the whole Earth at the Passion of our Lord, namely from the 6th to the 9th hours of the era 1237, there occurred a sign such has never happened since the Passion of our Lord until the present day. There was indeed night between the 6th and 9th hours and the Sun became as black as pitch and the Moon (*sic*) and many stars appeared in the sky. Then the receding of the darkness of night was followed by the receding and recovering of the Sun's original clarity. Many men and women assembled in the Church of the Holy Cross in Coimbra... everywhere the rays of the Sun penetrated into some hole (*foramen*).

[*Chronicon Conimbricense*, III; Florez (1747, vol. II, p. 336).]

Here we have a vivid eyewitness account in one of the chronicles of the city of Coimbra, then capital of Lusitania (Portugal). The Julian year is mistakenly given as 1237 rather than 1277, the scribe having written MCXXXVII rather than MCLXXVII. However, since the month and day are correct, the date is clearly AD 1239 Jun 3. Possibly the 'hole' refers to the prominence seen at Cesena, but the interpretation is obscure.

RESULTS

For totality at Coimbra, $-500 < \Delta T < 1450$ sec.
(NB local time of maximum phase = 11.2 h; solar altitude = 70 deg.)

(15) AD 1239 Jun 3 [Friday] (total, 1.08): Florence

1238. On the 3rd day of June, the whole of the Sun was obscured at the sixth hour and it remained obscured for several hours and from day it became night and the stars appeared; so that many people ignorant of the course of the Sun and the other planets marvelled greatly...

[*Storie Fiorentina*, VI; in Italian; quoted by Celoria (1877a).]

Note the error of a year. A slightly briefer account, giving the correct year, is to be found in the *Istoria Fiorentina* (*SRI*, VIII, 967).

RESULTS

For totality at Florence $-4100 < \Delta T < 3050$ sec.
(NB local time of maximum phase = 13.2 h; solar altitude = 65 deg.)

(16) AD 1239 Jun 3 [Friday] (total, 1.08): Montpellier (France)

> The King (James the Conqueror) entered the city of Montpellier on Thursday the 2nd of June of the year 1239; and on the next day, Friday, between midday and the ninth hour, the King writes that the Sun was eclipsed in a way people did not remember ever having seen before, because it was entirely covered by the Moon and the day grew so dark that one could see the stars in the sky.
>
> [Zurita, *Anales de la Corona de Aragon*, Lib. III, cap. 36; in Spanish; quoted by Celoria (1877a).]

There is a similar account in Latin, noting that the eclipse caused sudden darkness, in the *Indices rerum ab Aragoniae Regibus gestarum*; this is also cited by Celoria (1877a).

RESULTS

For totality at Montpellier, $-6050 < \Delta T < 1460$ sec.
(NB local time of maximum phase = 12.5 h; solar altitude = 69 deg.)

(17) AD 1239 Jun 3 [Friday] (total, 1.08): Siena (Italy)

> 1239, on Friday at the 6th hour, the Sun began to be obscured as if by a veil and was covered in a clear sky. At the ninth hour it was totally obscured, whence it gave no light; and as if a dark night arose with the result that a starry sky was seen, as on a clear night. People lit lamps in houses and shops. After some space of time it gradually became uncovered and restored to Earth, with the result that before the evening hour it was restored to its brilliance.
>
> [Archivio del Duomo di Siena, 106; *SRI*, XV, pp. 25–26.]

Note the excessive time-interval between first sighting of the eclipse and the onset of totality. This should have been no more than about one hour.

RESULTS

For totality at Siena, $-4700 < \Delta T < 3600$ sec.
(NB local time of maximum phase = 13.2 h; solar altitude = 65 deg.)

(18) AD 1239 Jun 3 [Friday] (total, 1.08): Split (Croatia)

At the same time, AD 1239 on the third day from the beginning of the month of June, a wonderful and terrible eclipse of the Sun occurred, for the entire Sun was obscured, and the whole of the clear sky was in darkness. Also stars appeared in the sky as if during the night, and a certain greater star shone beside the Sun on the western side. And such great fear overtook everyone, that just like madmen they ran about to and fro shrieking, thinking that the end of the world had come. However, it was a Friday, the 30th day of the (lunar) month. And although the same defection of the Sun appeared throughout the whole of Europe, it was not however spoken of in Asia and Africa.

[*Ex Thomae Historia Pontificum Salonitanorum et Spalatinorum*; *MGH*, XXIX, 584–585.]

This chronicle of events before and during the time of Archbishop Roger (AD 1249–1266) was compiled by Thomas, Archdeacon of Split (= Spalato) who died in 1268. Only two years after the eclipse, the same writer records another great eclipse, but although on this occasion (AD 1241 Oct 6) he makes no mention of the complete disappearance of the Sun, his account seems worth quoting in full. The time-interval between two such major events is remarkably short.

In this same year also, namely 1241 from the Incarnation, on the 6th day from the beginning of October, on Sunday, the Sun was again eclipsed and all the air was darkened. There was great terror among everyone, just as in that eclipse which happened three years previously, as we have attested above...

[*Ex Thomae Historia Pontificum Salonitanorum et Spalatinorum*; *MGH*, XXIX, 585.]

RESULTS

For totality at Split in AD 1239, $-3200 < \Delta T < 4600$ sec.

(NB local time of maximum phase = 13.7 h; solar altitude = 61 deg.)

In summary, only the observation from Coimbra provides reasonably narrow limits for ΔT in AD 1239 (i.e. $-500 < \Delta T < 1450$ sec). However, the upper limit is supported by the result for Montpellier ($\Delta T < 1460$). The various observing sites mentioned above for this eclipse are shown in figure 11.4 along with the computed belt of totality for $\Delta T = 0$.

In addition to the detailed records discussed above, Levi-Donati (1989) has drawn attention to a Latin inscription carved on a pillar at Marola, a village in the Appenine Mountains near Florence. This inscription, which is illustrated in figure 11.5, may be translated as follows:

1239. On the 3rd day before the Nones of June the Sun died at the ninth hour.

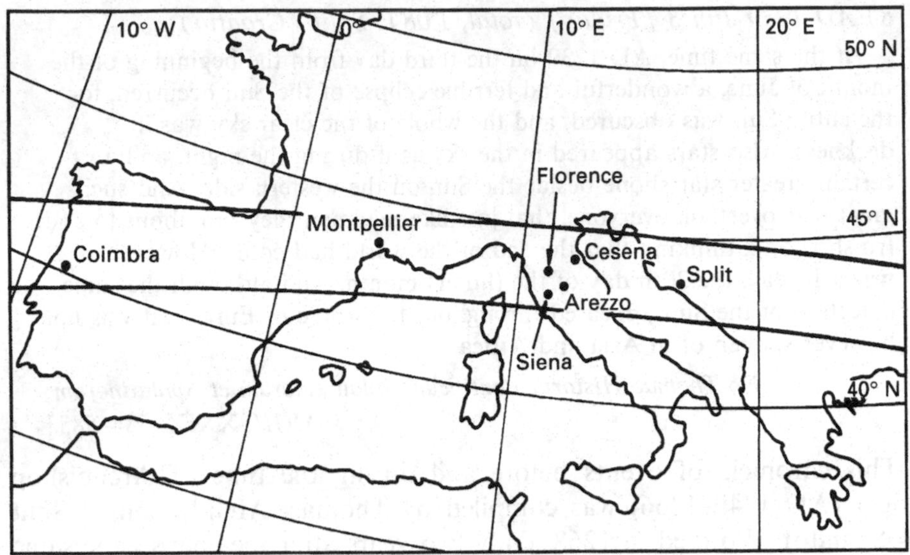

Fig. 11.4 The principal observing sites for the total solar eclipse of AD 1239 Jun 3.

Although less reliable than the accounts cited previously, totality at Marola is assured for any value of ΔT in the range $-2950 < \Delta T < 1610$ sec. This fully encompasses the range indicated by the careful observation of totality at Coimbra ($-500 < \Delta T < 1450$ sec). The computed local time of maximum (13.2 h) is actually towards the end of the 7th hour, but the quoted time is likely to be no more than a rough estimate.

(19) AD 1241 Oct 6 [Sunday] (total, 1.05): Reichersberg

> 1241. On the day before the Nones of October (Oct 6), the Sun while it was bright was suddenly covered with wonderful blackness a little after midday. As a result, no part of it could be seen and stars were seen as if during the night for about four hours.
>
> [*Chronicon Magni Presbyteri Continuatio*; *MGH*, XVII, 528.]

This continuation of the chronicle of Magnus of Reichersberg (see above) is contemporary with the events it describes.

RESULTS

For totality at Reichersberg, 450 sec $< \Delta T <$ 1400 sec.

(NB local time of maximum phase = 13.0 h; solar altitude = 32 deg.)

(20) AD 1241 Oct 6 [Sunday] (total, 1.05): Stade (Germany)

> 1241...There was an eclipse of the Sun on the Octave of St. Michael (Oct 6), namely the day before the Nones of October, on Sunday some

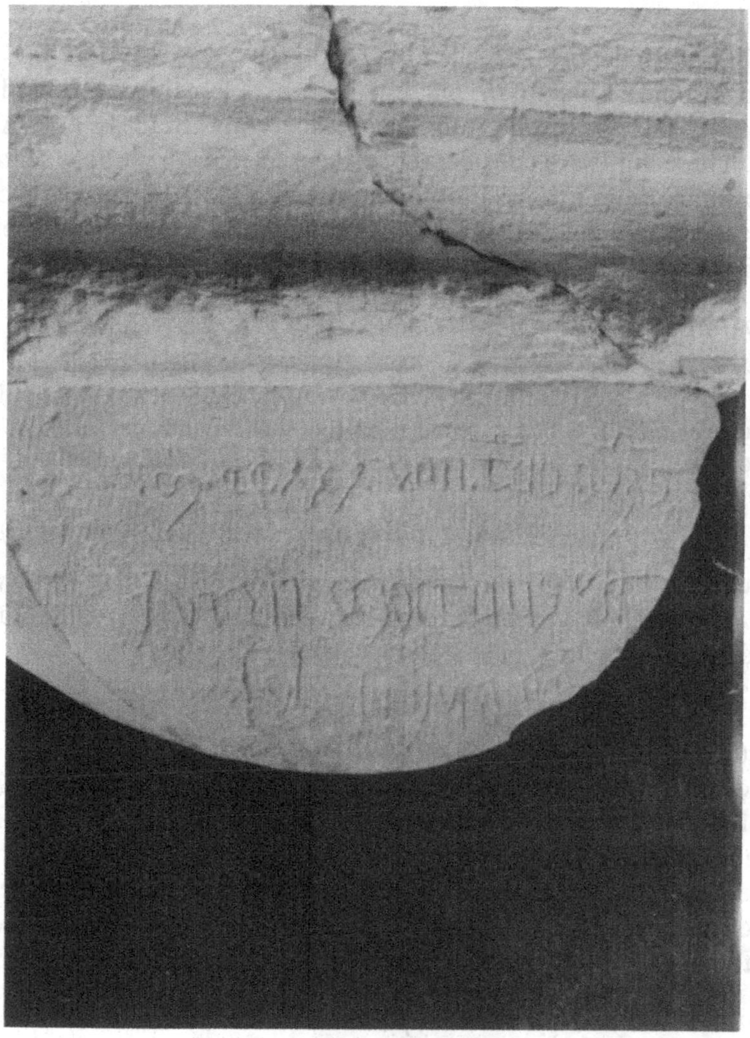

Fig. 11.5 Latin inscription from AD 1239 recording the 'death of the Sun', carved on a pillar at Marola in Italy. (Courtesy: Dr G.R. Levi-Donati.)

> time after midday. Stars appeared and the Sun was completely hidden from our sight. Yet the sky was so clear that no clouds appeared in the air.
>
> [*Annales Stadenses*, *MGH*, XVI, 368.]

This section of the chronicle of the monastery of St. Maria in Stade was compiled by Albertus who became Abbot there in 1232. Note Albertus' use of the first person plural, implying that he was an eye-witness. The date is precisely correct, including the weekday. Since the Feast of St Michael was on Sep 29, the Octave one week later (i.e. 8 days in inclusive counting) would be on Oct 6.

RESULTS

For totality at Stade, 630 sec < ΔT < 1630 sec.

(NB local time of maximum phase = 12.6 h; solar altitude = 28 deg.)

Combining the ΔT results from Reichersberg and Stade, 630 < ΔT < 1400 sec. These limits are considerably closer together than those obtained from the much more extensive observations in AD 1239, largely because the track of totality in 1241 was steeply inclined to the equator in Central Europe.

(21) AD 1263 Aug 5 [Sunday] (annular, 0.95): Orkney (Scotland)

> The eve of St Olaf was on a Sunday... When King Hacon lay in Rognvaldsvoe, a great darkness (*myrkur*) overtook the Sun, so that a little ring (*hringur litill*) was bright round it on the outside, and that lasted a while of the day... On St. Lawrence's day king Hacon sailed out of Rognvaldsvoe over the Pentland firth.
>
> [*Haconar Saga*; trans. Dasent (1894, pp. 346–7).]

The year can be firmly established from the context as AD 1263, while the date of the eclipse can be fixed as between the eve of St Olaf (Jul 29) – a Sunday – and St Lawrence's day (Aug 10).

RESULTS

For annularity in Orkney, −3500 < ΔT < −2150 sec.

(NB local time of maximum phase = 13.6 h; solar altitude = 43 deg.)

The above ΔT range is in marked conflict with roughly contemporaneous results, which indicate a value around +1000 sec. The eclipse could therefore have only been partial in Orkney. Even if it had been central there, no more than 0.95 of the solar diameter would have been covered so that the reference to 'great darkness' is much exaggerated.

(22) AD 1267 May 25 [Wednesday] (total, 1.02): Constantinople

> At that time the Moon obscured the Sun when it was in the 4th part (degree) of Gemini, at the 3rd hour before midday on the 25th day of May in the year 6775 (AD 1267). It was a total eclipse of about 12 digits or points. Also, such darkness arose over the Earth at the time of mid-eclipse that many stars appeared. No doubt this portended the very great and destructive calamities which were soon to be vented on the Romans by the Turks.
>
> [*Nicephori Gregorae Byzantinae Historiae*, Lib. IV, cap. 8; Migne (1865, CXLVIII, 246); in Greek, with a rendering into Latin by the editor.]

Nicephorus Gregoras, who himself lived at Constantinople, wrote about a century after the event. He does not directly name the place of observation but the above detailed account is included in his history of Constaninople. In reporting the partial eclipse of AD 1330 Jul 16 (Lib, IX, cap. 12), Nicephorus is careful to state that this was seen in the *parallel* of Constantinople (rather than the city itself), and specifies certain regions (e.g. Thrace) where observations were made. Although he assigns the correct date for the eclipse of AD 1267, Nicephorus is several degrees in error in his solar position; at the time of the eclipse the Sun would be in longitude 70.5, the 11th degree of Gemini.

RESULTS

For totality at Constantinople, $-850 < \Delta T < 820$ sec.
(NB local time of maximum phase = 11.5 h; solar altitude = 70 deg.)

(23) AD 1406 Jun 16 [Wednesday] (total, 1.06): Braunschweig (Germany)

1406. In this year there was an eclipse of the Sun so that the Sun stopped shining (*vorgingk or schyn*) before the Prime of the day (i.e. the Office held *c.* 6 a.m.) on St. Vitus' day (Jun 15); it was so dark that people could not recognise one another.

[*Bothonis Chronicon Brunsvicensis picturatum*; in German; quoted by Ginzel (1884a).]

RESULTS

For totality at Braunschweig, $-100 < \Delta T < 1450$ sec.
(NB local time of maximum phase = 6.8 h; solar altitude = 26 deg.)

(24) AD 1406 Jun 16 [Wednesday] (total, 1.06): Hamburg

In the year 1406, on the day of the Annunciation of the Virgin Mary (= Jun 16), in the morning at the 6th hour, the Sun was entirely (*ganz*) covered, and (there was) such a great eclipse that the people believed that it was the end of the world.

[*Adami Trazigeri Chronicon Hamburgensis*; in German; quoted by Zawilski (1994).]

RESULTS

For totality at Hamburg, $-830 < \Delta T < 740$ sec.
(NB local time of maximum phase = 6.8 h; solar altitude = 26 deg.)

(25) AD 1406 Jun 16 [Wednesday] (total, 1.06): Liege

1406. In that year on Jun 16, between the 6th and 7th hours before midday, there was a total eclipse of the Sun, with the result that the dawn, which was bright and clear, became as dark as at midnight.

[*Chronique Latine de Jean de Stavelot*; quoted by Zawilski (1994).]

This contemporary chronicle is specifically concerned with Liege.

RESULTS

For totality at Liege, $-550 < \Delta T < 900$ sec.

(NB local time of maximum phase = 6.5 h; solar altitude = 22 deg.)

(26) AD 1406 Jun 16 [Wednesday] (total, 1.06): Magdeburg (Germany)

> 1406. In the same year on St Vitus' day, at the beginning of the 9th hour in the morning, there was an eclipse here in the land, that is a darkening of the Sun, so that the Sun stopped shining (*schin vorging*) before the Prime of the day; it was so dark that there was no light and people could not recognise one another... People thought that the end of the world was coming.
>
> [*Magdeburger Schoppenchronik*; in German; quoted by Zawilski (1994).]

As noted above, in citing the entry from Braunschweig, St Vitus' Day is on Jun 15 rather than Jun 16. There are other similarities (as well as significant differences) between this account and that from Braunschweig. The two places are about 100 km apart so that inter-communication would be relatively easy.

RESULTS

For totality at Magdeburg, $+170 < \Delta T < 1700$ sec.

(NB local time of maximum phase = 6.9 h; solar altitude = 26 deg.)

Combining the various limits deduced above yields $170 < \Delta T < 740$ sec.

(27) AD 1415 Jun 7 [Friday] (total, 1.07): Neider Alteich (Germany)

> 1415. A dark eclipse of the Sun occurred with the result that the Sun entirely lost its light for twice the duration of the *Misere-mei* (Psalm 51) at the 6th hour of the day on the 7th day of June.
>
> [*Notae Altahenses*; *MGH*, XVII, 424.]

RESULTS

For totality at Neider Alteich, $-650 < \Delta T < 970$ sec.

Here we have the second of only two medieval allusions to the duration of totality (see also AD 1239, Arezzo). Near the central line, the total phase would last for 3 min 40 sec.

(NB local time of maximum phase = 7.5 h; solar altitude = 27 deg.)

(28) AD 1415 Jun 7 [Friday] (total, 1.07): Prague

> 1415. At that same time, on the 7th day of June, which was the sixth day of the week, after the Feast of S. Boniface (Jun 5), at the the 11th hour of the day, the whole Sun was eclipsed. As a result, the Mass could

> not be celebrated without lights...(This happened) on account of the death of Master Jan Hus (which occurred) soon afterwards...
>
> [Magister Laurentius de Brezina. *De gestis et variis accidentibus regno Boemiae*, II, part 1, 333; quoted by Ginzel (1884a).]

At this point in his chronicle, the author, who was a contemporary, is relating events which happened at Prague. Jan Hus, the prominent Czech religious reformer, was burnt at the stake on 6 July in AD 1415.

RESULTS

For totality at Prague, $-1160 < \Delta T < 670$ sec.
(NB local time of maximum phase = 7.1 h; solar altitude = 28 deg.)

(29) AD 1415 Jun 7 [Friday] (total, 1.07): Wroclaw (Poland)

> An eclipse of the Sun on the 6th day of the week (Friday). In the same year 1415, on the Octave of Corpus Christi, at the 12th hour (after sunset), at about the relighting of the Moon, there was a total eclipse of the Sun. This stood still for close to half an hour and there was terror and alarm among many people.
>
> [*Chronica Sigismundi Rosiczi; Scriptores rerum Silesicarum*, vol. 12; quoted by Zawilski (1994).]

The movable Feast of Corpus Christi, the first Thursday after Trinity, fell on May 30 in AD 1415 so that the Octave would be on Jun 6.

From Zawilski, the annalist Sigmund Rositz used notes preserved in Wroclaw; in particular the immediately preceding and following entries in his chronicle relate to (non-astronomical) events occurring in this city.

RESULTS

For totality at Wroclaw, $-1190 < \Delta T < 720$ sec.
(NB local time of maximum phase = 7.3 h; solar altitude = 30 deg.)

When the various ΔT ranges obtained for the eclipse of AD 1415 are combined, $-650 < \Delta T < 670$ sec.

(30) AD 1431 Feb 12 [Monday] (total, 1.04): Foligno (Italy)

> 1431. Memorandum, that on the 12th day of February at the 21st hour, which was the Monday of the Carnival, the Sun was so greatly obscured that it became as dark as night and the eye of the Sun became black like charcoal. And on that day the Moon turned (i.e. was new)...
>
> [Pietruccio di Giacomo degli Unti: *Memoriale*; in Italian; quoted by Levi-Donati (1987).]

This contemporary chronicle is concerned with events in the city of Foligno. The local time was evidently expressed relative to sunset. Feb 12, the day of the eclipse, was indeed a Monday. The time was apparently measured from sunset on the previous evening; see also the following entry.

RESULTS

For totality at Foligno, $10 < \Delta T < 870$ sec.

(NB local time of maximum phase = 15.3 h; solar altitude = 20 deg.)

(31) AD 1431 Feb 12 [Monday] (total, 1.04): Perugia (Italy)

> 1431. On February 12 at about the 21st or 22nd hour, the Sun was completely obscured and in front of the Sun was placed a black circle like a little wheel (*rotella*). It became as dark as night and the sky revealed the stars. The birds went to roost as they usually do at night. Everyone was feeling ill at ease as a result of this event. It began half an hour before the Sun was covered over. It gradually lost its light even to the hour stated above...
>
> [Antonio dei Veghi: *Diario dall'anno 1423 al 1491*; in Italian; quoted by Levi-Donati (1987).]

Antonio dei Veghi lived at Perugia. A copy of the manuscript version of the above text is illustrated in figure 11.6.

Once again the local time was apparently estimated from sunset.

RESULTS

For totality at Perugia, $-160 < \Delta T < 700$ sec.

(NB local time of maximum phase = 15.3 h; solar altitude = 20 deg.)

Combining the results from the observations at Foligno and Perugia leads to a value for ΔT at the epoch AD 1415 in the range $10 < \Delta T < 700$ sec.

(32) AD 1485 Mar 16 [Wednesday] (total, 1.05): Augsburg (Germany)

> In the year of salvation 1485, in the month of January, according to the ancient custom, the consuls of Augsburg ... were elected. On the 16th day of March, at the 3rd hour, during meal-time, the Sun was totally eclipsed. This produced such horrid darkness on our horizon for the space of half an hour that stars appeared in the sky. Crazed birds fell from the sky and bleating flocks and fearful herds of oxen unexpectedly began to return from their pastures to their stables.
>
> [Achilli Pirmini Gassari: *Annales Augustburgenses*; quoted by Zawilski (1994).]

In the longitudes of Central Europe, the track of totality ran almost parallel to the equator and hence this eclipse would be total at Augsburg for a wide range of ΔT.

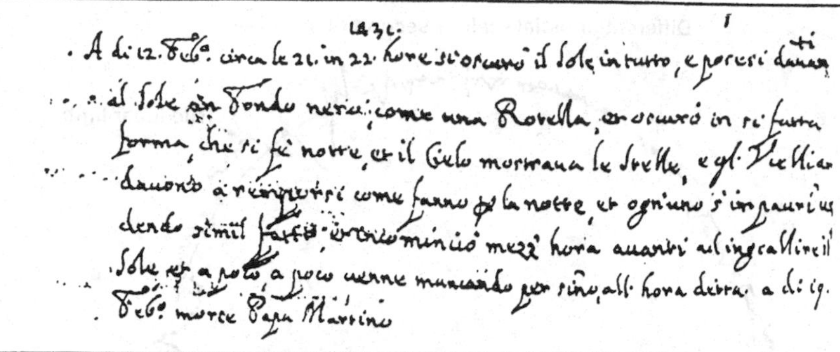
1431.

A di 12. Febr.o circa le 21. in 22. hore si oscurò il Sole in tutto, e pareci dinanzi al Sole un Tondo nero; come una Rotella, et oscurò in sì fatta forma, che si fè notte, et il Cielo mostrava le Stelle, e gl' Uccelli davano a ricoverarsi come fanno per la notte, et ogniuno s'impauriva vedendo simil fatto; e cominciò mezz' hora avanti al ingecallirsi il Sole et a poco a poco venne mancando per sino all' hora detta. a di 19. Febr.o morse Papa Martino

Fig. 11.6 A copy of the manuscript account of the total solar eclipse of AD 1431 Feb 12 in the *Diario dall'anno 1423 al 1491* by Antonio dei Veghi. (Courtesy: Dr G. R. Levi-Donati.)

RESULTS

For totality at Augsburg, $-5500 < \Delta T < 780$ sec.

(NB local time of maximum phase = 16.5 h, solar altitude = 16 deg.)
The time of day, expressed relative to noon, was only crudely estimated.

(33) AD 1560 Aug 21 [Wednesday] (total, 1.05): Coimbra

> I shall cite two remarkable eclipses of the Sun, which happened in my own time and thus not long ago. One of these I observed about midday at Coimbra in Lusitania in the year 1559 (*sic*), in which the Moon was placed between my sight and the Sun with the result that it covered the whole Sun for a considerable length of time. There was darkness in some manner greater than night; neither could one see clearly where one stepped. Stars appeared in the sky and (marvellous to behold) the birds fell down from the sky to the ground in terror of such horrid darkness...
>
> [Clavius (1593, p. 508).]

Clavius has mistaken the year; the eclipse of AD 1560 Aug 21 was the only one total in Portugal for many years around this time. Clavius, who was then aged 23, was studying at the University of Coimbra.

RESULTS

For totality at Coimbra, $-490 < \Delta T < 220$ sec.

NB for a duration of totality greater than one minute (i.e. 'a considerable length of time'), this range should be narrowed to $-475 < \Delta T < 205$ sec. This latter interval would seem more appropriate.

(NB local time of maximum phase = 11.3 h; solar altitude = 57 deg.)

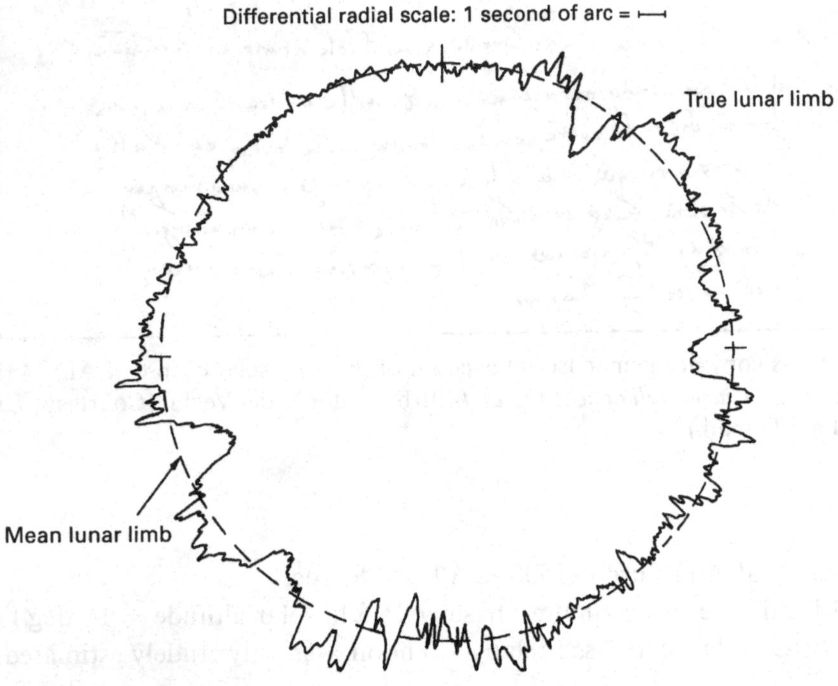

Differential radial scale: 1 second of arc = ⊢⊣

True lunar limb

Mean lunar limb

Fig. 11.7 The lunar limb profile (exaggerated by a factor of 60) at the solar eclipse of AD 1567 Apr 9. (After Duncombe, 1973).

(34) AD 1567 Apr 9 [Wednesday] (annular-total, 1.003): Rome

> ...The other (eclipse) I saw in Rome in the year 1567 also about midday in which although the Moon was placed between my sight and the Sun it did not obscure the whole Sun as previously but (a thing which perhaps never before occurred at any other time) a certain narrow circle was left on the Sun, surrounding the whole of the Moon on all sides.
>
> [Clavius (1593, p. 508).]

This is a continuation of the previous account (i.e. from AD 1560). Clavius had now moved to Rome, where he was teaching mathematics at the Collegio Romano.

The eclipse was neither total nor annular at Rome; the angular diameters of both Sun and Moon as seen from Rome were virtually identical. Even where the eclipse was central, part of the solar photosphere would be visible down the deep clefts at the Moon's edge – see figure 11.7, which is based on the lunar limb charts of Duncombe (1973). The approximate topocentric librations – as defined by Duncombe – are respectively: $L = -5$ deg, $B = 0$ deg.

RESULTS

Making a careful allowance for limb profile, Stephenson *et al.* (1997) have estimated a value of ΔT in the narrow range $145 < \Delta T < 165$ sec from this observation.

(NB local time of maximum phase = 12.2h; solar altitude = 59 deg.)

(35) AD 1601 Dec 24 [Monday] (annular, 0.90): near Bergen (Norway)

> Those who had seen the above mentioned eclipse described it with very great astonishment: 'But in the same form, the Sun had encompassed the Moon centrally within its own circumference in such a way that its own light, having been diffused up to the edge on all sides, more or less shone out equally by $1\frac{1}{2}$ digits' (i.e. $1\frac{1}{2}$ twelfths of the apparent solar diameter).
>
> [Longomontanus (1622).]

The above account of the ring phase, by fishermen on the sea coast near Bergen, was communicated to Longomontanus by Andreas Fosse, Bishop of Bergen, who was also a mathematician. The width of the annulus was considerably overestimated, probably on account of irradiation.

RESULTS

For annularity at Bergen, $-1460 < \Delta T < 3170$ sec.

By this late period, only a few years before the telescopic era, ΔT can be estimated fairly accurately as around 100 sec using extrapolation from the results obtained from telescopic observations. For any acceptable value of ΔT, Bergen would lie near the centre of this very wide zone of annularity. Hence the observation does not enable a useful result to be derived, although the fishermen deserve commendation for providing one of the few careful descriptions of the ring phase from the whole of the pre-telescopic period.

Annularity would last for more than eight minutes. On account of the low solar altitude (only 3 deg at mid-eclipse), the ring phase would probably be very distinct.

(NB local time of maximum phase = 13.8 h; solar altitude = 3 deg.)

(36) AD 1605 Oct 12 [Wednesday] (total, 1.03): Naples?

> I add a fourth argument (that the heavens are subject to change) from the eclipse of the Sun which we saw in the month of October. At Naples, indeed, the whole Sun appeared hidden, but this was entirely the same as that which Plutarch himself saw 1500 years ago; whose words from page 318 of my *Optics* are as follows: 'A certain splendour shines forth around the circumference, preventing the shadow from becoming profound and too great'. At Naples, indeed, this happened in the above mentioned year (AD 1605): the whole Sun was accurately hidden, but at any rate it did not last for a very long time. In the centre, where the Moon was, there was

Table 11.5 ΔT limits deduced from medieval European observations of central solar eclipses.

Year	ΔT Range (sec) LL	UL
+840	1 610	6 800
+968	1 580	2 600
1124	960	2 700
1133	540	1 150
1147	300	1 190
1176	−190	1 600
1185	−2 200	10 500
1239	−500	1 450
1241	630	1 400
1267	−850	820
1406	170	740
1415	−650	670
1431	10	700
1485	−5 500	780
1560	−475	205
1567	145	165
1601	−1 460	3 170

the appearance as if of a black cloud; all around it was a red and fiery splendour, of equal breadth on all sides, which occupied a good part of the sky. Out of the region of the Sun, towards the north, the sky was entirely dark, and as profound as night; stars, however, were not seen.

[Kepler (1605).]

RESULTS

For totality at Naples, $1400 < \Delta T < 2590$ sec.

Totality on this occasion cannot have been observed from the *city* of Naples; the indicated range of ΔT is much too high at such a recent epoch; as in the previous entry (AD 1601), ΔT can be estimated as around 100 sec from telescopic observations. This result gives a magnitude of only 0.97 in the city of Naples. However, the eclipse would certainly be total in the extreme south of Italy (Calabria), which was then part of the *kingdom* of Naples. The precise place of observation cannot be established; Kepler

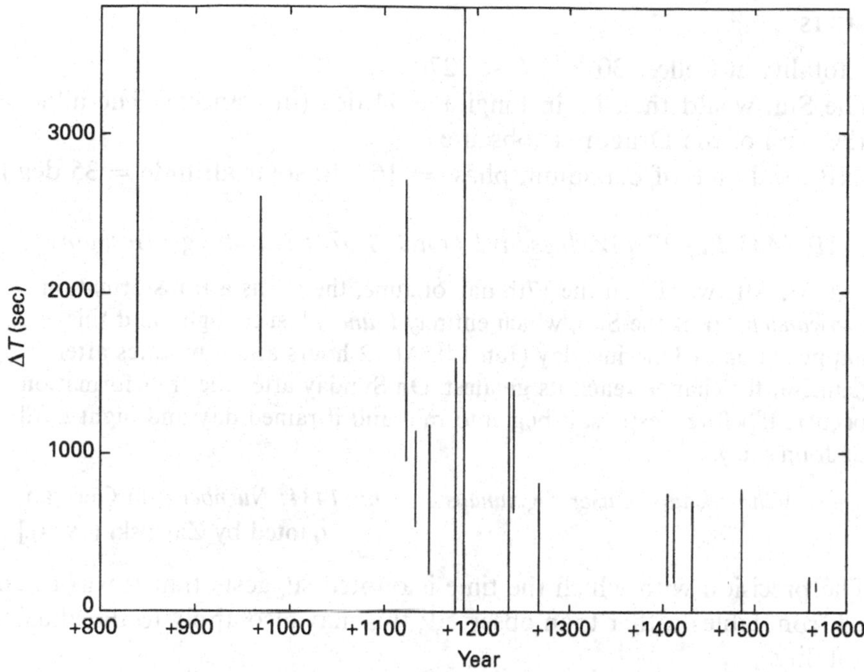

Fig. 11.8 ΔT limits obtained from central eclipses recorded in medieval Europe.

is surprisingly vague on this issue. On the other hand, he gives a careful description of the corona, which had not been previously referred to directly since AD 968 (see above).

(NB local time of maximum phase = 14.8 h, solar altitude = 30 deg.)

Table 11.5 summarises the ΔT limits derived in this section, apart from the anomalous results from AD 1263 and 1605. When more than one observation was made at the same eclipse, I have combined the various individual ΔT ranges, as already noted above. These limits are plotted in figure 11.8.

11.6.2 *Brief reports of total eclipses*

As noted above, although ΔT limits will be derived from the following brief records, they will not be included in subsequent analysis since the reliability of the observations is in doubt.

(1) AD 1433 Jun 17 [Wednesday] (total, 1.07): Celle (Germany)

1433. At Celle in the Duchy of Braunschweig, after midday the whole (*totus*) Sun was eclipsed by the intervention of the Moon in the Tail of the Dragon (*Cauda Draconis*)

[Heinrich Bunting: *Chronologia Catholica*; quoted by Zawilski (1994).]

RESULTS

For totality at Celle, $530 < \Delta T < 2270$ sec.

The Sun would then be in longitude 94 deg (in Cancer). The allusion to the 'Tail of the Dragon' is obscure.

(NB local time of maximum phase = 16.1 h; solar altitude = 35 deg.)

(2) AD 1433 Jun 17 [Wednesday] (total, 1.07): Nurnberg (Germany)

> 1433. Afterwards, on the 17th day of June, there was a transformation (*verwandelung*) of the Sun, which entirely (*gancz*) lost its light, and this happened on St Ullerius' day (Jun 17). At 12 hours and 4 minutes after (sunrise) the change reach its greatest. On Sunday after the transformation occurred, before Vespers, it began to rain and it rained day and night until St John's day.
>
> [*Chronik aus Kaiser Sigmundus Zeit bis 1434: Nurnberg*; in German; quoted by Zawilski (1994).]

The precision with which the time is quoted suggests that it was calculated from tables rather than observed; this may also apply to the allusion to totality.

Rain started to fall on Sunday Jun 21 and continued unabated until the following Wednesday (Jun 24 – i.e. the Nativity of St John the Baptist).

RESULTS

For totality at Nurnberg, $-530 < \Delta T < 1070$ sec.

(NB local time of maximum phase = 16.3 h; solar altitude = 34 deg.)

(3) AD 1433 Jun 17 [Wednesday] (total, 1.07): Karlstejn (Czech Republic)

> 1433. In this same year on the 4th day of the week (Wednesday), after St Vitus' Day (Jun 15), at about the 21st hour (after sunset), the Sun was eclipsed totally in its body. And in the same week and for two weeks afterwards there were great inundations of water.
>
> [*Kronika Bartoska z Drahonic*; *Fontes rerum Bohemicarum*, vol. 5; quoted by Zawilski (1994).]

This record, which contains only minimal astronomical details, is probably from Karlstejn, where Bartosek was writing his chronicle (Zawilski, 1994).

RESULTS

For totality at Karlstejn $420 < \Delta T < 2040$ sec.

(NB local time of maximum phase = 16.5 h; solar altitude = 32 deg.)

(4) AD 1485 Mar 16 [Wednesday] (total, 1.05): Melk (Austria)

1485. On the 16th day of March at the 4th hour there was a complete eclipse of the Sun.

[*Annales Mellicenses*; *MGH*, IX, 524.]

The value of this brief record in the annals of the monastery of Melk is reduced by the lack of any descriptive details.

RESULTS

For totality at Melk, $-4400 < \Delta T < 2100$ sec.

(NB local time of maximum phase = 16.8 h; solar altitude = 13 deg.)

(5) AD 1485 Mar 16 [Wednesday] (total, 1.05): Salzburg (Austria)

1485. In this same year, The Sun suffered a total eclipse on St Gertrude's Day (Mar 16) during Lent.

[*Chronicon Salisburgense*; quoted by Zawilski (1994).]

RESULTS

For totality at Salzburg, $-1000 < \Delta T < 2160$ sec.

(NB local time of maximum phase = 16.6 h; solar altitude = 15 deg.)

(6) AD 1485 Mar 16 [Wednesday] (total, 1.05): Vienna

1485. Mar 16. And on that day the Sun was totally eclipsed. As I observed, the eclipse in fact occurred an hour earlier (than expected).

[*Chronicle* of the Viennese physician Johannes Tichtel; quoted by Zawilski (1994).]

RESULTS

For totality at Vienna, $-4500 < \Delta T < 1780$ sec.

(NB local time of maximum phase = 16.9 h; solar altitude = 12 deg.)

(7) AD 1605 Oct 12 [Wednesday] (total, 1.03): Marseilles

Wendelin at Forcalquier in Provence saw the whole Sun hidden apart from a very narrow thread towards the north, which he ascribed to the illuminated atmosphere. He added that namely in Marseilles the whole Sun (*totum solem*) appeared obscured with dense darkness.

[Riccioli (1665, Book II, chap. 2); quoted by Zawilski (1994).]

The observation of a partial eclipse at Forcalquier will be considered in section 11.7.

RESULTS

For totality at Marseilles, $650 < \Delta T < 1700$ sec.

(NB local time of maximum phase = 13.7 h; solar altitude = 34 deg.)

The indicated range of ΔT at this late epoch is far outside the figure of approximately 100 sec derived from extrapolation of the telescopic data. Presumably the sparse information which Wendelin appears to have obtained at second hand from Marseilles was unreliable.

11.7 Eclipses which were definitely partial, yet very large

Each of the following observations suggests that the observer was not far outside the zone of totality or annularity.

(1) AD 1033 Jun 29 [Friday] (annular, 0.994): Cluny (France)

> AD 1033... On the 3rd day before the Kalends of July (Jun 29), the sixth day of the week (Friday) and the 28th of the Moon, an eclipse or defection of the Sun occurred from the 6th hour to the 8th; it was extremely terrible. Indeed, the Sun became sapphire in colour, showing in its upper part the appearance of the crescent Moon on the 4th day from its relighting...
>
> [*Glabri Rudolphi Cluniacenses monachi Historiarum sui temporis*; quoted by Ginzel (1884a).]

Glaber, a monk at Cluny, was a contemporary writer. His chronicle extends to 1044.

This is one of the few early accounts of a large partial solar eclipse to indicate whether the upper or lower portion of the Sun was obscured at maximum phase. In the longitudes of France, the narrow track of annularity ran almost parallel to the equator and in fact did not quite reach as far north as the latitude of Cluny (+46.42 deg) on any part of the Earth's surface (northern limit 46.35 deg) – see figure 11.1. Hence no value of ΔT will render the eclipse central at this site; calculation merely confirms the observation of a large partial eclipse in which the upper portion of the solar disc remained unobscured. At Cluny, the computed magnitude cannot have exceeded 0.992. Because of irradiation, the width of the solar crescent would probably appear much enhanced so that it seemed to resemble the crescent Moon considerably after new.

(NB local time of maximum phase = 11.7h ; solar altitude = 66 deg.)

Several other accounts of this same eclipse from France and Belgium also assert that the Sun became like the crescent Moon. However, the various sites are still further north than Cluny, so that the eclipse magnitudes must have been smaller.

(2) AD 1133 Aug 2 [Wednesday] (total, 1.07): Vysehrad (Prague)

> 1133. On the 4th day before the nones of August (Aug 2), an eclipse of the Sun appeared in a wonderful manner. This defection gradually diminished so much that a crown like a crescent Moon proceeded to the

south part, afterwards turning round to the east, henceforth to the west. At length it was transformed to its original state.

[*Canonici Wissegradensis continuatio Cosmae*; *MGH*, IX, 938.]

This continuation of the chronicle of Cosmas of Prague between 1125 and 1142 was by an anonymous canon of Vysehrad – a suburb of the city of Prague. He had an unusual interest in astronomy, and recorded many celestial phenomena (especially aurorae) which he presumably witnessed. Nevertheless, it is clear from the descriptions which he gives that he possessed only a very limited knowledge of astronomy. There can be little doubt that the above account relates to an eclipse which was not quite total. The author is mainly concerned with describing the changing aspect of the solar crescent and does not suggest that the Sun ever completely disappeared. However, the text is garbled. If the southern part of the Sun had remained unobscured, the crescent would have appeared to rotate from the south-east to the south-west.

RESULTS

For a partial eclipse at Vysehrad in which the southern limb of the Sun remained visible, $\Delta T > 2750$ sec.

The above result is in conflict with those derived from several observations of totality made elsewhere in Europe at the same eclipse (see section 11.6); these all indicate that ΔT was less than about 1500 sec (see above). However, if it be supposed that the *northern* part of the Sun remained uncovered at Vysehrad, this discord would be removed; a value of $\Delta T < 1470$ sec would be indicated. A possible explanation is that the Sun was viewed by reflection in water (a known practice in medieval times) but the author omitted to allow for this when recounting his experiences. However, this is no more than conjecture.

(NB local time of maximum phase = 12.8 h; solar altitude = 55 deg.)

(3) AD 1147 Oct 26 [Sunday] (annular, 0.95): Braunschweig (Germany)

1147. On the feast of St Simon and St Jude (Oct 28), the Sun was obscured with the result that it resembled a sickle.

[*Bothonis Chronicon Brunsvicensis picturatum*; in German; quoted by Ginzel (1884a).]

Note the error of two days in the date of the eclipse both in this and the following entry.

RESULTS

For a partial eclipse at Braunschweig, either $\Delta T < 1680$ or > 2640 sec.

(NB local time of maximum phase = 10.8h; solar altitude = 21 deg.)

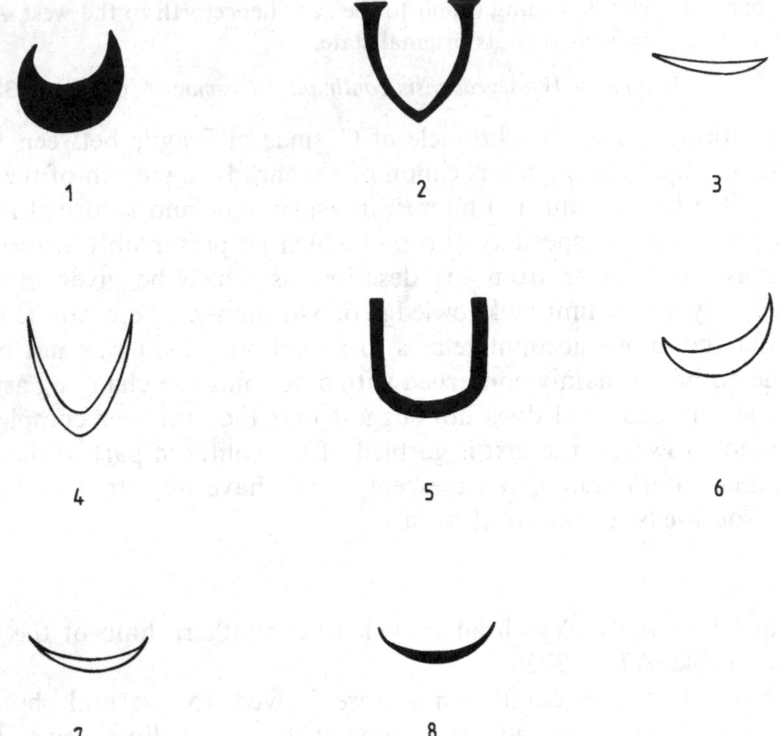

Fig. 11.9 Sketches of the eclipsed Sun in AD 1153 Jan 26 as found in the chronicles of Erfurt (Germany). (Holder-Egger, 1899.)

(4) AD 1147 Oct 26 [Sunday] (annular, 0.96): Magdeburg (Germany)

1147. In this same year on the 5th day before the Kalends of November (Oct 28), about midday there was an eclipse of the Sun and the Earth was covered with horrible darkness. Indeed a circle like a crescent was seen...

[*Annales Magdeburgenses*; *MGH*, XVI, 188.]

RESULTS

For a partial eclipse at Magdeburg, either $\Delta T < 1920$ sec or > 2870 sec. (NB local time of maximum phase = 10.9 h; solar altitude = 22 deg.)

Combining the limits from the two observations of partial eclipses in AD 1147 leads to either $\Delta T < 1680$ sec or > 2870 sec. Much more restricted ΔT limits at this date are yielded by the observation of annularity from Brauweiler – see section 11.6.

(5) AD 1153 Jan 26 [Monday] (annular, 0.93): Erfurt (Germany)

1153. A sign appeared in the Sun on the 7th day before the Kalends of February (Jan 26) in this manner...

[*Annales Sancti Petri Erphesfurdenses*; *MGH*, XVI, 21.]

This and several other Erfurt chronicles contain sketches showing the Sun as a crescent with the cusps pointing upwards (Holder-Egger, 1899, XLVII, 19, 56, 57 and 178) – see figure. 11.9.

RESULTS

For a partial eclipse at Erfurt with the southern limb of the Sun uncovered, $\Delta T < -50$ sec.

This result is clearly discordant when compared with those derived from a number of roughly contemporaneous observations – which indicate a value for ΔT of at least 500 sec (see section 11.6). If the sketches are taken to represent the Sun as seen by reflection in water (as in the case of the observation at Vysehrad in 1133), a more reasonable result of $\Delta T >$ 1370 sec is indicated instead. However, this is no more than speculation; unfortunately, none of the surviving accounts of the eclipse from Erfurt (all of which are very brief) state which limb of the Sun was obscured.

(NB local time of maximum phase = 12.9 h; solar altitude = 21 deg.)

(6) AD 1178 Sep 13 [Wednesday] (total, 1.05): Vigeois (France)

> In the year from the Incarnation of Our Lord 1178, on the 4th day of the week (Wednesday), the Ides of September (Sep 13), the 28th day of the Moon, on a clear day, about the 5th hour the Sun suffered an eclipse; and its disc began to be covered from the east (*sic*) until it was like a two- or three-day-old Moon. The star Venus was seen to the north. After 6 (i.e. the 6th hour) the brightness returned from the east (*sic*), in the order in which it was blackened, until the Sun was fully illuminated. Then we saw each others' faces as they appear beside a glowing furnace.
>
> [*Ex Chronico Gaufredi Vosiensis*; Bouquet (1781, p. 447).]

Gaufredus, the author of this eyewitness description, was prior of the monastery of Vosium (now Vigeois) at the time of the eclipse. Although he gives a careful description of a partial eclipse, he has mistaken the limb of the Sun at which it began to be covered and uncovered; in each case this would be the western edge. The reference to the apparent reddening of the observers' faces perhaps suggests that Gaufredus and his colleagues had been looking at the Sun for too long!

RESULTS

For a partial eclipse at Vigeois, either $\Delta T < 1140$ sec or > 2320 sec.

(NB local time of maximum phase = 11.6 h; solar altitude = 46 deg.)

(7) AD 1312 Jul 5 [Wednesday] (total, 1.01): Lanercost (England)

> AD 1312. On the third of the Nones of July (Jul 5), on the vigil of the octave of the Apostles Peter and Paul, was a new Moon (the thirtieth lunation), and an eclipse of the Sun about the first hour of the day, and

> the Sun appeared like a horned Moon, which was small at first and then larger, until about the third hour it recovered its proper and usual size; though sometimes it seemed green, but sometimes of the colour which it usually has.
>
> [*The Chronicle of Lanercost*; trans. Maxwell (1913, p. 198).]

Unlike most eclipse records from English monastic annals, this seems original. The chronicle, which is contemporary, is frequently concerned with local events. Since the Feast of the Apostles Peter and Paul was on Jun 29, the Vigil of the Octave would be 6 days later or Jul 5. This eclipse was recorded as producing intense darkness in Iceland (see section 11.8).

RESULTS

For a partial eclipse at Lanercost, either $\Delta T < 8700$ sec or > 8900 sec.

Since other roughly contemporaneous observations indicate a value for ΔT of less than about 1000 sec, the observation of a partial eclipse, although confirmed, is redundant.

(NB local time of maximum phase = 6.7 h; solar altitude = 24 deg.)

(8) A.D 1330 Jul 16 [Monday] (total, 1.02): Zbraslav (Czech Republic)

> 1330. In this same year on the Ides of July (Jul 15) at the 8th hour of the day, the Sun was so greatly obscured that of its great body only a small extremity like a three-night-old Moon was seen.
>
> [*Chronicon Aulae regiae*; quoted by Ginzel (1884a).]

The town in which the monastery of Aula Regia is situated is now known as Zbraslav; this is located about 10 km to the south of Prague.

RESULTS

For a partial eclipse at Zbraslav, either $\Delta T < 890$ sec or > 1210 sec.

(NB local time of maximum phase = 17.2 h; solar altitude = 23 deg.)

(9) AD 1333 May 14 [Friday] (annular, 0.997): Coimbra (Portugal)

> In the (Julian) year 1371, on the 14th day of the month of May, there was an eclipse of the Sun and the Sun became so diminished that it resembled nothing so much as the new Moon, very small in appearance. It increased in size and returned to its normal state; and as it grew it went through many colours, in such a way that the day was very dark, deprived of its brightness. This occurred at the hour of midday, and the Sun remained thus for one hour and a half.
>
> [*Chronicon Conimbricense*; in Portuguese; Florez (1747, vol. XXIII, p. 344).]

As noted in an earlier record from Coimbra – AD 1239 (see section 11.6.1) – the year is reckoned from the first year of the Julian calendar.

RESULTS

For a partial eclipse at Coimbra, either $\Delta T < 4150$ sec or > 4250 sec.

Because the belt of annularity was so narrow, only a very restricted range of ΔT is disallowed. Many roughly contemporaneous observations indicate a value of ΔT at this date of considerably less than about 1000 sec so that the above limits, although not in discord with other results, are redundant.

(NB local time of maximum phase = 14.2 h; solar altitude = 55 deg.)

(10) AD 1354 Sep 17 [Wednesday] (total, 1.004): Perugia (Italy)

> In this year on 17 September that novelty appeared. The Sun became dark on a Wednesday at about the third hour and it lasted for the space of two hours. Above the Sun and Moon, which were joined together – that is, the Moon was covering the Sun – there appeared a very large star with fiery rays like a torch... Many people viewed the rays of the small Sun by reflection in a mirror or in clear water. And the rays of the Sun were so small and so dark, on account of the Moon covering the Sun, that there did not remain unobscured as much as 3 fingers of the Sun... Everyone appeared deathly pale.
>
> [*Memorie di Perugia dall'anno 1351 al 1438*; in Italian; quoted by Levi-Donati (1987).]

This account of a partial eclipse is clearly much exaggerated. However, the estimate of the magnitude in digits is interesting. The 'very large star with fiery rays like a torch' seen above the Sun was possibly a bright comet, although East Asian records make no mention of such an object. It was not Venus, which was about 18 deg to the east of the Sun and in mid-morning would be at a rather lower altitude.

RESULTS

For a partial eclipse at Perugia either $\Delta T < 4200$ sec or > 4320 sec.

As in AD 1333, the belt of totality was extremely narrow and this leads to a very narrow excluded zone, only 120 sec wide.

(NB local time of maximum phase = 9.4 h; solar altitude = 34 deg.)

(11) AD 1605 Oct 12 [Wednesday] (total, 1.03): Forcalquier (France)

> Wendelin at Forcalquier in Provence saw the whole Sun hidden apart from a very narrow thread towards the north, which he ascribed to the illuminated atmosphere.
>
> [Riccioli (1665, Book II, chap. 2); quoted by Zawilski (1994).]

Presumably Wendelin attributed the narrow thread of light to scattering in the supposed lunar atmosphere.

Table 11.6 ΔT limits deduced from medieval European observations of large partial solar eclipses.

	ΔT Range (sec)	
Year	LL	UL
1147	2870	1680
1178	2320	1140
1330	1210	890
1333	4250	4150
1354	4320	4200
1605	—	1060

RESULTS

For a partial eclipse at Forcalquier in which the southern portion of the solar disc was obscured, $\Delta T < 1060$ sec. At this late date, immediately before the telescopic era, this result – although sound – is by no means critical.

(NB local time of maximum phase = 13.7 h; solar altitude = 35 deg.)

The ΔT limits derived in this section – apart from the discordant results from AD 1133 and 1153 – are listed in table 11.6. In each case, a set of ΔT values is excluded. Hence in this table, as in table 8.8, the lower limit of one acceptable range of ΔT is actually higher than the upper limit which bounds another acceptable range of ΔT.

11.8 Detailed records which are deficient in some way

Although alleging a very large eclipse, the following records are lacking in important details. Either (i) they do not necessarily imply totality or annularity or (ii) the place of observation is in doubt. They will thus not be used to determine ΔT. I have included the observations because they give some additional indication of the remarkable variety of the descriptions which are preserved in medieval European texts – as well as further illustrating the effects of a major solar eclipse on an unsuspecting observer. In each case I have used a preliminary value of ΔT (determined from equation (8.1)) to deduce the approximate magnitude and local time of greatest phase at the assumed place of observation.

(1) AD 733 Aug 14 [Friday] (annular, 0.98): northern England

In the year 733 an eclipse of the Sun occurred on the 19th day before the Kalends of September (i.e. Aug 14), about the third hour of the day, with the result that almost the whole of the Sun's disc seemed to be covered by a black and horrid shield.

[*Bedae Continuatio*; ed. Plummer (1896, vol I, p. 361)]

This accurately dated account is one of the earliest observations of an eclipse recorded in an English chronicle. However, it is not clear from the text whether the ring phase was witnessed. Possibly a very large partial obscuration of the Sun would suffice.

Calculations based on equation (8.1) ($\Delta T = 3100$ sec) indicate that in northern England (lat. = 55 deg, long. = +1.5 deg) – the most likely source of information for the continuation of Bede's history – the eclipse would reach a magnitude of 0.91 at about 8.8 h. Although small, this may have been sufficient to produce the observed effects. The track of annularity itself did not reach further north than latitude 52.7 deg N at any point on the Earth's surface and would cross S. England.

(2) AD 812 May 14 [Friday] (total, 1.04): Syria?

In the year 1123 (Seleucid), on the 14th of Ayyar (= May) there was a total eclipse of the Sun from the ninth to the 11th hours. The darkness was as profound as night; the stars were seen and people lit torches. The Sun eventually reappeared over about an hour.

[*Chronicle* of Michael the Syrian, Book XII, chap. 7; in Syriac; translated from the rendering into French by Chabot (1905, vol. III, p. 26).]

This is one of the earliest *detailed* accounts of a total solar eclipse from any part of the world. The event occurred more than three centuries before the chronicler's own era. Although the date is accurately recorded (1123 Seleucid = AD 811/2), the place of observation is unfortunately not specified.

(3) AD 1133 Aug 2 [Wednesday] (total, 1.07): Kerkrade (Netherlands)

In the year from the Incarnation of our Lord 1133, darkness occurred over the whole Earth at the time of an eclipse of the Sun about midday. It lasted for almost a whole hour of the day, the Moon being on the 17th day, on the 4th day before the Nones of August (Aug 2). For then, just as at night, stars appeared, and the birds in the sky flew away, and the ground was wet with dew. Men were struck with very great terror; they thought that the last day had come.

[*Annales Rodenses*, *MGH*, XVI, 710.]

Although the date on the Julian calendar is correctly given, it seems likely that the reported '17th day' of the Moon's age is a scribal error for the 27th day. This is one of the very few early records to mention the formation of dew resulting from the fall in temperature during a large eclipse. Calculations using equation (8.1) ($\Delta T = 1200$ sec) indicate that at Kerkrade the magnitude would be about 0.995 at 12.0 h – close to the reported time.

(4) AD 1133 Aug 2 [Wednesday] (total, 1.07): Zwiefalten? (Germany)

> 1133. On the third day before the Nones of August... an eclipse of the Sun occurred. Concerning this same eclipse, we read in a Zwiefalten parchment. After meal-time there, at the beginning of midday, the Sun was suddenly covered with darkness (so it is written); people working in the houses, since they could not see to work, hurried outside in order to see, but things had suddenly changed. Then, with no clouds appearing, the disc of the Sun was very clearly without light. After half an hour, a small portion (*particulam*) began to appear, like the one-day old Moon. Afterwards the Sun gradually emerged, eventually with its usual splendour.
>
> [Footnote to *Bertholdi Zwifaltensis Chronicon*; *MGH*, X, 120; quoted by Zawilski (1994).]

The above passage, which is not in the *Bertholdi Zwifaltensis Chronicon* itself, is from the *Annales Suevici*, a chronicle compiled around AD 1590 by Martinus Crusius. Crusius, who frequently abstracted from the *Bertholdi Zwifaltensis Chronicon*, came across the parchment describing the eclipse while searching for material for his own chronicle. However, it would appear that Bertholdus, who was a monk at the monastery of Zwiefalten at the time of the eclipse and became abbot there in AD 1139, was unaware of this document. The parchment may well have been acquired by the monastery from some other source long after the time of Bertholdus.

With $\Delta T = 1200$ sec (equation (8.1)), the computed magnitude at Zwiefalten is 0.975 at 12.4 h.

(5) AD 1140 Mar 20 [Wednesday] (total, 1.06): Malmesbury (England)

> 1140. During this same year in Lent, on the 13th day before the Kalends of April (March 20), at the 9th hour, on the 4th day of the week (Wednesday), there was an eclipse of the Sun throughout the whole of England, as I have heard. With us, certainly, and with all our neighbours, the solar eclipse was so remarkable that men seated at tables, as almost everywhere they were at the time, for it was Lent, feared the primeval chaos; but soon learning what it was, they went outside and saw stars around the Sun. It was thought and said by many, not mistakenly, that the King (Stephen) would not continue to reign for a year without loss.
>
> [*Willelmi monachi Malmesburiensis Historia Novella*, lib. II; Potter (1955, pp. 42–43).]

William was librarian at the monastery of Malmesbury. His 'New History', which covers the interval from AD 1126 to 1142, is an important source for the period. In February of 1141, less than a year after the eclipse, King Stephen was imprisoned following the defeat of his army.

From equation (8.1) (ΔT = 1200 sec) it may be computed that the eclipse was fully total at Malmesbury at 14.6 h. This evidently was the time when men were dining, probably after the midday service (Sext).

(6) AD 1230 May 14 [Tuesday] (total, 1.05): Belvoir? (England)

> In the same year (1230) an extraordinary eclipse of the Sun occurred, in the very early morning immediately after sunrise, on the day before the Ides of May (May 14) in Rogationtide, namely the third day of the week (i.e. Tuesday). As a result, the workers in the fields and many others, leaving their morning's work on account of the excessive darkness, decided to return to bed and go back to sleep. But at length, after the space of one hour, to the astonishment of many, the Sun regained its usual brightness.
>
> [Roger of Wendover: *Flores Historiarum*; Hewlett (1887, vol. II, p. 384).]

Around this time, Roger of Wendover was prior of Belvoir in Leicestershire. Although his *Flores Historiarum* ('Flowers of History') is a chronicle of events in the whole of England, he may have witnessed the phenomenon personally. Calculations based on equation (8.1) ($\Delta T = 900$ sec) indicate that this eclipse would be total throughout much of eastern England (including Belvoir) at sunrise, the local time being about 4.1 h. This was indeed in the 'very early morning' and emphasises the long working day of agricultural labourers during the summer months.

(7) AD 1312 Jul 5 [Wednesday] (total, 1.01): Eastern Iceland

> 1312. A solar eclipse occurred in the month of March; there was such great darkness in the eastern fjords that one could not find one's way about, neither on the water nor indeed on land. Afterwards came a great mortality of men.
>
> [*Oddaverja annal*; in Icelandic;
> trans. into German by Thorkellson (1933).]

The month is incorrectly given as March. Since the eastern fjords cover almost the whole of the east coast of Iceland, the exact place of observation cannot be located, but the mean co-ordinates are: lat. $\sim +65$ deg, long. $\sim +14$ deg. With ΔT = 800 sec (from equation (8.1)), the eclipse would be indeed total here at 6.3 h.

(8) AD 1415 Jun 7 [Friday] (total, 1.07): Kobrin/Mytho (Belarius)

> 1415. An astonishing eclipse of the Sun. When King Vladislav was riding from Kobrin to Mytho, on the 6th day of the week, after the Octave of Corpus Christi (= Oct 6) at the 3rd hour of the day, a notable eclipse of the Sun appeared. As a result of this unexpected and unfamiliar event, King Vladislav and his followers were at first astonished and bewildered, later full of reverence. For it was so notable that the birds, terrified by the sudden darkness, fell to the ground, and the stars were shining as if by

night. The same King Vladislav was forced to stop on account of the darkness and could not proceed until the obscuration of the Sun passed.

[Joannis Dlugossi: *Annales seu Chronicae inditii Regni Poloniae*; quoted by Zawilski (1994).]

Despite the description of the sudden fall of darkness, the complete disappearance of the Sun is not directly alluded to. The chronicler, Jan Dlugosz, who was himself born in 1415, is clearly quoting an eye-witness account. Kobrin and Mytho (near Lida) are about 100 km apart. It can only be presumed that the observation was made somewhere between these towns.

With $\Delta T = 400$ sec (from equation (8.1)), the eclipse would be total at both Kobrin and Mytho at 8.0 h.

(9) AD 1476 Feb 25 [Sunday] (total, 1.04): Pskov (Russia)

(6984.) During the same month (i.e. February), on the 25th day, in Butter week, at the second hour of the day, as the Sun was rising and people were going to markets and elsewhere in the town, it suddenly began to grow darker and the darkness lasted for a little while, less than an hour; people could not see one another on the market nor anywhere else in the town, and they were frightened; and again God gave us light as before.

[*Pskovskaya I Letopis*; in Russian; trans. Vyssotsky (1949, p. 20).]

Here we have another very large eclipse at sunrise. The year 6984 is expressed relative to the Byzantine era, and thus corresponds to AD 1475/6. From equation (8.1) ($\Delta T = 300$ sec), totality would occur soon after sunrise (6.9 h), the solar altitude then being only about 2 deg.

(10) AD 1544 Jan 24 [Thursday] (total, 1.004): Neider Alteich

In the year of Our Lord 1544, on the Feast of St. Timothy the Apostle (i.e. Jan 24), there was an eclipse of the Sun before breakfast at the 9th hour, and there was darkness over the whole world (*in universa terra*) as if it was entirely night, with the result that people could not see one another well. Such an eclipse was not in the memory of men.

[*Notae Altahenses*; *MGH*, XVII, 426.]

Using the extrapolated ΔT value of 120 sec based on early telescopic data (rather than equation (8.1)) for this relatively recent date leads to a magnitude at Neider Alteich of 0.997 at 9.5 h. Evidently the recorded local time was expressed relative to midnight. In order for this eclipse to have been total (which is not specifically alleged) at Neider Alteich $-290 < \Delta T < -40$ sec. These values are in discord with roughly contemporaneous figures.

11.9 Timed solar eclipse contacts

Goldstein (1979) has drawn attention to a number of timed eclipse contacts measured by the astronomer Levi ben Gerson in southern France between AD 1321 and 1339, and also by Johannes de Muris in Northern France in the years 1333 and 1337. By this period, ΔT would be fairly small (only 750 sec according to equation (8.1)), and it seems doubtful whether timings of eclipses made with the unaided eye can yield a useful result for this parameter. However, the measurements by de Muris, although few in number, seem much more careful than those of ben Gerson and I have considered them in detail below.

The observations by Johannes de Muris were discovered in a Latin manuscript by Beaujouan (1974). The main objective of de Muris in making the measurements seems to have been to compare them with calculations based on the Alfonsine tables. He also made several determinations of the meridian altitude of the Sun. I have translated the following accounts from the texts published by Beaujouan.

(1) AD 1333 May 14: Evreux (France)

> Three brothers and I, in the presence of the Queen of Navarre, observed the beginning of this eclipse at Evreux near St Germain. And the altititude of the Sun at the moment of the beginning of the eclipse was close to 50 deg, and the altitude at the end of the eclipse was 33 deg. The magnitude of the eclipse, according to our estimate, was 10 digits... And thus the eclipse began after our midday by 2 hours and 20 minutes....

Johannes de Muris noted that the eclipse took place about 17 minutes earlier than as predicted by the Alfonsine tables. This same eclipse was observed to be partial in Coimbra – see section 11.7.

Results

(i) Assuming a solar altitude of 49.5 deg (in the west) at first contact, equivalent LT = 14.40 h, UT = 14.24 h. Computed TT = 13.89 h. Hence $\Delta T = -1250$ sec.

(ii) For a solar altitude of 33 deg (in the west) at last contact, equivalent LT = 16.18 h, UT = 16.03 h. Computed TT = 16.30 h. Hence $\Delta T = 1000$ sec.

(2) AD 1337 Mar 3: St Germain des Pres (France)

> In the current year of our Lord 1337, on the 3rd day of March after sunrise, a Monday, we saw the beginning of the eclipse, with the altitude of the Sun 10 deg. And already part was sensibly eclipsed, whereby we concluded that the edges of the luminaries could have touched at the altitude of 9 deg. Similarly we saw the exit of the Moon from contact with the Sun, as far as was possible, with the Sun at an altitude of 27 deg and

about 30 min...(The magnitude) was only 5 digits. In this experiment there were ten of us present and several had good astrolabes.

Johannes de Muris remarked that the eclipse occurred about 16 minutes earlier than expected from the Alfonsine tables.

RESULTS

(i) For a solar altitude of 9 deg (in the east) at first contact, equivalent LT = 7.20 h, UT = 7.22 h. Computed TT = 7.49 h. Hence ΔT = 950 sec.

(ii) For a solar altitude of 29 deg (in the east) at last contact, equivalent LT = 9.61 h, UT = 9.63 h. Computed TT = 9.71 h. Hence ΔT = 300 sec.

The ΔT result obtained from the first contact measurement in AD 1333 is in poor agreement with the other values derived in this section. Although it may well be that the recorded altitude of 'nearly 50 deg' at first contact on this occasion represents a scribal error, the other measurements are too few in number for it to be rejected as discordant. It would thus appear that the whole set of results is inadequate to obtain a refined value for ΔT. Incidentally, the two magnitude estimates are remarkably accurate. Using equation (8.1) (ΔT = 600 sec), the calculated magnitude in AD 1333 would be 0.82 – very close to the observed figure of 0.83. Similarly, in AD 1337, the calculated magnitude would be 0.42 – identical with the observed value. However, at this late epoch it is most unlikely that these unaided-eye estimates would themselves yield meaningful results for ΔT.

Rather more than a century later, the German astronomers Regiomontanus and Bernard Walther also made careful measurements of eclipse times (mainly of lunar obscuration) at Nurnberg. These observations appear to be of similar precision to those of Johannes de Muris. For details, see Zinner (1990). However, by this period ΔT would be so small (less than about 200 sec) that the observations are of mainly historical interest. Accordingly, they will not be analysed here.

11.10 An eighth century occultation of Jupiter by the eclipsed Moon

Unique in ancient and medieval history is an account of an occultation of the planet Jupiter by the eclipsed Moon in AD 755. The report of this compound event, which is due to the twelfth century English annalist Symeon of Durham, may be translated as follows:

AD 756...Also the Moon was covered over with the redness of blood on the 8th day before the Kalends of December (Nov 24), in the 15th day of its age, that is, at full Moon. And the darkness gradually decreasing, it returned to its original light. Further, astonishingly, a bright star following that same Moon, and passing through it, preceded the luminary by as great a space as it had followed it before it was obscured.

[*Symeonis Monachi Historia Regum*; Arnold (1885, p. 41).]

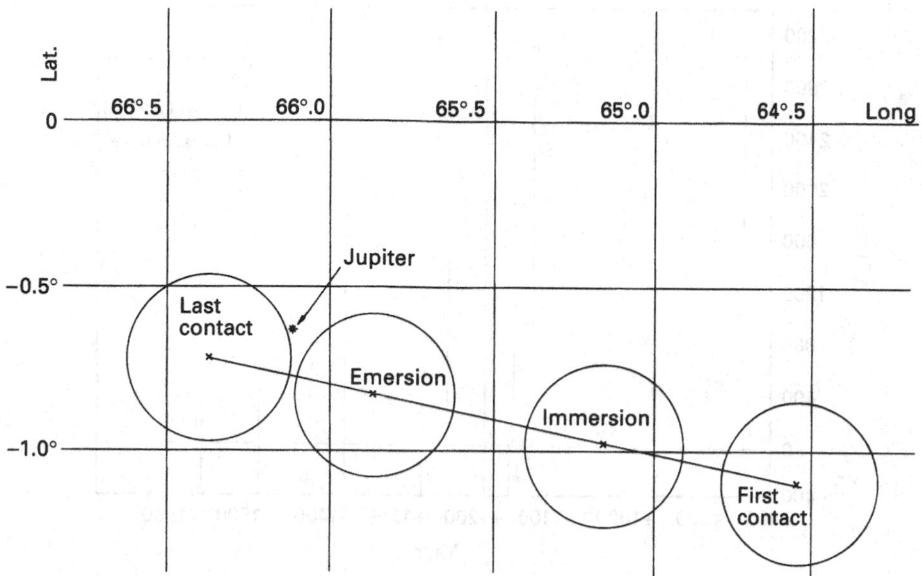

Fig. 11.10 Calculated position of the Moon relative to the planet Jupiter at each phase of the total lunar eclipse of AD 755 Nov 23/24.

The dates of the birth and death of Symeon, who was the precentor of Durham Cathedral in the early twelfth century, are uncertain. However, he is known to have written his history between 1104 and 1129. He seems to have taken his eighth century material from a contemporary chronicle, which because of the frequent references to York is sometimes referred to as the 'York Annals' (Blair, 1963).

Since celestial bodies appear to move from east to west on account of the diurnal rotation of the Earth, the text may be understood to imply that the bright star was first seen to the east of the Moon and afterwards appeared to the west of it. There was a lunar eclipse on Nov 11 in AD 756 but this was only partial and there was no bright star or planet nearby (Jupiter would be about 14 deg to the east of the Moon on that date). However, on the evening of Nov 23/24 in AD 755 the totally eclipsed Moon was very close to the planet Jupiter and this must be the correct identification.

Although fairly rare in other civilisations, many medieval European accounts of total lunar eclipses allude to the blood-red colour of the Moon; possibly the annalists were inspired by the Old Testament reference in *Joel* II, 31 – repeated in *Acts* II, 20.

Calculations based on equation (8.1) (ΔT = 2950 sec) show that from the viewpoint of York, Jupiter would be occulted by the Moon during the closing stages of the eclipse of AD 755 Nov 23/24, as the record asserts (see figure 11.10).

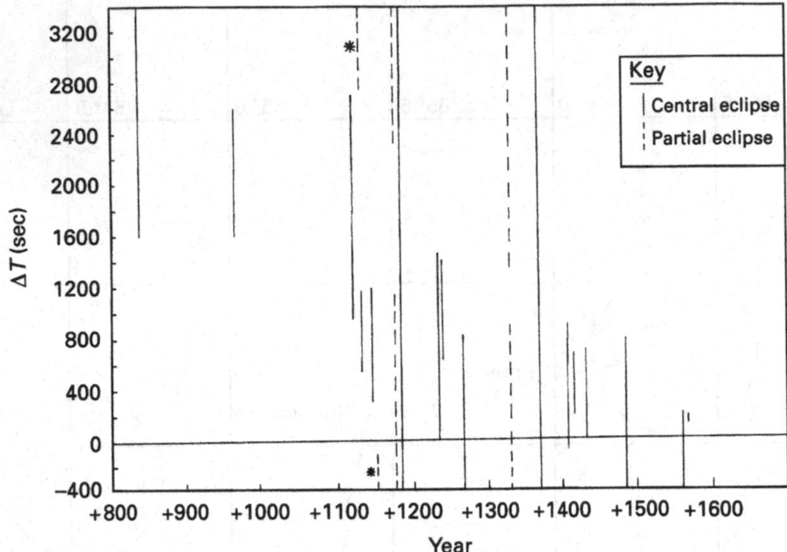

Fig. 11.11 ΔT limits obtained from European observations of both central and partial solar eclipses. Discordant results are indicated by an asterisk.

The computed local times of the eclipse contacts are as follows: beginning at 16.9 h; immersion at 18.0 h; emersion at 19.5 h; and end at 20.7 h. The Moon would start to occult Jupiter around 19.8 h local time (soon after emersion) and the occultation would end at 20.7 h (just before last contact). Jupiter would appear to pass about midway between the lunar centre and its upper limb. Although the observation is too imprecise for a value of ΔT to be deduced, it is of considerable historical interest because of its uniqueness.

11.11 Conclusion

In figure 11.11 are plotted the ΔT limits obtained in this chapter from both central solar eclipses (section 11.6) and partial eclipses (section 11.7). The anomalous results obtained from the partial eclipse observations in 1133 (Vysehrad) and 1153 (Erfurt) – are indicated by asterisks. Apart from these, it can be seen that there is generally good mutual accord between the various ΔT limits, especially bearing in mind that the observers were not astronomers. It is possible to trace a gradual decline in ΔT from about 2000 sec around the epoch +900 to less than 200 sec near +1600.

12
Solar and lunar eclipses recorded in medieval Arab chronicles

12.1 Introduction

Arab observations of solar and lunar eclipses which are of value in the study of the Earth's past rotation originate exclusively from the medieval period. These records are found in two main sources: (i) chronicles, and (ii) compendia on astronomy. As might be expected, the observations reported in *chronicles* are essentially qualitative; measurements of any kind are fairly rare. Eclipses and other celestial phenomena (such as bright comets and meteor showers) were mainly noted on account of their spectacular nature. Yet chronicles contain some important astronomical records. The present chapter will be devoted to eclipses described in these works. Arab compendia on astronomy contain many careful measurements of the times of the various phases for both solar and lunar eclipses. These observations will be discussed in chapter 13.

The Arabic names for the Moon and Sun are respectively *al-Qamar* and *al-Shams.* In general, medieval Arab chronicles use the term *khusuf al-Qamar* for an eclipse of the Moon and *kusuf al-Shams* for an eclipse of the Sun. These designations are still in use today, both among astronomers and the general public. However, medieval Muslim astronomers mostly used *kusuf* for both types of eclipse, adding the appropriate term for the Moon or Sun. The word *khusuf* means 'sinking', but in describing a lunar eclipse it came to mean a failing of the Moon's light. By contrast, *kusuf* means a 'cut' – i.e. in the solar (or lunar) limb. Totality of an eclipse was, and still is, usually indicated by expressions such as *tamm*, *kamil* or *kulliyy.*

Dates of eclipses are normally carefully recorded in Arab chronicles, but when the local time of occurrence is given, this is nearly always crudely expressed – rarely to better than the nearest hour. As remarked in previous chapters, such informal estimates of time are of negligible value in the determination of ΔT. Several detailed accounts of total solar eclipses are preserved from the medieval Arab world and these merit careful investigation. There are also a few allusions to very large partial eclipses of the Sun, but no references to central annular obscurations. Apart from the records of large solar eclipses (both total and partial), the only viable observations for the determination of ΔT are instances where the Sun or Moon rose or set eclipsed. It should be emphasised that the occasional estimates of the magnitudes of partial solar eclipses are too rough to be of value.

12.2 Chronicles and chroniclers

Arab chronicles cover much the same period as their European counterparts: mainly from about AD 800 to 1500. However, whereas vast numbers of European annals have been collected together and published (see chapter 11), relatively few Arabic texts of this nature have appeared in print. This is probably the main reason why only a relatively small number of reports of eclipses from the medieval Arab world have so far come to light. To date, searches for eclipse observations have concentrated exclusively on published chronicles. It may well be that many further observations are contained in manuscripts, but at present it is impossible to assess the potential of this material – much of which is probably widely scattered.

Until very recently, researchers had shown little interest in the eclipse records in Arab chronicles. However, in 1989, the author along with Dr Said S. Said and Dr Wafiq S. Rada, published the results of an extensive literature search through printed annals for references to eclipses of the Sun (Said *et al.*, 1989); this work will subsequently be abbreviated to SSR. Many records of lunar eclipses have since been uncovered in these and other printed sources (Stephenson and Said, 1997). All of the observations discussed in this chapter are taken from these two compilations. Printed editions of the Arab chronicles referred to below are listed in the appendix at the end of this chapter (section 12.10), rather than placed in the main list of references. I have not considered it necessary to include translations of the names of various book titles, some of which are quite lengthy.

As was also true of medieval European annalists, interest in celestial phenomena among Arab chroniclers evidently varied very much from one individual to another. Several writers seem to have had a special interest in astronomical matters, recording many eclipses and other occurrences. However, other chroniclers passed over such events in silence. Sometimes an eclipse was regarded as foretelling the death of a ruler. Not all such predictions proved to be successful! Thus, in reporting the lunar eclipse of AD 1487 Feb 7/8 the contemporary annalist Ibn Iyas relates the following:

> (892 AH.) In (the month of) Safar, the Moon's body was eclipsed and the Earth was darkened. It remained in eclipse for about 50 degrees (i.e. 3

Table 12.1 The principal Arab chroniclers recording eclipses of the Sun and Moon.

Chronicler	AD dates	Main domicile
al-Tabari	839–923	Baghdad
Ibn Hayyan	987–1076	Cordoba
Ibn al-Jawzi	1116–1201	Baghdad
Ibn al-Athir	1160–1233	Mosul
Abu Shama	1199–1266	Damascus
al-Maqrizi	1367–1442	Cairo
al-'Asqalani	1372–1449	Cairo
Ibn Taghri Birdi	1411–1470	Cairo
Ibn Iyas	1448–1524	Cairo
Ibn Tulun	1475–1546	Damascus
al-'Umari	? –1816+	Mosul

hours 20 min). People were saying that the demise of the Sultan was coming near. Nothing of what they said happened and the Sultan stayed (in power) for a long time after that.

[*Bada'i' al-Zuhur fi Waqa'i' al-Duhur* (vol III, p. 238); trans. Stephenson and Said (1997).]

The elderly Sultan (al-Malik al-Ashraf Qaytabay) lived for a further nine years after the eclipse.

Ibn Iyas had the unusual habit for an annalist of recording time intervals – both astronomical and in everyday life – in degrees (each equivalent to 4 minutes), as in the above account. The computed duration of the above event is 51 deg (3 h 23 m) so that on this occasion Ibn Iyas' figure is remarkably accurate.

The inclusive dates and main residence of each of the annalists who recorded the eclipses which are investigated in this chapter are listed in table 12.1. Exact dates are known for all of these men except al-'Umari, for whom neither the date of birth nor death is established; he is only known to have died some time after AD 1816.

In Arab as well as European chronicles, eclipse observations are cited alongside mainly non-astronomical matters, purely in order of occurrence. Generally, Arab annalists were in the habit of reporting material from a wide geographic area; few of their writings are comparable with the town and monastic chronicles of medieval Europe. Hence unless the location where an eclipse was seen is clearly stated or there is good reason for believing that the writer witnessed the event himself, the true place

Table 12.2 Geographic co-ordinates of places of observation for eclipses noted in Arab chronicles.

Place	Latitude (deg)	Longitude (deg)
Aleppo	36.23	−37.17
Baghdad	33.33	−44.43
Cairo	30.05	−31.25
Cizre	37.35	−42.22
Cordoba	37.88	+4.77
Cueva de la Mora	40.33	+3.92
Damascus	33.50	−36.32
Granada	37.17	+3.58
Mosul	36.35	−43.13
Olmos	40.15	+3.98

of observation may be in doubt. Fortunately, some annalists are quite specific regarding these details.

The modern names and geographic co-ordinates (in degrees and decimals) of the various sites where the eclipses mentioned in this chapter were observed are listed in table 12.2. As elsewhere in this book, the adopted sign convention is that longitudes west of the Greenwich meridian are positive, as are latitudes north of the equator.

12.3 Date conversion

Eclipse dates in Arab chronicles and astronomical compendia are mainly expressed in terms of the Islamic lunar calendar. On this calendar, years are numbered from *al-Hijrah* – the migration of the Prophet Muhammad from Mecca to Medina in AD 622. Years reckoned from this epoch are customarily designated by AH (*Anno Hegirae*). The Islamic year consists of 12 lunar months, each of either 29 or 30 days. From the time of the Prophet himself, intercalation has never been practised so that as a period of 12 lunar months is about 11 days shorter than the orbital period of the Earth around the Sun, the beginning of each year regresses through the seasons in approximately 33 years. In particular, a Julian or Gregorian century is closely equal to 103 Islamic years.

Whenever possible, each lunar month should commence with the sighting of the lunar crescent. At all significant centres of population a watch is kept for the young crescent at the close of the 29th day of every month. Should the crescent prove to be invisible – for example on account of cloud – an extra day is added to that month. Since meteorological condi-

Table 12.3 Lunar months of the Islamic year.

No.	Month	Days
1	Muharram	30
2	Safar	29
3	Rabi' al-Awwal	30
4	Rabi' al-Akhir	29
5	Jumada al-Ula	30
6	Jumada al-Ukhra	29
7	Rajab	30
8	Sha'ban	29
9	Ramadan	30
10	Shawwal	29
11	Dhu al-Qa'dah	30
12	Dhu al-Hijjah	29/30

tions are a major factor in determining crescent visibility, it is possible for adjacent communities to begin the month one day apart. Observation of the crescent is especially important in the last five months of the Islamic year. These are Sha'ban, Ramadan and Shawwal (to fix the month of Fasting - i.e. Ramadan), and also Dhu al-Qa'dah and Dhu al-Hijjah (to determine the dates of Pilgrimage to Mecca).

Although date conversion can be effected using computation of the visibility of the crescent based on a series of observational boundary conditions (see for example the details in SSR), it is more convenient to employ tables based on a 'standardised' Islamic calendar. The 12 lunar months of the Islamic year along with their tabular lengths are listed in table 12.3.

It should be noted that the following month names are occasionally confused in Arabic literature: (i) Rabi' al-Awwal (often abbreviated to Rabi I) and Rabi' al-Akhir (Rabi II); (ii) Jumada al-Ula (Jumada I) and Jumada al-Ukhra (Jumada II). Because the first part of the name is identical in each case, a careless scribe might easily make a mistake. NB the terms al-Awwal and al-Akhir (both masculine) mean respectively first and last, as do al-Ula and al-Ukhra (both feminine).

In both this and the succeeding chapters, interconversion of dates between the Islamic and Julian calendars has been achieved using a computer program modelled on the well-known tables of Freeman-Grenville (1977). These tables are based on the following assumptions:

(i) The first day of Muharram in the year in which the Hijra took place was Friday July 16 in AD 622. (NB some authorities prefer the alternative of July 15.)

(ii) The Islamic year consists of alternate lunar months of 30 and 29 days duration in the first eleven months (from Muharram to Dhu al-Qa'dah) and either 29 or 30 days in the twelfth month (Dhu al-Hijjah). The Islamic year thus contains either 354 or 355 days.

(iii) The number of days assigned to the twelfth month follows a regular pattern based on a 30-year cycle commencing in AD 622: it contains 30 days in the years 2, 5, 7, 10, 13, 16, 18, 21, 24, 26 and 29 of every cycle and 29 days in other years. The total number of days in a cycle of 360 months is thus 10 631 – an average length for the lunar month of 29.530 56 days. This compares very favourably with the true synodic month length of 29.530 59 days.

As is true for any lunar calendar based on first visibility of the crescent (for example that of ancient Babylon), fully accurate conversion of dates to the Western calendar can never be guaranteed unless independent information is supplied – for example the weekday. However, errors seldom exceed a single day, which is unimportant for events as rare as eclipses.

Many Arab accounts of eclipses specify the day of the week. From Sunday (the first day of the week), to Thursday (the fifth day of the week), each day has a name which simply indicates the appropriate number. For example Sunday is *yawm al-Ahhad*, meaning 'Day One'. Friday is called *yawm al-Jumu'ah*, meaning the 'Day of Assembly'. Saturday is known as *yawm al-Sabt* - meaning the 'Day of Rest'; this is presumably derived from the Hebrew Sabbath (*Shabbat*). When the weekday is specified, it is usually possible to derive the exact date on the Julian or Gregorian calendar and in such cases there is usually precise accord with calculated eclipse dates.

One of the earliest Arab eclipse records containing accurate dates originates from AD 882. This is cited by the contemporary Baghdad writer al-Tabari as follows:

> (269 AH.) In (the month of) Muharram in this year, the Moon was eclipsed on the night of the 14th (day) and set eclipsed. The Sun was eclipsed at the time of sunset on Friday, when two nights remained to the completion of Muharram, and set eclipsed. So in this month there were both lunar and solar eclipses.
>
> [al-Tabari, *Tarikh al-Tabari: Tarikh al-Rusul wa'l Muluk*, vol IX, p. 613; trans. SSR.]

In the Islamic calendar, lunar eclipses consistently take place on or about the 14th day of a month and solar eclipses around the 28th day. When the two dates in the above example are converted to the Julian calendar using the scheme outlined above, they correspond respectively

Table 12.4 Daily prayer times in Islam.

Prayer	Time
Fajr	Dawn
Zuhr	Noon
‘Asr	Afternoon
Maghrib	Sunset
‘Isha	Nightfall

to Aug 2/3 and Aug 17 in AD 882. In particular, Aug 17 was indeed a Friday. On each of these dates it may be calculated that an eclipse did occur, indicating that al-Tabari's chronology is very accurate.

The Islamic day begins at sunset, roughly 6 hours before the start of the civil day (i.e. local midnight). Hence in converting the recorded date of a lunar eclipse which began before local midnight to the Julian (or Gregorian) calendar, one day must be subtracted to allow for this discrepancy. In the case of a lunar obscuration commencing after midnight – and for all solar eclipses – the Islamic and civil dates are identical so that this correction is unnecessary. The practice adopted throughout this book of using double dates for all lunar eclipses avoids such confusion.

Local times of eclipses are occasionally expressed relative to the five Islamic prayer times. These are listed in table 12.4. For example, in AD 1377 it is recorded that an obscuration of the Sun lasted from *Zuhr* (the noon prayer) until after *‘Asr* (the afternoon prayer). Similarly, in AD 1399 a lunar eclipse was said to begin after *‘Isha* (the nightfall prayer) and to last until midnight.

Theoretically, the dawn prayer begins when the Sun's depression below the eastern horizon is about 18 deg while the nightfall prayer starts when the Sun has reached a similar depression below the western horizon. As it happens, these angular values correspond to the limit of astronomical twilight in modern convention; when the Sun is beyond this limit there is negligible sunlight glow. In high latitudes, of course, the Sun never attains a depression of 18 deg during the summer months, but in the low latitudes occupied by most Muslim countries this problem does not arise. The sunset prayer closely follows sunset – normally a well-defined reference moment. However, to fix the time of the noon and afternoon prayers may require the use of a sundial, as found at many mosques.

The noon prayer begins a little *after* the Sun crosses the meridian (in principle, when the length of the shadow of a gnomon begins to show a noticeable increase after the noon minimum). By comparison, according to majority opinion the afternoon prayer commences when the shadow length equals its noon length plus the height of the gnomon itself. For a discussion of these various definitions, see Said and Stephenson (1991).

One early eclipse (occurring in AD 1241) is dated relative to the Coptic calendar which numbers years from the persecution of Christians by the Roman Emperor Diocletian in AD 284. This scheme, which has twelve months of 30 days followed by 5 or 6 intercalary days, is still adopted by the Coptic church today. The twelve month-names are derived from ancient Egyptian names. Each year, which normally begins on Aug 29 of the Julian calendar, commences with the month Tut.

12.4 Records of total and near-total solar eclipses

In this section I shall only discuss those few records for which the place of observation is fairly well established and either totality or a very large partial eclipse is clearly described. At the head of each entry below is given the tabular Julian date of the eclipse and the calculated weekday. These are followed by the type of eclipse, computed magnitude in the central zone and place of observation.

All translations of Arabic texts in this and the following three sections (12.5 to 12.7) are by SSR, with minor emendations. After each record are given a few brief comments, followed by the indicated ΔT range. For general reference, the local time of maximum phase and the solar altitude at that time are calculated using equation (8.1).

(1) AD 912 Jun 17 [Wednesday] (total, mag. = 1.06): Cordoba

> (299 AH.) In this year, the Sun was eclipsed and all (*jami'*) of it disappeared on Wednesday when one night remained to the completion of (the month of) Shawwal. The stars appeared and darkness covered the horizon. Thinking it was sunset, most of the people prayed the *Maghrib* (Sunset) Prayer. Afterwards, the darkness cleared and the Sun reappeared for half an hour and then set.
>
> [Ibn Hayyan: *al-Muqtabis fi Tarikh al-Andalus*, vol. III, p. 147.]

The recorded date (including the weekday), is correct. Although Ibn Hayyan wrote rather more than a century after the event, this section of *al-Muqtabis* is a chronicle of the reign of 'Abd Allah b. Muhammad, the Umayyad Caliph, in Cordoba and frequently mentions events in the city. Ibn Hayyan himself was born in Cordoba and spent most of his life there. It thus seems highly likely that the place of observation was Cordoba.

Results

For totality at Cordoba, $880 < \Delta T < 2600$ sec.

(NB computed local time of maximum phase = 19.2h, solar altitude = + 3 deg.)

(2) AD 1061 Jun 20 [Wednesday] (total, mag. = 1.06): Baghdad

> (453 AH.) On Wednesday, when two nights remained to the completion of (the month of) Jumada al-Ula, two hours after daybreak, the Sun was eclipsed totally. There was darkness and the birds fell whilst flying. The astrologers claimed that one-sixth of the Sun should have remained (uneclipsed) but nothing of it did so. The Sun reappeared after four hours and a fraction (of an hour). The eclipse was not in the whole of the Sun in places other than Baghdad and the provinces.
>
> [Ibn al-Jawzi: *al-Muntazam fi Tarikh al-Muluk wa'l-Umam*, vol. III, p. 221.]

A copy of the printed Arabic text of a page containing this record is illustrated in figure 12.1.

The Baghdad chronicler Ibn al-Jawzi wrote a century after the eclipse. He is, however, careful to specify the precise place of observation and his account is clearly based on an eyewitness description. The date which he gives is exactly correct.

RESULTS

For totality at Baghdad, $800 < \Delta T < 2140$ sec.
(NB local time of maximum phase = 7.7 h; solar altitude = 33 deg.)

(3) AD 1176 Apr 11 [Sunday] (total, mag. = 1.06): Cizre

> (571 AH.) In this year the Sun was eclipsed totally and the Earth was in darkness so that it was like a dark night and the stars appeared. That was the forenoon (*Duha*) of Friday the 29th of (the month of) Ramadan at Jazirat Ibn 'Umar (Ibn 'Umar's island), when I was young and in the company of my arithmetic teacher. When I saw it I was very much afraid; I held on to him and my heart was strengthened. My teacher was learned about the stars and told me, 'Now you will see that all of this will go away', and it went quickly.
>
> [Ibn al-Athir: *al-Kamil fi al-Tarikh*; vol. IX, p. 138.]

For a copy of the printed Arabic text of this account, see figure 12.2.

This eclipse was also observed to be total in Antioch (= Antakya) – as discussed in chapter 11. Although the year, month and day of the month are correctly given in the above account, Ibn al-Athir has mistaken the weekday. Nevertheless, he is careful to tell his readers precisely where he was when he saw the eclipse; this is now known as Cizre, a Turkish town on the frontier with Syria. At the time of this event, Ibn al-Athir was aged 16. He gives a particularly appealing account of the effect of a total eclipse on a young and unsuspecting eye-witness.

كتاب المنتظم ٢٢١ ج – ٨

والخليفة يحتمل ويصبر وجاء يوما الى الديوان بثياب بيض وتوسط الامر قاضى القضاء الدامغانى وابو منصور بن يوسف واستقر الأمر على ان كتب الخليفة لعميدالملك انا قد استخلفناك على هذا الامر ورضينا بك فيما تفعله مما يعود بمرضاتنا ومرضاة ركن الدين فاعمل فى ذلك برأيك الصائب الموفق تزجية للحال ودفعا بالأيام وترقبا لأحد امرين اما قناعة السلطان بهذا الامر اوطلب الاتمام فلا يمكن المخالفة ثم دخل عميدالملك يوما الى الخليفة ومعه قاضى القضاة وجماعة من الشهود وقال اسأل مولانا امير المؤمنين التطول بذكر ماشرف به ركن الدين الخادم الناصح فيما رغب فيه وسمت نفسه اليه ليعرفه الجماعة من رأيه الكريم وأراد ان يقول الخليفة ما يلزمه به الحجة بالاجابة ففطن لذلك فقال قد شرط فى المعنى ما فيه كفاية والحال عليه جارية فانصرف مغتاظا ورحل فى عشية يوم الثلاثاء السادس والعشرين من جمادى الآخرة ورد المال والجواهر والآلات الى همذان وبقى الناس وجلين من هذه المنازعة .

وفى يوم الاربعاء لليلتين بقيتا من جمادى الاولى على ساعتين منه انكسفت الشمس جميعها واظلمت الدنيا كلها وسقطت الطيور فى طيرانها وكان المنجمون قد زعموا أنه يبقى سدسها فلم يبق منها شىء وكان انجلاؤها على اربع ساعات وكسر ولم يكن الكسوف فى غير بغداد واقطارها عاما فى جميع الشمس .

وفى رجب ورد رسول من عميد الملك يذكر ان كتاب السلطان ورد عليه بان الخليفة ان لم يجب الى الوصلة التى سألناها فطالبه بتسليم ارسلان خاتون اليك واعدها معك لأسير بنفسى وا تولى الخطاب على هذا وانه أراد العود من الطريق لفعل مارسم له من هذا فخاف ان لا ينضبط له العسكر اذا عادوا الى بغداد يقول انى قد اعدت هذا الرسول لحمل ارسلان خاتون الى دار المملكة الى حين اجتماعى بالسلطان واصلاح هذه القصة وكاتب ارسلان بمثل ذلك وبانتقالها عن الدار فتجدد الانزعاج والخوف وذافع الخليفة عن الجواب وتبسط اصحاب فى اشياء توجب خرق الحشمة فاظهر الخليفة الخروج من بغداد وتقدم باصلاح

Fig. 12.1 Printed Arabic text showing the report of the solar total eclipse of AD 1061 Jun 20 by Ibn al-Jawzi (Baghdad).

RESULTS

For totality at Cizre, $600 < \Delta T < 2350$ sec.

(NB local time of maximum phase = 7.1 h; solar altitude = 25 deg.)

Combining the above limits with those set by the observation of the same eclipse at Antioch (i.e. $-190 < \Delta T < 1600$ sec) yields a revised range $600 < \Delta T < 1600$ sec. The southern edge of the computed track of totality calculated with $\Delta T = 600$ sec (passing through Cizre) and the northern edge calculated with $\Delta T = 1600$ sec (passing through Antioch) are delineated in figure 12.3. For any value of ΔT in the range $600 < \Delta T < 1600$ sec, the eclipse would be total at both places in accordance with the records.

مكة فحاربوهم وقتل من الفريقين جماعة وصاح الناس الغزاة إلى مكة فهجموا عليها فهرب أمير مكة مكثر فصعد إلى القلعة التى بناها على جبل أبى قبيس فحصروه بها ففارقها وسار عن مكة وولى أخوه داود الإمارة ونهب كثيرا من الحاج وأخذوا من أموال التجارة المقيمين بها شيئا كثيرا وأحرقوا دورا كثيرة ومن أعجب ما جرى فيها أن إنسانا زراقا ضرب دار بقارورة نفط فأحرقها وكانت لأيتام فأحرقت ما فيها ثم أخذ قارورة أخرى ليضرب بها مكانا آخر فأتاه حجر فأصاب القارورة فكسرها فاحترق هو بها فبقى ثلاثة أيام يعذب بالحريق ثم مات .

﴿ذكر عدة حوادث﴾ فى هذه السنة فى شهر رمضان انكسفت الشمس جميعها وأظلمت الارض حتى بقى الوقت كأنه ليل مظلم وظهرت الكواكب وكان ذلك ضحوة النهار يوم الجمعة التاسع والعشرين منه وكنت حينئذ صبيا بظاهر جزيرة ابن عمر مع شيخ لنا من العلماء أقرأ عليه الحساب فلما رأيت ذلك خفت خوفا شديدا وتمسكت به فقوى قلبى وكان عالما بالنجوم أيضا وقال لى الآن ترى هذا جميعه انصرف فانصرف سريعا ، وفيها ولى الخليفة المستضىء بأمر الله حجبة الباب أبا طالب نصر بن على الناقد وكان يلقب فى صغره قنبرا فصاروا يصيحون به ذلك إذا ركب بأمر الخليفة أن يركب معه جماعة من الأتراك ويمنعون الناس من ذلك فامتنعوا فلما كان قبل العيد خلع عليه ليركب فى الموكب فاشترى جماعة من أهل بغداد من القنابر شيئا كثيرا وعزموا على إرسالها فى الموكب إذا رأوا ابن الناقد فأنهى ذلك إلى الخليفة وقيل له يصير الموكب ضحكة فعزله وولى ابن المعوج وفيها فى ذى الحجة يوم العيد وقعت فتنة ببغداد بين العامة وبين الأتراك بسبب أخذ جمال النحر فقتل بينهم جماعة ونهب شىء كثير من الأموال ففرق الخليفة أموالا جليلة فيمن نهب ماله . وفيها زلزلت بلاد العجم من جهة العراق إلى ماوراء الرى وهلك فيها خلق كثير وتهدمت دور كثيرة وأكثر ذلك كان بالرى وقزوين . وفيها فى ربيع الآخر استوزر سيف الدين غازى صاحب الموصل جلال الدين أبا الحسن ابن جمال الدين محمد بن على وكان جمال الدين وزير البيت الأتابكى وقد تقدمت أخباره وهو المشهور بالجود والإفضال ولما ولى جلال الدين الوزارة ظهرت منه كفاية عظيمة ومعرفة تامة بقوانين الوزارة وله مكاتبات وعهود حسنة مدونة مشهورة وكان جوادا فاضلا خيرا وكان عمره لما ولى الوزارة خمسا وعشرين سنة . وفيها فى ذى الحجة استناب سيف الدين أيضا عنه بقلعة الموصل مجاهد الدين قايماز وفوض اليه الأمور وكان قبل ذلك اليه الأمر بمدينة إربل وأعمالها وكان رحمه الله من صالحى الأمراء وأرباب المعروف بنى كثيرا من الجوامع والخانات فى الطرق والقناطر على الأنهار والربط وغير ذلك من أبواب البر وكان دائم الصدقة كثير الإحسان عادل السيرة رحمه الله . وفيها قبض الخليفة على سنجر المقتفوى أستاذ الدار ورتب مكانه أبا الفضل هبة الله بن على بن هبة الله بن الصامت . وفيها فى رمضان قدم شمس الدولة تورانشاه بن أيوب الذى ملك اليمن إلى دمشق ولما سمع أن أخاه صلاح الدين ملكها حن إلى الوطن والأتراب ففارق اليمن وسار إلى الشام وأرسل من الطريق إلى أخيه صلاح الدين يعلمه بوصوله وكتب فى الكتاب شعرا من قول ابن المنجم المصرى :

وإلى صلاح الدين أشكو أننى ٭ من بعده مضنى الجوانح مولع
جزعا لبعد الدار منه ولم أكن ٭ لولا هواه لبعد دار أجزع
فلأركبن اليـــه متن عزائمى ٭ ويخب بى ركب الغرام ويوضع

Fig. 12.2 Printed Arabic text showing the report of the solar total eclipse of AD 1176 Apr 11 by Ibn al-Athir (Cizre).

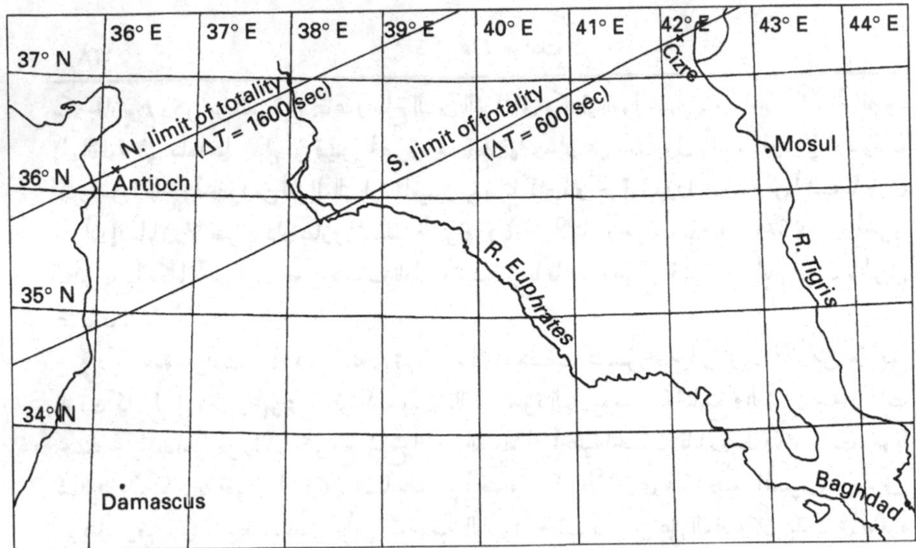

Fig. 12.3 Edges of the computed track of totality in AD 1176 for the two ΔT limits set by the observations at Cizre and Antioch (southern limit for $\Delta T = 600$ sec, northern limit for $\Delta T = 1600$ sec).

(4) AD 1433 Jun 17 [Wednesday] (total, mag. = 1.06): Cairo

(836 AH.) On Wednesday the 28th of (the month of) Shawwal, the Sun was eclipsed by about two-thirds in the sign Cancer more than one hour after the *'Asr* (Afternoon) Prayer. As the Sun was setting, the eclipse started to clear. During the eclipse there was darkness and some stars appeared.

[Al-Maqrizi: *al-Suluk fi Ma'rifat Duwal al-Muluk*, vol. IV, p. 892.]

Al-Maqrizi was a contemporary writer living in Cairo. It is possible that he witnessed the eclipse himself. An eclipse of magnitude only $\frac{2}{3}$ certainly would not bring out 'some of the stars' but the effect of irradiation may well have made the degree of obscuration of the Sun seem smaller than it actually was. Hence the record may best be interpreted as denying totality. Both Venus and Jupiter were well placed to the east of the Sun, which was indeed in Cancer (longitude 94 deg), but unless the eclipse was extremely large, only Venus would be likely to be seen.

RESULTS

For a partial eclipse at Cairo, $\Delta T < -4000$ or > -2460 sec.

At this relatively recent epoch, extrapolation of telescopic results indicates that ΔT would be fairly small. In consequence, the above limits are of little significance. They will not be considered further.

(NB local time of maximum phase = 18.2 h; solar altitude = 9 deg.)

Table 12.5 ΔT limits derived from medieval observations of total solar eclipses in Arabic chronicles.

	ΔT Range (sec)	
Year	LL	UL
+912	880	2600
1061	800	2140
1176	600	1600

The ΔT ranges derived in this section are summarised in table 12.5. This table gives the year (+ rather than AD), lower limit (LL) and upper limit (UL) to ΔT for each date. Since there were two separate observations of totality in 1176 (at Antioch – see chapter 11 – and Cizre), the two sets of ΔT limits are combined.

12.5 Other records of large solar eclipses

Although the following records of large eclipses are deficient in some way (e.g. the place of observation is uncertain or the precise magnitude is in doubt), they seem interesting enough to render them worth quoting in full. They will not, however, be used to fix ΔT. For each observation, I have used equation (8.1) to calculate the apparent magnitude and local time of greatest phase.

(1) AD 939 Jul 19 [Friday] (total, mag. = 1.07): Olmos/Cueva del la Mora (Spain)

> (327 AH.) The Caliph al-Nasir li Din Allah advanced (northwards from Cordoba) heading for his holy battle (*jihad*) until he reached Toledo on Thursday, when seven nights (*sic*) remained to the completion of (the month of) Ramadan. He stayed there for six days and left on Thursday, when two nights remained to the completion of Ramadan, for Welmish fortress and on Friday to Khalifah Castle. During the forenoon (*Duha*) of that day (Friday) the Sun was eclipsed totally and its disk became dark except for a slight portion as seen by eye.
>
> [Ibn Hayyan, *al-Muqtabis fi Tarikh al-Andalus*, vol V, p. 434.]

Ibn Hayyan is here quoting from the contemporary writer al-Razi, who probably resided at Cordoba. It would appear that the eclipse was witnessed while the Caliph and his army were travelling between Welmish and Khalifah. I learn from Dr Edward Cooper of London Guildhall

University that the present-day names for Welmish and Khalifah are respectively Olmos and Cueva de la Mora. Olmos is situated approximately 30 km north of Toledo, while Cueva de la Mora is some 20 km further north than Olmos.

It is difficult to reconcile the two descriptions (i) 'the Sun was eclipsed totally' and (ii) 'its disk became dark except for a slight portion as seen by eye'. Possibly the Sun was reduced to a small point of light or a bright prominence appeared at the solar limb. However, the true phase must remain in doubt. The value of the observation is further weakened by the uncertainty in the precise place of observation.

Calculations using equation (8.1) (ΔT = 1980 sec) indicate that at both Olmos and Cueva de la Mora the eclipse would be fully total at 7.2 h.

(2) AD 993 Aug 20 [Sunday] (total, mag. = 1.06): Cairo?

> (383 AH.) In this year the Sun was eclipsed totally at the end of (the month of) Jumada al-Ukhra. It was so dark that the stars appeared and people could not see the palms of their hands. The eclipse cleared at the end of the day.
>
> [Al-Maqrizi: *Itti'az al-Hunafa bi Akhbar al-A'imma al-Fatimiyin al-Khulafa*, vol. I, p. 280.]

Although al-Maqrizi spent most of his life in Cairo, he wrote more than four centuries after the eclipse. Hence the place of observation must be regarded as very doubtful; there is nothing in the text itself to indicate a Cairo source. Although al-Maqrizi appears to quote from an eye-witness description of totality, careful observations made in Cairo and recorded by the contemporary astronomer ibn Yunus indicate that the eclipse was only partial there and ended in mid-morning (see chapter 13). Al-Maqrizi may have obtained his information from a source much to the east of Cairo.

Calculations using equation (8.1) (ΔT = 1750 sec) indicate that at Cairo the magnitude would be 0.96 at 8.9 h.

(3) AD 1176 Apr 11 (total, 1.06): River Orontes

> (570 AH.) In the last days of (the month of) Shawwal, I remember we crossed the River al- 'Asi (Orontes) on our return (from Hamah to Damascus). The Sun was eclipsed and it became dark in the day time. People were very frightened and stars appeared. Then we arrived at Hams...
>
> [Al-Katib al-Isfahani, as quoted by Abu Shama: *Kitab al-Rawdataim fi Akhbar al-Dawlatain*, vol. I, p. 250.]

Al-Katib al-Isfahani (AD 1125–1201) accompanied Saladin (the Sultan Yusuf b. Ayyub Salah al-Din) on many campaigns. However, the date which he gives is incorrect. This corresponds to within a day or two of AD 1175 May 21 but there was no solar eclipse visible at the time. The nearest eclipse which was large in Syria occurred almost a year later, AD 1176 Apr 11 (28 Ramadan in 571 A.H.). Thus we must assume a serious dating error here, as was suggested by Lane-Poole (1926, p. 143). Although the place of observation can be accurately deduced – the point where the road from Hamah to Hams crosses the River Orontes (Syria: lat. = 37.35 deg, long. = −42.18 deg) – the text fails to describe the complete disappearance of the Sun.

According to computation using equation (8.1) (ΔT = 1050 sec), at the Orontes crossing the eclipse would be fully total at 7.6 h.

(4) AD 1241 Oct 6 [Sunday] (total, mag. = 1.05): Nile Delta

> On the 9th of the month of Babah of the year 958...in the reign of al-Malik al-Salih Najm al-Din Ayyub (King Ayyub the Good) and under the presidency of Patriarch Cyril, something strange and wonderful occurred in the world such that everyone who saw it or heard about it was astonished. Namely, the Sun was gradually darkened until it became completely dark. Also the day became dark like night. Some people saw the stars and people lit lamps and people were very frightened and prayed... The darkness cleared instantly and then the Sun appeared as usual and lit up the world, and lamps were extinguished. The duration of the solar eclipse was about one hour – from the middle of the 8th to the middle of the 9th hour...
>
> [*Synaxarium Alexandrinum.*]

Observations of this same eclipse were also made in Europe – see chapter 11. The *Synaxarium Alexandrinum* is a contemporary Coptic religious yearbook written in Arabic. It is mainly a compilation of the lives of martyrs, saints, etc. of the Coptic church. The eclipse is the only astronomical event mentioned in the entire work. The year, 958 relative to the Diocletian era, corresponds to AD 1241/2. Babah, the second month of the year, is a version of the ancient Egyptian month Phaophi. This month began on Sep 28 so that Babah 9 would be equivalent to Oct 6. Hence the recorded date is exactly correct. The stated local time corresponds to between about 13.5 and 14.5 h. According to Ginzel (1918), the place of observation was either Cairo (the residence of the Coptic patriarchs) or elsewhere in the Nile Delta – where the other Coptic communities were mainly to be found. However, this is an expansive area, extending roughly 200 km north–south and east–west. The *Synaxarium* itself was compiled under the direction of Amba Michael, Bishop of Atrib, a town in the Nile Delta. A translation of this yearbook into Latin has been published by Forget (1963).

Because of the uncertainty in the precise place of observation, *firm* limits to the value of ΔT (necessary at such a late date) cannot be set.

Using equation (8.1) ($\Delta T = 850$ sec), the eclipse would be total at Cairo and throughout much of the Nile Delta at 15.1 h.

(5) AD 1431 Feb 12 [Monday] (total, mag. = 1.04)

Two separate accounts of this eclipse – which was also observed to be total in Italy (see chapter 11) – are preserved by contemporary Arab writers: al-Maqrizi and al-'Asqalani, both of whom lived in Cairo.

(i) (834 AH.) In (the month of) Jumada al-Ukhra, the astrologers warned that the Sun would be eclipsed and in Cairo there were callings to the people that they should pray and do good deeds. However, the eclipse did not occur and those who gave the warnings were denounced. Then news arrived from al-Andalus (Islamic Spain) of the occurrence of an eclipse there covering all of the Sun's body except one-eighth of it. That was after midday (*nisf al-nahar*) on the 28th of the month.

[Al-Maqrizi: *al-Suluk fi Ma'rifat Duwal al-Muluk*, vol. IV, p. 855.]

(ii) (834 AH.) In (the month of) Jumada al-Ula it was known that the calendar experts agreed that the Sun was to be eclipsed on the 28th of the month after the *Zawal* (i.e. after the Sun had crossed the meridian). The Sultan and the people were prepared for it and were watching the Sun until it set but nothing of it had changed at all.

[Al-'Asqalani: *Inba' al-Ghumr bi 'Bna' al-'Umr*, vol. VIII, p. 179.]

The correct lunar month is Jumada al-Ula rather than the following month Jumada al-Ukhra. The most likely place of observation in Spain – where the eclipse was said to be fairly large – is the region of Granada, the capital of the Muslim territories there. Using equation (8.1) ($\Delta T = 400$ sec), it may deduced that the magnitude at Granada was 0.90 at 13.8 h. Despite the negative observation in Cairo, similar calculations indicate that as much as 0.43 of the solar diameter would be covered there in the late afternoon (16.8 h). The failure of so many potential observers to notice it is indeed surprising, since neither text makes any mention of unfavourable weather.

(6) AD 1433 Jun 17 [Wednesday] (total, mag. = 1.06): Aleppo

There are two records of this eclipse from Aleppo.

(i) (836 AH.) On the 28th of (the month of) Shawwal, the Sun was eclipsed after the *'Asr* (Afternoon) Prayer and continued until the time of sunset. It cleared up after the conclusion of the eclipse prayer, which I led in the Great Mosque. Then the Sun set and we prayed the *Maghrib* (Sunset) Prayer in the mosque. When the eclipse prayer was concluded, I sent a witness to ascend the minaret of the mosque to see if the Sun had cleared. He returned, saying that it had cleared completely.

[Al-'Asqalani: *Inba' al-Ghumr bi 'Bna' al-'Umr*, vol. VIII, p. 280.]

The editor of the above text refers to a gloss (marginal note) in one of the manuscripts of the chronicle of al-'Asqalani, presumably written by a student or companion of his. In this note, we find the following statement:

(ii) The eclipse was dense and it became dark such that we thought that (the time for) the *Maghrib* (Sunset) Prayer had arrived. Then we reckoned that it was still afternoon. I looked at the Sun and found that it was eclipsed and that the eclipse was great (*'azim*). We accompanied the author (i.e. al-'Asqalani) to the Great Mosque and prayed after him until it cleared.

We know from his chronicle that at the time of the eclipse al-'Asqalani was in Aleppo, having travelled there from Cairo with the Sultan. The 'Great Mosque' is a well known mosque in Aleppo, which confirms the location. Although the writer of the gloss (ii) gazed directly at the Sun, he did not give a clear description of the degree of obscuration. His remark that the eclipse was 'great' could either imply totality or a very large partial phase. From equation (8.1) ($\Delta T = 380$ sec), the magnitude at Aleppo was 0.998 (i.e. on the verge of totality) at 18.4 h, the solar altitude then being only 9 deg.

(7) AD 1463 May 18 [Wednesday] (annular, mag. = 0.96): Cairo

(i) (867 AH.) In the month of Sha'ban, the Sun was eclipsed and the eclipse was excessive (*fahish*) from late forenoon (after *Duha*) almost to the afternoon such that the Earth was darkened in the eyes of the people.

[Ibn Iyas: *Bada'i' al-Zuhur fi Waqa'i' al-Duhur*, older edition, vol II, p. 75.]

The above record is taken from an edition of Ibn Iyas dating from 1894. A more recent edition of the same work, published between 1960 and 1975, gives the following account:

(ii) (867 AH.) In the month of Sha'ban, there was a complete (*tamm*) eclipse of the Sun such that the Earth was darkened. The eclipse continued for about 40 degrees (i.e. 2 hours and 40 minutes).

[Ibn Iyas: *Bada'i' al-Zuhur fi Waqa'i' al-Duhur*, new edition, vol. II, p. 447.]

A contemporary writer, Ibn Iyas lived at Cairo. In his chronicle, he often omitted to state the day of the month (which on this occasion would be the 28th of Sha'ban).

Why the above accounts differ so much is unknown. In describing the eclipsed Sun, the term 'excessive' in the older edition could mean either extremely large or total, but the newer edition implies totality. As it happens the eclipse was only annular, no more than 0.96 of the solar diameter being covered where the central phase would be visible. At Cairo, the computed magnitude based on equation (8.1) (ΔT = 300 sec) is only 0.85 at 12.5 h. The calculated duration there is as much as 55 deg (3h 40 m), considerably more than that reported by Ibn Iyas. To achieve *annularity* at Cairo (if this is a plausible interpretation of the record), a range of ΔT between 2360 and 3030 sec would be required. This is an order of magnitude too great at such a recent epoch. Both this and the following observation recorded by Ibn Iyas (which also alleges totality) would appear to be unreliable.

(8) AD 1491 May 8 [Sunday] (annular, mag. = 0.94): Cairo

> (896 AH.) In the month Jumada al-Ukhra, the Sun was eclipsed totally. The Sun stayed eclipsed for about 30 degrees (i.e. two hours).
>
> [Ibn Iyas: *Bada'i' al-Zuhur fi Waqa'i' al-Duhur*, vol. III, p. 282.]

The calculated date of this event is equivalent to the 28th of the month Jumada al-Ukhra. As in the previous entry, although this eclipse was described by Ibn Iyas (a contemporary) as total, it was in fact generally annular. The record is lacking both in descriptive detail and any reference to the place of observation.

The configuration of this eclipse track is such that only impossibly high values of ΔT (between about 28 000 and 30 000 sec) could lead to annularity at Cairo. Using equation (8.1) (ΔT = 280 sec), only 0.40 of the solar diameter would be covered at maximum phase around 17.1 h. Possibly Ibn Iyas is quoting the results of prediction rather than observation. As it happens, the computed duration of the eclipse at Cairo is 2h 0m (30 deg), identical to the reported figure.

(9) AD 1513 Mar 7 [Monday] (annular, mag. = 0.998): Damascus

> (918 AH.) In the afternoon of Monday the 29th of (the month of) Dhu al-Hijjah, the Sun was eclipsed for 13 deg (i.e. 52 minutes). There was darkness and some shopkeepers lit lamps in their shops. In the mean time there were clouds. After the *Jumu'ah* (Friday) Prayer in the Umayyad Mosque, a *khutbah* (sermon) was delivered about the eclipse and the eclipse prayer was performed.
>
> [Ibn Tulun: *Mufakahat al-Khullan fi Hawadith al-Zaman*, part I, p. 375.]

The contemporary writer Ibn Tulun lived in Damascus, the site of the Umayyad Mosque. In the narrow central zone the loss of daylight would be considerable. However, the degree of obscuration of the Sun at Damascus is not directly specified. As noted elsewhere in this chapter, it was not unusual for Muslims to assemble at the mosque during a significant eclipse in order to recite a special prayer – a practice which still continues. At the eclipse of AD 1513, the computed magnitude would be as much as 0.98 at 15.6 h (assuming a value for ΔT of 240 sec). The computed duration of the eclipse is 2h 35 m (38 deg). This is much greater than the recorded figure of 13 deg, but cloud may have considerably limited the period of visibility.

12.6 Sun eclipsed near sunrise or sunset

There are a number of examples in Arab chronicles where the Sun is said to have risen or set whilst eclipsed. Whether such observations are as reliable as those made by astronomers is doubtful. In particular, the precise place of observation is seldom stated directly. Additionally, if a chronicler makes a simple statement such as 'the Sun was eclipsed at sunrise' – as was the case in AD 1174 – it is conceivable that he may simply mean that the event occurred in the early morning. Until AD 1398 there are only two records which assert that the Sun rose or set whilst eclipsed (AD 882 and 1178). By AD 1398, ΔT would be so small (probably less than about 300 sec) that it is debatable whether the rather crude observations found in chronicles are of much value in this context.

In view of these cautionary remarks, I have only analysed the observations in AD 882 and 1178. In each case the author was a contemporary, living in Baghdad. The following translations are by SSR. I have made no allowance for horizon profile in computing local times of sunrise or sunset: Baghdad is situated in a very flat plain near sea-level. Although a further observation in which the Sun set after the end of totality was reported from Cordoba in AD 912 (see section 12.4), this will not be utilised to determine limits to ΔT. The horizon towards the north-west of Cordoba is quite hilly, and thus materially affects the local time of sunset in summer.

(1) AD 882 Aug 17 [Friday]: Baghdad

> (269 AH.) In (the month of) Muharram in this year... the Sun was eclipsed at the time of sunset on Friday, when two nights remained to the completion of Muharram, and set eclipsed....
>
> [Al-Tabari: *Tarikh al-Tabari: Tarikh al-Rusul wa'l Muluk*, vol. IX, p. 613.]

The complete record, which notes eclipses of both Sun and Moon, is cited in section 12.3.

Al-Tabari settled in Baghdad around AD 875 and spent the rest of his life there. Hence Baghdad is the most likely place of observation.

Unfortunately, how far the eclipse was advanced by sunset is not clear from the record. Hence it will only be assumed that the Sun reached the horizon between first and last contact.

RESULTS

(i) First contact before sunset at Baghdad on Aug 17. LT of sunset = 18.58 h. Hence LT of contact < 18.58 h, UT < 15.65 h. Computed TT = 15.99 h, thus $\Delta T > 1250$ sec.

(ii) Last contact after sunset. LT of contact > 18.58 h, UT > 15.65 h. Computed TT = 17.60 h, thus $\Delta T < 7000$ sec.

Combining these limits yields $1250 < \Delta T < 7000$ sec.

(2) AD 1178 Sep 13 [Wednesday]: Baghdad

> (574 AH.) The Moon was eclipsed after the last third of the night in the middle of (the month of) Rabi' al-Awwal, and stayed in that state until it set after sunrise. Also the Sun was eclipsed on Wednesday the 29th of Rabi' al-Awwal in the afternoon and stayed like that until near sunset.
>
> [Ibn al-Jawzi: *al-Muntazam fi Tarikh al-Muluk wa'l-Umam*, vol. X, p. 283.]

The lunar observation will be considered in section 12.7. In the case of the solar eclipse, the text seems to imply that the eclipse ended before the Sun had set.

RESULTS

Last contact before sunset at Baghdad on Sep 13. LT of sunset = 18.10 h. Hence LT of contact < 18.10 h, UT < 15.02 h. Computed TT = 14.12 h, thus $\Delta T > -3250$ sec.

This single limit is in keeping with all realistic values of ΔT and will not be considered further. However, the inference that the eclipse ended before sunset is confirmed.

The results deduced in this section are listed in table 12.6 along with the ΔT limits deduced from lunar observations in section 12.7. It should be noted that the ΔT range derived from the solar eclipse of AD 882 is much narrower than for the lunar eclipse in the same year.

12.7 Moon eclipsed near moonrise or moonset

Records of lunar eclipses occurring near moonrise or moonset present similar difficulties to the solar observations discussed in the last section. In particular, Ibn al-Athir (AD 1160–1233) notes the occurrence of three moonrise or moonset eclipses which took place more than two centuries before his own time. The dates of these events are AD 951, 966 and 969. Since Ibn al-Athir wrote so long afterwards, it is no more than a presumption that Mosul, where he spent most of his life, was the place of observation. In the following century, the Nestorian Bishop Elias of Nisibis, who frequently abstracted from Arab chronicles, made a list of as many as 17 observations of lunar eclipses which were 'made in our own time'. The chronicle of Elias, which is in Syriac, has been translated into French by Delaporte (1910). The eclipses which Elias cites range in date from AD 1005 to 1030; he died in 1049. In several instances the Moon was said to rise or set eclipsed. Elias may well have witnessed some of these events at Nisibis (now known as Nusaybin: lat. = +37.08 deg, long. = −41.18 deg) but in no case is the source of the original record preserved by him. Even if we could be sure that Nisibis was the place of observation, the fact that this site is in a mountainous area would necessitate careful allowance for horizon profile.

If the above observations are rejected, there are only two seemingly useful records before AD 1400 – the rather arbitary cutoff date adopted in section 12.6. In both cases (AD 882 and 1178) the author was a contemporary. As in the previous section, no allowance for horizon profile is necessary since the two events were reported from Baghdad.

Following the usual policy, a double date (e.g. Aug 2/3) is given for lunar eclipses. The calculated weekday refers to the day beginning at sunset – i.e. the second of the pair of dates. In each case the computed magnitude of the eclipse is also given for reference.

(1) AD 882 Aug 2/3 [Friday] (mag. = 1.43): Baghdad

> (269 AH.) In (the month of) Muharram this year... the Moon was eclipsed on the night of the 14th (day) and set eclipsed.
>
> [Al-Tabari: *Tarikh al-Tabari: Tarikh al-Rusul wa'l Muluk*, vol. IX, p. 613.]

An observation of a solar eclipse, recorded two weeks later by al-Tabari, has already been discussed in section 12.6. There is no mention of totality, but since the record is so brief, it cannot be assumed that the Moon set before the onset of the total phase. Hence it will merely be supposed that the Moon reached the western horizon between first and last contact.

RESULTS

(i) First contact before moonset at Baghdad on Aug 3. LT of moonset = 5.13 h. Hence LT of contact < 5.13 h, UT < 2.23 h. Computed TT = 1.07 h, thus $\Delta T > -4200$ sec.

(ii) Last contact after moonset. LT of moonset = 5.30 h. Hence LT of contact > 5.30 h, UT > 2.41 Computed TT = 1.07 h, thus $\Delta T < 7550$ sec.

Combining these limits yields $-4200 < \Delta T < 7550$ sec.

Table 12.6 ΔT limits from solar and lunar eclipses occurring near rising or setting.

Year	Type	ΔT Range (sec)	
		LL	UL
+882	Moon	−4200	7550
+882	Sun	1250	7000
1178	Moon	−5650	3450

(2) AD 1178 Aug 29/30 [Wednesday] (mag. = 0.46): Baghdad

(574 AH.) The Moon was eclipsed after the last third of the night in the middle of (the month of) Rabi' al-Awwal, and stayed in that state until it set after sunrise...

[Ibn al-Jawzi: *al-Muntazam fi Tarikh al-Muluk wa'l-Umam*, vol. X, p. 283.]

An observation of a solar eclipse recorded in this same passage has already been discussed in the previous section. Ibn al-Jawzi was living in Baghdad at the time. His remark that the Moon set after sunrise suggests unusually careful observation.

RESULTS

(i) First contact before moonset at Baghdad on Aug 3. LT of moonset = 5.65 h. Hence LT of contact < 5.65 h, UT < 2.65 h. Computed TT = 1.08 h, thus $\Delta T > -5650$ sec.

(ii) Last contact after moonset. LT of moonset = 5.74 h. Hence LT of contact > 5.74 h, UT > 2.74 h. Computed TT = 3.70 h, thus $\Delta T < 3450$ sec.

Combining these limits yields $-5650 < \Delta T < 3450$ sec.

NB the computed LT of sunrise is 5.63 h, so that on any reasonable value for ΔT the Moon would set a little after sunrise, as the text alleges.

The few ΔT results derived in this and the previous section are assembled in table 12.6.

12.8 Selected records of total lunar eclipses

In this section, a few examples of records of total lunar eclipses are given, as an illustration of the kind of material available in Arabic chronicles. Translations are by Stephenson and Said (1996). It should be emphasised that the observations are of no value for determining ΔT.

Most Arab records of lunar eclipses, although often distinguishing between total and partial obscurations, contain little descriptive information.

One of the most detailed accounts of an eclipse in which the Moon completely disappeared is given by the fifteenth century Cairo annalist Ibn Taghri Birdi. This relates to the event of AD 1461 Jun 22/23:

> (865 AH.) On the night of Tuesday, the 14th of (the month of) Ramadan, the whole of the Moon's body was eclipsed and disappeared in the eclipse for about seventy degrees. The stars in the sky became as though it was the 29th of the month (i.e. on a moonless night). Perhaps the occurrence of such an eclipse is very rare.
>
> [Ibn Taghri Birdi: *Hawadith al-Duhur fi Mada 'l -Ayyam wa'l Shuhur*, extract from vol. VII, part 4, pp. 741–2.]

Ibn Taghri Birdi gives the correct weekday and day of the lunar month. Computation indicates that the true duration was only about 3.9 hours (58 degrees), so that the reported duration was too long.

In medieval Islamic history, there seems to be no definite instance of the Moon described as turning red during totality. The earliest reliable observation of this kind is relatively recent. It is reported by the eighteenth century historian al-'Umari of Mosul, who describes an event occurring during his own time:

> (1186 AH.) In this year, the Moon was totally eclipsed in the first half of the night. All of its light disappeared and it became like a copper disc.
>
> [Al-'Umari: *al-'Athar al-Jaliyyah fi 'l -Hawadith al-Ardiyyah*, pp. 136–7.]

Only the year is given, corresponding to AD 1772–3. Either the total lunar eclipse of 17/18 April or that of 11/12 October (both in AD 1772) could be intended.

Although several *medieval* annals describe alleged eclipses in which the Moon appeared red, all events of this kind can be explained by atmospheric phenomena such as dust storms. For example, the Damascus chronicler Abu Shama records the following observations, made in his own time. These began on a date corresponding to AD 1256 Jul 11:

> (654 AH.) On the night of the 16th of the month Jumada al-Ukhra, the Moon was eclipsed at the beginning of the night and it was extremely red. It then reappeared. The Sun was eclipsed the following day and it became reddish at both the time of its rising and setting. It remained like this for (several) days...
>
> [Abu Shama: *Tarajum Rijal al-Qarnain al-Sadis wa'l Sabi'*, also known as *al-Dhail 'ala al-Rawdatain*, pp. 189–191.]

No solar eclipse would be visible near Damascus around this time. Although there was a lunar obscuration on the night of 8/9 July – i.e. two days earlier than the date reported by Abu Shama – this was only partial (magnitude 0.19).

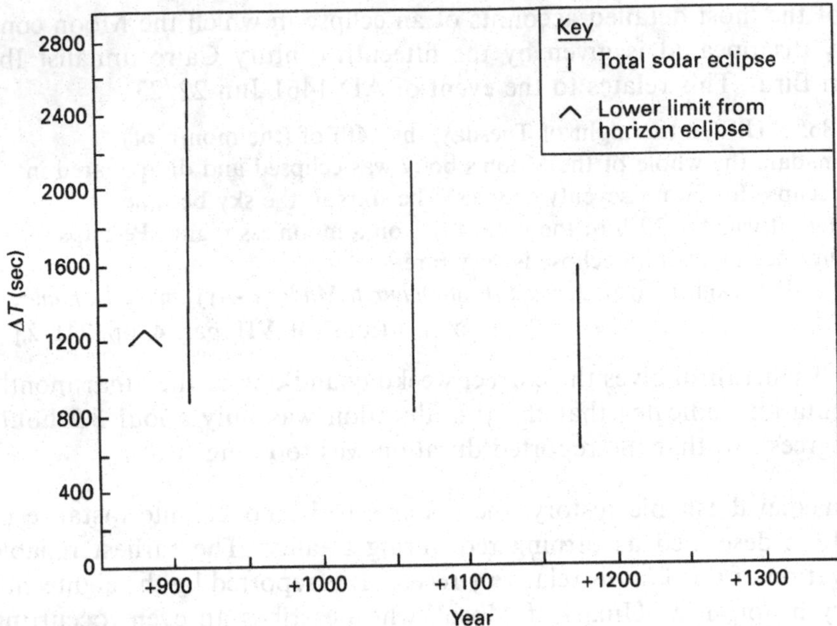

Fig. 12.4 ΔT limits derived from eclipse observations recorded in Arab chronicles.

12.9 Conclusion

The few ΔT limits obtained in this chapter are shown diagrammatically in figure 12.4. Although the observations of total solar eclipses yield valuable limits for ΔT, the number of results is disappointing.

12.10 Appendix: Arab chronicles consulted

Abu Shama: *Kitab al-Rawdataim fi Akhbar al-Dawlatain* (2 vols.). Cairo, 1870–71.

Abu Shama: *Tarajum Rijal al-Qarnain al-Sadis wa'l Sabi'*, also known as *al-Dhail 'ala al-Rawdatain.* Cairo, 1947.

al-'Asqalani: *Inba' al-Ghumr bi 'Bna' al-'Umr* (8 vols.). Hyderabad-Deccan, 1967–75.

al-Maqrizi: *al-Suluk fi Ma'rifat Duwal al-Muluk* (4 vols.). Cairo, 1934–71.

al-Maqrizi: *Itti'az al-Hunafa bi Akhbar al-A'imma al-Fatimiyin al-Khulafa* (3 vols.). Cairo, 1967–73.

al-Tabari: *Tarikh al-Tabari: Tarikh al-Rusul wa'l Muluk* (10 vols.). Cairo, 1960–69.

al-'Umari: *al-'Athar al-Jaliyyah fi 'l -Hawadith al-Ardiyyah.* Al-Najaf, 1973.

Ibn al-Athir: *al-Kamil fi al-Tarikh* (9 vols.). Cairo, 1929–38.

Ibn al-Jawzi: *al-Muntazam fi Tarikh al-Muluk wa'l-Umam* (10 vols.: only vol. V, part 2 to vol. X accessible). Hyderabad-Deccan, 1938–42.

Ibn Hayyan: *al-Muqtabis fi Tarikh al-Andalus* (10 vols.: only vols. III and V accessible). Vol. III: Paris, 1937; vol. V: Madrid, 1979.

Ibn Iyas: [*Bada'i' al-Zuhur fi Waqa'i' al-Duhur*. New edition (5 vols.), Wiesbaden, 1960–75. Older edition (3 vols.), Cairo, 1894–5.

Ibn Taghri Birdi: *Hawadith al-Duhur fi Mada 'l -Ayyam wa'l Shuhur*. Only extracts from vol. 7, parts 2 and 4 have been published, by the University of California, 1931 and 1942.

Ibn Tulun: *Mufakahat al-Khullan fi Hawadith al-Zaman* (2 parts). Cairo, 1962–4.

13

Observations of eclipses by medieval Arab astronomers

13.1 Introduction

The eclipse observations made by medieval Arab astronomers are among the most accurate and reliable data from the whole of the pre-telescopic period. Careful records of both solar and lunar eclipses are contained in a number of compendia – some known as *zijes* (astronomical handbooks containing various tables along with explanatory text). These include measurements of the times of occurrence and other details such as magnitude estimates. Although the main emphasis in this chapter will be on timed data, solar magnitude estimates and horizon observations of eclipses will also be considered.

Many of the observations discussed below were investigated by Newcomb (1878) and Newton (1970). However, these authors relied on published translations which sometimes contained significant errors, while their own interpretations are occasionally suspect. Furthermore, in neither case was a direct solution made for ΔT.

13.2 Sources of data

Most of the accessible eclipse observations by medieval Arab astronomers are contained in a single treatise – the *zij* compiled by the great Cairo astronomer Ibn Yunus, who died in AD 1009 (his date of birth is unknown). A few eclipses are also recorded in works by al-Battani (who lived between AD 850 and 929) and al-Biruni (AD 973–1048).

Ibn Yunus cites reports of some thirty solar and lunar eclipses from between AD 829 and 1004. His treatise, dedicated to Caliph al-Hakim, is entitled *al-Zij al-Kabir al-Hakimi.* Not all of Ibn Yunus' text survives today, but portions of it are extant in manuscripts at Leiden and Oxford; a further manuscript at Paris contains abridged versions of certain sections of the text (King, 1976). Only the manuscript which is preserved in the library at Leiden University (cat. no. Or 143) contains the eclipse observations; this also notes some planetary conjunctions. This text was published in its original Arabic, along with a translation into French, by Caussin (1804). Although Caussin's translation is generally very sound, it contains several errors. Recently Dr Said S. Said, in conjunction with the author, has made a detailed study of a microfilm of the manuscript supplied by Leiden University, and has re-translated all of the eclipse records (Said and Stephenson, 1997). These translations, which preserve the original astronomical terminology of the record as closely as possible, form the basis of much of the present chapter.

The observations compiled by Ibn Yunus fall into three distinct groups: AD 829–866; 923–933; and finally 977–1004. Most of the eclipses in the first set were witnessed by al-Mahani of Baghdad, while the principal observer in the second group was Ali ibn Amajur, also of Baghdad. Most of the eclipses in the third set were observed by the Cairo astronomer Ibn Yunus himself, as is clear from his use of the first person in the text. Sometimes, the place where the eclipse was seen is directly specified in the record; this is particularly true of the Cairo observations. However, more often it may be inferred from the reference to the astronomer's name. There can be little doubt that all of the observations from between AD 829 and 933 were made at Baghdad, while the remaining observations were made in Cairo (al-Qahirah), which was founded in AD 969.

In assembling these various accounts, Ibn Yunus had two main motives. Firstly, he wished to emphasise the need for improved planetary tables by illustrating how poorly computations made with existing tables were supported by observation and secondly he simply wanted to list these observations for the benefit of future astronomers (see King, 1976).

Four eclipses witnessed by al-Battani and his colleagues are reported in his treatise entitled *al-Zij al-Sabi'*, written *c.* AD 910. This work has been published along with a translation into Latin by Nallino (1899). The places of observation were al-Raqqah (in present-day Syria) and Antakyah (in Turkey). About a century later, al-Biruni also recorded four eclipses, as seen from Jurjan (in present-day Iran), Jurjaniyyah (in Turkmenistan) and Ghaznah (Afghanistan). These observations are contained in al-Biruni's *Kitab Tahdid al-Amakin li-Tashih Masafat al-Masakin* (dating from AD 1025) and his *al-Qanun al-Mas'udi* (composed in AD 1030). The former work has been translated into English by Ali (1967). The main aim of both al-Battani and al-Biruni in observing eclipses appears to have been the determination of longitude difference between selected sites (see section 13.3).

The eclipse reports cited by al-Battani and al-Biruni probably represent only a small proportion of the number actually available to them. Summaries of a few additional observations are in fact recorded by al-Biruni in his *Kitab Tahdid* but unfortunately he does not give any useful timed

Table 13.1 Locations from which eclipses were reported by medieval Muslim astronomers.

Original name	Modern name	N lat. (deg)	E long. (deg)
al-Qahirah	Cairo	30.05	31.25
al-Raqqah	Raqqa	35.94	39.02
Antakyah	Antakya	36.20	36.17
Baghdad	Baghdad	33.34	44.40
Ghaznah	Ghazni	33.55	68.43
Jurjan	Gorgan	36.83	54.48
Jurjaniyyah	Kunya-Urgench	42.32	59.16
Nishapur	Neyshabur	36.21	58.83

information. Further observations by other medieval Arab astronomers may well survive in unpublished texts. However, at present the range of source material is rather restricted.

A careful account of a central annular eclipse seen in AD 873 at Nishapur (in present-day Iran) is given by al-Biruni in his *al-Qanun al-Mas'udi*, although no times are reported. This observation will also be considered below.

Geographical co-ordinates (degrees and decimals) of the places of observation discussed in this chapter are listed in table 13.1. This table gives both original names and their present-day equivalents. The locations of the various sites are also shown in figure 13.1. Throughout this chapter I have normally used the original names except for Cairo, for which it would seem pedantic to use al-Qahirah.

13.3 Motives for observation

Some of the main reasons why medieval Arab astronomers made careful observations of eclipses have already been briefly mentioned in the previous section. Further comments now seem desirable. Ibn Yunus compared some of the observations made in Baghdad with tables produced around AD 810 by Yahya ibn Abi Mansur. These were contained in the latter writer's *al-Zij al-Mumtahan* ('Tables verified by observation') and were developed from Ptolemy's methods as laid down in his *Almagest*. Ibn Yunus was able to demonstrate that the times of eclipses calculated from the tables of Yahya frequently deviated from the observed times by half an hour or more. As is apparent from the records of the eclipses of AD 923 to 933 cited by Ibn Yunus, the contemporary astronomer Ibn Amajur had already shown that the tables of Habash al-Hasib, produced around

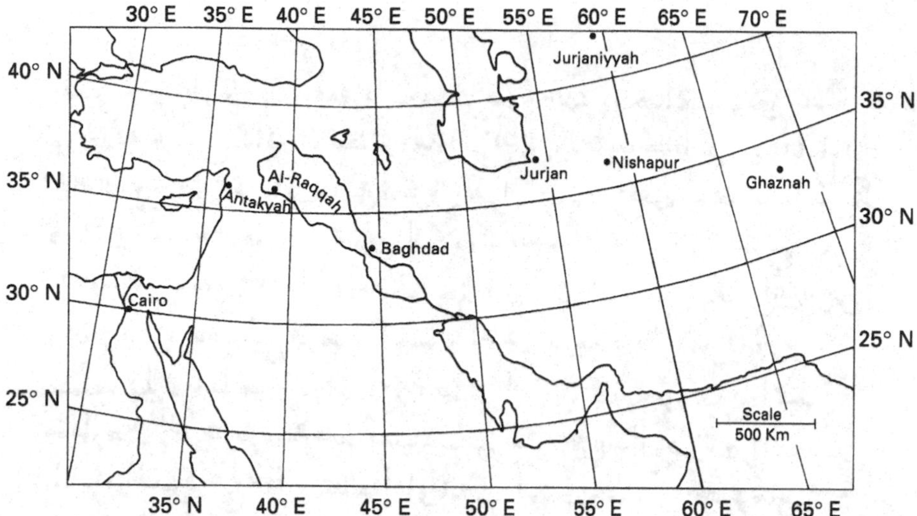

Fig. 13.1 Map showing the locations of the various sites where eclipses were observed by medieval Muslim astronomers.

AD 840, were in error by similar amounts. These tables, also entitled *al-Zij al-Mumtahan*, were based on observations made by Habash. For instance, the account of the solar eclipse of AD 923 Nov 11 – as quoted by Ibn Yunus – contains the following details:

> (This) solar eclipse was calculated and observed by Abu al-Hasan Ali ibn Amajur, who used the *al-Zij al-Arabi* of Habash... We as a group observed and clearly distinguished it... We observed this eclipse at several sites on the *Tarmah* (an elevated platform on the outside of the building)... According to calculation from the conjunction tables in the Habash *Zij* the middle was at 0;31 h (i.e. 31 min) and its clearance at 0;44 hours (i.e. 44 min), calculation being in advance of observation.
>
> [Trans. Said and Stephenson (1997).]

The Arabic text of the above record (from the Leiden manuscript of Ibn Yunus) is shown in figure 13.2. This figure includes the report of the lunar eclipse of AD 923 Jun 1/2.

A more extensive translation of the above record is given in section 13.7 below. It is apparent that eclipse predictions – although of limited accuracy – were evidently sufficiently reliable to enable careful advance planning of observations. This was particularly important when eclipses were used to determine the difference in longitude between two separate sites.

The technique of determination of the longitude difference between two cities by using simultaneous timings of lunar eclipse – widely practised

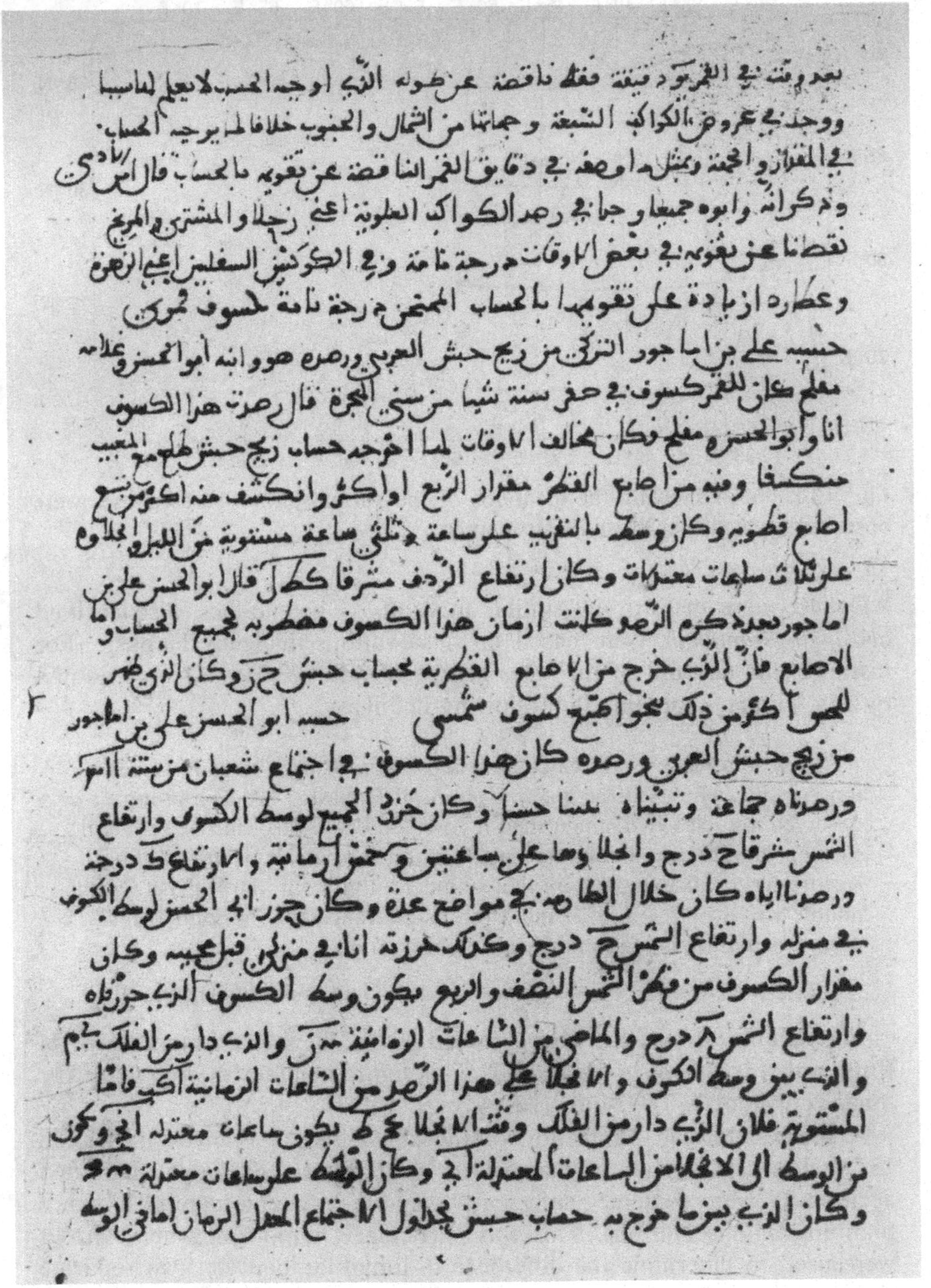
بعد وقت في القمر بو دقيقة فقط ناقصة عن طوله الذي اوجبه الحساب لا يعلم لما سببا
ووجد في عروض الكواكب السبعة وجماتها من الشمال والجنوب خلافا لما يوجبه الحساب
في المقدار والجهة ومثل ما وصفه في دقايق القمر الناقصة عن تقويمه بالحساب قال ابن الادمي
وذكر انه وابوه جميعا وجدا في رصد الكواكب العلوية اعني زحلا والمشتري والمريخ
نقصانا عن تقويمه في بعض الاوقات درجة تامة وفي الكوكبين السفليين اعني الزهرة
وعطارد ازيادة على تقويمهما بالحساب الممتحن درجة تامة كسوف قمري
حسبه علي بن اماجور التركي من زيج حبش العربي ورصده هو وابنه ابو الحسن علي
مفلح كان للقمر كسوف في صفر سنة ثلثمائة عشر من سني الهجرة قال رصدت هذا الكسوف
انا وابو الحسن مفلح فكان يخالف الاوقات لما اخرجه حساب زيج حبش طلع مع المغيب
منكسفا وفيه من اصابع القطر مقدار الربع او اكثر وانكشف منه اكثر من تسع
اصابع قطرية وكان وسطه بالتقريب على ساعة وثلثي ساعة مستوية من الليل وانجلاؤه
على ثلاث ساعات معتدلات وكان ارتفاع الردف مشرقا كط ل قال ابو الحسن علي بن
اماجور وبعد كرم الرصد كانت ازمان هذا الكسوف مضطربة بجميع الحساب واما
الاصابع فان الذي خرج من الاصابع القطرية بحساب حبش ح ز وكان الذي ظهر
للبصر اكثر من ذلك بنحو اصبع كسوف شمسي حسبه ابو الحسن علي بن اماجور
من زيج حبش العربي ورصده كان هذا الكسوف في اجتماع شعبان من سنة اثم
ورصدناه جماعة وتبيناه بيانا حسنا وكان حزر الجميع لوسط الكسوف وارتفاع
الشمس شرقا ح درج وانجلاوها على ساعتين وخمس زمانية وارتفاع ك درجة
ورصدناه اياه كان خلال الطاهر في مواضع عدة وكان حزر ابي الحسن لوسط الكسوف
في منزله وارتفاع الشمس ح درج وكذلك حزرته انا في منزلي قبل مجيئه وكان
مقدار الكسوف من قطر الشمس النصف والربع يكون وسط الكسوف الذي حزرناه
وارتفاع الشمس ح درج والماضي من الساعات الزمانية حب والذي دار من الفلك يم
والذي بين وسط الكسوف والانجلاء على هذا الرصد من الساعات الزمانية اكثر فاما
المستوية فلان الذي دار من الفلك وقت الانجلاء كح ط يكون ساعات معتدلة اجو وكون
من الوسط الى الانجلاء من الساعات المعتدلة اج وكان الوسط على ساعات معتدلة
وكان الذي بين ما خرج به حساب حبش لاول الاجتماع المحقق الزمان اماني اوله

Fig. 13.2 Arabic text of eclipse records from AD 923 Jun 1/2 (lunar) and 923 Nov 11 (solar) in the Ibn Yunus manuscript (Leiden University library: Or. 143).

in Europe during the eighteenth and early nineteenth centuries – was pioneered by medieval Arab astronomers. Hipparchus in the second century BC was already well aware of the value of lunar eclipses for measuring longitude, as noted by Strabo:

> In like manner, we cannot accurately fix points that lie at varying distances from us, whether to the east or to the west, except by a comparison of the eclipses of the Sun and Moon. That, then, is what Hipparchus says on the subject.
>
> [Strabo: *Geography*, I, 1, 13; trans. Jones (1917, pp. 23–25).]

However, 'the total absence in antiquity of any scientific organisation deprived the whole method of its practical importance' (Neugebauer, 1975, p. 667). Hipparchus would have been unable to obtain the observations which he needed for such a goal. Similar remarks apply to Ptolemy in the second century AD. Hence the technique remained no more than an idealistic concept until it was put to use by the Arabs.

Al-Biruni, in his *Kitab Tahdid*, gives the following outline of the method as applied in his time:

> If we know beforehand of the formation of a lunar eclipse and we wish to determine the longitudinal difference between two towns, we make arrangements beforehand for someone in each town who can measure the times accurately by instruments, to obtain as accurately as possible the times of the beginning of an eclipse and its end and those of the beginning of clearance and its end.
>
> [Trans. Ali (1967, p. 130).]

Using lunar eclipse observations, al-Biruni determined the longitude difference between Jurjan and Ghaznah as (2;21) minutes of day and between Jurjaniyyah and Ghaznah as (1;42,12) minutes of day. Since one minute of day corresponded to 1/60 day or 6 deg, and al-Biruni followed the standard practice of using sexagesimal arithmetic, these results were equivalent to 14.1 deg and 10.22 deg respectively. The true values are respectively 14.0 and 9.4 deg, representing tolerable precision at this epoch.

A diagram by al-Biruni explaining the formation of lunar eclipses is illustrated in figure 13.3.

13.4 Calendrical remarks

Dates of the various observations cited by Ibn Yunus, al-Battani and al-Biruni are usually expressed in terms of the Islamic lunar calendar. Years are numbered from *al-Hijrah*, the migration of the Prophet Muhammad to Medina in AD 622. The operation of this calendar and conversion of

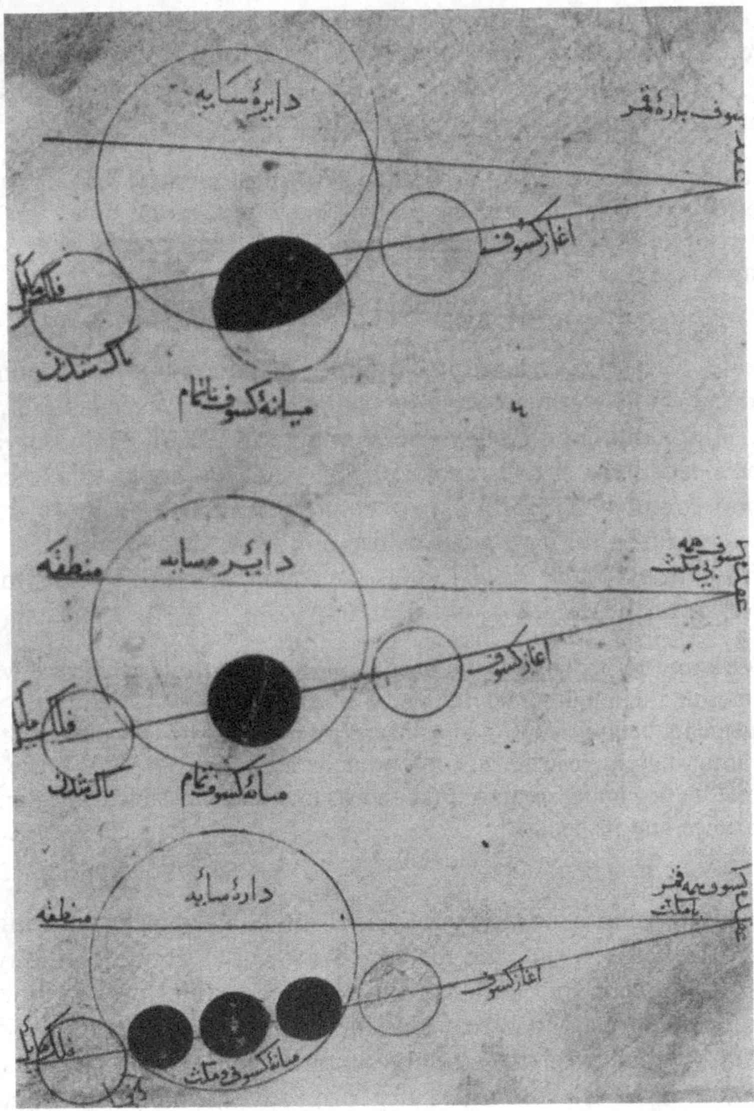

Fig. 13.3 Diagram by al-Biruni explaining the formation of lunar eclipses (Nasr, 1976, Fig. 40).

dates to the Julian or Gregorian calendars have already been discussed in chapter 12. However, some of the dates cited by Ibn Yunus are also expressed in terms of other systems: these are the Persian, Syrian and Coptic calendars. In giving translations of the various eclipse records below, I have – for the sake of brevity – normally given only the date according to the Muslim calendar in order to avoid cumbersome entries; certain texts cite all four dating schemes mentioned above! Invariably, when the Muslim date is converted to the Julian calendar there is either exact agreement with the calculated date of an eclipse or the discrepancy between them is unimportant. This is also true of dates given on the other three calendars.

The Syrian and Coptic calendars have already been briefly discussed (chapters 11 and 12). Some remarks on the Persian calendar are as follows. The Persian calendar resembles to some extent the ancient Egyptian calendar. Years consist of 12 months, each of 30 days, with five *epagomenal* (additional) days placed after the end of the eighth month. Years are counted from the era of Yazdijerd, the accession of King Yazdijerd III: AD 632 Jun 16. Because of its simple rules, the Persian calendar – like the ancient Egyptian scheme – is convenient for counting days between two epochs.

13.5 Observational techniques

Medieval Arab astronomers were well aware of the hazards of observing solar eclipses with the naked eye. As a result, they were in the habit of viewing the Sun by reflection in water in order to reduce its glare. Thus al-Biruni in his *Kitab Tahdid* makes the following remarks:

> The faculty of sight cannot resist it (the Sun's rays), which can inflict a painful injury. If one continues to look at it, one's sight becomes dazzled and dimmed, so it is preferable to look at its image in water and avoid a direct look at it, because the intensity of its rays is thereby reduced... Indeed such observations of solar eclipses in my youth have weakened my eyesight.
>
> [Trans. Ali (1967, p. 131).]

Eclipse magnitudes were frequently estimated by the Muslim astronomers. These were usually expressed as a fraction of the diameter of the luminary obscured, but occasionally were quoted in terms of surface area. Often these estimates were expressed to the nearest digit or twelfth of the solar or lunar diameter, following ancient Babylonian and Greek practice (see chapter 3). On a few occasions it is reported that the Moon or Sun was eclipsed when it rose or set, sometimes the proportion of the disk obscured at the time being used. However, the numerous timed data are the most valuable for determining ΔT.

For partial solar and lunar eclipses, medieval Arab astronomers timed three separate phases: 'beginning' (i.e. first contact), 'middle' and 'clearance' (last contact). No total solar eclipse is known to have been recorded by them (unlike in the case of chroniclers), but total lunar obscurations were frequently reported. Here two extra phases were timed: 'beginning of staying' (second contact) and 'end of staying' or 'beginning of clearance' (third contact). In such cases, the time of mid-eclipse was merely the average of the measurements for second and third contact.

Owing to the mediocre quality of clocks at this early period, eclipse times were usually determined indirectly by the Arab astronomers – by measuring the altitude of the Sun in the case of a solar eclipse and of either the Moon or a selected bright star for a lunar obscuration. As noted above (section 13.3), often several astronomers would make independent observations at the same site.

The type of instrument used for determining altitudes is hardly ever mentioned in the text, although it is most likely to have been a hand-held astrolabe. Among astronomical instruments, the astrolabe was the most widely used in both the Arab dominions and Europe during the Middle Ages. (For a discussion of its construction and use, see North (1974); a photograph of a tenth century example is shown in figure 13.4.) Of ancient Greek origin, the astrolabe served two main purposes: (i) as an observational device for the determination of the altitudes of celestial bodies; and (ii) as an analogue computing device, particularly for the determination of local time.

Most altitude measurements relating to lunar and solar eclipses were made to the nearest degree or half degree, probably the best that could be achieved with an astrolabe. On one occasion (AD 928) the altitude of the Sun at an eclipse contact was recorded with significantly higher precision, but this seems to be unique:

> We (found) that it cleared and nothing of the eclipse remained and we distinguished the (full) circle of the Sun's body in water; (that was) when the altitude was 12 deg in the east, less $\frac{1}{3}$ of a division of the *al-halaqa* (i.e. the ring), which is graduated in thirds (of a degree), that is (less by) $\frac{1}{9}$ degree....
>
> [Trans. Said and Stephenson (1997).]

By comparison, certain altitude determinations, although quoted to the nearest degree or even half degree, are said to be only approximate. Just how inferior these are to normal measurements is not clear. Evidently, medieval Arab astronomers were capable of making altitude measurements of very high accuracy. A recent analysis of many determinations of the meridian altitude of the Sun made by Muslim astronomers between AD 832 and 1018 revealed that several observers achieved consistent precision using a fixed instrument of around 1 arcmin (Stephenson and Said, 1991). Presumably owing to the large changes in azimuth of the Sun or Moon (or clock star) during the course of an eclipse – and the mediocre definition of the contacts using the unaided eye – it was normally considered adequate to use a hand-held instrument for determining altitudes under these circumstances.

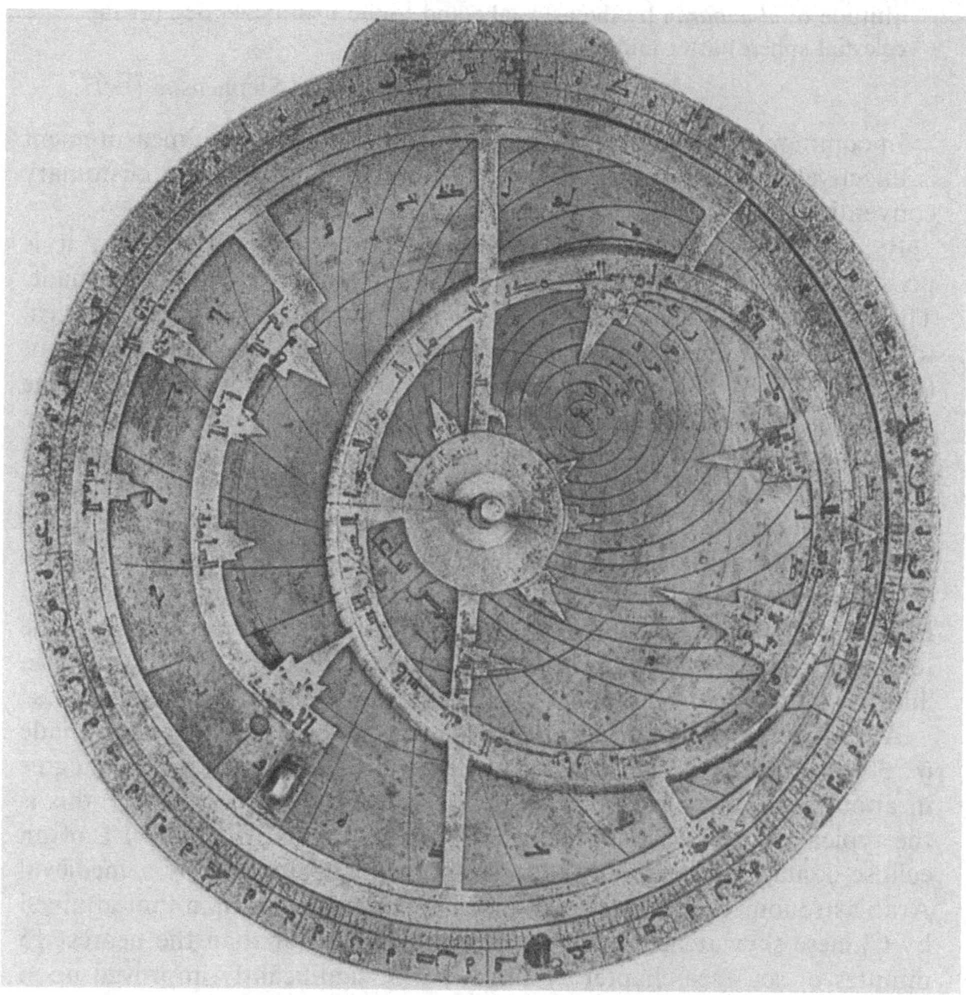

Fig. 13.4 Tenth century Arab astrolabe (Courtesy: Dar al-Athar al-Islamiyyah, Ministry of Information, Kuwait).

Altitudes taken during an eclipse were afterwards reduced to local time (LT) by the observers either with the aid of tables or an astrolabe. Detailed examples of the use of tables by medieval astronomers to convert from solar altitudes to LT are given by King (1973). The text of Ibn Yunus notes application of an astrolabe for conversion of star altitudes on two separate occasions, as in the following quotation from al-Mahani on a date corresponding to AD 854 Aug 11/12:

> ...It was found by observation that the time of beginning of the eclipse was when the altitude of (the star) *al-dabaran* (Aldebaran: α Tau) was 45;30 deg in the east...We determined the time of the beginning from the

altitude of *al-dabaran* by the astrolabe and found it to be 44 deg (of the celestial sphere) after midnight....

[Trans. Said and Stephenson (1997).]

In common with other altitude determinations, the above measurement is effectively expressed in degrees and arcmin. Following the customary convention, I have rendered this as 45;30 deg.

By computing the LT direct from the altitude determination, it is possible to assess the accuracy with which the above reduction was made. The estimate of 44 degrees (i.e. 2h 56m) after midnight is virtually identical to the LT which may be calculated directly from the altitude measurement (2h 55.6m). By chance this result is unusually precise. Among the reductions reported by Ibn Yunus, the average error – whether using the astrolabe or tables – can be shown to be about 4 minutes (Stephenson and Said, 1991). This still represents tolerable precision.

On some dates, only contact timings are available, the original altitude measurements being no longer preserved. Eclipse times were normally expressed in terms of hours and minutes, both equal hours and seasonal hours being used with comparable frequency. Usually it is clear from the text which system was adopted. For instance, if a text quotes times in 'hours of daytime' or 'hours of night', these are evidently seasonal hours.

At the relatively low latitudes where most observations were made (*c.* 30 to 35 deg) the altitude of a celestial body changes by one degree in about 5 minutes (except when near the meridian). In principle, this is the typical accuracy with which it is possible for us to fix the LT of an eclipse contact when an altitude measurement is reported by a medieval Arab astronomer. Such precision is significantly greater than that attained by Chinese skywatchers using a clepsydra (no better than the nearest 15 minutes or so – see chapter 9) and was not significantly improved upon in Europe until the sixteenth century.

When recording first contact of an eclipse, the Baghdad astronomers simply noted the time of the 'beginning' of the event. However, at Cairo Ibn Yunus on two occasions recorded both this moment and the time when an eclipse was first 'perceived'. Evidently he was aware of the delay in detecting the start of an eclipse owing to the limited resolution of the unaided eye. His contemporary al-Biruni offered the following specific remarks on this problem for the case of lunar eclipses:

A (lunar) eclipse does not become noticeable to the observer until the segment removed from it (i.e. from the Moon) according to some authors of *zijes* reaches a limit of one digit; I mean one part in twelve of its body (i.e. disk). A limit is also set to its time, which is 1;49 *azman* (time degrees) or 0;7,16 (equal) hours. By this amount of time the true beginning of the eclipse is in advance of the apparent beginning, and the true completion of

> clearance delays behind the apparent completion of clearance. I would think that the amount of a digit in this respect is (too) large because even though the first contact between the shadow (of the Earth) and the Moon is not perceptible, a small indentation (on the Moon's edge) could be seen....
>
> [*Kitab Tahdid*; trans. Said and Stephenson (1997).]

Obviously, the effect of visual acuity on the detection of eclipse contacts is a very subjective issue. In practice, recorded observations only allude to the delay in detection of first contact, never making reference to an advance of last contact. When an observation contains estimates of both the time when an eclipse was first 'perceived' and the time of 'beginning' I have given it special consideration. However, if only one measurement is reported for first contact, I have accepted this without comment.

13.6 References to large solar eclipses in astronomical treatises

Only two observations of this kind are known. A description of the ring phase in AD 873 was quoted by al-Biruni, while in 1004 Ibn Yunus witnessed a very large partial eclipse which was not quite annular. These two observations will be considered before the other types of data.

(1) AD 873 Jul 28 [Tuesday] (annular: mag. = 0.94), Nishapur

> This solar eclipse was observed by Abu al-'Abbas al-Iranshahri at Nishapur early in the morning on Tuesday the 29th of the month of Ramadan in the year 259 of *al-Hijrah*...(date on Persian calendar)...He mentioned that the Moon's body (i.e. disk) was in the middle of the Sun's body. The light from the remaining uneclipsed portion of the Sun surrounded it (i.e. the Moon). It was clear from this that the Sun's diameter exceeded in view that of the Moon.
>
> [*al-Qanun al-Mas'udi*; trans Said and Stephenson (1997).]

Attention was drawn to this observation by Goldstein (1979). Although the eclipse occurred long before al-Biruni's own time, he cited it as evidence that the angular diameter of the Sun could exceed that of the Moon. Ptolemy (*Almagest*, V, 14) had implied that annular eclipses could not take place because he believed that the apparent lunar diameter at apogee was equal to the apparent solar diameter (assumed to be fixed). This issue was still being debated in seventeenth century Europe – see chapter 11.

Goldstein notes that none of the works of al-Iranshahri survive, but this author is mentioned by al-Biruni in several other places.

RESULTS

For annularity at Nishapur, $1820 < \Delta T < 3750$ sec.

(NB computed LT of maximum phase = 5.6 h; solar altitude = 7 deg.)

(2) AD 1004 Jan 24 [Monday] (annular, mag. = 0.98): Cairo

> This solar eclipse was in the sign of Aquarius and was in the late afternoon of Monday the 29th of the month of Rabi' al-Awwal in the year 394 of *al-Hijrah*...(date on Persian, Syrian and Coptic calendars)...The Sun was eclipsed until what remained of it resembled the crescent Moon on the first night of the month. I estimated the eclipsed portion of the Sun to be 11 digits (i.e. of surface). The altitude of the Sun when the eclipse became noticeable in it (i.e. on its disk) was $16\frac{1}{2}$ deg in the west; thus I estimated (the altitude) at the beginning to be $18\frac{1}{2}$ deg. About quarter of the diameter was eclipsed when the altitude was 15 deg, and half of the diameter was eclipsed when the altitude was 10 deg. The eclipse was complete (i.e. maximum phase) when the altitude was 5 deg.
>
> [*Al-Zij al-Kabir al-Hakimi*; trans. Said and Stephenson (1997).]

The Arabic text of the above record from the Ibn Yunus manuscript is illustrated in figure 13.5. This figure also includes the records of the solar eclipse of AD 993 Aug 20 and the lunar eclipses of AD 1001 Sep 5/6 and 1002 Mar 1/2.

The solar altitude measurements will be considered in section 13.7. Although the eclipse was generally annular, it is clear from the text that the observed phase at Cairo was only partial. The progress of the eclipse was carefully followed, and it thus seems very unlikely that the observers overlooked the annular phase.

RESULTS

For a partial eclipse at Cairo, either $\Delta T < 1770$ or $\Delta T > 1940$ sec.

Since the belt of annularity was rather narrow, only a restricted range of ΔT is ruled out by the observation of a partial eclipse. It is unfortunate that Ibn Yunus does not state whether the upper or lower part of the Sun remained visible. If he had done so, one of the above ΔT ranges would have been excluded. His remark that what remained of the eclipsed Sun 'resembled the crescent Moon on the first night of the month' may simply be an indication of the narrowness of the solar crescent, not necessarily the direction in which it was facing.

(NB computed LT of maximum phase = 16.7 h; solar altitude = 7 deg.)

13.7 Solar eclipse times recorded by Ibn Yunus

Although Ibn Yunus lists the various observations which he cites purely in chronological order (i.e. without regard to type), he records so many observations that I have decided to group the solar and lunar eclipses separately. Lunar eclipse timings reported in his treatise will be discussed in section 13.8. Except where otherwise stated, all translations in this and the following sections are by Said and Stephenson (1997), with occasional

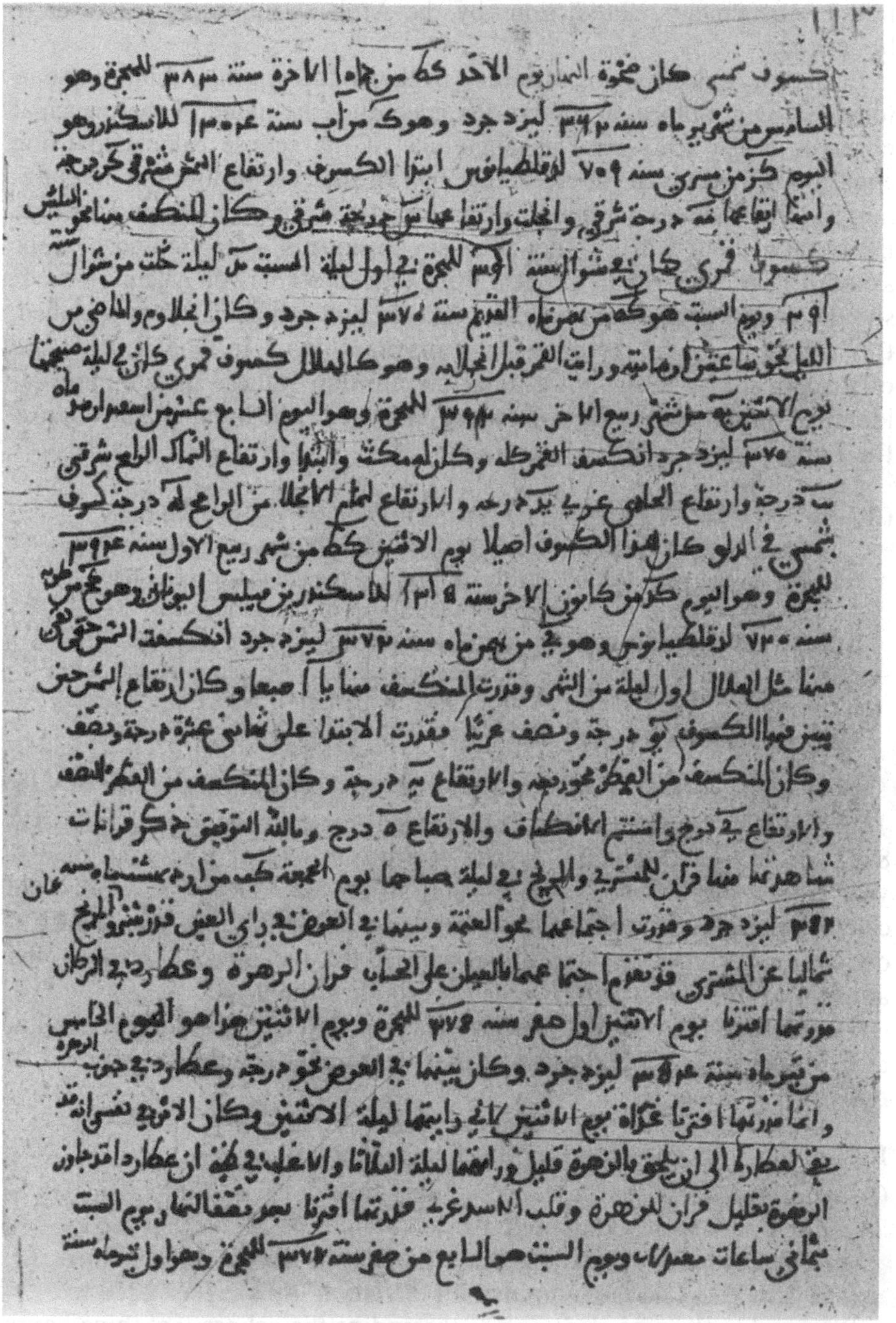

Fig. 13.5 Arabic text of eclipse records of AD 993 Aug 20 (solar), 1001 Sep 5/6 (lunar), 1002 Mar 1/2 (lunar) and 1004 Jan 24 (solar) in the Ibn Yunus manuscript (Leiden University library: Or. 143).

minor amendments. Predictions by the Arab astronomers will usually be ignored.

Each entry below begins with the calculated date on the Julian calendar, followed by the place of observation, including the name of the principal observer, where known. The magnitude of the eclipse calculated using equation (8.1) is given for reference only. In the case of altitude measurements, allowance for refraction is unnecessary since no measured altitude is less than about 5 deg. Under these circumstances, refraction corrections amount to no more than about 0.1 deg (see the tables of Allen, 1976). It should be noted that although the time of mid-eclipse is often specified, this will be taken as the moment of maximal phase instead. In the case of a solar eclipse – but not a lunar obscuration – these instants are not identical owing to the irregular rate of advance of the lunar shadow across the Earth's surface (see chapter 3).

(1) AD 829 Nov 30 [Tuesday] (computed mag. = 0.56): Baghdad

> Ahmad b. 'Abd Allah known as Habash said: 'There was a lunar eclipse... in the year 198 of Yazdijerd... As for the solar eclipse, which (occurred) in this year at the end of the month of Ramadan, all calculations were in error. The altitude of the Sun at the beginning of the eclipse was 7 deg as they (the astronomers) claim. The eclipse ended when the altitude of the Sun was about 24 deg, as though it was 3 hours of day (after sunrise)'.

Here we have a combination of a Persian year and a Muslim month. The year 198 of Yazdijerd (= 214/215 AH) covered the period from AD 829 Apr 28 to 830 Apr 27. A date at the very end of Ramadan in that year closely corresponds (within a day or so) to AD 829 Nov 30 – the calculated day of a solar eclipse visible in Baghdad. No observational details are recorded for the lunar eclipse; for the present purpose, the allusion to this event merely serves to establish the year.

The remark that the astronomers *claimed* that the altitude of the Sun at first contact was 7 deg suggests that the measurement was regarded as only approximate.

RESULTS

(i) Solar altitude at first contact ~ 7 deg (in the east). Equivalent LT ~ 7.68 h, UT ~ 4.60 h. Computed TT = 4.85 h, thus ΔT ~ 900 sec.

(ii) Solar altitude at last contact ~ 24 deg (in the east). Equivalent LT ~ 9.52 h, UT ~ 6.44 h. Computed TT = 7.06 h, thus ΔT ~ 2250 sec.

Presumably the recorded time of the end of the eclipse (i.e. three hours after sunrise) was derived from the solar altitude measurement. The fact that the time was expressed in 'hours of day' implies seasonal hours. Since the LT of sunrise = 6.87 h, 1 seasonal hour = 0.855 h. Hence the estimated

LT of last contact = 9.44 h. This compares fairly well with the above result deduced from the altitude measurement of 9.52 h.

(2) AD 866 Jun 16 [Sunday] (mag. = 0.66): Baghdad

This solar eclipse was mentioned by al-Mahani. He said: 'The Sun is to be eclipsed on Sunday the 28th of (the month of) Jumada al-Ula in the year 252 of *al-Hijrah*... (date on Persian calendar)... It was found (by observation) that this eclipse began (a little) more than a third of an hour after *Zawal*; the middle of the eclipse, as we estimated, was at 7 hours and $\frac{1}{3}$ and $\frac{1}{10}$ (i.e. 7;26 h after sunrise); then the eclipse cleared at 8 hours $\frac{1}{2}$ (i.e. 8;30 h)... (calculated details)... The eclipsed part of the Sun's diameter, as we estimated, was more than 7 digits and less than 8 digits'.

The month should be Jumada al-Ukhra, rather than the previous month Jumada al-Ula; otherwise the recorded Islamic date (including the week-day) is correct – as is the date on the Persian calendar. Elsewhere in the above text, it is stated that times are expressed in seasonal hours. No altitude measurements are preserved, only the reduced times. *Zawal* means the moment when the shadow of a gnomon begins to noticeably increase after the Sun transits the meridian and is thus a few minutes after noon. It will be assumed that the estimated time of first contact ('more than a third of an hour after *Zawal*') means roughly half an hour after midday.

RESULTS

(i) First contact ~ 0.52 h after midday. LT of contact ~ 12.50 h, UT ~ 9.51 h. Computed TT = 10.12 h, thus ΔT ~ 2200 sec.

(ii) Maximum phase 7.43 seasonal hours after sunrise. LT of sunrise on Jun 16 = 4.82 h, thus 1 seasonal hour = 1.20 equal hours. LT of observation = 13.71 h, UT = 10.73 h. Computed TT = 11.43 h, thus ΔT = 2500 sec.

(iii) Last contact 8.5 seasonal hours after sunrise. LT of contact = 14.99 h, UT = 12.01 h. Computed TT = 12.70 h, thus ΔT = 2450 sec.

These three ΔT results are fairly self-consistent.

(3) AD 923 Nov 11 [Tuesday] (mag. = 0.80): Baghdad

This solar eclipse was calculated and observed by Abu al-Hasan Ali ibn Amajur, who used the *al-Zij al-Arabi* of Habash. 'This eclipse was at the conjunction (i.e. new Moon) of (the month of) Sha'ban in the year 311 (AH). We as a group observed (this eclipse) and clearly distinguished it. The estimate of all (observers) for the middle of the eclipse was that it occurred when the altitude of the Sun was 8 deg in the east; its clearance was at $2\frac{1}{5}$ seasonal hours (after sunrise), when the altitude of the Sun was 20 deg. We observed this eclipse at several sites on the *Tarmah* (an elevated platform on the outside of the building). The estimate of Abu al-Hasan for the middle of the eclipse at his house was when the altitude of the Sun was 8 deg, as I estimated myself at my house before he arrived. The magnitude of the eclipse was $\frac{1}{2}$ and $\frac{1}{4}$ (i.e. $\frac{3}{4}$) of the Sun's diameter; the middle of the eclipse, which we estimated when the Sun's altitude was 8 deg, would be when the elapsed time (after sunrise) was 0;50 seasonal hours, and the (celestial) sphere had revolved (through) 10;40 deg. (The interval) between the middle of the eclipse and its clearance in this observation was 1;22 seasonal hours...(alternative times in equal hours)...'.

The statement that the eclipse 'was at the conjunction of (the month of) Sha'ban' implies that it occurred at the new Moon of Sha'ban – i.e. the very end of the previous lunar month Rajab.

RESULTS

(i) Solar altitude at maximum phase = 8 deg (in the east). Equivalent LT = 7.56 h, UT = 4.37 h. Computed TT = 4.91, thus ΔT = 1950 sec.

(ii) Solar altitude at last contact = 20 deg (in the east). Equivalent LT = 8.72 h, UT = 5.53 h. Computed TT = 5.98 h, thus ΔT = 1600 sec.

NB the LTs derived in (i) and (ii) above agree well with the times deduced by the observers from their own altitude determinations (i.e. 7.51 h and 8.70 h respectively).

(4) AD 928 Aug 18 [Monday] (mag. = 0.25): Baghdad

This solar eclipse was calculated and observed by Ali ibn Amajur. (According to calculation), the beginning was to be at...(calculated details)...on Monday. He said: 'I observed this eclipse with my son Abu al-Hasan and Muflih and (found) that the Sun rose (already) eclipsed by less than one $\frac{1}{4}$ of its surface. The eclipse continued to increase by an amount that we could perceive until $\frac{1}{4}$ (of its surface) was eclipsed. We observed the Sun distinctly (by reflection) in water. We (found) that it cleared and nothing of the eclipse remained and we distinguished the (full) circle of the Sun's body in water; (that was) when the altitude was 12 deg in the east, less $\frac{1}{3}$ of a division of the *al-halaqa* (i.e. the ring), which is graduated in thirds (of a degree), that is (less by) $\frac{1}{9}$ degree.... (comparison with calculation)...'.

Although the full date is not stated, it can fairly readily be established. The observation is cited between the records of two lunar eclipses: those of AD 927 Sep 13/14 and 929 Jan 27/28 (entries 6 and 7 of section 13.8 below). Only one solar obscuration would be visible at Baghdad during this interval – that of AD 928 Aug 18, which indeed occurred on a Monday. Hence the date is firmly established.

RESULTS

Solar altitude at last contact = 11.9 deg (in the east). Equivalent LT = 6.44 h, UT = 3.50 h. Computed TT = 4.01 h, thus ΔT = 1800 sec.

(5) AD 977 Dec 13 [Thursday] (mag. = 0.60): Cairo

This solar eclipse was in the early morning of Thursday the 28th of the month of Rabi ‘al-Akhir, in the year 367 of *al-Hijrah*... (date on Persian calendar)... We, a group of scholars (ten names are given), attended at al-Qarafah (a district of Cairo) in the Mosque of Abu Ja‘far Ahmad ibn Nasr al-Maghribi to watch this eclipse. Everyone waited for the beginning of this eclipse. It began to be perceived when the altitude of the Sun was more than 15 deg but less than 16 deg. (Those) present all agreed that about 8 digits of the Sun's diameter were eclipsed, that is (a little) less than 7 digits of surface. The Sun completely cleared when its altitude was more than 33 deg by about $\frac{1}{3}$ of a degree, as estimated by me, and agreed by all those present... (calculated details)...

RESULTS

(i) Solar altitude at first contact = 15.5 deg (in the east). Equivalent LT = 8.38 h, UT = 6.29 h. Computed TT = 6.80 h, thus ΔT = 1800 sec.

(ii) Solar altitude at last contact = 33.3 deg (in the east). Equivalent LT = 10.69 h, UT = 8.60 h. Computed TT = 9.15 h, thus ΔT = 2000 sec.

(6) AD 978 Jun 8 [Saturday] (mag. = 0.50): Cairo

This solar eclipse was on Saturday the 29th of (the month of) Shawwal in the year 367 of *al-Hijrah*... (date on Persian, Syrian and Coptic calendars)... A maximum of $5\frac{1}{2}$ digits of the Sun's diameter were eclipsed, according to estimation, that is 4 digits 10 minutes (i.e. 4 $\frac{1}{6}$ digits) of surface. The altitude of the Sun when a portion of the eclipse began to be perceived was 56 deg approximately. The completion of the clearance was when the altitude of the Sun was 26 deg or about so... (calculated details)...

RESULTS

(i) Solar altitude at first contact ~ 56 deg (in the west). Equivalent LT ~ 14.49 h, UT ~ 12.37 h. Computed TT = 12.73 h, thus ΔT ~ 1300 sec.

(ii) Solar altitude at last contact ~ 26 deg (in the west). Equivalent LT ~ 16.82 h, UT ~ 14.70 h. Computed TT = 15.26 h, thus ΔT ~ 2000 sec.

Although the direction in which the Sun was located is not stated, it is obvious from the context that at last contact the solar altitude was declining. For first contact, a solar altitude of 56 deg leads to a LT of either 9.51 h or 14.49 h. The former would imply an impossible duration of more than 7 hours, so that a westerly azimuth at this phase may also be inferred.

(7) AD 979 May 28 [Wednesday] (mag. = 0.45): Cairo

This solar eclipse was in the late afternoon of Wednesday the 23rd [read: 28th] of (the month of) Shawwal in the year 368 of *al-Hijrah*... (date on Persian, Syrian and Coptic calendars)... The eclipse was perceptible when the altitude of the Sun was $6\frac{1}{2}$ deg. About $5\frac{1}{2}$ digits of the Sun's diameter were eclipsed, as I estimated, that is 4;10 digits of surface. The Sun set eclipsed... (comparison with magnitude of previous solar eclipse)...

The 23rd of a lunar month is too early for a solar eclipse, implying a minor scribal error. The day of the month – like the other numbers – is expressed using numerals based on the letters of the Arabic alphabet (similar to the practice of the ancient Greeks). The presence or absence of diacrytical points (dots) can lead to widely differing numbers. In this particular instance, the numerals for 23 and 28 are very similar.

RESULTS

Solar altitude at first contact = 6.5 deg (in the west). Equivalent LT = 18.37 h, UT = 16.21 h. Computed TT = 16.61 h, thus ΔT = 1450 sec.

(8) AD 985 Jul 20 [Monday] (mag. = 0.30): Cairo

This solar eclipse was in the late afternoon on Monday at the end of (the month of) Safar in the year 375 of *al-Hijrah*. The altitude of the Sun when I perceived its eclipse by eye was 23 deg approximately. The altitude was 6 deg when nothing of its eclipse remained to be perceived by the eye. A maximum of $\frac{1}{4}$ of the Sun's diameter was eclipsed.

As noted in the comment on the previous entry, the Arabic numeral for 23 closely resembles that for 28. Hence the altitude of the Moon at first contact could well have been 28 deg. A mean of 25.5 deg will be assumed.

RESULTS

(i) Solar altitude at first contact ~ 25.5 deg (in the west). Equivalent LT ~ 16.74 h, UT ~ 14.73 h. Computed TT = 15.15 h, thus ΔT ~ 1500 sec.

(ii) Solar altitude at last contact = 6 deg (in the west). Equivalent LT = 18.30 h, UT = 16.29 h. Computed TT = 16.51 h, thus ΔT = 750 sec.

(9) AD 993 Aug 20 [Thursday] (mag. = 0.96): Cairo

This solar eclipse was in the forenoon of Sunday the 29th of (the month of) Jumada al-Ukhra in the year 383 of *al-Hijrah*... (date on Persian, Syrian and Coptic calendars)... The eclipse began when the altitude of the Sun was 27 deg in the east and was complete (i.e. reached its maximum) when the altitude was 45 deg in the east. The Sun cleared when its altitude was 60 deg in the east. About $\frac{2}{3}$ of it (i.e. the surface) was eclipsed.

Use of the provisional ΔT result obtained from equation (8.1) (1750 sec) leads to a calculated magnitude of 0.96. The observed magnitude (equivalent to 0.59 in terms of the solar diameter) would thus appear to represent a considerable underestimate (see also section 13.10). The eclipse was independently reported to be total by the Cairo chronicler, al-Maqrizi (see chapter 12). However, al-Maqrizi lived more than four centuries after the event and his source is unknown.

RESULTS

(i) Solar altitude at first contact = 27 deg (in the east). Equivalent LT = 7.67 h, UT = 5.60 h. Computed TT = 6.16 h, thus ΔT = 2000 sec.

(ii) Solar altitude at maximum phase = 45 deg (in the east). Equivalent LT = 9.08 h, UT = 7.01 h. Computed TT = 7.40 h, thus ΔT = 1400 sec.

(iii) Solar altitude at last contact = 60 deg (in the east). Equivalent LT = 10.35 h, UT = 8.28. Computed TT = 8.72 h, thus ΔT = 1600 sec.

NB for this eclipse to be partial at Cairo, in accordance with the observation recorded by Ibn Yunus, either $\Delta T < -2800$ or $\Delta T > -750$ sec. Neither of these limits are at all critical and will be disregarded in further investigation.

(10) AD 1004 Jan 24 [Monday] (mag. = 0.98): Cairo

> This solar eclipse was in the sign of Aquarius and was in the late afternoon of Monday the 29th of the month of Rabi' al-Awwal in the year 394 of *al-Hijrah*... (date on Persian, Syrian and Coptic calendars)... The Sun was eclipsed until what remained of it resembled the crescent Moon on the first night of the month. I estimated the eclipsed portion of the Sun to be 11 digits (i.e. of surface). The altitude of the Sun when the eclipse became noticeable in it (i.e. on its disk) was $16\frac{1}{2}$ deg in the west; thus I estimated (the altitude) at the beginning to be $18\frac{1}{2}$ deg. About quarter of the diameter was eclipsed when the altitude was 15 deg and half of the diameter was eclipsed when the altitude was 10 deg. The eclipse was complete (i.e. maximum phase) when the altitude was 5 deg.

The observation that the eclipse fell only a little short of annularity has already been discussed in section 13.6. Since the altitude at greatest phase was only 5 deg, the Sun would set still partially obscured; this explains the lack of any reference to last contact. Ibn Yunus' adjustment of as much as 2 deg to the measured altitude at the start suggests that the eclipse was not noticed until it was already fairly well advanced. The altitude correction represents a delay in sighting the eclipse by as much as 0.15 h; possibly intermittent cloud was responsible. Presumably the observer made an empirical correction based on the phase when the eclipse was first seen. It will be assumed that the amended altitude (i.e 18.5 deg) is accurate.

The determinations of solar elevation when it was estimated that respectively one quarter and half of the solar diameter was covered are probably also worth considering; the phase was changing fairly rapidly – by about 1 digit every 5 or 6 minutes. However, these observations are likely to lead to less accurate ΔT results than those derived from the measurements at first contact and maximum phase.

RESULTS

(i) Solar altitude at first contact = 18.5 deg (in the west). Equivalent LT = 15.68 h, UT = 13.85 h. Computed TT = 14.25 h, thus ΔT = 1450 sec.

(ii) Solar altitude when ~ 0.25 of solar diameter covered = 15 deg (in the west). Equivalent LT = 16.00 h, UT = 14.17 h. Computed TT = 14.53 h, thus ΔT ~ 1300 sec.

(iii) Solar altitude when ~ 0.50 of solar diameter covered = 10 deg (in the west). Equivalent LT = 16.45 h, UT = 14.62 h. Computed TT = 14.81 h, thus ΔT ~ 700 sec.

(iv) Solar altitude at maximum phase = 5 deg (in the west). Equivalent LT = 16.88 h, UT = 15.05 h. Computed TT = 15.37 h, thus ΔT = 1150 sec.

The various ΔT results derived in this section are assembled in table 13.2.

The ΔT values listed in table 13.2 are plotted in figure 13.6. The scale of this diagram has been chosen to enable ready comparison with later ΔT plots in this chapter (figures 13.8, 13.9 and 13.10).

13.8 Lunar eclipse times recorded by Ibn Yunus

For all lunar eclipses, calculated double dates (centred on local midnight) are used systematically in this chapter – as elsewhere in this book. However, I have computed weekdays for the second of each pair of dates since the Islamic day begins at sunset; as a result direct comparison can be made with the recorded weekday. The form of the various entries follows much the same pattern as in the previous section. It should be noted that whereas the Baghdad observers regularly measured altitudes of selected clock stars rather than the Moon itself, the Cairo astronomers generally preferred to determine the lunar elevation.

Among the lunar records cited by Ibn Yunus from al-Mahani is an account of an eclipse whose magnitude was so small that the observers believed it to be penumbral. Calculations from tables had indicated a likely magnitude of 1.5 digits. However, al-Mahani stated that although the brightness at the northern edge of the Moon appeared diminished, no part of the lunar disk was lost to view. He did not note the time of occurrence so that the record cannot be utilised to determine ΔT. The Islamic date of this event (252 AH, Dhu al-Qa‘dah 15) corresponds to AD 866 Nov 25/26, on which day the computed magnitude was only 0.03. A truly penumbral eclipse was seen by Babylonian astronomers in 188 BC (see chapter 6).

Table 13.2 ΔT results from solar eclipse timings recorded by Ibn Yunus.

Year[a]	Ct[b]	ΔT[c]
+829	1	~900
+829	4	~2250
+866	1	~2200
+866	M	2500
+866	4	2450
+923	1	1950
+923	4	1600
+928	4	1800
+977	1	1800
+977	4	2000
+978	1	~1300
+978	4	~2000
+979	1	1450
+985	1	~1500
+985	4	750
+993	1	2000
+993	M	1400
+993	4	1600
1004	1	1450
1004	E	~1300
1004	E	~700
1004	M	1150

[a] Year
[b] Contact, etc. (M = maximal phase; E = estimated phase – in AD 1004 only).
[c] Derived ΔT value.

(1) AD 854 Feb 16/17 [Saturday] (mag. = 0.92): Baghdad

This lunar eclipse was mentioned by al-Mahani. 'There was an eclipse of the Moon in the month of Ramadan in the year 239 of *al-Hijrah* on the night of Saturday, the middle of the month. It was found by observation that the beginning of this eclipse was at 10 hours and something like half of one-tenth of an hour (i.e. 10;03 h) after midday of Friday. We did not determine its times apart from the beginning. It was found that the uneclipsed part of its body was (a little) more than 1/10... (calculated details)...'.

The fact that only the time of first contact is reported is surprising; as an

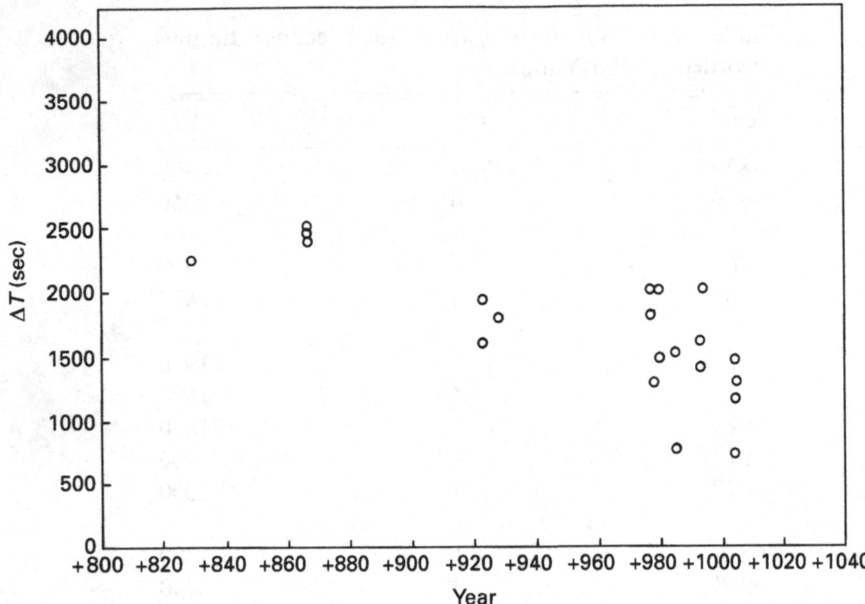

Fig. 13.6 ΔT values derived from solar eclipse timings recorded by Ibn Yunus.

evening eclipse, the entire course of events should have been visible unless unfavourable weather intervened. Although the text does not specify whether equal or unequal hours were used, since part of the measured interval of 10;03 hours after midday was in daylight and part in darkness, only equal hours would be meaningful.

Results

LT of first contact ~ 22.05 h on Feb 16. UT ~ 19.35 h. Computed TT = 20.22 h, thus ΔT ~ 3150 sec.

(2) AD 854 Aug 11/12 [Sunday] (mag. = 1.14): Baghdad

This lunar eclipse was mentioned by al-Mahani. 'The Moon was eclipsed on the night of Sunday 13th of the month of Rabi' al-Awwal in the year 240 of *al-Hijrah.* It was found by observation that the time of beginning of the eclipse was when the altitude of (the star) *al-dabaran* (Aldebaran: α Tau) was 45;30 deg in the east. We did not find its times (accurately) except this time, which was exact and precise. We measured the time of the completion of (the first phase of) the eclipse, which is the time of the beginning of the staying (*al-makth*) (in totality) and found it (to be) when the altitude of (the star) (*al-shi'ra*) *al-shamiyyah* (Procyon: α CMi) was between 22 and 23 deg in the east. This (latter) measurement is not exact but approximate. We determined the time of the beginning from the altitude of *al-dabaran* by the astrolabe and found it to be 44 deg (of the celestial sphere) after midnight...(calculated details)...We (also) determined the time (of the beginning of) the stay by the astrolabe, taking the altitude of *al-shamiyyah* as 23 deg and found it to be $23\frac{1}{2}$ parts (i.e. degrees) of the celestial sphere after the beginning (of the eclipse)'.

RESULTS

(i) Altitude of α Tau at first contact = 45.5 deg in the east. Equivalent LT (Aug 12) = 2.92 h, UT = 0.01 h. Computed TT = 0.71 h, thus ΔT = 2500 sec.

(ii) Altitude of α CMi at second contact ~ 22.5 deg in the east. Equivalent LT ~ 4.45 h, UT ~ 1.53 h. Computed TT = 1.91 h, thus ΔT ~ 1350 sec.

The LTs deduced from the measured altitudes of both α Tau (2.92 h) and α CMi (4.45 h) agree well with the astrolabe reductions (respectively 2.93 and 4.50 h).

(3) AD 856 Jun 21/22 [Monday] (mag. = 0.59): Baghdad

(This is) the third lunar eclipse mentioned by al-Mahani. 'There was an eclipse of the Moon on the night of Monday the middle of (the month of) Safar in the year 242 of *al-Hijrah*...(date on the Persian Calendar)...It was found by observation that the beginning of the eclipse was when the altitude of (the star) *al-dabaran* (Aldebaran: α Tau) was 9;30 deg in the east; the amount of the revolution of the (celestial) sphere from midnight to this time, as we determined (from this measurement) with the astrolabe was 50 deg. We did not determine its times except for the beginning. It was found (by observation) that the uneclipsed part of its body (i.e. disk) was more than one-quarter and less than one-third.... (calculated details)...'.

RESULTS

Altitude of α Tau at first contact = 9.5 deg in the east. Equivalent LT (Jun 22) = 3.39 h, UT = 0.43 h. Computed TT = 1.08 h, thus ΔT = 2350 sec.

Once again, the astrolabe reduction (3.33 h) is fairly precise.

(4) AD 923 Jun 1/2 [Monday] (mag. = 0.66): Baghdad

This lunar eclipse was calculated by Ali ibn Amajur al-Turki from the *al-Zij al-'Arabi* of Habash and observed by him and his son Abu al-Hasan and his freedman Muflih. (He said:) 'There was an eclipse of the Moon in (the month of) Safar in the year 311 of *al-Hijrah*...(calculated details)...The Moon rose at sunset (already) eclipsed by $\frac{1}{4}$ or (a little) more of the digits of the diameter (i.e. 3 digits or a little more). The Moon was eclipsed by (a little) more than 9 digits of diameter. The middle of the eclipse was at 1 and $\frac{2}{3}$ of equal hours of night (i.e. after sunset). The clearance of the eclipse was at 3 equal hours (after sunset) and (that was) when the altitude of (the star) *al-ridf* (Deneb: α Cyg) was 29;30 deg in the east...(calculated details)...'.

The alternative Arabic name for *al-ridf* is *Dhanab*, from which the modern name Deneb is derived.

RESULTS

(i) Mid-eclipse 1.67 h after sunset on Jun 1. LT of sunset = 19.14 h. Hence LT of observation = 20.81 h, UT = 17.78 h. Computed TT = 18.12 h, thus ΔT = 1200 sec.

(ii) Altitude of α Cyg at last contact = 29.5 deg in the east. Equivalent LT = 21.97 h, UT = 19.94 h. Computed TT = 19.50 h, thus ΔT = 2000 sec.

(5) AD 925 Apr 11/12 [Tuesday] (mag. = 1.08): Baghdad

> This lunar eclipse was calculated and observed by Abu al-Hasan ibn Amajur. This eclipse was on the night of Tuesday, the 15th of (the month of) Muharram, year 313 of *al-Hijrah.* He mentioned that the Moon was totally eclipsed and reported its five times (as found by calculation but not given in the text). Then he said: 'I observed this eclipse. The beginning was when the altitude of (the star) (*al-simak*) *al-ramih* (Arcturus: α Boo) was 11 deg in the east. The end of clearance was when the altitude of (the star) *al-nasr al-waqi'* (Vega: α Lyr) was 24 deg'. He then said: 'The beginning of the eclipse would be when the time elapsed from the start of night (i.e. from sunset) was 0;55 seasonal hours; the observation was later by 0;23 seasonal hours than that calculated from the *al-Zij al-Mumtahan* of Habash. The end of clearance by observation would be at 4;36 seasonal hours; observation was (again) later than calculation by 0;17 seasonal hours'.

There must be an error in the preserved altitude of α Boo at first contact. This star would be 11 deg above the eastern horizon at about 55 minutes *before* sunset rather than after it but the text clearly implies an observation. Presumably a scribe made an error in reporting the elevation of the star, but there is no obvious alternative reading here. The reported time of 55 minutes after sunset will be adopted instead. When Ibn Amajur compared this time with the result based on the tables of Habash, it showed much the same discrepancy as for the reported time of last contact.

It is not stated whether the azimuth of α Lyr was east or west of the meridian at last contact. However, it is clear from the reported time of 4 h 36 m after sunset (LT = 22.72 h) that the star was to the east of the meridian; an elevation of 24 deg would correspond to a rather similar LT of 22.77 h.

RESULTS

(i) First contact 0.92 h after sunset on Apr 11. LT of sunset = 18.52 h. Hence LT of observation = 19.44 h, UT = 16.46 h. Computed TT = 17.13 h, thus ΔT = 2400 sec.

(ii) Altitude of α Lyr at last contact = 24 deg in the east. Equivalent LT (Apr 11) = 22.77 h, UT = 19.74 h. Computed TT = 20.44 h, thus ΔT = 2350 sec.

These two results for ΔT are almost identical.

(6) AD 927 Sep 13/14 [Friday] (mag. = 0.22): Baghdad

This lunar eclipse was calculated and observed by Ali ibn Amajur, who used the *al-Zij al-'Arabi* of Habash. This eclipse was on the night of Friday, in the year 315 of *al-Hijrah*...(calculated details)...' He said: 'This eclipse was observed by my son Abu al-Hasan. The beginning of the eclipse was when the altitude of (the star) *al-shi'ra al-yamaniyyah* (Sirius: α CMa) was 31 deg in the east; the part of the celestial sphere which has revolved between sunset and the beginning of the eclipse is 148 deg plus a third of a degree and this is (equivalent to) 9;52 equal hours, which is 10;0 seasonal hours. The estimated digits of the eclipse was more than $\frac{1}{4}$ but less than $\frac{1}{3}$, as though it was $3\frac{1}{2}$ digits...(calculated details...)'.

Although the month (and day of the month) is not cited, no other lunar eclipse apart from that identified above occurred in 315 AH (AD 927 Mar – 928 Feb). Confirmation of the calculated date is provided by the recorded weekday (Friday). In addition, the estimate of magnitude is in tolerable accord with calculation for this eclipse.

RESULTS

Altitude of α CMa at first contact = 31 deg in the east. Equivalent LT (Sep 14) = 3.98 h, UT = 0.91 h. Computed TT = 1.73 h, thus ΔT = 2950 sec.

(7) AD 929 Jan 27/28 [Wednesday] (mag. = 1.19): Baghdad

This lunar eclipse was calculated and observed by Abu al-Hasan Ali ibn Amajur. This eclipse was at the opposition (i.e. full Moon) of (the month of) Dhu al-Hijjah, year 316 of *al-Hijrah*...(calculated details)...He said: 'I observed this eclipse at its beginning when the altitude of (the star) (*al-simak*) *al-ramih* (Arcturus: α Boo) was 18 deg in the east; the time elapsed from the start of the night (i.e after sunset) (to the beginning) was 5 hours seasonal, as required by calculation from the (*al-Zij*) *al-Mumtahan*'.

RESULTS

Altitude of α Boo at first contact = 18 deg in the east. Equivalent LT (Jan 27) = 22.79 h, UT = 20.10 h. Computed TT = 21.77 h, thus ΔT = 6000 sec.

Although the LT derived from the altitude measurement is in fair accord with the stated time of 5 unequal hours after sunset (equivalent to a LT of 22.88 h), and was said to confirm calculation from tables, the resulting ΔT value is very discordant compared with roughly contemporaneous results. Hence the observation will be rejected.

(8) AD 933 Nov 4/5 [Tuesday] (mag. = 1.42): Baghdad

> This lunar eclipse was calculated and observed by Ali ibn Amajur al-Turki. He said: '(According to calculation) from the *al-Zij al-Mumtahan al-'Arabi* of Habash, the eclipse opposition was on the night of Tuesday the 13th of (the month of) Dhu al-Qa'dah, year 321 of *al-Hijrah*...(calculated details)...' He said 'I observed this eclipse when (the Moon) entered (the shadow) and (that was) when the altitude of (the star) (*al-simak*) *al-ramih* (Arcturus: α Boo) was 15 deg in the east and when the time elapsed from the start of the night was 9;56 seasonal hours...(calculated details)...'.

Results

Altitude of α Boo at first contact = 15 deg in the east. Equivalent LT (Nov 5) = 4.56 h, UT = 1.36 h. Computed TT = 1.99 h, thus ΔT = 2250 sec.

The LT deduced from the altitude of α Boo agrees reasonably well with the stated time after sunset (corresponding to a LT of 4.41 h).

(9) AD 979 May 14/15 [Thursday] (mag. = 0.70): Cairo

> This lunar eclipse was in (the month of) Shawwal in the year 368 of *al-Hijrah*. The Moon rose eclipsed on the night whose morning was Thursday...(date on Persian, Syrian and Coptic calendars)...More than 8 digits but less than 9 of the Moon's diameter were eclipsed...(calculated details)...The eclipse cleared when about one equal hour and a fifth of night had elapsed, as I estimated...(calculated details)....

Results

Last contact $\sim$ 1.20 h after moonrise on May 14. LT of moonrise = 18.87 h. Hence LT of contact $\sim$ 20.07 h, UT $\sim$ 17.90. Computed TT = 18.31 h, thus $\Delta T \sim$ 1650 sec.

(10) AD 979 Nov 6/7 [Friday] (mag. = 0.84): Cairo

> This lunar eclipse was in the month of Rabi 'al-Akhir in the year 369 (of *al-Hijrah*), on the night whose morning was Friday the 13th of the month...(date on Persian, Syrian and Coptic calendars)...A group of scholars gathered to observe this eclipse. They estimated that what was eclipsed of the Moon's circular surface was 10 digits. The altitude of the Moon when they perceived the eclipse was $64\frac{1}{2}$ deg in the east. The altitude when its clearance completed was 65 deg in the west...(calculated details).

Although the Moon was high in the sky at both contacts, each altitude was sufficiently below the meridian elevation (77 deg) for the rate of change with time to be significant.

RESULTS

(i) Altitude of Moon at first contact = 64.5 deg in the east. Equivalent LT (Nov 6) = 22.40 h, UT = 20.08 h. Computed TT = 20.61 h, thus ΔT = 1900 sec.

(ii) Altitude of Moon at last contact = 65 deg in the west. Equivalent LT = 1.62 h, UT = 23.30 h. Computed TT = 23.67 h, thus ΔT = 1300 sec.

(11) AD 980 May 2/3 [Monday] (mag. = 1.61): Cairo

> The Moon was totally eclipsed in (the month of) Shawwal in the year 369 of *al-Hijrah* on the night whose morning was Tuesday...(date on Persian calendar)...A group of scholars gathered to observe this eclipse. They perceived the trace of the eclipse when the altitude of the Moon was $47\frac{2}{3}$ deg. The eclipse cleared when about $\frac{3}{5}$ of an equal hour remained to the end of the night (i.e. before sunrise). Our gathering to observe this eclipse was in the Mosque of Ibn Nasr (al-Maghribi) at al-Qarafa.

The day of the week is in error; it should be Monday rather than Tuesday.

There is an obvious error in the recorded lunar altitude when the eclipse began; this significantly exceeds the meridian altitude of the Moon (42.5 deg). A *possible* alternative reading in place of 47 is 40 (the figures are similar), leading to an altitude at first contact of $40\frac{2}{3}$ deg. However, even if this restoration is correct, the Moon would be too close to the meridian for the rate of change of altitude to be significant. Hence only the timing of fourth contact is usable.

RESULTS

Last contact ~ 0.60 h before sunrise on May 3. LT of sunrise = 5.25 h. Hence LT of contact ~ 4.65 h, UT ~ 2.48 h. Computed TT = 3.07 h, thus ΔT ~ 2100 sec.

(12) AD 981 Apr 21/22 [Friday] (mag. = 0.18): Cairo

> This lunar eclipse was in the month of Shawwal in the year 370 of *al-Hijrah* on the night whose morning was Friday...(date on Persian, Syrian and Coptic calendars)...We gathered to observe this eclipse at al-Qarafa in the Mosque of Ibn Nasr al-Maghribi. We perceived the beginning of this eclipse when the altitude of the Moon was approximately 21 deg. About one-quarter of the Moon's diameter was eclipsed. The Moon cleared completely when about $\frac{1}{4}$ of an hour remained to sunrise.

As the eclipse ended shortly before sunrise, the Moon must have been to the west of the meridian when its altitude was measured. Although it is not stated whether the time interval of one-quarter of an hour before sunrise for last contact was in equal or seasonal hours, the latter would only be 0.02 h less. A mean of 0.24 h will be assumed.

RESULTS

(i) Altitude of Moon at first contact ~ 21 deg in the west. Equivalent LT (Apr 22) ~ 3.53 h, UT ~ 1.39 h. Computed TT = 1.97 h, thus ΔT ~ 2000 sec.

(ii) Last contact ~ 0.24 h before sunrise on Apr 22. LT of sunrise = 5.39 h. Hence LT of contact ~ 5.15 h, UT ~ 3.01 h. Computed TT = 3.67 h, thus ΔT ~ 2350 sec.

(13) AD 981 Oct 15/16 [Sunday] (mag. = 0.36): Cairo

> The Moon was eclipsed in the month of Rabi' al-Akhir, in the year 371 of *al-Hijrah* on the night whose morning was Sunday. About 5 digits of diameter were eclipsed. The altitude of the Moon at the contact from outside (external or first contact) was about 24 deg, as I estimated... (calculated details)...

It is not specified whether the Moon was to the east or west of the meridian at first contact. The stated altitude of 24 deg corresponds to a LT of either about 19.56 h (UT = 17.66 h) or 4.44 h (UT = 2.11 h) depending on whether the Moon was in an easterly or westerly azimuth. Since the computed TT is 2.54 h, only the latter alternative would lead to a meaningful result for ΔT (i.e. ~ 1500 sec rather than ~ 32 000 sec). Hence it will be assumed that the Moon was to the west of the meridian.

RESULTS

Altitude of Moon at first contact ~ 24 deg in the west. Equivalent LT (Oct 16) ~ 4.44 h, UT ~ 2.11 h. Computed TT = 2.54 h, thus ΔT ~ 1500 sec.

(14) AD 983 Mar 1/2 [Friday] (mag. = 1.07): Cairo

> This lunar eclipse was in the month of Ramadan in the year 372 of *al-Hijrah* on the night whose morning was Friday, the 15th of the month... (date on Persian calendar)... The Moon was totally eclipsed. The altitude of the Moon when the eclipse became perceivable was 66 deg. The altitude when the Moon cleared completely was 35 plus $\frac{1}{2}$ and $\frac{1}{3}$ (i.e. 35;50 deg); the Moon was totally dark for about an hour... (calculated details)...

The recorded altitude of the Moon at first contact slightly exceeds the meridian altitude of 65.8 deg. However, if the Moon was near the meridian at first contact, this could still represent a good measurement, although of course a value of ΔT cannot be derived.

RESULTS

Altitude of Moon at last contact = 35.8 deg in the east. Equivalent LT (Mar 2) = 3.43 h, UT = 1.55 h. Computed TT = 1.87 h, thus ΔT = 1200 sec.

(15) AD 986 Dec 18/19 [Sunday] (mag. = 0.91): Cairo

> This lunar eclipse was on the night whose morning was Sunday, the 15th of (the month of) Sha'ban in the year 376 of *al-Hijrah*. The eclipse became noticeable when the altitude of the Moon was 24 deg in the west. I estimated the (first) contact was when the altitude was $50\frac{1}{2}$ deg. About 10 digits of the Moon's diameter were eclipsed. The observation was in the Mosque of Abu Ja'far Ahmad ibn Nasr al-Maghribi at al-Qarafa in the presence of Abu Ahmad ibn 'Asim and 'Abd al-Rahman ibn 'Isa ibn Tabyan. The Moon set eclipsed.

In using the Arabic alphabet for numerals to express altitude in degrees, the 50 of the text could possibly be read as 30, which is much closer to the altitude when the eclipse first 'became noticeable'. If so, there would still be a delay of about half an hour between true first contact and the observed start. Although this could be explained by cloud, there is no mention of unfavourable weather in the record. It thus seems best to consider only the original measurement of 24 deg. Since Ibn Yunus judged that he was late in detecting first contact, it will be assumed that this latter value is only approximate.

RESULTS

Altitude of Moon at first contact ~ 24 deg in the west. Equivalent LT (Dec 19) ~ 4.89 h, UT ~ 2.84 h. Computed TT = 3.07 h, thus $\Delta T \sim 800$ sec.

NB an earlier moment for the detection of the eclipse would, of course, lead to a higher result for ΔT.

(16) AD 990 Apr 12/13 [Sunday] (mag. = 0.74): Cairo

> This lunar eclipse was on the night whose morning was Sunday the 16th of (the month of) Muharram in the year 380 of *al-Hijrah*. $7\frac{1}{2}$ digits of the Moon's diameter were eclipsed, as I guessed. The Moon cleared when the ascendant was the beginning of Aquarius. The altitude of the Moon when the eclipse began, I mean at the time of contact, was 38 deg.

It is not mentioned whether the Moon was east or west of the meridian at first contact. For an altitude of 38 deg, the corresponding LTs are respectively 21.80 h and 2.20 h. Since the latter time is later than the time of last contact as inferred from the rising of the beginning of Aquarius (approximately 1.3 h) the Moon must have been in an easterly azimuth when the eclipse began.

The way in which the time of end is expressed is without parallel among medieval Arabic eclipse records and presumably was derived from a separate measurement. Since the nature of this measurement cannot be established, and the phrase 'the beginning of Aquarius' may represent only an approximate celestial longitude, this result will not be used to determine ΔT. (NB Aquarius extends from long. 300 to 330 deg.)

RESULTS
Altitude of Moon at first contact = 38 deg in the east. Equivalent LT (Apr 12) = 21.80 h, UT = 19.70 h. Computed TT = 20.69 h, thus ΔT = 3500 sec.

(17) AD 1001 Sep 5/6 [Saturday] (mag. = 0.86): Cairo

> This lunar eclipse was in (the month of) Shawwal in the year 391 of *al-Hijrah* at the start of the night of Saturday, the 14th of the month... (date on Persian calendar)... The Moon cleared when about 2 seasonal hours of night had elapsed (i.e. after sunset). I saw the Moon before its clearance and it was like a crescent.

It is not clear whether the estimate of 2 hours is only approximate, but this will be assumed.

RESULTS
(i) Last contact ~ 2.00 h after sunset on Sep 5. LT of sunset = 18.24 h. 2 h unequal = 1.92 h. Hence LT of contact ~ 20.12 h, UT ~ 18.01 h. Computed TT = 18.17 h, thus ΔT ~ 600 sec.

(18) AD 1002 Mar 1/2 [Monday] (mag. = 1.44): Cairo

> This lunar eclipse was on the night whose morning was Monday the 15th of the month of Rabi‘ al-Akhir in the year 392 of *al-Hijrah*... (date on Persian calendar)... The Moon was totally eclipsed and had a staying (*al-makth*: i.e. in totality or in darkness). The eclipse began when the altitude of (the star) *(al-simak) al-ramih* (Arcturus: α Boo) was 12 (or 52) deg east and when the altitude of (the star) *al-hadi* (Capella: α Aur) was 14 deg in the west. The altitude of (the star) (*al-simak*) *al-ramih* at the complete clearance was 35 deg.

There is evidence of scribal errors in recording both altitudes of α Boo. The elevation of this star at first contact (implying an LT of 20.25 h) is incompatible with that of α Aur at the same moment (LT = 23.55 h). However, the symbols for the numbers 12 and 52 are so similar that confusion is not uncommon – as noted by Caussin (1804). Reading 52 deg for the altitude of α Boo leads to a LT of 23.48 h, which is in fairly good accord with that derived from the α Aur measurement. This altitude will thus be adopted.

Assumption of an altitude for α Boo of 35 deg in the west at the end of the eclipse implies an LT of 6.41 h. This is about 40 minutes after sunrise, so that the star would be invisible; further, an excessive duration for the eclipse of nearly 7 h would be implied. In this case, it is not possible to satisfactorily restore the altitude measurement at last contact.

Table 13.3 ΔT results from lunar eclipse timings recorded by Ibn Yunus.

Year	Ct	ΔT
+854a	1	~3150
+854b	1	2500
+854b	2	~1350
+856	1	2350
+923	M	1200
+923	4	2000
+925	1	2400
+925	4	2350
+927	1	2950
+933	1	2250
+979a	4	~1650
+979b	1	1900
+979b	4	1300
+980	4	~2100
+981a	1	~2000
+981a	4	~2350
+981b	1	~1500
+983	4	1200
+986	1	~800
+990	1	3500
1001	4	~600
1002	1	1750
1002	1	1950

RESULTS

(i) Altitude of α Aur at first contact = 14 deg in the east. Equivalent LT (Mar 1) = 23.55 h, UT = 21.67 h. Computed TT = 22.15 h, thus ΔT = 1750 sec.

(ii) Altitude of α Boo at first contact = 52 deg in the east. Equivalent LT = 23.48 h, UT = 21.61 h. Computed TT = 22.15 h, thus ΔT = 1950 sec.

The ΔT results obtained in this section are listed in table 13.3, and are plotted in figure 13.7.

Comparison between figure 13.7 and figure 13.6 shows that the scatter is considerably greater for the lunar data. A partial explanation is poorer

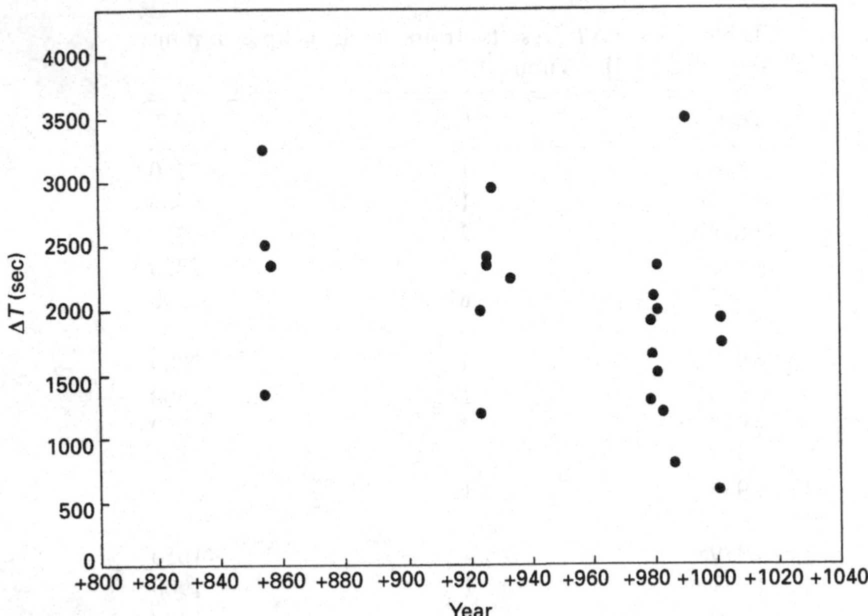

Fig. 13.7 ΔT values derived from lunar eclipse timings recorded by Ibn Yunus.

resolution of the contacts, perhaps combined with the greater difficulty of reading an instrument at night.

13.9 Solar and lunar eclipse timings recorded by al-Battani and al-Biruni

The following translations are from Said and Stephenson (1997) except where otherwise stated. Observations by al-Battani (AD 883–901) are taken from his *al-Zij al-Sabi*; those by al-Biruni are extracted from both his *al-Qanun al-Mas'udi* (AD 1003–4) and his *Kitab Tahdid al-Amakin li-Tashih Masafat al-Masakan* (AD 1019 only). Apart from the very last observation, only the times of maximal phase are preserved. Dates cited by al-Battani are given in terms of the Syrian calendar, each year of which began on 1 October of the Julian calendar.

(1) AD 883 Jul 23/24 [Wednesday] (lunar, mag. = 0.95): al-Raqqah

> This lunar eclipse was observed by us at the city of al-Raqqah on the 23rd of (the month of) Tammuz in the year 1194 of *Dhu al-Qarnayn* ('the Two-horned', i.e. Alexander IV of Macedonia), which is the year 1206 of *al-Iskandar* (i.e. Alexander the Great). The middle of the eclipse was at a little more than 8 equal hours after midday. A little more than $\frac{1}{2}$ plus $\frac{1}{3}$ (i.e. $\frac{5}{6}$ or 10 digits) of the Moon's diameter was eclipsed... (calculated details)...
>
> [Al-Battani.]

The reference to Alexander IV of Macedonia is evidently an indirect allusion to the Seleucid era. Both the year 1194 from the Seleucid era (adopted as 1 Oct in BC 312) and the year 1206 from the death of Alexander the Great (323 BC) correspond to AD 883/4. Tammuz (equivalent to July on the Julian calendar) was the tenth month of the year.

The 'little more than 8 equal hours' will be assumed to be 8.1 h. Al-Battani expresses most times to the nearest $\frac{1}{4}$ or $\frac{1}{3}$ of an hour.

RESULTS

LT of mid-eclipse ~ 20.10 h, UT ~ 17.57 h. Computed TT = 17.82 h, thus ΔT ~ 900 sec.

(2) AD 891 Aug 8 [Sunday] (solar, mag. = 0.89): al-Raqqah

> This solar eclipse was observed by us at the city of al-Raqqah on the 8th of (the month of) Ab in the year 1202 of *Dhu al-Qarnayn*, which is the year 1214 after the death of *al-Iskandar*. The middle of the eclipse was at one seasonal hour after midday. (A little) more than $\frac{2}{3}$ of the Sun (i.e of surface) was eclipsed in view...(calculated details)...
>
> [Al-Battani.]

Ab (corresponding to August) was the 11th month of the year.

RESULTS

1 unequal hour = 1.14 h. Hence LT of mid-eclipse ~ 13.14 h, UT ~ 10.59 h. Computed TT = 11.06 h, thus ΔT = 1700 sec.

(3) AD 901 Jan 23 [Friday] (solar, mag. = 0.67): Antakyah

> This solar eclipse was observed by us at the city of Antakyah on the 23rd of (the month of) Kanun al-thani in the year 1212 of *Dhu al-Qarnayn* which is the year 1224 after the death of *al-Iskandar*. The middle of the eclipse was about $3\frac{2}{3}$ equal hours before midday. (A little) more than $\frac{1}{2}$ of the Sun (i.e Sun's surface) in sight was eclipsed.... (calculated details)...
>
> [Al-Battani.]

Kanun al-thani (corresponding to January) was the fourth month of the year.

RESULTS

LT of mid-eclipse ~ 8.33 h, UT ~ 6.17 h. Computed TT = 6.63 h, thus ΔT ~ 1650 sec.

(4) AD 901 Jan 23 [Friday] (solar, mag. = 0.68): al-Raqqah

> ...This eclipse was observed by someone on our behalf at the city of al-Raqqah (on the same date as above). The middle of the eclipse was (a little) less than $3\frac{1}{2}$ equal hours before midday. (A little) less than $\frac{2}{3}$ of the Sun (i.e. Sun's surface) in view was eclipsed...(calculated details)...
>
> [Al-Battani.]

This is a continuation of the previous text. The interval of '3 hours and less than half an hour equal' will be assumed to be 3.4 h.

RESULTS

LT of mid-eclipse ~ 8.60 h, UT ~ 6.26 h. Computed TT = 6.69 h, thus ΔT ~ 1550 sec.

(5) AD 901 Aug 2/3 [Sunday] (lunar, mag. = 1.06): Antakyah

> This lunar eclipse was observed by us at the city of Antakyah on the 2nd of (the month of) Ab in the year 1212 of *Dhu al-Qarnayn*, which is the year 1224 from the death of *al-Iskandar*. The middle of the eclipse was at approximately 15 plus $\frac{1}{3}$ (i.e 15;20 h) equal hours after midday. The Moon was eclipsed by less than its diameter by a small amount...(calculated details)...
>
> [Al-Battani.]

Although this eclipse was recorded as marginally partial, it was actually total according to computation.

RESULTS

LT of mid-eclipse ~ 3.33 h on Aug 3, UT ~ 0.98 h. Computed TT = 1.16, thus ΔT ~ 650 sec.

(6) AD 901 Aug 2/3 [Sunday] (lunar, mag. = 1.06): al-Raqqah

> This eclipse was observed (by someone on our behalf) at the city of al-Raqqah (at the same date as above). The middle of the eclipse was approximately 15 plus $\frac{1}{3}$ and $\frac{1}{4}$ (i.e. 15;35 equal hours) after midday...(calculated details)...
>
> [Al-Battani.]

This is a continuation of the preceding account.

RESULTS

LT of mid-eclipse ~ 3.58 h on Aug 3, UT ~ 1.04 h. Computed TT = 1.16, thus ΔT ~ 450 sec.

(7) AD 1003 Feb 19/20 [Saturday] (lunar, mag. = 0.14): Jurjan

> This lunar eclipse was on the night of Saturday the 14th of the month of Rabi' al-Akhir in the year 393 (of *al-Hijrah*). I observed the beginning and clearance at Jurjan by the altitude of the (two stars) *al-Shi'rayan* (i.e. *al-shi'ra al-yamaniyyah* – Sirius: α CMa – and *al-shi'ra al-shamiyyah* – Procyon: α CMi). The Moon was eclipsed by $\frac{1}{4}$ of its diameter by estimate. The longitude difference between Jurjan and Ghaznah is 2;21 minutes of day. The middle of the eclipse at *it* (presumably at Ghaznah) was 19;11 (minutes of day) after midday of Friday... in the year 1751 of Buktinassar (i.e. Nabonassar).
>
> [Al-Biruni: *al-Qanun*.]

In the following observation (AD 1003 Aug 14/15), al-Biruni quoted only the measured time of mid-eclipse at Ghaznah and it will be presumed that he did the same in this case.

A minute of day (*daqa'iq al-Ayyar*) – i.e. 1/60 of a day – was equivalent to 0.4 of an equal hour. Hence mid-eclipse was 7.67 h after noon at Ghaznah.

Results

LT of mid-eclipse at Ghaznah = 19.67 h on Feb 19, UT = 15.35 h. Computed TT = 15.75 h, thus ΔT = 1450 sec.

NB If the LT of mid-eclipse (7.67 h after noon) related to Jurjan, the corresponding value for ΔT would be −1900 sec, an obviously errant result.

(8) AD 1003 Aug 14/15 [Sunday] (lunar, mag. = 0.14): Jurjan

> This lunar eclipse was on the night of Sunday the 13th of (the month of) Shawwal in the year 393 (of *al-Hijrah*). I observed it at Jurjan by the altitudes of (the two stars) *al-nasran* (i.e. 'the two eagles': *al-nasr al-ta'ir* – Altair: α Aql – and *al-nasr al-waqi'* – Vega: α Lyr) and (the star) *al-'ayyuq* (i.e. Capella: α Aur). The middle of the eclipse at Ghaznah occurred when more than $\frac{1}{4}$ of its diameter was eclipsed and that was 31;21 (minutes of day) after midday of Saturday... in the year 1751 (of Nabonassar).
>
> [Al-Biruni: *al-Qanun*.]

It seems curious that despite al-Biruni's mention of the star altitudes which he made at Ghaznah, he does not quote any LTs for that city in both this and the previous entry.

Mid-eclipse was 12.54 h after noon at Ghaznah.

Results

LT of mid-eclipse at Ghaznah = 0.54 h on Aug 15, UT = 20.01 h. Computed TT = 20.20 h, thus ΔT = 700 sec.

(9) AD 1004 Jul 4/5 [Wednesday] (lunar, mag. = 0.17): Jurjaniyyah

> This lunar eclipse was on the night of Wednesday the 14th of the month of Ramadan in the year 394 (of *al-Hijrah*). I observed its middle at Jurjaniyyah of Khwarizm and found it to be 36;32 (minutes of day) after midday of Tuesday...in the year 1752 (of Nabonassar). Ghaznah is east of Jurjaniyyah by 1;42,12 (minutes of day).
>
> [Al-Biruni: *al-Qanun*.]

In this instance, al-Biruni did not mention the clock stars which he used to determine the LT of observation.

It seems most likely that in this instance the place to which the LT refers is Jurjaniyyah; the allusion to Ghaznah would appear to be incidental. (The longitude difference between Ghaznah and Jurjaniyyah according to al-Biruni is equivalent to 0.68 h.) Mid-eclipse was observed 14.61 h after noon at Jurjaniyyah.

RESULTS

LT of mid-eclipse at Jurjaniyyah = 2.61 h on Jul 5, UT = 22.72 h. Computed TT = 23.45 h, thus ΔT = 2600 sec.

(10) AD 1019 Sep 16/17 [Thursday] (lunar, mag. = 0.76): Ghaznah

> This lunar eclipse was (seen) at Ghaznah in (the month of) Jumada al-Ula, in the year 410 (of *al-Hijrah*). I observed it and (found that) at the moment when the indentation at (the edge of) the full Moon became noticeable the altitude of (the star) *al-'ayyuq* (Capella: α Aur) from the east was slightly less than 66 deg, that of (the star) *al-shi'ra al-yamaniyyah* (Sirius: α CMa) was 17 deg, that of (the star) *al-shamiyyah* (Procyon: α CMi) was 22 deg, and that of (the star) *aldabaran* (Aldebaran: α Tau) was 63 deg; all in the east. All these (altitude measurements) necessitate that the beginning of the eclipse would be when approximately 8 hours of night had elapsed (i.e. after sunset)...(calculated details)...It was clear to the sight that the world was lit up, the stars had disappeared, the Sun was about to rise, and the Moon was about to set behind the mountains which screened it. A small portion of the eclipse (still) remained in its body (i.e. disk) and I was unable to observe it (the time of completion of clearance) exactly.
>
> [Al-Biruni: *Kitab*.]

All four star altitudes refer to first contact. It will be assumed that the altitude of Capella (α Aur) was 65.7 deg. Since visibility of the eclipse was eventually interrupted by mountains, last contact could not be observed from Ghaznah.

Table 13.4 ΔT results from solar and lunar eclipse timings recorded by al-Battani and al-Biruni.

Year	Type	Ct	ΔT
+883	Moon	M	~900
+891	Sun	M	1700
+901	Sun	M	~1650
+901	Sun	M	~1550
+901	Moon	M	~650
+901	Moon	M	~450
1003a	Moon	M	1450
1003b	Moon	M	700
1004	Moon	M	2600
1019	Moon	1	1900
1019	Moon	1	1700
1019	Moon	1	1800
1019	Moon	1	1600

RESULTS

(i) Altitude of α Aur at first contact = 65.7 deg in the east. Equivalent LT = 2.29 h, UT = 21.60 h. Computed TT = 22.13 h, thus ΔT = 1900 sec.

(ii) Altitude of α CMa at first contact = 17 deg in the east. Equivalent LT = 2.34 h, UT = 21.65 h. Computed TT = 22.13 h, thus ΔT = 1700 sec.

(iii) Altitude of α CMi at first contact = 22 deg in the east. Equivalent LT = 2.32 h, UT = 21.63 h. Computed TT = 22.13 h, thus ΔT = 1800 sec.

(iv) Altitude of α Tau at first contact = 17 deg in the east. Equivalent LT = 2.38 h, UT = 21.69 h. Computed TT = 22.13 h, thus ΔT = 1600 sec.

These four results are fairly self-consistent.

The various ΔT values from this section are assembled in table 13.4. These values are plotted in figure 13.8, which also shows the limits obtained later in this chapter from horizon observations – see section 13.11. It is evident that apart from the individual measurements made at the eclipse of AD 1019, the scatter in these results is significantly greater than for the data from Ibn Yunus.

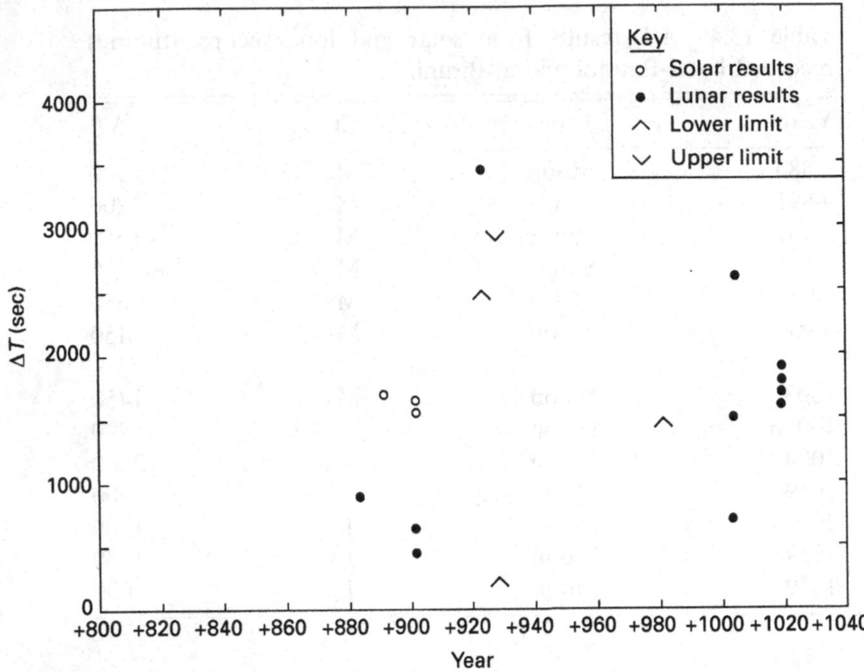

Fig. 13.8 ΔT values derived from eclipse timings recorded by al-Battani and al-Biruni along with ΔT limits derived from horizon observations.

13.10 Solar magnitude estimates

The solar eclipse records cited in sections 13.7 and 13.9 above contain a number of estimates of magnitude. Most of these are expressed as a fraction of the apparent diameter of the Sun, sometimes accompanied by the surface equivalent, but in one case (AD 928) only a surface magnitude appears to be given. In this year it is stated that:

> ...The Sun rose (already) eclipsed by less than one $\frac{1}{4}$ of its surface. The eclipse continued to increase by an amount that we could perceive until $\frac{1}{4}$ (of its surface) was eclipsed...

It seems a reasonable presumption that the *peak* magnitude was also expressed in terms of the visible surface of the Sun. From table 3.1 (chapter 3), it may be seen that the equivalent linear magnitude would be 0.36.

Where both linear and surface magnitudes are quoted, I have adopted only the former result. At the eclipses of AD 891 and 901 (two separate observations were made in the latter year), the magnitude was said to be (i) $> \frac{2}{3}$; (ii) $> \frac{1}{2}$; and (iii) $< \frac{2}{3}$ (see section 13.9). As the observers were familiar with the use of digits, I have taken these estimates to be

Table 13.5 ΔT results from medieval Arab estimates of solar eclipse magnitude.

Date AD	Place	Mag. est.	ΔT
891 Aug 8	al-Raqqah	$>\frac{2}{3}$	50
901 Jan 23	Antakyah	$>\frac{1}{2}$	5500
901 Jan 23	al-Raqqah	$<\frac{2}{3}$	5850
923 Nov 11	Baghdad	$\frac{3}{4}$	2700
928 Aug 18	Baghdad	$\frac{1}{4}$ surf.	—
977 Dec 13	Cairo	$\frac{8}{12}$	2900
978 Jun 8	Cairo	$\frac{5.5}{12}$	500
979 May 28	Cairo	$\frac{5.5}{12}$	—
985 Jul 20	Cairo	$\frac{1}{4}$	850
993 Aug 20	Cairo	$\frac{2}{3}$ surf.	−8050

as follows: (i) between 8 and 9 twelfths (i.e. between $\frac{2}{3}$ and $\frac{3}{4}$); (ii) and (iii) $\frac{7}{12}$ (i.e. between $\frac{1}{2}$ and $\frac{2}{3}$). I have not made any amendments to the observed magnitudes to allow for possible systematic errors. There is sound evidence for such errors in the case of lunar obscurations – see chapter 7 – but not for solar eclipses.

As discussed in chapter 3, unless it is clear whether the upper or lower limb of the Sun was covered, in principle an observation of magnitude can lead to two separate values for ΔT – typically some 20 000 sec apart. Since numerous contemporary measurements (eclipse timings, etc.) indicate that ΔT at this period was of the order of 2000 or 3000 sec, usually only one solution need be considered.

The observations and ΔT results are summarised in table 13.5. Full (AD) dates are cited in each case since translations of individual records are not given in this section.

No value of ΔT can satisfy the estimates of magnitude in AD 928 (0.37) or 979 (0.46). The geographical circumstances are such that the actual magnitudes cannot have exceeded 0.33 and 0.42 respectively. The ΔT value of −8050 sec obtained in AD 993 (the only possible result on this date for a magnitude of 0.67) is clearly anomalous. Not only is the ΔT figure itself highly discordant, it also leads to a computed solar altitude at greatest phase of 60 deg rather than the observed figure of 45 deg. Clearly the scatter of the remaining ΔT values in table 13.5 is considerably larger than that obtained from timed data (tables 13.2 to 13.4). It is apparent that the use of magnitude estimates to determine ΔT is scarcely justifiable at this relatively late period (see also chapters 5 and 9). Accordingly, the results in table 13.5 will not be considered in subsequent analysis.

13.11 Rising and setting phenomena

Among the observations discussed in sections 13.7 to 13.9 are six instances where it is stated that the Sun or Moon rose or set whilst eclipsed. Five of these observations, ranging in date from AD 923 to 986, were made at either Baghdad or Cairo. In Baghdad, the terrain is flat and the horizon is extremely level. This is true also of Cairo except towards the south-east, where the horizon is screened by hills. A remarkably vivid description of a moonset eclipse in AD 1019 was reported from Ghaznah by al-Biruni (see section 13.9). Unfortunately, this city is located in a very mountainous region so that an accurate horizon profile would be needed to make proper use of the observation. Hence only the five earlier records from Baghdad and Cairo will be considered here, along with a sixth observation from Cairo in AD 981 which states that a lunar eclipse ended not long before sunrise.

In two instances (AD 923 and 928), an estimate of the degree of obscuration of the Sun or Moon when on the horizon is clearly given. On the former occasion, it is stated that the Moon rose already rather more than $\frac{1}{4}$ covered, the magnitude eventually reaching 9 digits. Hence this observation can be utilised to obtain an independent value for ΔT. In AD 928, the fraction of the solar disk said to be covered at sunrise (less than $\frac{1}{4}$) is very close to the observed peak magnitude of $\frac{1}{4}$. Hence on this occasion a useful result for ΔT cannot be obtained.

Extracts from the relevant texts – all quoted in full in sections 13.7 and 13.8 – are given below along with comments and the derived ΔT limits. All six records are taken from the *zij* of Ibn Yunus.

(1) AD 923 Jun 1/2 (lunar, mag. = 0.66): Baghdad

> ...The Moon rose at sunset (already) eclipsed by $\frac{1}{4}$ or (a little) more of the digits of the diameter (i.e. 3 digits or a little more). The Moon was eclipsed by (a little) more than 9 digits of diameter...

In order to determine limits to ΔT, it will be assumed that the Moon rose between first contact and mid-eclipse. However, the estimate of the fraction of the Moon obscured as it reached the horizon leads to a separate determination of ΔT.

RESULTS

(i) First contact before moonrise on Jun 1. LT of moonrise = 19.07 h. Hence LT of contact < 19.07 h, UT < 16.04 h. Computed TT = 16.74 h, thus $\Delta T > 2500$ sec.

(ii) Mid-eclipse after moonrise. LT of moonrise = 19.13 h. Hence LT of contact > 19.13 h, UT > 16.10 h. Computed TT = 18.12 h, thus $\Delta T < 7250$ sec.

Combining these limits yields $2500 < \Delta T < 7250$ sec.

(iii) If one assumes between 3 and 4 twelfths – a mean of 0.29 – and adjusts for the likely systematic error made by the unaided eye in assessing the magnitude of a lunar eclipse (see chapter 7), the estimated proportion of the Moon's diameter actually in shadow at moonrise is 0.22. For 0.22 of the lunar diameter to be obscured at moonrise (phase increasing), $\Delta T = 3450$ sec.

(2) AD 928 Aug 18 (solar, mag. = 0.25): Baghdad

> ...The Sun rose (already) eclipsed by less than one $\frac{1}{4}$ of its surface. The eclipse continued to increase by an amount that we could perceive until $\frac{1}{4}$ (of its surface) was eclipsed...

It may be inferred that the Sun rose at some time between first contact and maximal phase.

RESULTS

(i) First contact before sunrise on Aug 18. LT of sunrise = 5.44 h. Hence LT of contact < 5.44 h, UT < 2.50 h. Computed TT = 2.57 h, thus $\Delta T > 250$ sec.

(ii) Maximal phase after sunrise. Hence LT of contact > 5.44 h, UT > 2.50 h. Computed TT = 3.31 h, thus $\Delta T < 2900$ sec.

Combining these limits yields $250 < \Delta T < 2900$ sec.

(3) AD 979 May 14/15 (lunar, mag. = 0.70): Cairo

> ...The Moon rose eclipsed...More than 8 digits but less than 9 of the Moon's diameter were eclipsed...The eclipse cleared when about one equal hour and a fifth of night had elapsed, as I estimated....

As the horizon at Cairo is somewhat hilly towards the south-east, accurate limits to ΔT cannot be derived without considering the horizon profile in detail. However, making no allowance for this effect leads to such wide limits for ΔT ($-4100 < \Delta T < 5750$ sec) that attempts to refine this solution would not be worthwhile. The observation will thus not be considered further.

(4) AD 979 May 28 (solar, mag. = 0.45): Cairo

...The eclipse was perceptible when the altitude of the Sun was $6\frac{1}{2}$ deg. About $5\frac{1}{2}$ digits of the Sun's diameter were eclipsed, as I estimated, that is 4;10 digits of surface. The Sun set eclipsed...

The computed magnitude is very close to the observed value (0.46). The fact that the solar altitude was estimated as only 6.5 deg (in the west) at first contact suggests that this eclipse had not reached maximum phase when the Sun set. This altitude measurement leads to a very average figure for ΔT of 1450 sec (as deduced in section 13.7) and on this assumption (or the ΔT value of 1720 sec indicated by equation (8.1)) the computed depression of the Sun below the horizon at last contact would be about 12 deg. Hence it seems reasonable to assume that the Sun reached the western horizon at some time between first contact and maximal phase.

RESULTS

(i) First contact before sunset on May 28. LT of sunset = 18.95 h. Hence LT of contact < 18.95 h, UT < 16.80 h. Computed TT = 16.52 h, thus $\Delta T > -1000$ sec.

(ii) Maximum phase after sunset. LT of maximum > 18.95 h, UT > 16.80 h. Computed TT = 17.48 h, thus $\Delta T < 2450$ sec.

Combining these limits yields $-1000 < \Delta T < 2450$ sec.

NB the latter figure can be only approximate.

(5) AD 981 Apr 21/22 (lunar, mag. = 0.18): Cairo

...About one-quarter of the Moon's diameter was eclipsed. The Moon cleared completely when about $\frac{1}{4}$ of an hour remained to sunrise.

Sunrise (LT = 5.40 h) would be marginally later than moonset. Hence the observation that last contact occurred before the Moon reached the horizon (implicit in the text) yields a slightly more critical limit for ΔT.

RESULTS

Last contact before moonset on Apr 22. LT of moonset = 5.39 h. Hence LT of contact < 5.39 h, UT < 3.25 h. Computed TT = 3.67 h, thus $\Delta T > 1500$ sec.

Only a single (lower) limit to ΔT may be derived on this occasion.

(6) AD 986 Dec 18/19 (lunar, mag. = 0.91): Cairo

...The eclipse became noticeable when the altitude of the Moon was 24 deg in the west...About 10 digits of the Moon's diameter were eclipsed...The Moon set eclipsed.

Table 13.6 ΔT limits from eclipses observed by Arab astronomers.

Year	Type	Observation	ΔT Range (sec) LL	UL
+837	Sun	annular	1820	3750
+923	Moon	moonrise	2500	7250
+928	Sun	sunrise	250	2900
+979	Sun	sunset	−1000	2450
+981	Moon	moonset	1500	—
+986	Moon	moonset	−6450	5450
1004	Sun	partial	1940	1770

The text does not state clearly whether the eclipse reached a maximum before moonset or if the phase was still growing when the Moon reached the horizon. Moonset at some time between first and last contact will be assumed.

RESULTS

(i) First contact before moonset on Dec 19. LT of moonset = 6.92 h. Hence LT of contact < 6.92 h, UT < 4.87 h. Computed TT = 3.07 h, thus $\Delta T > -6450$ sec.

(ii) Last contact after moonset. LT of moonset = 7.01 h. Hence LT of contact > 7.01 h, UT > 4.97 h. Computed TT = 6.48 h, thus $\Delta T < 5450$ sec.

Combining these limits yields $-6450 < \Delta T < 5450$ sec.

The ΔT limits (lower = LL, upper = UL) obtained from the five selected eclipses are listed in table 13.6. This table also includes the ΔT ranges derived earlier (section 13.6) from the observation of an annular solar eclipse in AD 873 and a large partial eclipse in AD 1004.

The ΔT limits in table 13.6 lying between 0 and +4000 sec, along with the single ΔT value obtained from the estimate of phase at moonrise in AD 923 (ΔT = 3450 sec), are plotted in figure 13.8. This diagram also displays the few ΔT results obtained from solar and lunar eclipse timings recorded by al-Battani and al-Biruni (section 13.9).

13.12 Conclusion

With the ΔT results obtained in this chapter from observations by medieval Arab astronomers, we come to the end of our compilation of data. Apart from solar eclipse magnitudes – discussed in section 13.10 – the sets

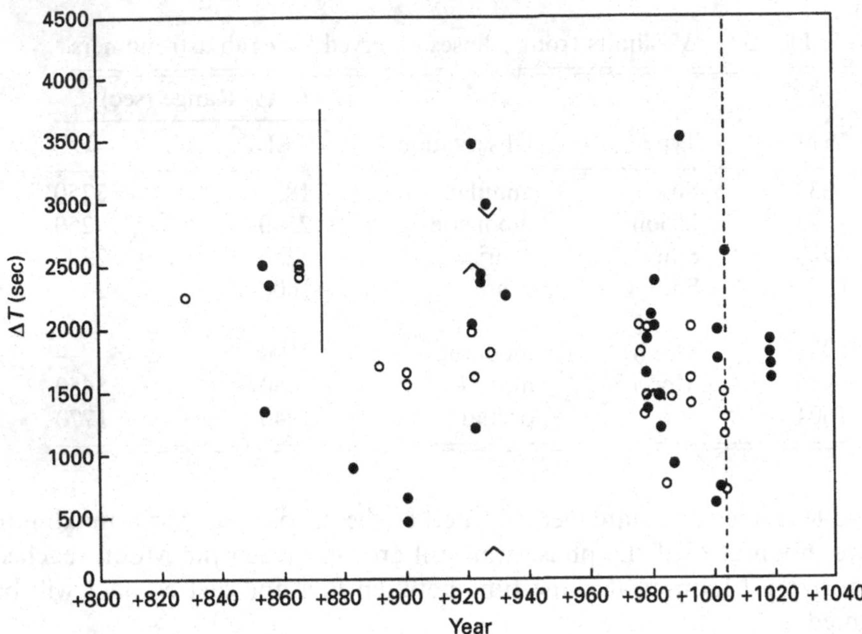

Fig. 13.9 ΔT values and limits derived from medieval Arab eclipse observations.

of both individual ΔT values and limits derived from this material are unusually important for the study of the Earth's past rotation. The various results in tables 13.2, 13.3, 13.4 and 13.6 – along with an isolated value for ΔT derived in section 13.11 from an estimate of phase at moonrise (i.e. 3450 sec at AD 923) are plotted in figure 13.9.

In this diagram, as previously, open circles represent solar eclipse timings and shaded circles lunar eclipse timings. Arrow heads denote upper and lower limits to ΔT derived from rising and setting events. The full vertical line indicates the range in ΔT determined from the annular eclipse of AD 873 and the broken line the two possible ranges obtained from the partial eclipse of AD 1004.

14
Determination of changes in the length of the day

14.1 Introduction

In chapters 4–13, more than 400 timed and untimed observations of both solar and lunar eclipses from the pre-telescopic period (date range −700 to +1600) have been investigated in detail. ΔT values or limits (depending on whether the observations were timed or untimed) have been derived in almost every case. The fundamental objective of the present chapter is to use these results to obtain the best-fitting ΔT curve to the data and hence to determine changes in the length of the day (LOD) over the historical period. In addition, the various geophysical mechanisms responsible for the observed variations will also be discussed. I am grateful to Dr L. V. Morrison of the Royal Greenwich Observatory for undertaking the data analysis which forms the basis of much of this chapter and producing several of the diagrams. This chapter is essentially an enlargement of section 5 of the paper by Stephenson and Morrison (1995). Although several tens of further historical data have been added (notably medieval Chinese timing – see chapter 9), the basic conclusions obtained in that paper remain unchanged.

All of the ΔT results derived in chapters 4–13 are summarised in tabular form in the Appendices for ready reference. For each individual observation, only the year (− or +) and appropriate ΔT value or limits are tabulated. Appendix A contains timed data and Appendix B untimed material, each being divided into sub-groups depending on the source and type of observation it contains. An outline of the contents of each sub-group is as follows:

A1. Babylonian timings of solar and lunar eclipses (intervals < 25 deg).

A2. Babylonian timings of solar eclipses (intervals > 25 deg).

A3. Babylonian timings of lunar eclipses (intervals > 25 deg or approximate measurements < 25 deg + estimates of phase on horizon).

A4. Chinese measurements of solar eclipse times to $\frac{1}{100}$ day (+ estimates of phase on horizon).

A5. Chinese measurements of lunar eclipse times to $\frac{1}{5}$ night-watch (+ estimates of phase on horizon).

A6. Chinese measurements of lunar eclipse times to $\frac{1}{100}$ day.

A7. Ancient Greek timings of lunar and solar eclipses.

A8. Arab timings of solar eclipses.

A9. Arab timings of lunar eclipses.

B1. Untimed total and annular solar eclipses.

B2. Untimed partial solar eclipses (ΔT limits reversed).

B3. Sunrise/sunset or moonrise/moonset observations.

Estimates of the degree of obscuration of the Sun or Moon when on the horizon are ranked with timed observations since they yield results of similar form). The total numbers of timed and untimed data listed are respectively about 305 and 105.

14.2 Timed and untimed observations

As these two groups form independent data sets, the ΔT values determined from them can be considered separately. The results in Appendix A are largely derived from individual timings by astronomers of the various stages of solar or lunar eclipses. Since each measurement leads to a specific value for ΔT (rather than a limit), the results can be treated statistically. In the analysis of this material, the ΔT values in each group will be assigned unit weight except for data in categories A1 and A8. The results derived from Babylonian records in which the measured time-interval was less than 25 deg (group A1) and also Arab timings of solar eclipses (A8) show an unusually high degree of self-consistency even when the observations were said to be only approximate. Hence they will be allotted double weight. Although the Arab observations originated from several independent sources, it seems best to regard them as a unit.

None of the data in Appendix B are based on measurement. Here the observers, many of whom were not astronomers, simply described what they saw – sometimes in graphic terms. Untimed observations set limits on ΔT. Unless the observations are questionable in any way, a viable ΔT curve should not infringe these limits. Nevertheless, the form of the results in each of the three categories B1, B2 and B3 have certain mutual differences. Data in group B1 define ranges (solution space) of ΔT anywhere within which the true value of this parameter is equally likely to lie. Most ΔT ranges in this class are fairly narrow, seldom exceeding 2000 sec in width. Furthermore, the limits are extremely sharp; a total solar eclipse, in particular, is a remarkably well-defined event. Observations in category B2 define the inverse of the solution space in category B1; in each case an extensive range of ΔT is indicated (several tens of thousands of arcsec in width) apart from a fairly narrow gap – up to about 2000 sec wide – in which the solution cannot lie; otherwise the eclipse would be fully total or annular instead of partial.

ΔT ranges in group B3 have similar properties to those in B1, but are much broader – typically between about 5000 and 10 000 sec in width. Because of this feature, many data in this category prove to be redundant. Additionally, some limits are rather poorly defined – especially when it is inferred that the Moon or Sun reached the horizon at mid-eclipse.

In figure 14.1 are plotted the ΔT ranges obtained from the untimed data, as listed in Appendix B. A different symbolism from that adopted in earlier chapters is used, as explained in the key to the diagram. Solution space derived from a total or annular solar eclipse (B1) is denoted by a heavy vertical line. In the case of a partial solar obscuration (B2), the solution space at that date is indicated by a pair of lines of medium width – separated by a gap within which the value of ΔT must not lie. Extending these lines to the upper and lower edges of the diagram tends to give them undue prominence. Hence beyond the excluded zone only short lines terminated by arrow heads are used; beyond the arrow heads any value of ΔT is still acceptable. Finally, solution space deduced from a rising or setting event (B3) is represented by a fine vertical line, with an arrow head where appropriate. These various symbols replace respectively the full lines, broken lines and isolated arrow heads of earlier diagrams. For most of the partial solar obscurations, the solution space on *one* side of the belt of totality or annularity proves to be redundant; it is obviated by other roughly contemporaneous observations. These redundant solutions are not shown in figure 14.1 in order to simplify the diagram. Certain solar observations which lead to extremely narrow forbidden ΔT ranges (in the years −187, −79, −1, +1333 and +1354) are entirely omitted, as are limits obtained from horizon observations which do not contribute useful constraints to the solution space for ΔT.

Figure 14.2 is similar to figure 14.1 but here the individual ΔT values obtained from the timed data (Appendix A) are shown. Once again a revised set of symbols is used, as defined in the key to the diagram; this distinguishes observations in the various categories. In particular, double-weighted results (from sub-groups A1 and A8 of Appendix A) are represented by heavy black symbols.

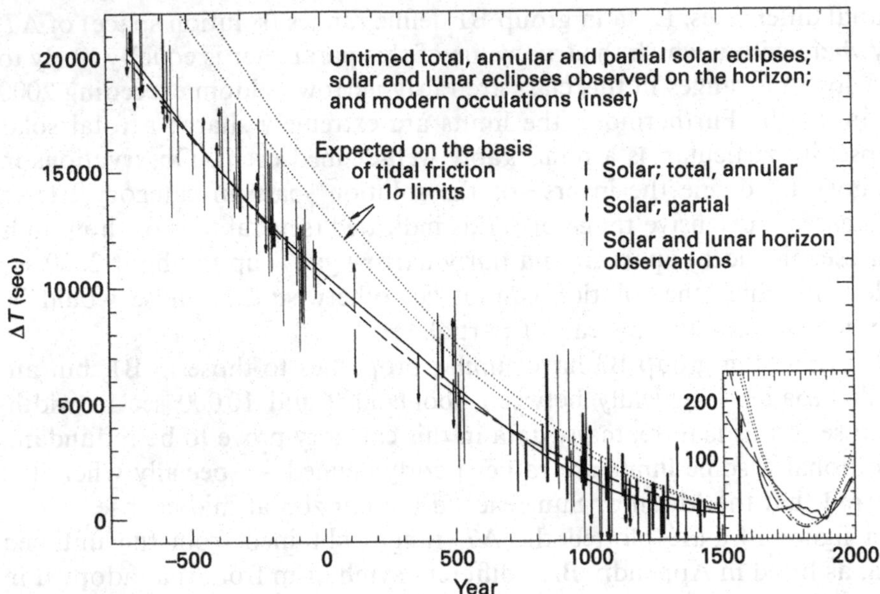

Fig. 14.1 ΔT ranges obtained from untimed total, annular and partial solar eclipses and also solar and lunar eclipses on the horizon: −800 to +1600. Also shown (inset) is the ΔT curve obtained from modern occultations. (Courtesy: Dr L. V. Morrison.)

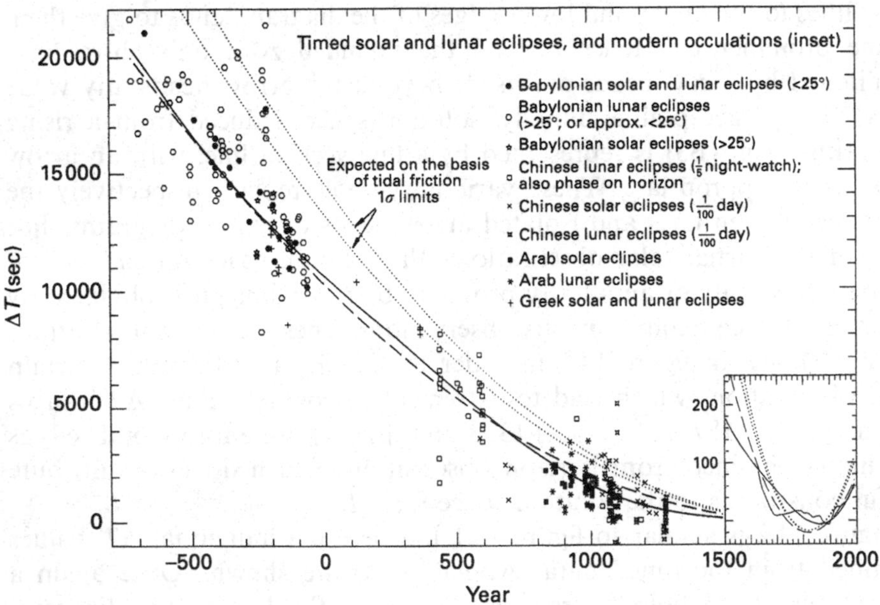

Fig. 14.2 ΔT values obtained from timed solar and lunar eclipses. Also shown (inset) is the ΔT curve obtained from modern occultations. (Courtesy: Dr L.V. Morrison.)

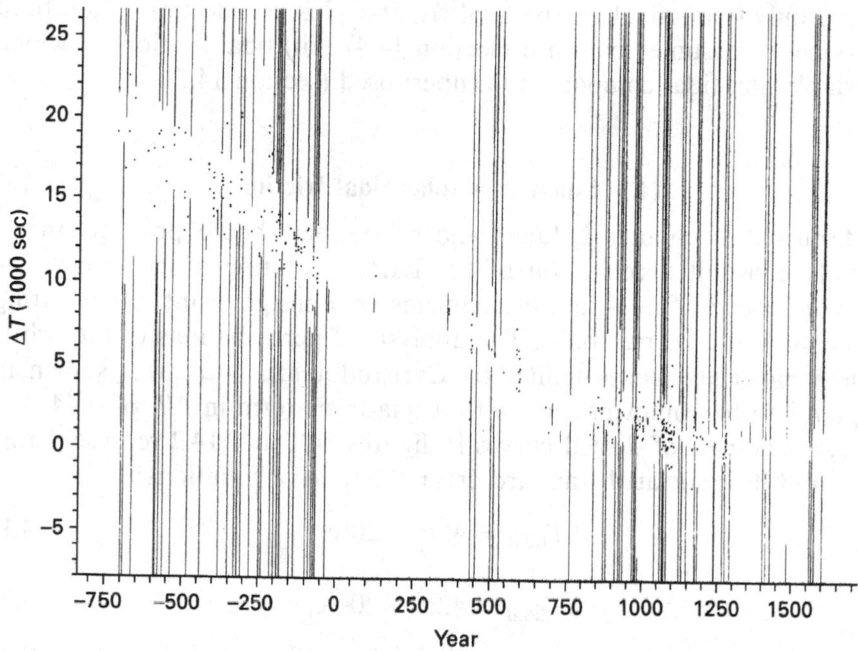

Fig. 14.3 ΔT ranges and individual values obtained from untimed and timed solar and lunar eclipses: −800 to +1600. In this diagram, each allowed ΔT range is indicated by a gap between two vertical lines; this convention in the opposite to that adopted in figures 14.1, 14.4 and 14.5. (Courtesy: Dr R. S. Roberts).

Also shown in both figure 14.1 and figure 14.2 (inset, on a much larger scale) is the ΔT curve since the year +1600, obtained mainly from telescopic observations of occultations (Stephenson and Morrison, 1984). This is depicted as an irregular full line. The nature of the four smooth curves will be discussed below.

It is obvious from comparison of figures 14.1 and 14.2 that the untimed and timed data define similar trends in ΔT over the past 2700 years. This is particularly clear from figure 14.3, which displays both the timed and untimed data. In this diagram, each individual ΔT value is represented by a dot of equal size. Each allowed ΔT range as indicated by a total or annular solar eclipse or a horizon observation is denoted by a gap between two vertical lines. In the case of a partial solar eclipse, almost the entire solution space is permitted except for that indicated by a short vertical line. This method of depicting ΔT limits was suggested to me by Dr R. S. Roberts of the University of Durham, to whom I am grateful for producing figure 14.3. The convention adopted is the reverse of that followed in figure 14.1.

In the following sections, the timed and untimed data will be used to assess the viability of the following assumptions: (i) lunar and solar tidal friction only (section 14.3); (ii) tidal friction plus an additional non-tidal component – assumed constant (section 14.4); (iii) tidal friction on which a variable non-tidal component is superposed (section 14.5).

14.3 Lunar and solar tidal friction

As discussed in chapter 2, lunar and solar tides should produce an essentially constant deceleration of the Earth's rotation over the last few millennia. Hence if non-tidal mechanisms were insignificant, the resulting ΔT curve would be parabolic. The analysis of perturbations of the orbits of near-Earth artificial satellites by Christodoulidis *et al.* was shown in chapter 2 to be commensurate with a quadratic term in ΔT of $+44 \pm 2$ sec/cy^2. The pair of dotted curves in figures 14.1 and 14.2 represent this result and its associated standard error. They have equations:

$$\Delta T_{\text{tidal}} = 46t^2 - 20\text{sec}, \tag{14.1}$$

$$\Delta T_{\text{tidal}} = 42t^2 - 20\text{sec}, \tag{14.2}$$

where t is measured in Julian centuries from the epoch +1820 (rather than +1800, as defined earlier). This is the approximate date at which the average LOD is equal to the standard LOD on the TT scale (i.e. 86 400 SI sec), as discussed in chapter 1. The constant (–20 sec) is included to obtain a satisfactory fit to the telescopic curve shown in the insets to figures 14.1 and 14.2; over the pre-telescopic period, this term is, of course, negligible.

It is clear from figures 14.1 and 14.2 that the curve of best fit to the pre-telescopic data (whether timed or untimed) lies far below the tidal range; the discrepancy in ΔT amounts to about one hour by the epoch +150 and two hours by –550. There is thus evidence of marked non-tidal behaviour. The existence of this was first clearly shown by Stephenson and Morrison (1984) from the analysis of smaller data sets than those displayed in figures 14.1 and 14.2.

14.4 Constant non-tidal component

A much better approximation to the observational data in the pre-telescopic period is provided by a solution which incorporates a quadratic non-tidal term in ΔT in addition to the tidal parabola. On the (provisional) assumption that the rate of increase in the LOD – both tidal and non-tidal – has been constant over the past 2700 years, the parabola represented by a broken line in figures 14.1 and 14.2 has been chosen to satisfy the greatest number of boundary conditions set by the eclipse limits. This parabola has the equation:

$$\Delta T = 31t^2 - 20\text{sec.} \tag{14.3}$$

Although the parabola is a remarkably good fit to most of the data (including the telescopic observations) there are important discrepancies – especially in the period from +700 to +1400. During this interval, the eclipses of +761, +1133, +1221 and +1267 were all clearly described as total and that of +1147 was recorded as annular. Yet in each case the ΔT values yielded by equation (14.3) render the phase as only partial at the appropriate place of observation. Indeed the observational limits to ΔT all lie substantially below the parabola, as shown in figure 14.1. The discrepancies (in chronological order) are −170, −290, −190, −110 and −110 sec. There are also three marked discrepancies among the results obtained from partial eclipses: in +1178, +1330 and +1361. In each case, use of equation (14.3) would indicate a total eclipse but the record specifically denies this. Furthermore, the ΔT discrepancies (respectively −120, −170 and −150 sec) are of the same sign as for the central eclipses listed above.

Strong evidence for a systematic discrepancy between a parabolic representation and the results of observation during the same interval is provided by two independent sets of timed data. These are derived from Arab and Chinese measurements in the approximate date range from +800 to +1300. In this interval more than 80 per cent of all data points lie below the parabola (see figure 14.2). The situation is more clearly shown in figure 14.4, which depicts on a larger scale than in figure 14.2 both the timed and untimed data between the years +400 and +1600.

Around the epoch +500, there is a discrepancy between observation and the mean parabolic fit (equation (14.3)) of opposite sign to that noted above. Amounting to some +350 sec, this is mainly demonstrated by the total solar eclipse of +454; a value for ΔT of at least 6130 sec would be required to render this eclipse total at the Chinese capital of Chien-k'ang, where, on account of the darkness 'all the constellations were brightly lit' (see chapter 8). A figure for ΔT of only 5780 sec is yielded by equation (14.3) at this date.

Near this same date, the Greek and Chinese timed observations (covering the period from +364 to +596 indicate a solution above the parabola, most ΔT values lying above +5500 sec. Since there are only a few observations and the scatter in ΔT is rather large, the evidence is somewhat weak. However, it should be mentioned that the two low data points (ΔT = +1750 and +2800 sec) are obtained from two phases of the same lunar eclipse – that of +434 – and are thus are not independent. Both results are about an hour less than the mean for that date derived from the remainder of the observations in this group. Possibly on this occasion the clepsydra was malfunctioning or careless readings were taken.

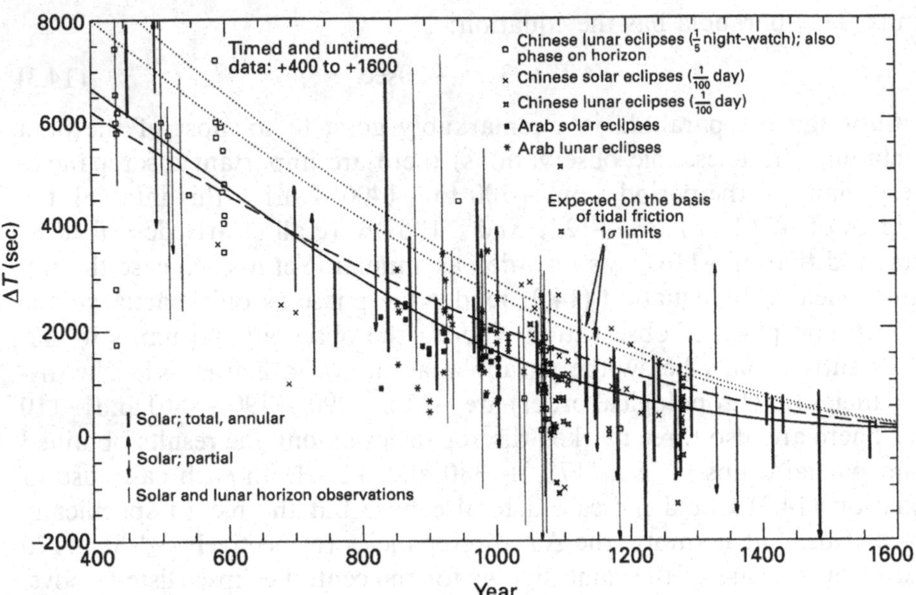

Fig. 14.4 Combined timed and untimed data for the period from +400 to +1600. (Courtesy: Dr L. V. Morrison.)

The accord between the parabola represented by equation (14.3) and the observational results is closer in ancient times, but fewer ΔT limits are critical. Figure 14.5 displays the limits set by the untimed data over the period from −800 to +500 on an enhanced scale.

By choosing a different origin for the cusp of the parabola and varying the quadratic term, it is possible to obtain a better fit to the set of pre-telescopic data. However, this is at the expense of the modern observations. Stephenson and Morrison (1995) found that the following parabola fitted the ancient and medieval observations more closely than equation (14.3):

$$\Delta T = 35\tau^2 - 20\text{sec},$$

with τ measured in centuries from a cusp at the epoch +1735. However, they remarked that this parabola does not 'remotely resemble the ΔT curve between +1600 and the present'. In addition, it does not meet the condition that the cusp lie near the beginning of the nineteenth century, in keeping with the requirement that the average LOD at that epoch should be equal to the standard LOD on the TT scale – see chapter 1.

It is thus apparent that no single parabola can satisfactorily represent both the pre-telescopic and telescopic data, implying that the non-tidal component of the Earth's spin is variable with time. The degree of variability will be investigated in the following section by fitting a more flexible curve than a parabola to the data.

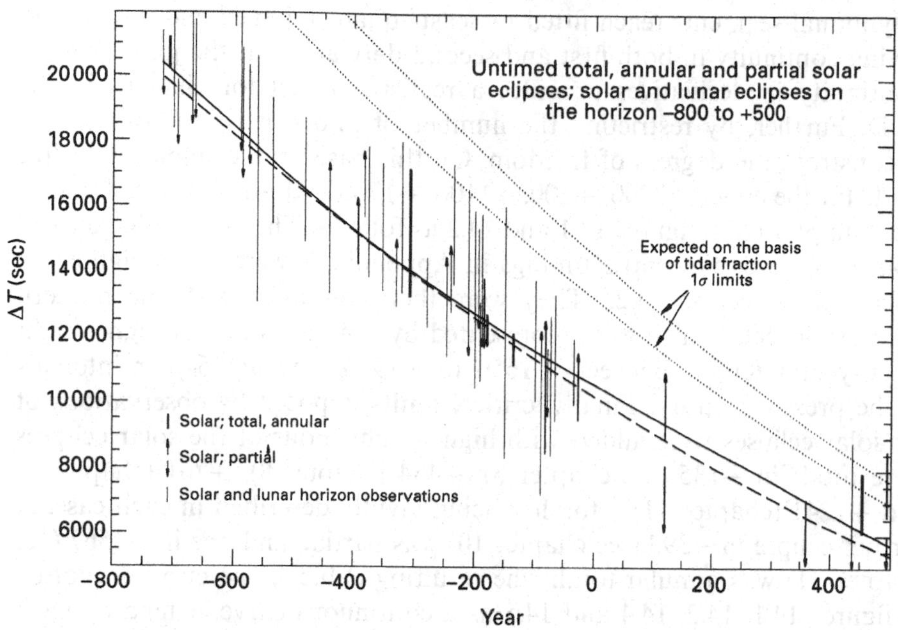

Fig. 14.5 ΔT limits obtained from untimed total, annular and partial solar eclipses and also solar and lunar eclipses on the horizon: −800 to +700. (Courtesy: Dr L. V. Morrison.)

14.5 Variable non-tidal component

Before attempting to obtain a better fit to both the pre-telescopic and telescopic results, the following limitations should be noted. In particular, it is clear from figure 14.3 that the temporal distribution of the pre-telescopic data is far from uniform. In summary, there are many observations (both timed and untimed) between −600 and −50, scarcely any between −50 and +400, a small but useful number of data from +400 to +800, and many from +800 to +1300. Between +1300 and +1600 – a period when good data would have been especially helpful in extending the well-defined telescopic curve backwards in time – there is only a mediocre selection of observations (all untimed). Further, in the whole of the period back to −500, the parabola represented by equation (14.3) does not deviate by more than about 500 sec from observation. As a result of these circumstances, it is only feasible to use the pre-telescopic material to investigate minor departures from a parabola on a time-scale of several centuries to millennia. Hence in fitting a more flexible curve than a parabola to the data, it should be tightly constrained to avoid introducing spurious short-term fluctuations.

Under these circumstances, cubic splines provide the most effective method of curve-fitting. Cubic splines consist of a number of cubic polynomial segments (each fitted by least-squares), joined end to end and having continuity in both first and second derivatives at the joins (knots). Continuity in derivatives provides a realistic model for changes in the LOD. Further, by restricting the number of knots, it is readily possible to constrain the degrees of freedom. On this basis, cubic splines with five knots (at the epochs −200, +300, +1100, +1700 and +1990) were fitted to the data plotted in figures 14.1 and 14.2 as follows. The ΔT results derived from pre-telescopic eclipse timings in Appendix A were first weighted as discussed in section 14.2. They were then combined with the modern occultation data which were represented by normal points of high weight at 50-year intervals between +1650 to +1900, and at 25-year intervals in the present century. Finally, critical limits imposed by observations of six solar eclipses were added with high weight. Four of the solar eclipses were total: in −135 (see chapter 5), +454 (chapter 8), +761 (chapter 8) and +1241 (chapter 11) – totality being vividly described in each case. A further eclipse in −393 (see chapter 10) was partial, and one in +1567 (see chapter 11) was annular-total. The resulting cubic spline curve is plotted in figures 14.1, 14.2, 14.4 and 14.5 as a continuous curve, extending back to the epoch −500.

The total eclipse in +454 and an observation of a partial eclipse in +360 set close limits to the form of the ΔT curve at this epoch. The spline cannot pass much above the lower boundary of +454 without infringing the upper limit in +360. The account of the eclipse of +360 asserts that at maximum phase the Sun had the appearance of a hook (see chapter 8). Hence it may be concluded that the observers did not witness the annular phase but only saw a large partial eclipse.

Despite the small number of critical limits, the spline curve since −500 satisfies the numerous constraints imposed by all the untimed solar and lunar eclipses with only three exceptions: horizon observations of lunar eclipses in −382, −239 and +1067. In −382, only a Greek transcription of a Babylonian record is available (see chapter 4). This may have become corrupted. In −239, Babylonian astronomers recorded a lunar eclipse which began 'at 3 deg (i.e. 12 min) before sunrise,' adding that 'it set eclipsed' (see chapter 7). For any value of ΔT less than 13 650 sec, the Moon would set at Babylon before the eclipse began, but the spline fit passes some 300 sec below this limit. Possible explanations are confusion of the deep penumbral shadow with the umbra or anomalous refraction. In +1067 Chinese astronomers recorded a lunar eclipse of 'more than 8 divisions. It set eclipsed...' (see chapter 9). As this eclipse was, in fact, total, it would appear that as seen from the capital (Pien) the Moon set before totality. The indicated range of $\Delta T < +750$ sec corresponds to the Moon setting just before the total phase. This result is in dispute with the roughly contemporaneous lower limits set by the well-observed total solar eclipses of +1061 (see chapter 12) and +1124 (chapter 11), and would severely distort the form of the ΔT curve around this date. Possibly cloud interfered with observation, although this is not mentioned in the record. It should be stressed that the limits set by horizon observations are not as reliable as for central and partial solar eclipses.

14.6 Further discussion of timed observations

In figures 14.6a and b are plotted histograms showing the residuals from the spline curve for the Arab timings (date range +829 to +1019) and the Chinese timings (+1040 to +1280). In figure 14.6b, five highly discordant Chinese results are not represented. A bin of width 100 sec is used in each diagram. The arrows denote the mean position of the parabola $31t^2 - 20$ sec (equation (14.3)). Clearly, the distribution of the Arab results (standard deviation $\sigma = 550$ sec) is significantly narrower than for the Chinese data ($\sigma = 700$ sec). However, the parabola is at variance with both datasets by almost identical amounts (−500 sec: in the sense observed–computed).

In principle, this discrepancy of 500 sec could be explained by a systematic delay by this amount in the Arab and Chinese timings of both solar and lunar eclipses. Nevertheless it seems most unlikely that it arises from a data artefact. The Arab and Chinese astronomers employed quite different techniques for measuring time (respectively altitude determinations and water clocks). Some delay in detecting the start of an eclipse might be expected owing to the limited resolution of the unaided eye, although since both sets of astronomers made approximate eclipse predictions, they would be able to keep a watch for these events (see chapters 9 and 13). In 500 sec, the depth of the indentation at the solar or lunar limb increases by up to 4 arcmin – far greater than the unaided eye limit (typically 1 arcmin). In addition, at last contact the opposite effect would be expected. Since the observations from both Arab and Chinese sources are roughly equally distributed between first contact, last contact and other phases (principally mid-eclipse), any residual bias should be minimal. A real variation in the Earth's rate of rotation is thus indicated.

Most of the Arab altitude measurements (whether of the Sun, Moon or a reference star) were made to the nearest degree. Except when a celestial body is near the meridian, at mid-latitudes its altitude changes by this amount in about 6 min. For the solar eclipses in figure 14.6a, the standard deviation is 6 min, so that these measurements are fairly accurate. However, in the case of lunar eclipses σ is as much as 11 min. In addition to the poorer definition of the contacts for a lunar eclipse, the fact that scales would then have to be read at night may have contributed to this significant discrepancy. In the case of the Chinese measurements (figure 14.6b), $\sigma = 12$ min (ignoring the five severely deviant results mentioned

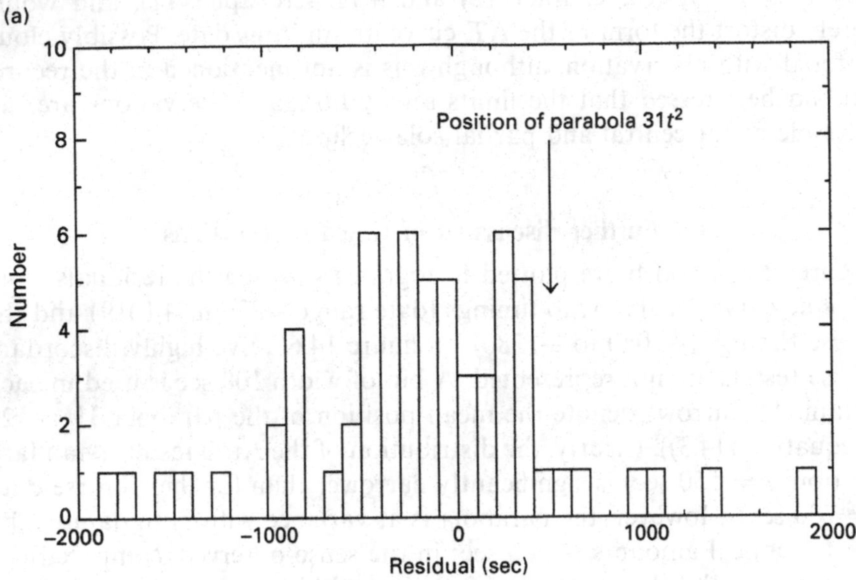

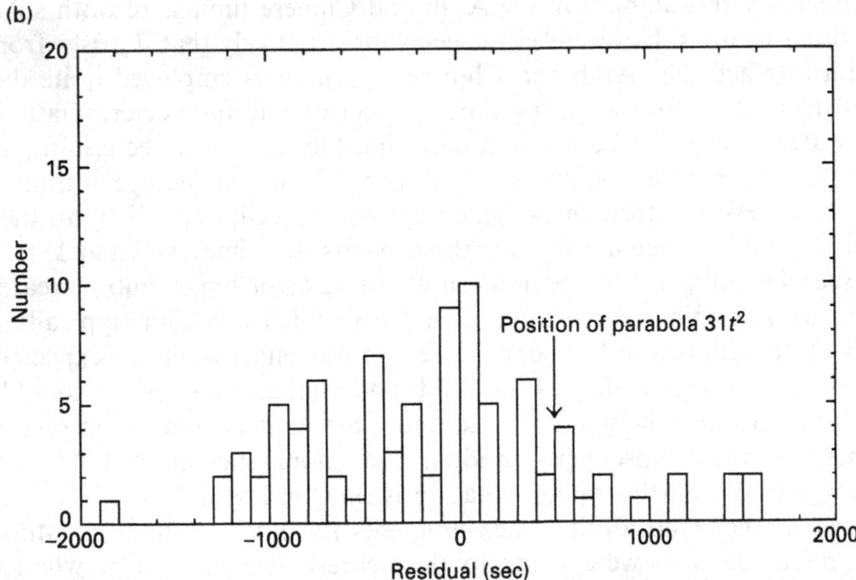

Fig. 14.6 Histograms of the residuals of the medieval timed data from the spline curve: (a) Arab observations from +829 to +1019; (b) Chinese observations from +1046 to +1280. The arrow indicates the mean position of the best-fitting parabola $31t^2$. (Courtesy: Dr L. V. Morrison.)

above). This represents tolerable precision since times were only estimated to the nearest 15 to 30 min.

From the frequency distribution of the residuals for all the Babylonian timings of eclipses (not shown diagrammatically), the estimate for the standard deviation is about 30 min. However, in the case of the restricted set of lunar eclipses which occurred within 25 deg (1 h 40 min) of sunset/sunrise (shown by black dots in figure 14.2), the standard deviation is much smaller: 13 min, after the rejection of one outlier. These various measurements were probably made made with a clepsydra, which although presumably showing a considerable drift over several hours might be expected to be more precise in measuring shorter intervals of time. All but the very earliest lunar eclipses were timed to the nearest degree (4 min) – see chapter 4 – so that the real accuracy of even the best measurements was relatively poor. Nevertheless, the precision attained by the Babylonian astronomers was not much inferior to that achieved more than 1000 years later by the Arabs and Chinese.

14.7 Changes in the LOD

By taking the first derivative along the cubic spline in figures 14.1 and 14.2, changes in the LOD (ΔLOD) relative to the standard reference day of 86 400 SI sec) may be derived. The resulting LOD variation is represented by a continuous curve in figure 14.7. The irregular decade fluctuations in the period from 1830 to the present are also shown; these are taken from Jordi *et al.* (1994). Similar variations have been detected using earlier telescopic observations (Stephenson and Morrison, 1984) but because of the lower precision of measurement these are less well defined. It may be presumed that decade fluctuations on a similar scale occurred throughout the entire period covered by the historical data but the integrated effect of these variations (i.e. their contribution to ΔT) is too small to be detectable using observations made with the unaided eye.

In figure 14.7 are also shown the straight lines corresponding to the effect of tidal friction alone (represented by a dotted line) and the mean observed long-term trend in the LOD (broken line). The rate of increase in the LOD owing to lunar and solar tides is $+2.3 \pm 0.1$ milliseconds per century (ms/cy). As noted in chapter 1, there is no evidence to suggest that tidal dissipation has altered significantly over the past 2700 years. The observed mean rate of increase in the LOD (corresponding to the parabolic fit to ΔT in figures 14.1 and 14.2) is substantially less: +1.7 ms/cy. In order to obtain an indication of the uncertainty in this latter result, comparison may be made between the spline curve and the mean parabola in figures 14.1 and 14.2. In the whole of the period covered by reliable data (since −500), the two curves do not deviate from one

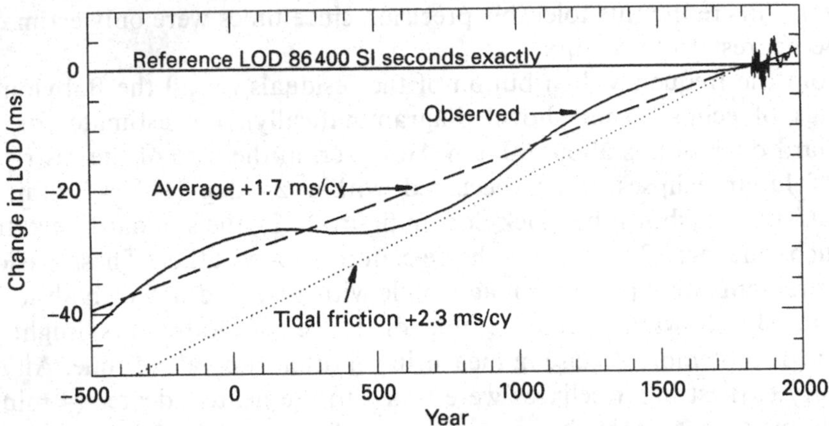

Fig. 14.7 Changes in the length of the mean solar day (LOD) from −500 to +1990 obtained by taking the first derivative along the spline curves shown in figures 14.1, 14.2, 14.4 and 14.5. (Courtesy: Dr L. V. Morrison). The decade fluctuations in the LOD from +1830 to +1990 are taken from Jordi *et al.* (1994).

another by more than about 500 sec, the mean deviation being about half of this amount. Taking the average value of t^2 over this date range of 250 cy^2, the estimated error in the coefficient of t^2 is 0.10 ms/cy^2. The corresponding uncertainty in the rate of change in the LOD is 0.05 ms/cy, a result of $+1.70 \pm 0.05$ (ms/cy).

Comparison of the tidal and observed mean rates of increase in the LOD indicates that in opposition to the tides there is an accelerative component producing an average change in the LOD of -0.6 ± 0.1 ms/cy. On this residual non-tidal term are superposed oscillations of amplitude ~4 ms and periodicity ~1500 years. These oscillations are of similar amplitude to the decade fluctuations in the LOD observed in recent centuries, although of much longer period. The scale of the long-term variations is such that for several centuries around the epoch +300 there was negligible mean change in the LOD (the tidal term was effectively annulled) while at other periods (around −500 and +1000) the average rate of increase in the LOD was roughly twice the tidal figure.

In table 14.1 are shown side by side the ΔT and ΔLOD (relative to the reference LOD) values at 50-year intervals from the epoch −500 to +1600. (A lunar acceleration of −26 arcsec/cy^2 is assumed.) Before −500 the data are too sparse to derive reliable results for both parameters. More precise values since +1600 are tabulated by Stephenson and Morrison (1984). It is hoped that the ΔT values in this table will be of interest to historians of astronomy as well as geophysicists. An accurate knowledge of ΔT is, of course, essential in the calculation of historical eclipses.

Table 14.1 ΔT and ΔLOD values from the spline curve in figures 14.1, 14.2, etc. (NB lunar accel. = $-26''/\text{cy}^2$.)

Year	ΔT (s)	ΔLOD (ms)
−500	16 800	−42
−450	16 000	−40
−400	15 300	−38
−350	14 600	−36
−300	14 000	−35
−250	13 400	−33
−200	12 800	−32
−150	12 200	−30
−100	11 600	−29
−50	11 100	−29
0	10 600	−28
+50	10 100	−27
+100	9 600	−27
+150	9 100	−27
+200	8 600	−26
+250	8 200	−27
+300	7 700	−27
+350	7 200	−27
+400	6 700	−27
+450	6 200	−27
+500	5 700	−27
+550	5 200	−27
+600	4 700	−26
+650	4 300	−25
+700	3 800	−24
+750	3 400	−23
+800	3 000	−22
+850	2 600	−20
+900	2 200	−19
+950	1 900	−17
1000	1 600	−15
1050	1 350	−14
1100	1 100	−12
1150	900	−10
1200	750	−9

Table 14.1 *continued.*

Year	ΔT (s)	ΔLOD (ms)
1250	600	–7
1300	470	–6
1350	380	–5
1400	300	–4
1450	230	–3
1500	180	–3
1550	140	–2
1600	110	–2

14.8 Geophysical discussion

It has been shown above that over the past 2500 years there is a variable non-tidal decrease in the LOD at an average rate of –0.6 ms/cy in opposition to the steady tidal increase of +2.3 ms/cy. The non-tidal acceleration and the fluctuations which occur on a time-scale of 1500 years may have separate geophysical origins. The most plausible cause of the former would appear to be post-glacial uplift. Following the deglaciation after the last ice-age, there has been a gradual decrease in the terrestrial oblateness owing to viscous rebound of the solid Earth. Artificial satellite measurements of the present-day rate of change in the Earth's zonal harmonic J_2 lead to similar results to that obtained above for the mean non-tidal rate of decrease in the LOD. By analysing the acceleration of the node of the orbit of the Starlette satellite, Cheng *et al.* (1989) obtained a value for $\dot{J}_2$ of $-(2.5 \pm 0.3) \times 10^{-11}$/y, corresponding to a rate of decrease in the LOD of -0.44 ± 0.05 ms/cy. This result was fairly close to that obtained previously by Yoder *et al.* (1983) based on orbital studies of the Lageos satellite: equivalent to –0.6 ms/cy.

If reliable historical data had extended much further back in time, it might have proved possible to detect a gradual decrease in $\dot{J}_2$ – as expected from the theory of post-glacial rebound. However, the time-scale of 2700 years is too short to achieve this objective, bearing in mind the mediocre quality of the early observational data.

With regard to the quasi-periodic fluctuations on a time-scale of some 1500 years, a possible mechanism would appear to be electromagnetic coupling between the core and mantle of the Earth. This is the most likely cause of the decade fluctuations (Lambeck, 1980, p. 247). Since the observed peak angular acceleration – relative to the mean non-tidal term – is $\sim \pm 2$ ms/cy ($\sim \pm 5 \times 10^{-22}$ rad/sec^2), it is readily calculated that the required torques operating on the mantle are $\sim \pm 4 \times 10^{16}$ Nm. The decade fluctuations observed over the last few centuries require torques an order of magnitude greater – but, of course, for a much shorter period of time. The possibility of appreciable changes in the moment of inertia of the Earth caused by global sea-level variations cannot be ruled out. Significant long-term alterations in climate have been detected in the last few millennia (Lamb, 1982). However, at present the associated changes in the LOD cannot be quantified owing to the lack of reliable sea-level measurements on a global scale (Pirazolli, 1991).

The investigation of long-term changes in the rate of rotation of the Earth using historical observations thus raises a number of geophysical issues which await further development.

14.9 Concluding remarks

The study of Earth's past rotation is just one of several modern scientific disciplines in which ancient and medieval astronomical observations play a major role. It is particularly satisfying that observations from a wide variety of early cultures made using different techniques can be integrated into a viable entity.

Appendix A Timed data

Table A1 Babylonian timings of solar and lunar eclipses (intervals < 25 deg).

Year	ΔT (sec)
–665	21 050
–536	18 800
–482	17 300
–420	15 500
–407	15 250
–406	16 200
–405	16 500
–352	14 550
–352	15 250
–321	14 150
–316	15 550
–307	14 100
–280	12 950
–239	14 200
–214	17 750
–211	11 800
–193	13 650
–188	11 250
–169	12 300
–142	12 550
–135	12 600
–133	10 950
–128	12 600
–108	12 050
–66	10 150

Table A2 Babylonian timings of solar eclipses (intervals > 25 deg).

Year	ΔT (sec)
–253	11 450
–253	11 500
–253	11 650
–248	13 850
–189	12 900
–189	13 400
–189	14 150
–169	12 200
–135	12 100
–135	12 250

Table A3 Babylonian timings of lunar eclipses (intervals > 25 deg or approximate measurements < 25 deg + estimates of phase on horizon).

Year	ΔT (sec)
−720	21 650
−719	19 000
−694	16 750
−685	22 500
−684	19 000
−666	17 250
−620	17 800
−600	18 000
−598	14 400
−586	18 850
−579	19 050
−572	18 050
−561	15 950
−536	18 200
−522	19 250
−501	18 950
−500	14 950
−500	15 300
−500	14 950
−500	13 000
−490	13 800
−423	16 700
−423	17 000
−423	17 050
−407	14 300
−406	14 950
−406	13 850
−396	14 150
−382	17 750
−381	16 400

Table A3 *–continued*

Year	ΔT (sec)
−381	16 000
−377	16 000
−377	16 200
−377	15 750
−377	14 950
−370	12 750
−366	19 550
−352	15 950
−352	14 450
−316	16 550
−316	16 100
−316	19 950
−316	20 200
−238	8 250
−225	17 550
−225	17 550
−225	18 950
−225	19 350
−225	15 450
−214	16 500
−214	17 650
−214	16 850
−214	14 150
−211	21 350
−193	14 000
−189	12 700
−188	10 750
−162	9 700
−159	14 050
−153	13 150

Table A3 *–continued*

Year	ΔT (sec)
–142	12 900
–142	12 750
–135	11 800
–128	11 850
–128	12 050
–119	12 500
–119	11 700
–119	10 900
–119	11 950
–98	13 050
–95	13 150
–79	10 350
–79	11 450
–79	10 800
–79	11 300
–79	12 050
–66	11 400
–65	9 900
–65	10 450

Table A4 Chinese measurements of solar eclipse times to $\frac{1}{100}$ day + estimates of phase on horizon.

Year	ΔT (sec)
+586	3 650
+680	2 500
+691	1 000
+702	2 350
+937	2 100
1040	2 850
1042	1 200
1052	2 900
1054	1 900
1059	2 400
1066	1 700
1068	1 600
1068	1 200
1068	850
1069	500
1080	0
1094	1 500
1094	2 650
1100	–1 000
1107	1 050
1107	1 400
1107	1 150
1173	900
1173	2 000
1173	1 400
1183	800
1195	850
1202	700
1202	1 700
1216	400
1243	350
1245	1 350
1277	300
1277	500
1277	600

Table A5 Chinese measurements of lunar eclipse times to $\frac{1}{5}$ night-watch (+ estimates of phase on horizon).

Year	ΔT (sec)
+434	1 750
+434	2 800
+437	7 450
+437	6 000
+437	6 550
+438	5 800
+440	6 200
+440	8 100
+503	6 000
+513	5 750
+543	4 600
+585	6 000
+585	5 700
+585	7 200
+592	5 900
+593	4 650
+595	5 450
+595	4 050
+595	3 500
+596	4 200
+596	4 800
+948	4 450
1042	700
1067	100
1068	950
1071	1 300
1073	1 850
1074	1 600
1074	500
1081	100
1168	−2 300
1185	100

Table A6 Chinese measurements of lunar eclipse times to $\frac{1}{100}$ day.

Year	ΔT (sec)
1052	2 100
1069	2 650
1069	1 750
1069	800
1071	500
1071	300
1071	950
1073	900
1073	1 400
1073	800
1078	300
1078	1 200
1081	1 350
1082	1 650
1082	1 150
1082	1 200
1085	150
1085	50
1085	200
1088	650
1088	−900
1088	150
1089	1 100
1092	700
1092	−1 100
1092	1 050
1097	1 050
1099	5 100
1099	3 850
1099	650

Table A6 *–continued*

Year	ΔT (sec)
1099	1 150
1099	700
1099	250
1106	200
1106	950
1168	0
1168	100
1270	−1 300
1270	−450
1270	−200
1272	−200
1272	−750
1272	−750
1277	650
1277	−50
1277	900
1277	200
1279	450
1279	50
1279	−400
1279	700
1279	900
1279	550
1280	1 050

Table A7 Ancient Greek timings of lunar and solar eclipses.

Year	ΔT (sec)
−200	12 400
−199	11 800
−199	12 650
−199	13 550
−173	11 000
−173	10 750
−140	8 550
+125	10 350
+364	8 100
+364	8 300
+364	8 400

Table A8 Arab timings of solar eclipses.

Year	ΔT (sec)
+829	900
+829	2 250
+866	2 200
+866	2 500
+866	2 450
+891	1 700
+901	1 650
+901	1 550
+923	1 950
+923	1 600
+928	1 800
+977	1 800
+977	2 000
+978	1 300
+978	2 000
+979	1 450
+985	1 500
+985	750
+993	2 000
+993	1 400
+993	1 600
1004	1 450
1004	1 300
1004	700
1004	1 150

Table A9 Arab timings of lunar eclipses.

Year	ΔT (sec)
+854	3 150
+854	2 500
+854	1 350
+856	2 350
+883	900
+901	650
+901	450
+923	1 200
+923	2 000
+923	3 450
+925	2 400
+925	2 350
+927	2 950
+933	2 250
+979	1 650
+979	1 900
+979	1 300
+980	2 100
+981	2 000
+981	2 350
+981	1 500
+983	1 200
+986	800
+990	3 500
1001	600
1002	1 750
1002	1 950
1003	1 450
1003	700
1004	2 600
1019	1 900
1019	1 700
1019	1 800
1019	1 600

Appendix B Untimed data

Table B1 Untimed total and annular solar eclipses.

Year	ΔT range (sec) LL	UL
−708	20 230	21 170
−309	13 300	17 200
−180	11 800	12 720
−135	11 210	12 140
+454	6 130	7 900
+761	1 720	3 290
+840	1 610	6 800
+873	1 820	3 750
+912	880	2 600
+968	1 580	2 600
+975	1 230	4 480
1061	800	2 140
1124	960	2 700
1133	540	1 150
1147	300	1 190
1176	600	1 600
1185	−2 200	10 500
1221	—	980
1239	−500	1 450
1241	630	1 400
1267	−850	820
1275	−700	1 300
1292	−50	1 830
1406	170	740
1415	−650	670
1431	10	700
1485	−5 500	780
1560	−475	205
1567	145	165
1575	−2 150	1 390
1601	−1 460	3 170

Table B2 Untimed partial solar eclipses (ΔT limits reversed).

Year	ΔT range (sec) LL	UL
−430	12 650	11 800
−393	14 250	12 710
−187	14 140	13 830
−79	8 520	8 200
−27	9 530	8 090
−1	2 150	2 050
+120	8 970	8 150
+360	9 400	7 120
+494	6 600	5 980
+702	2 760	1 460
+729	1 190	420
+822	—	4 020
1004	1 940	1 770
1135	3 810	1 840
1147	2 870	1 680
1178	2 320	1 140
1330	1 210	890
1333	4 250	4 150
1354	4 320	4 200
1361	1 760	500
1605	—	1 060

Table B3 Sunrise/sunset or moonrise/moonset observations.

Year	ΔT range (sec) LL	UL
−719	—	21 400
−701	18 350	25 000
−694	9 700	19 850
−666	11 400	23 500
−665	—	21 750
−590	12 550	24 700
−587	11 000	20 850
−576	8 250	20 300
−562	13 050	20 600
−536	15 450	19 300
−482	14 900	18 700
−441	13 350	24 650
−382	15 700	21 450
−381	—	17 850
−352	13 450	17 350
−330	13 000	25 050
−321	13 400	18 000
−280	12 350	14 450
−247	11 450	23 100
−240	12 800	—
−239	13 650	17 300
−211	6 800	16 550
−200	10 500	15 900
−194	—	15 350
−189	10 700	13 100
−188	6 900	15 800
−149	8 350	16 050
−98	10 100	13 800
−88	7 250	—
−79	6 850	11 700

Table B3 *–continued*

Year	ΔT range (sec) LL	UL
−66	4150	15350
−65	9500	13350
−34	10000	12000
+438	4800	8850
+503	5000	9800
+513	2250	6300
+514	−250	5500
+532	2400	6850
+882	1250	7000
+937	500	3500
+923	2500	7250
+928	250	2900
+979	−1000	2450
+981	1500	—
+986	−6450	5450
1042	−850	2050
1067	−3150	750
1068	−200	3400
1071	−1150	2600
1073	−1850	3100
1081	−850	2750
1178	−5650	3450

References

Aaboe, A., Britton, J. P., Henderson, J. A., Neugebauer, O. and Sachs, A. J., 1991. *Trans. Amer. Phil. Soc.*, **81**, No. 6, 1–75.

Adams, J. C., 1853. *Phil. Trans. R. Soc. Lond.*, **143**, 397–406.

Adams, J. C., 1859. *Comptes Rendus des Seances de l'Academie des Sciences*, **48**, 247–8.

Airy, G., 1853. *Phil. Trans. R. Soc. Lond.*, **143**, 179–200.

Airy, G., 1857. *Mem. R. Astr. Soc.*, **26**, 131–49.

Ali, J., 1967, *The Determination of the Co-ordinates of Positions for the Correction of Distances between Cities*. Beirut.

Allen, C. W., 1976. *Astrophysical Quantities*. London.

Arnold, T., 1885. *Symeonis Monachi Opera Omnia*. Longman and Co., London.

Aston, W. G., 1972. *Nihongi: Chronicles of Japan from the Earliest Times to AD 697*. Tokyo.

Barron, J. P. and Easterling, P. E., 1985. In *The Cambridge History of Classical Literature*, vol. I: *Greek Literature* (ed. by P. E. Easterling and B. M. W. Knox). Cambridge University Press, pp. 117–28.

Beaujouan, G. 1974. In *Actes du XVIe Congres International d'Histoire des Sciences*, vol. 2, pp. 27–30. Tokyo–Kyoto.

Beijing Observatory, 1988. *Zhongguo Gudai Tianxiang Jilu Zongji* (A Union Table of Ancient Chinese Records of Celestial Phenomena). Kexue Jishi Chubanshe, Kiangxu.

Berger, A. L., 1977. *Celestial Mechanics*, **15**, 53–74.

Bey, M., 1860. *Comptes Rendus*, **51**, 684.

Bickerman, E. J., 1980. *Chronology of the Ancient World*. Cornell University Press, Ithaca, N. Y.

Bielenstein, H., 1979. *Bull. Museum Far Eastern Antiquities*, Stockholm, **51**, 3–300.

Blair, P. H., 1963. In *Studies on the Early British Border* (ed. by N. K. Chadwick). Cambridge, pp. 63–118.

Bouquet, M. (ed.), 1781. *Recueil des Historiens des Gaules et de la France*, vol 10. Paris.

Bowra, C. M., 1964. *Pindar*. Clarendon Press, Oxford.

Bretagnon, P., 1982. *Astron. Astrophys.*, **114**, 278–88.

Brinkman, J. A., 1968. *A Political History of Post-Kassite Babylonia* (*Analecta Orientalia*, 43). Pontifical Biblical Institute, Rome.

Britton, J. P., 1985. *The Quality of Ptolemy's Solar and Lunar Observations and Parameters.* (Unpublished manuscript.)

Brown, E. W., 1915a. *Mon. Not. R. Astr. Soc.*, **75**, 508–16.

Brown, E. W., 1915b. *Report of the 84th Meeting of the British Association for the Advancement of Science*, 311–21.

Brown, E. W., 1919. *Tables of the Motion of the Moon*, Yale University Press, New Haven, Conn., vol. I.

Brown, E. W., 1926. *Trans. Astr. Obs. Yale Univ.*, **3**, 205–35.

Budge, W. A., 1925. *The Rise and Progress of Assyriology*. Hopkinson and Co., London.

Calame, O. and Mulholland, D., 1978. *Science*, **199**, 977–78.

Carter, T. F., 1925. *The Invention of Printing in China and its Spread Westwards.* New York.

Caussin C., 1804. *Le Livre de la Grande Table Hakemite par ebn Iounis.* Paris.

Celoria, G., 1877a. *Memorie del Reale Istituto Lombardo di Scienze e Letteri, Classe di Scienze, Mathematiche e Naturali*, **13**, 275–300.

Celoria, G., 1877b. *Memorie del Reale Istituto Lombardo di Scienze e Letteri, Classe di Scienze, Mathematiche e Naturali*, **13**, 367–82.

Chabot, J. B., 1905. *Chronique de Michel le Syrien.* 4 vols. Leroux, Paris.

Chapront-Touze, M. and Chapront, J., 1983. *Astron. Astrophys.*, **124**, 50–62.

Chen Jiujin, 1983a. *Zhi-ran Ke-xue Shi-yan-jiu* (Studies in the History of Natural Sciences), **2**, No. 2, 118–32.

Chen Jiujin, 1983b. *Zhi-ran Ke-xue Shi-yan-jiu* (Studies in the History of Natural Sciences), **2**, No. 4, 303–15.

Cheng, M. K., Eanes, R. J., Shum, C. K., Schutz, B. E. and Tapley, B. D., 1989. *Geophys. Res. Lett.*, **16**, 393–6.

Cherniss, H. and Helmbold, W.C., 1957. *Plutarch's Moralia*, vol. XII. Heinemann, London.

Chou Hung-hsiang, 1979. *Scientific American*, **240**, No. 4, 134–49.

Christodoulidis, D. C., Smith, D. E., Williamson, R. G. and Klosko, S. M., 1988. *J. Geophys. Res.*, **93**, 6216–36.

Clark, D. H. and Stephenson, F. R., 1977. *The Historical Supernovae.* Pergamon Press, Oxford.

Clavius, C., 1593. *In sphaeram Ioannis de Sacrobosco, Commentarius.* Sumptibus Fratrum de Gabiano, Lugduni.

Clemence, G. M., 1948. *Astron. J.*, **53**, 169–79.

Clemence, G. M., 1971. *J. Hist. Astr.*, **2**, 73–9.

Cohen A. P. and Newton, R. R., 1981–83. *Monumenta Serica*, **35**, 347–430.

Corcoran, T. H., 1971. *Seneca: Naturales Quaestiones*. Heinemann, London.

Cowell, P. H., 1905. *Mon. Not. R. Astr. Soc.*, **66**, 3–5.

Cowell, P. H., 1906a. *Mon. Not. R. Astr. Soc.*, **66**, 352–5.

Cowell, P. H., 1906b. *Mon. Not. R. Astr. Soc.*, **66**, 523–41.

Cowell, P. H., 1907. *Mon. Not. R. Astr. Soc.*, **68**, 18–9.

D'Elia, P. M., 1960. *Galileo in China*. Cambridge, Mass.

Dakyns, H. G., 1892. *The Works of Xenophon*, vol II. Macmillan and Co., London.

Danjon, A., 1929. *l'Astronomie*, **43**, 13–22.

Darwin, G. H., 1880. *Phil. Trans. R. Soc. Lond.*, **171**, 713–891.

Dasent, G. W., 1894. *The Saga of Hacon and a Fragment of the Saga of Magnus*. H.M. Stationery Office, London.

de Sitter, W., 1927. *Bull. Astr. Inst. Neth.*, **4**, 21–38.

Delaporte, L. J., 1910. *Chronographie de Mar Elie bar Sinaya, Metropolitaine de Nisibe*. Paris.

Delaunay, C.-E., 1859. *Comptes Rendus des Seances de l'Academie des Sciences*, **48**, 817–27.

Delaunay, C.-E., 1860a. *Comptes Rendus des Seances de l'Academie des Sciences*, **51**, 695–702.

Delaunay, C.-E., 1860b. *Connaissance des Temps, Additions pour l'Annee 1862*, 1–58.

Delaunay, C.-E., 1865. *Comptes Rendus des Seances de l'Academie des Sciences*, **61**, 1023–32.

Dicke, R. H., 1966. In *The Earth–Moon System* (ed. by B. G. Marsden and A. G. W. Cameron). Plenum Press, New York, pp. 98–163.

Dickey, J. O., Williams, J. G. and Yoder, C. F., 1982. *IAU Colloquium No. 63* (ed. by O. Calame). D. Reidel Publishing Company, Dordrecht, p. 217.

Dickey, J. O., Bender, P. L., Faller, J. E. *et al.* 1994. *Science*, **265**, 482–490.

Dicks, D. R., 1960. *The Geographical Fragments of Hipparchus*. University of London Press.

Dubs, H. H., 1938, 1944, 1955. *History of the Former Han Dynasty*. Harvard University Press, 3 vols.

Dubs, H. H., 1947. *Harvard Journal of Asiatic Studies*, **10**, 162.

Duncombe, J. S., 1973. *U.S. Naval Observatory Circular*, No. 141. Washington, DC.

Dunthorne, R., 1749. *Phil. Trans. R. Soc. Lond.*, **46**, 162–71.

Essen, L. and Parry, J.V.L., 1957. *Phil. Trans. R. Soc. Lond., A,* **250**, 45–69.

Ferrari, A. J., Sinclair, W. S., Sjogren, W. L., Williams, J. G., and Yoder, C. F., 1980. *J. Geophys. Res.,* **85**, 3939–51.

Ferrel, W., 1865. *Proc. Amer. Acad. Arts and Sciences,* **6**, 379–83.

Fitzgerald, C. P., 1976. *China, a Short Cultural History.* Cresset Press, London.

Florez, H., 1747. *Espana Sagrada.* Marin, Madrid.

Foley, N. B. F. and Stephenson, F. R., 1997. In preparation.

Forget, I., 1963. *Synaxarium Alexandrinum.* Louvain.

Fotheringham, J. K., 1908. *Mon. Not. R. Astr. Soc.,* **69**, 26–30.

Fotheringham, J. K., 1909. *Mon. Not. R. Astr. Soc.,* **69**, 204–10.

Fotheringham, J. K., 1915. *Mon. Not. R. Astr. Soc.,* **75**, 377–94.

Fotheringham, J. K., 1920a. *Mon. Not. R. Astr. Soc.,* **80**, 578–81.

Fotheringham, J. K., 1920b. *Mon. Not. R. Astr. Soc.,* **81**, 104–26.

Fotheringham, J. K., 1920c. *Observatory,* **43**, 189–91.

Fotheringham, J. K., 1935. *Mon. Not. R. Astr. Soc.,* **95**, 719–23.

Freeman-Grenville, G. S. P., 1977. *The Muslim and Christian Calendars.* 2nd ed. Oxford University Press.

Geer, R. M., 1954. *The Library of History of Diodorus Siculus,* Books XIX–XX. Heinemann, London.

Gingerich, O., 1986. *Scientific American,* **251**, No. 4, cover and pp. 74–83.

Ginzel, F. K., 1884a. *Sitzungsberichte der Kaiserlichen Akademie des Wissenschaften in Wien, Math.-naturwiss. Kl., Abt. II,* **88**, 629–755.

Ginzel, F. K., 1884b. *Sitzungsberichte der Kaiserlichen Akademie des Wissenschaften in Wien, Math.-naturwiss. Kl., Abt. II,* **89**, 491–559.

Ginzel, F. K., 1899. *Spezieller Kanon der Sonnen-und Mondfinsternisse,* Mayer und Muller, Berlin.

Ginzel, F. K., 1918. *Abhandlungen der Koniglichen Akademie der Wissenschaften, Berlin, Phys.-Math. Kl.,* **4**, 3–46.

Glauert, H., 1915. *Mon. Not. R. Astr. Soc.,* **75**, 489–95.

Goad, C. and Douglas, B., 1978. *J. Geophys. Res.,* **83**, 2306–10.

Godley, A. D., 1971. *Herodotus,* Books V–VII. Heinemann, London.

Godley, A. D., 1975. *Herodotus,* Books I and II. Heinemann, London.

Goldstein, B. R., 1979. *Archives Internationales d'Histoire des Sciences,* **29**, 101–56.

Grayson, A. K., 1975. *Assyrian and Babylonian Chronicles.* J. J. Augustin, New York.

Halley, E., 1693. *Phil. Trans. R. Soc. Lond.,* **17**, 913–21.

Halley, E., 1695. *Phil. Trans. R. Soc. Lond.*, **19**, 106–75.

Halley, E., 1715. *Phil. Trans. R. Soc.*, **29**, 245–262 and 314–16.

Han, Yu-shan, 1955. *Elements of Chinese Historiography.* Hawley, Hollywood.

Hansen, P. A., 1857. *Tables de la Lune construites d'apres le Principe Newtonien de la Gravitation Universelle.* H. M. Stationery Office, London.

Heath, T. L., 1932. *Greek Astronomy.* Dent and Sons, London. Reprinted 1991 by Dover Publications Inc., Mineola, NY.

Heiskanen, W., 1921. *Ann. Acad. Scient. Fennicae, A*, **18**, 1–84.

Herschel, W., 1781. *Phil. Trans. R. Soc. Lond.*, **71**, 115–21.

Hewlett, H. G. 1887. *The Flowers of History by Roger of Wendover.* H. M. Stationery Office, London.

Hide, R., Birch, N. T., Morrison, L. V., Shea, D. J. and White, A. A., 1980. *Nature*, **286**, 114–17.

Hilton, J. L., Seidelmann, P. K. and Liu Ciyuan, 1992. *Astron. J.*, **104**, 2250–2.

Ho Peng Yoke, 1966. *The Astronomical Chapters of the Chin Shu.* Mouton, Paris.

Holder-Egger, O. (ed.), 1899. *Scriptores Rerum Germanicarum*, XLVII (1899), pp. 19, 56, 57 and 178.

Hsueh Chung-san and Ou-yang I, 1956. *A Sino-Western Calendar for Two Thousand Years.* San-lien Shu-tien, Beijing.

Huber, P. J., 1973. *Babylonian Eclipse Observations from 750 BC to 0.* (Unpublished manuscript.)

Hughes, D. W., Yallop, B. D. and Hohenkerk, C. Y., 1989. *Mon. Not. R. Astr. Soc.*, **238**, 1529 35.

Hunger, H., 1992. *Astrological Reports to Assyrian Kings* (*State Archives of Assyria*, vol. VIII). Helsinki University Press.

Hunger H., 1995. Pre-print.

Hunger, H. and Pingree, D., 1989. *MUL.APIN: an Astronomical Compendium in Cuneiform.* Ferdinand Berger, Horn (Austria). (*Archiv fur Orientforschung*, **24**).

IAU, 1954. *Trans. Int. Astr. Un.*, **8**, 80.

IAU, 1968. *Trans. Int. Astr. Un.*, *XIIIB*, 48.

ILE, 1954. *Improved Lunar Ephemeris: 1952–1959*, US Naval Observatory, Washington, DC.

Innes, R. T. A., 1925. *Astr. Nachr.*, **225**, 109–10.

Jeffreys, H., 1920. *Phil. Trans. R. Soc., A*, **221**, 239–64.

Jeffreys, H., 1976. *The Earth: Its Origin, History and Physical Constitution* (sixth edition). Cambridge University Press.

Jeffreys, H., 1982. *Geophys. J.*, **71**, 555–66.

Jones, H. L. (ed.), 1917. *The Geography of Strabo*. Heinemann, London, vol. I, pp. 23–5.

Jones, H. S., 1926. *Mon. Not. R. Astr. Soc.*, **87**, 4–31.

Jones, H. S., 1932. *Ann. Cape Obs.*, **13**, part 3.

Jones, H. S., 1939. *Mon. Not. R. Astr. Soc.*, **99**, 541–58.

Jordi, C., Morrison, L. V., Rosen, R. D., Salstein, D. A. and Rossello, G., 1994. *Geophys. J. Int.*, **117**, 811–8.

Kanda Shigeru, 1934. *Nihon Temmon Shiryo*. Tokyo.

Kanda Shigeru, 1935. *Nihon Temmon Shiryo Soran*. Tokyo.

Kant, I., 1754. *Wochentliche Frag- und Anzeigungs-Nachrichten* Nos. 23 and 24. (See Ley, 1968.)

Karlgren, B., 1950. *The Book of Odes*. Stockholm, 1950.

Keightley, D. N. 1978a, *Sources of Shang History: The Oracle-bone Inscriptions of Bronze-age China*. Berkeley and Los Angeles.

Keightley, D. N., 1978b. *Harvard Journal of Asiatic Studies*, **38**, 423–38.

Kelvin, Lord, 1894. *Popular Lectures and Addresses*. Macmillan and Co. Ltd, London, vol. II.

Kelvin, Lord and Tait, P. G., 1883. *Treatise on Natural Philosophy*. Cambridge University Press.

Kepler, J., 1605. *De Stella Nova in Pede Serpentarii*, chap. XXIII, 4. Reprinted in *Joannis Kepleri Astronomi Opera Omnia* (ed. by C. Frisch), vol. II, p. 696. Frankfurt, 1859.

King, D. A., 1973. *Archive for History of Exact Science*, **10**, 342–94.

King, D. A., 1976. Contribution to *Dictionary of Scientific Biography*, vol. 14, pp. 574–80. New York.

Kugler, F. X., 1907, 1909–24, 1913–14. *Sternkunde und Sterndienst in Babel*, vols. I, II, Erganzungen. Munster.

Lagrange, J.-L., 1792. *Nouveaux Memoires de l'Academie Royale des Sciences et Belles Lettres de Berlin, Oeuvres*, **5**, 687–98.

Lalande, J.-J. 1757. *Memoire de l'Academie pour l'annee 1757.*

Lamb, H. H., 1982. *Climate History and the Modern World*. Methuen, London.

Lambeck, K., 1977. *Phil. Trans. R. Soc. Lond., A*, **287**, 545–94.

Lambeck, K., 1980. *The Earth's Variable Rotation*. Cambridge University Press.

Lane-Poole, S., 1926. *Saladin and the Fall of the Kingdom of Jerusalem*. London.

Laplace, P.-S., 1787a. *Memoire de l'Academie pour l'annee 1786.*

Laplace, P.-S., 1787b. *Connaissance des Temps, Additions pour l'Annee 1790*, 291–5.

Laskar, J., 1986. *Astron. Astrophys.*, **157**, 59–70.

Legge, J., 1872. *The Chinese Classics*, vol. 5. Hong Kong. Reprinted Hong Kong, 1960.

Levi-Donati, G. R., 1987. *Bolletino della Deputazione di Storia Patri per l'Umbria*, **84**, 89–107.

Levi-Donati, G. R., 1989. *l'astronomia*, no. 89, 24–9.

Ley, W. 1968. *Kant's Cosmogony*. Greenwood Publishing Company, New York.

Liais, E., 1866. *Memoires de l'Academie*, **28**, 1119–21.

Liu Baolin and Fiala, A. D., 1992. *Canon of Lunar Eclipses: 1500 BC – AD 3000*. Willmann-Bell, Inc., Richmond, Virginia.

Liu Chao-yang, 1944. *Chung-kuo Wen-hua Yen-ch'iu Hui-k'an* (*China Cultural Research Quarterly*), **4** (2), 39.

Liu Chao-yang, 1945. *Yu Chou* (The Universe), **15**, 15–16.

Livingstone, R. W., 1949. *Thucydides: the History of the Peloponnesian War*. Oxford University Press, London.

Longomontanus, C. S., 1622. *Astronomica Danica*. Amsterdam.

Macey, S. L. (ed.), 1994. *Encyclopedia of Time*. Garland Publishing, Inc., New York, p. 443.

Markowitz, W., Hall, R. G., Essen, L. and Parry, J. V. L., 1958. *Phys. Rev. Lett.*, **1**, 105–6.

Maspero, H., 1939. *Melanges Chinois et Bouddhiques*, **6**, 183–382 (esp. 212–3).

Maxwell, H., 1913. *The Chronicle of Lanercost*. James Maclehose and Sons, Glasgow.

Mayer, T., 1753. *Novae Tabulae Motuum Solis et Lunae*. Gottingen.

Meeus, J., 1982. *J. Brit. Astr. Assn.*, **92**, 124–6.

Migne, J. P. (ed.), 1865. *Patrologiae Graecae*, vol. 148. Paris.

Millard, A. R., 1994. *The Eponyms of the Assyrian Empire: 910–612 BC* University of Helsinki. (*State Archives of Assyria Studies*, vol. II).

Morrison, L. V., 1979, *Mon. Not. R. Astr. Soc.*, **187**, 41–82.

Morrison, L. V. and Stephenson, F. R., 1981. In *Reference Coordinate Systems for Earth Dynamics* (ed. by E. M. Gaposchkin and B. Kolaczek). D. Reidel Publishing Company, Dordrecht, pp. 181–5.

Morrison, L. V. and Ward, C. G., 1975. *Mon. Not. R. Astr. Soc.*, **173**, 183–206.

Muller, P. M., 1975. PhD Thesis, University of Newcastle upon Tyne.

Muller, P. M., 1976. Report SP 43-26. Jet Propulsion Laboratory, Pasadena, California.

Muller, P. M. and Stephenson, F. R., 1975. In *Growth Rhythms and the History of the Earth's Rotation* (ed. by G. D. Rosenberg and S. K. Runcorn). John Wiley and Sons, London, pp. 459–534.

Munk, W. H. and MacDonald, G. J. F., 1960. *The Rotation of the Earth.* Cambridge University Press.

Muratori, L. A. (ed.), 1723– . *Scriptores Rerum Italicarum* (25 vols.), Milan.

Nakayama Shigeru, 1969. *A History of Japanese Astronomy.* Harvard University Press, Cambridge, Mass.

Nakiboglu, S. M. and Lambeck K., 1991. In *Glacial Isostacy, Sea-level and Mantle Rheology* (ed. by R. Sabadini, K. Lambeck and E. Boschi). Kluwer, Dordrecht, pp. 237–58.

Nallino, C. A., 1899. *Al-Battani sive Albatenii Opus Astronomicum.* Milan.

Nasr, S. H., 1976. *Islamic Science: an Illustrated Study.* World of Islam Publishing Co. Ltd., London.

Needham, J., Lu Gwei-djen, Combridge, J. H. and Major, J. S., 1986. *The Hall of Heavenly Records.* Cambridge University Press.

Needham, J., Wang Ling and Price, D deS., 1986. *Heavenly Clockwork.* Cambridge University Press.

Neugebauer, O., 1947. *Isis*, **37**, 37–43.

Neugebauer, O., 1952. *The Exact Sciences in Antiquity.* Princeton, New Jersey.

Neugebauer, O., 1955. *Astronomical Cuneiform Texts*, I–III. London.

Neugebauer, O., 1975. *A History of Ancient Mathematical Astronomy.* Springer-Verlag, Berlin.

Nevill, E., 1906a. *Mon. Not. R. Astr. Soc.*, **66**, 404–20.

Nevill, E., 1906b. *Mon. Not. R. Astr. Soc.*, **67**, 2–13.

Newcomb, S., 1870. *Amer. J. Sci.*, Ser. II, **50**, 183–94.

Newcomb, S., 1878. *Researches on the Motion of the Moon, Washington Observations for 1875*, Appendix II. US Naval Observatory, Washington, DC.

Newcomb, S., 1895a. *Astr. Pap. Amer. Eph.*, **6**, Washington, DC.

Newcomb, S. 1895b. *The Elements of the Four Inner Planets and the Fundamental Constants of Astronomy.* Washington, DC.

Newcomb, S., 1898. *Popular Astronomy.* Macmillan and Co. Ltd, London.

Newcomb, S., 1909. *Mon. Not. R. Astr. Soc.*, **69**, 164–9.

Newcomb, S., 1912. *Researches on the Motion of the Moon and Related Astronomical Elements, Astr. Pap. Amer. Eph.*, **9**, part I. Washington DC.

Newhall, X X, Standish, E. M. and Williams, J. C., 1983. *Astron. Astrophys.*, **125**, 150–67.

Newton, R. R., 1968. *Geophys. J. R. Astr. Soc.*, **14**, 505–39.

Newton, R. R., 1970. *Ancient Astronomical Observations and the Accelerations of the Earth and Moon.* Johns Hopkins University Press, Baltimore.

Newton, R. R., 1972a. *Mem. R. Astr. Soc.*, **76**, 99–128.

Newton, R. R., 1972b. *Medieval Chronicles and the Rotation of the Earth.* Johns Hopkins University Press, Baltimore.

Niebuhr, B. G. (ed.), 1828. *Corpus Scriptorum Historiae Byzantinae*, vol. 33. Weber, Bonn.

North, J. D., 1974. *Scientific American*, **230**, No. 1, 96–106.

O'Keefe, J. A., 1970. *J. Geophys. Res.*, **75**, 6565–74.

Oates, J., 1986. *Babylon* (revised edition). Thames and Hudson, London.

Panchenko, D., 1994. *J. Hist. Astr.*, **25**, 275–88.

Pang, K. D., Yau, K. K. C., Chou Hung-hsiang and Wolf, R., 1988a. *Vistas in Astronomy*, **31**, 833–47.

Pang, K. D., Chou, Hang-hsiang and Yau, K. K. C., 1988b. *Bull. Amer. Astr. Soc.*, **21**, 753.

Park, Seong-rae, 1979. *J. Social Sciences and Humanities*, **49**, 53–117.

Parker, R. A., 1974. In *The Place of Astronomy in the Ancient World* (ed. by D. G. Kendall and S. Piggott). Oxford University Press, pp. 51–65. Reprinted in *Phil. Trans. R. Soc. Lond. A*, **276**, 51–65.

Parker, R. A. and Dubberstein, W. H., 1956. *Babylonian Chronology: 626 BC – AD 75.* Brown University Press, Providence, R.I.

Parkinson, J. H., Stephenson, F. R. and Morrison, L. V., 1988. *Nature*, **331**, 421–3.

Parpola, S., 1983. *Letters from Assyrian Scholars to the Kings Esarhaddon and Assurbanipal.* Part 2: *Commentary and Appendices. Alter Orient und Altes Testament*, vol 5/1. Butzon und Bercker Kevelaer, Neukirchen–Vluyn.

Perrin. B., 1961. *Plutarch's Lives*, vol. V, Heinemann, London.

Pertz, G. H. (ed.), 1826– . *Monumenta Germaniae Historica, Scriptores* (32 vols.). Hahn, Hannover.

Pingre, A.-G., 1901. *Annales Celestes du Dix-Septieme Siecle* (ed. by M. G. Bigourdan). Paris.

Pirazzoli, P. A., 1991. In *Glacial Isostacy, Sea-level and Mantle Rheology* (ed. by R. Sabadini, K. Lambeck and E. Boschi). Kluwer, Dordrecht, pp. 259–70.

Plechl, H. (ed.), 1972. *Orbis Latinus.* Kinkhardt and Biermann, Braunschweig.

Plummer, C., 1896. *Venerabilis Bedae Opera Historica.* Clarendon Press, Oxford.

Potter, K. R., 1955. *The Historia Novella by William of Malmesbury.* Nelson and Sons, London.

Rabi, I. I., 1945. Richtmeyer Lecture to the American Physical Society, New York. (Unpublished.)

Race, W. H., 1986. *Pindar.* Twayne, Boston, Mass.

Rackham, H., 1938. *Pliny: Natural History*, Books I–II. Heinemann, London.

Rankin, H. D., 1977. *Archilochus of Paros.* Noyes, Park Ridge (New Jersey).

Ranyard, A. C., 1879. *Mem. R. Astr. Soc.*, **41**.

Rassam, H., 1897. *Asshur and the Land of Nimrod.* Eaton and Mains, New York.

Ravn, O. E., 1942. *Herodotus' Description of Babylon.* Copenhagen.

Reade, J. E., 1986. In E. Leichty, *Catalogue of the Babylonian Tablets in the British Museum*, vol VI, pp. xiii–xxxvi. British Museum Press, London.

Riccioli, G. B., 1665. *Astronomiae Reformatae tomi duo.* Bologna.

Rolfe, J. C., 1946. *Quintus Curtius.* Heinemann, London.

Rosan, L. J., 1949. *The Philosophy of Proclus.* Cosmos, New York.

Rowton, M. B., 1946. *Iraq*, **8**, 94–110.

Sachs, A. J., 1948. *J. Cuneiform Stud.*, **2**, 271–90.

Sachs, A. J., 1967. *Archaeologia*, **15**, Mars–Avril, 12–19.

Sachs, A. J., 1974. In *The Place of Astronomy in the Ancient World* (ed. by D. G. Kendall and S. Piggott), Oxford University Press, pp. 43–50. Reprinted in *Phil. Trans. R. Soc. Lond. A*, **276**, 43–50.

Sachs, A. J., 1976. In *Cuneiform Studies in Honor of Samuel Noah Kramer* (ed. by B. L. Eichler). *Alter Orient und Altes Testament Sonderreihe*, vol 25. Butzon und Bercker Kevelaer, Neukirchen-Vluyn.

Sachs, A. J. and Hunger, H., 1988, 1989, 1996. *Astronomical Diaries and Related Texts from Babylonia*, vols. I, II and III. Österreichische Akademie der Wissenschaften, Wien.

Sachs, A. J. and Schaumberger, J., 1955. *Late Babylonian Astronomical and Related Texts.* Brown University Press, Providence, R.I.

Sachs, A. J. and Wiseman, D. J., 1954. *Iraq*, **16**, 202–11.

Sage, E. T., 1936. *Livy*, vol. XI. Heinemann, London.

Said, S. S. and Stephenson, F. R., 1991. *J. Hist. Astr.*, **22**, 297–310.

Said, S. S. and Stephenson, F. R., 1997. In preparation.

Said. S. S., Stephenson, F. R., and Rada, W. S., 1989. *Bulletin of the School of Oriental and African Studies, London Univ.*, **52**, 38–64.

Sandys, J., 1978. *The Odes of Pindar.* Heinemann, London.

Sawyer, J. F. A., 1972. *J. Theological Studies*, **23**, 124–8.

Schaumberger, J., 1952. *Zeitschrift fur Assyriologie und Vorderasiatische Archaologie*, **50**, 214–29.

Schroeter, J. Fr., 1923. *Spezieller Kanon der Zentralen Sonnen- und Mondfinsternisse, welche innerhalb des Zeitraums von 600 bis 1800 n. Chr. in Europa sichtbar waren.* Jacob Dybwad, Kristiania.

Seidelmann, P. K. (ed.), 1992. *Explanatory Supplement to the Astronomical Almanac.* University Science Books, Mill Valley, California.

Standish, E. M., 1982. *Astron. Astrophys.*, **114**, 297–302.

Stephenson, F. R., 1974. In *The Place of Astronomy in the Ancient World* (ed. by D. G. Kendall and S. Piggott), Oxford University Press, pp. 118–21. Reprinted in *Phil. Trans. R. Soc. Lond. A*, **276**, 118–21.

Stephenson, F. R., 1992. *Quart. J. Roy. Astr. Soc.*, **33**, 91–8.

Stephenson, F. R., 1993. *Encyclopedia Britannica*, **17**, 872b–7b.

Stephenson. F. R., 1997. In preparation.

Stephenson, F. R. and Fatoohi, L. J., 1993. *J. Hist. Astr.*, **24**, 255–7.

Stephenson, F. R. and Fatoohi, L. J., 1994a. *Quart. J. R. Astr. Soc.*, **35**, 81–94.

Stephenson, F. R. and Fatoohi, L. J., 1994b. *J. Hist. Astr.*, **25**, 99–110.

Stephenson, F. R. and Fatoohi, L. J., 1995. *J. Hist. Astr.*, **26**, 227–36.

Stephenson, F. R. and Houlden, M. A., 1986. *Atlas of Historical Eclipse Maps: East Asia, 1500 BC to AD 1900.* Cambridge University Press.

Stephenson, F. R., Jones, J. E. and Morrison, L. V., 1997. *Astron. Astrophys.* In press.

Stephenson, F. R., Hunger, H. and Yau, K. K. C., 1985. *Nature*, **314**, 587–92.

Stephenson, F. R. and Morrison, L. V., 1984. In *Rotation in the Solar System* (ed. by R. Hide). The Royal Society, London. Reprinted in *Phil. Trans. R. Soc. Lond., A*, **313**, 47–70.

Stephenson, F. R. and Morrison, L. V., 1995. *Phil. Trans. R. Soc. Lond., A*, **351**, 165–202.

Stephenson, F. R. and Said, S. S., 1991. *J. Hist. Astr.*, **22**, 195–207.

Stephenson, F. R. and Said, S. S., 1997. In preparation.

Stephenson, F. R. and Walker, C. B. F. (eds.), 1985. *Halley's Comet in History.* British Museum Publications, London.

Stephenson, F. R. and Yau, K. K. C., 1985. *J. British Interplanetary Society*, **38**, 195–216.

Stephenson, F. R. and Yau, K. K. C., 1992a. *Quart. J. R. Astr. Soc.*, **33**, 91–8.

Stephenson, F. R. and Yau, K. K. C., 1992b. *Astron. Astrophys.*, **260**, 485–8.

Stier, H. E. and Kirsten, E., 1956. *Westermann's Atlas zur Weltgeschichte.* Westermann, Braunschweig.

Stubbs, W., 1867. *The Chronicle of the Reigns of Henry II and Richard I: AD 1169–1192.* H. M. Stationery Office, London.

Taylor, G. I., 1919. *Phil. Trans. R. Soc., A*, **220**, 1–33.

Thompson, J. E. S., 1974. In *The Place of Astronomy in the Ancient World* (ed. by D. G. Kendall and S. Piggott). Oxford University Press, pp. 83–98. Reprinted in *Phil. Trans. R. Soc. Lond. A*, **276**, 83–98.

Thompson, R. C., 1900. *The Reports of the Magicians and Astrologers of Nineveh and Babylon.* London, 2 vols. (reprinted New York, 1974).

Thorkelson, T., 1933. *Visindafelag Islendinga* (Reykjavik), **15**.

Tihon, A., 1976–77. *Bulletin de l'Institut Historique Belge de Rome*, **46–47**, 35–79.

Toomer, G. J., 1974. *Archive for History of Exact Sciences*, **14**, 126–42.

Toomer, G. J., 1984. *Ptolemy's Almagest: a Translation and Annotation.* Duckworth, London.

Toomer, G. J., 1988. In *A Scientific Humanist: Studies in Memory of Abraham Sachs* (ed. by E. Leichty, M. deJ. Ellis and P. Gerardi). Samuel Noel Kramer Fund, Philadelphia, pp. 353–62.

Tung Tso-Pin, 1945. *Yin Li P'u* (On the Calendar of the Yin Period). Lichuang.

Unger, E. 1931. *Babylon.* W. de Gruyter, Berlin.

Volland, H., 1990. In *Earth's Rotation from Eons to Days* (ed. by P. Brosche and J. Sundermann). Springer-Verlag, Berlin, pp. 127–40.

von Oppolzer, T. R., 1887. *Canon der Finsternisse.* Kaiserl. Akad. der Wiss., Wien. Reprinted 1962 as *Canon of Eclipses*, Dover, New York.

Vyssotsky, A. N., 1949. *Medd. fran. Lunds Astr. Obs. Historical Papers*, No. 22.

Waley, A., 1934. *Travels of an Alchemist.* Routledge, London.

Williams, J. G., Sinclair, W. S. and Yoder, C. F., 1978. *Geophys. Res. Lett.*, **5**, 943–6.

Wise, D. V., 1969. *J. Geophys. Res.*, **74**, 6034–45.

Wiseman, D. J., 1965. Contribution to *The Cambridge Ancient History.* Cambridge University Press, vol. II, chap. 31.

Xu Zhentao, Stephenson, F. R. and Jiang Yaotiao, 1995. *Quart. J. Roy. Astr. Soc.*, **36**, 397–406.

Xu Zhentao, Yau, K. K. C. and Stephenson, F. R., 1989. *Archaeoastronomy* (Supplement to *J. History of Astronomy*), no. 14, S61–72.

Yabuuchi Kiyoshi, 1979. *Acta Asiatica*, **36**, 7–48.

Yoder, C. F., Williams, J. G., Dickey, J. O., Schutz, B. E., Eanes, R. J. and Tapley, B. D., 1983. *Nature*, **303**, 757–62.

Zawilski, M., 1994. *The Collection of Texts Containing the Historical Observations for the Solar Eclipses for Europe and the Near East.* (Unpublished manuscript.)

Zinner, E., 1990. *Regiomontanus: his Life and Work.* Amsterdam, p. 142.

Acknowledgements

I am grateful to Dr Leslie V. Morrison of the Royal Greenwich Observatory for his continued help and advice. My thanks are also due to Professor Peter J. Huber (Bayreuth University), and Professor Hermann Hunger (University of Vienna) for permission to publish their translations of Babylonian texts. I wish to thank Mrs Pauline Russell of the Department of Physics, University of Durham for skilfully executing numerous diagrams. I would also like to acknowledge with gratitude my indebtedness to the late Professor S. Keith Runcorn, who introduced me to this subject and gave me much encouragement.

Much of the research for this book was undertaken at Jet Propulsion Laboratory, Pasadena during my tenure of a Senior Resident Research Associateship funded by the National Research Council, Washington DC. I am grateful for the opportunities provided by this award.

Index of eclipse records

Note: The letters in the 'Region' column refer to the following broad regions or cultures: A Arab; B Babylonian (including Assyrian); C Chinese (including Japanese and Korean); E European (including Alexandrian).

Principal solar eclipse records

	Julian date			Region	Page(s)
BC	763	Jun	15	(B)	125–7
	709	Jul	17	(C)	221, 226
	669	May	27	(B)	125
	657	Apr	15	(B)	125
	648	Apr	6 ?	(E)	338–42
BC	601	Sep	20	(C)	226
	585	May	28 ?	(E)	342–4
	549	Jun	19	(C)	226–7
	463	Apr	30 ?	(E)	344–6
	444	Oct	24	(C)	227
BC	431	Aug	30	(E)	346–8
	394	Aug	14	(E)	366
	369	Apr	11	(B)	138–9
	322	Sep	26	(B)	79, 131–2, 141–2
	310	Aug	15	(E)	348–51
BC	281	Jan	30	(B)	133, 142
	254	Jan	31	(B)	72, 133–4
	249	May	4	(B)	134, 139
	247	Sep	7	(B)	122
	241	Nov	28	(B)	142
BC	198	Aug	7	(C)	238
	195	Jun	6	(B)	134–5, 139
	190	Mar	14	(B)	121, 135–6, 139
	188	Jul	17	(C)	234
	188	Jul	17	(E)	367

Principal lunar eclipse records

	Julian date			Region	Page(s)
BC	731	Apr	9	(B)	122
	721	Mar	19/20	(B)	99
	720	Mar	8/9	(B)	100
	720	Sep	1/2	(B)	100
	702	Mar	19/20	(B)	194–5
BC	695	Apr	30/May 1	(B)	162, 195
	686	Apr	21/22	(B)	162
	685	Oct	3/4	(B)	163–4
	667	Oct	14/15	(B)	88, 164, 195–6
	666	Apr	10/11	(B)	151, 206
BC	621	Apr	21/22	(B)	100
	601	Apr	10/11	(B)	164
	599	Feb	19/20	(B)	165
	591	Mar	22/23	(B)	196
	588	Jan	18/19	(B)	196–7
BC	587	Jan	7/8	(B)	165
	580	Aug	14/15	(B)	165
	577	Jun	13/14	(B)	197
	573	Apr	1/2	(B)	166
	563	Sep	5/6	(B)	197
BC	562	Mar	2/3	(B)	152
	555	Oct	6/7	(B)	173–4
	537	Oct	16/17	(B)	152, 198, 208
	523	Jul	16/17	(B)	101, 166
	522	Jan	9/10	(B)	167
BC	502	Nov	19/20	(B)	101–2
	501	Nov	7/8	(B)	174
	491	Apr	25/26	(B)	102
	483	Nov	18/19	(B)	152–3, 198
	442	Mar	23/24	(B)	198–9
BC	424	Sep	28/29	(B)	168–9, 181
	421	Feb	2/3	(B)	153
	408	Oct	31/Nov 1	(B)	169
	407	Oct	20/21	(B)	174–5
	406	Oct	9/10	(B)	153

Index of places of observation

Name index

Subject index